Applied and Numerical Harmonic Analysis

Series Editor
John J. Benedetto
University of Maryland

Editorial Advisory Board

Akram Aldroubi
NIH, Biomedical Engineering/
Instrumentation

Ingrid Daubechies
Princeton University

Christopher Heil
Georgia Institute of Technology

James McClellan
Georgia Insitute of Technology

Michael Unser
NIH, Biomedical Engineering/
Instrumentation

M. Victor Wickerhauser
Washington University

Douglas Cochran
Arizona State University

Hans G. Feichtinger
University of Vienna

Murat Kunt
Swiss Federal Institute of
Technology, Lausanne

Wim Sweldens
Lucent Technologies
Bell Laboratories

Martin Vetterli
Swiss Federal Institute of
Technology, Lausanne

Applied and Numerical Harmonic Analysis

J.M. Cooper: *Introduction to Partial Differential Equations with MATLAB* (ISBN 0-8176-3967-5)

C.E. D'Attellis and E.M. Fernández-Berdaguer: *Wavelet Theory and Harmonic Analysis in Applied Sciences* (ISBN 0-8176-3953-5)

H.G. Feichtinger and T. Strohmer: *Gabor Analysis and Algorithms* (ISBN 0-8176-3959-4)

T.M. Peters, J.H.T. Bates, G.B. Pike, P. Munger, and J.C. Williams: *Fourier Transforms and Biomedical Engineering* (ISBN 0-8176-3941-1)

A.I. Saichev and W.A. Woyczyński: *Distributions in the Physical and Engineering Sciences* (ISBN 0-8176-3924-1)

R. Tolimierei and M. An: *Time-Frequency Representations* (ISBN 0-8176-3918-7)

G.T. Herman: *Geometry of Digital Spaces* (ISBN 0-8176-3897-0)

A. Procházka, J. Uhlíř, P.J.W. Rayner, and N.G. Kingsbury: *Signal Analysis and Prediction* (ISBN 0-8176-4042-8)

J. Ramanathan: *Methods of Applied Fourier Analysis* (ISBN 0-8176-3963-2)

A. Teolis: *Computational Signal Processing with Wavelets* (ISBN 0-8176-3909-8)

W.O. Bray and Č.V. Stanojević: *Analysis of Divergence* (ISBN 0-8176-4058-4)

G.T. Herman and A. Kuba: *Discrete Tomography* (ISBN 0-8176-4101-7)

J.J. Benedetto and P.J.S.G. Ferreira (Eds.): *Modern Sampling Theory* (ISBN 0-8176-4023-1)

A. Abbate, C.M. DeCusatis, and P.K. Das: *Wavelets and Subbands* (ISBN 0-8176-4136-X)

L. Debnath: *Wavelet Transforms and Time-Frequency Signal Analysis* (ISBN 0-8176-4104-1)

K. Gröchenig: *Foundations of Time-Frequency Analysis* (ISBN 0-8176-4022-3)

D.F. Walnut: *An Introduction to Wavelet Analysis* (ISBN 0-8176-3962-4)

Agostino Abbate
Casimer M. DeCusatis
Pankaj K. Das

Wavelets and Subbands
Fundamentals and Applications

With 234 Figures

Springer Science+Business Media, LLC

Agostino Abbate
Panametrics, Inc.
221 Crescent Street
Waltham, MA 02453
USA

Casimer M. DeCusatis
Department HHLA
IBM Corporation
522 South Road
Poughkeepsie, NY 12601
USA

Pankaj K. Das
Department of Electrical and
   Computer Engineering
University of California, San Diego
La Jolla, CA 92093
USA

Library of Congress Cataloging-in-Publication Data
Abbate, Agostino.
   Wavelets and subbands : fundamentals and applications / Agostino Abbate, Casimer M.
DeCusatis, Pankaj Das.
      p. cm.—(Applied and numerical harmonic analysis)
   Includes bibliographical references and index.
   ISBN 978-0-8176-4136-8   ISBN 978-1-4612-0113-7 (eBook)
   DOI 10.1007/978-1-4612-0113-7
   1. Signal processing.   2. Wavelets (Mathematics)   3. Frequency spectra.   I. DeCusatis,
Casimer.   II. Das, Pankaj K., 1937–   III. Title.   IV. Series.
TK5102.9 .A23 2001
621.382'.2—dc21   2001025371

Printed on acid-free paper.
© 2002 Springer Science+Business Media New York
Originally published by Birkhäuser Boston in 2002

ISBN 978-0-8176-4136-8   SPIN 10728139

Production managed by Louise Farkas; manufacturing supervised by Jeffrey Taub.
Camera-ready copy prepared from the authors' WordPerfect files.

9 8 7 6 5 4 3 2 1

To my friend, my guide, my companion, and my wife,
Josefina.

To my wife, Carolyn, my children, Anne and
Rebecca, and my parents for their continued
support.

To my wife, Virginia, and my granddaugther,
Rachel.

Contents

4. Discrete Wavelet Transform: From Frames to Fast Wavelet Transform

Preface

There are probably more than 100 books and monographs on the subject of wavelet/subband transforms. Many of them are good; some are excellent. So naturally, the question arises, Why another book on this topic? The easy answer is to say that we wanted to write our own book; however, there are other more practical reasons as well.

The subject of wavelets is quite diverse and extensive. Original and pioneering work was done in this field by mathematicians, electrical engineers, physicists, and also by experts in applied sciences such as geophysics and statistics. For this reason, a book written by a physicist will describe the wavelet as an integral transform, whereas the one written by an engineer would utilize a linear system filter bank description, and a mathematician would utilize the mathematical framework of multiresolution, and so on. We became interested in this subject around 1993 and had a difficult time trying to understand the different facets of wavelets and subbands, and the connection between them. We have since utilized wavelet and subband transform techniques in signal processing applications such as ultrasonic nondestructive testing, image processing using acousto-optic devices, and applications in wireless and spread spectrum communication.

The genesis of this book can be traced to 1997, when we were invited by the Technion-Israel Institute of Technology to teach the first of a series of short courses on wavelets. At that time, we were in need of a text capable of explaining in a simple way the linkage between the various representations of wavelet theory and of subband transform. This book is an expanded version of the viewgraphs and notes handed out to the students, and it is aimed at helping a newcomer to the subject to understand the basic principles of wavelet and subband transform and the applicability to his/her needs. The other objective of this book is to present some practical applications that a traditional electrical engineer might consider useful in signal processing, image processing, communication, and control and systems analysis.

We are thankful to the students who took these courses, as their probing questions made the deeper meaning of the subject clearer to us and helped us to decide what content should be included in this book. Also, we would like to thank Professor G. Saulnier, Dr. M. Medley, and Dr. K. Hetling for liberal use of materials in joint papers.

Waltham, Massachussets   Agostino Abbate
Poughkeepsie, New York   Casimer M. DeCusatis
LaJolla, California   Pankaj K. Das

Notation

Sets

$\mathbb{R}$	Real numbers
$\mathbb{R}^+$	Positive real numbers (including 0)
$\mathbf{Z}$	Integers
$\mathbb{N}$	Positive integers (including 0)
$\mathbb{C}$	Complex numbers
$\mathbf{H}$	Hilbert space

Operators

$h * g$	Convolution between h and g
$h \otimes g$	Correlation between h and g
$\langle h, g \rangle$	Inner product between h and g
$\| s \| = \langle s, s \rangle^{1/2}$	Norm of the function s
$\mathbb{I}$	Identity operator
$\mathscr{F}\{s\}$	Fourier transform of function s
s^*	Complex conjugate of the function s
$\mathbf{F}\{s\}$	Frame operator
$\mathbf{A}\{s\}$	Analytic operator

Variables

t	Continuous time
n	Discrete time
f	Frequency
$\omega = 2\pi f$	Angular frequency
$\langle t \rangle$	Average or mean time
$T = 2\sigma_t$	Signal duration
$\langle \omega \rangle$	Average or center angular frequency
$B = 2\sigma_\omega$	Signal frequency bandwidth
$\tau_g(\omega)$	Group delay
$\omega_i(t)$	Instantaneous frequency
a	Scaling parameter
b	Shift parameter
E	Energy of a signal
(x, y)	Rectangular cooordinates
(r, θ)	Polar coordinates

Special Functions

$\delta(t),\ \delta(n),\ \delta_n$	Delta function in continuous-time, discrete-time, and finite duration discrete-time, respectively
$u_T(t)$	Rectangular function
$\mathrm{sinc}(x) = \sin(x)/x$	Sinc function

Signals

$s(t)$	Continuous-time signal
$s(n)$	Discrete-time signal
s_n	Finite duration discrete-time signal
$S(\omega) = \mathscr{F}\{s\}$	Fourier Transform of the signal s
$h(t)$	Window function, Wavelet function

$S(\tau,\omega)$	Short-Time Fourier Transform, Windowed Fourier Transform or Gabor Transform
$P(\tau,\omega)= \lvert S(\tau,\omega)\rvert^2$	Spectrogram
$s_H(t)$	Hilbert transform
$\psi(t)$	Wavelet function (mother wavelet)
$\psi_{a,b}(t)$	Wavelet Basis functions (daughter wavelets)
$\psi_{m,n}(\tau)$	Wavelet frames
$\varphi(t)$	Basis function
$\phi(t)$	Scaling function
$W_s(a,b)$	Continuous wavelet transform
$W_{m,n} = W_s(m,n)$	Discrete wavelet transform.
h_n	Discrete-time filters coefficients or taps
$h_0,\ h_1,\ g_0,\ g_1$	Subband filters (time domain)
$H_0,\ H_1,\ G_0,\ G_1$	Subband Filters (frequency domain)

Chapter 1

Introduction

The concept of wavelets has been discussed in the literature for a very long time. It is based on fundamental ideas which were first expressed more than a century ago in a variety of forms. However, it is only recently that significant progress has been made in the application of wavelets to practical problems in signal processing. The wavelet transform has been proposed as a flexible tool for the multiresolution decomposition of continuous time signals. The pioneering work of Daubechies in the early 1980s has shown the linkage between the wavelet and subband transform theories. Since then, there has been an explosion of interest and a flurry of interdisciplinary research and development activities on wavelet and subband transforms, and their applications [Dau90].

Figure 1.1.1 shows the diversity of the wavelet/subband concept and its many applications.* Historically mathematics, physics, electrical engineering, and other applied sciences have contributed to the development of the wavelet theory. For example, significant practical applications of wavelets have been found in signal and image processing, spread-spectrum and wireless communications, and control system analysis, which are in the electrical engineering domain [Wor96], [Sch96a].

---

* The figure numbers in this book have three digits. The first two digits are the chapter and section, respectively, as in Figure 1.1.1; the last digit is the figure number in the section. The same applies for the equation numbers.

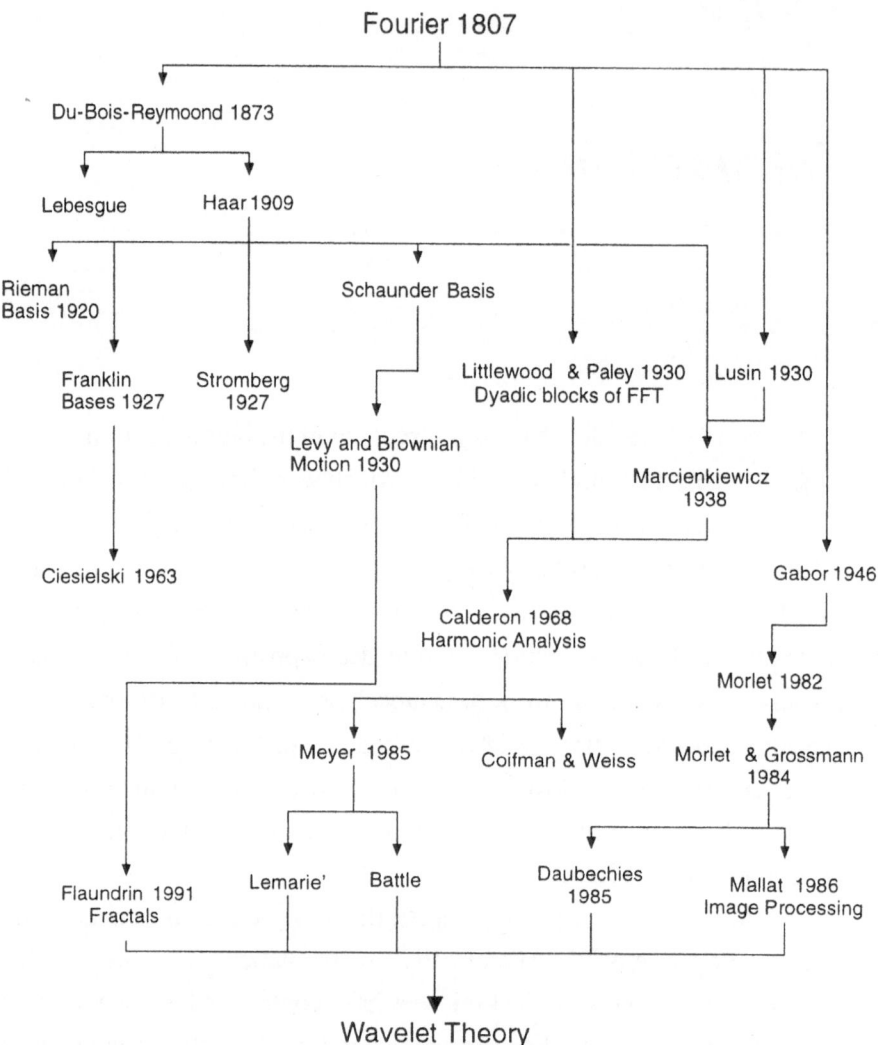

Figure 1.1.1 Historical tree of wavelet theory.

In the following, we will review the history of the development of wavelet and subband theory and their impact on the various areas of science and engineering.

1.1   Historical Review: From Fourier Analysis to Wavelet Analysis and Subband

The history of wavelets begins with the development of the traditional Fourier series in 1807, as shown in Figure 1.1.1. The techniques of Fourier series and Fourier transforms were pioneered by the French Physicist Jean Baptiste Joseph, Baron de Fourier (1768-1830). Although rejected by his contemporaries, Fourier's ideas have developed into one of the cornerstones of contemporary mathematics and engineering. The Fourier transform is very useful in many applications, but it had to be modified to deal with the case of singularities or sharp transient signals. This shortcoming of the Fourier transform was originally identified by Paul DuBois-Reymond in 1873. He also proposed a solution to the singularity problem that eventually was fully developed by Henri Lebesgue. A different solution was designed by Haar in 1909; he replaced the sine and cosine functions of the Fourier transform with another orthonormal basis, now commonly known as the Haar basis. The original idea of Haar has opened the door to the construction of an infinite number of other bases. Using a very simple function, Haar created a basis using dyadic scaling. The functions of the Haar basis suffer from a major disadvantage for many applications; as they are discontinuous, they are not optimal for approximating a continuous function $f(t)$, especially if the function has a continuous derivative. Nevertheless, the importance of the Haar decomposition lays in the development of an orthonormal basis based on dyadic sampling, which opened one of the routes leading to the concept of wavelets and, in particular, to what it is now called multiresolution analysis. The Haar basis is the simplest example to date of a wavelet basis which satisfies the multiresolution properties [Mey93].

The shortcomings of the Haar decomposition led to the development of other bases such as the Schander basis and the Riemann basis in the 1910-1920 decades. In the 1930s, a lot of interest was given to the study of Brownian motion. A lot of work was devoted to define the best possible representation for a signal representing this motion. The Fourier analysis was adequate for representing the spectral properties of the Browian motion, especially to extract any particular resonance frequency. Unfortunately, the randomness of the motion itself was not well described by the Fourier decomposition. It was then that Paul Levy decided to apply the Schaunder basis to describe the local regularity properties of the Brownian motion. This work has been recently (1991) extended by Patrick Flandrin for the analysis of fractal Brownian motion, following the noise models of Mandelbrot and Van Ness.

Also in the 1930s, Littlewood and Paley established a new approach to group the Fourier transform coefficients of a signal in order to extract

Multirate & Subband	
Speech & Music Coding	Crosier et al. 1976
Polyphase Filters	Bellanger et al. 1976
Quadrature Mirror Filters (QMF)	Nussbaumer 1977
M-Channel QMF	Estaban & Galand 1977
Perfect Reconstruction (PR)	Smith, Barnwell & Ramstead 1981
Laplacian Pyramid	Burt & Adelson 1983
ParaUnitary	Vaidyanathan
Cosine Modulated Lapped	Malvar 1990 Ramstead
Non-Uniform Bandwidth and decimated ratios Non-Uniform Sampling	Hoang & Vaidyanathan
2-D Case	Vetterli 1984
Image Coding	Wood & O'Neil 1986

Figure 1.1.2 History of subband transform.

rapidly and more efficiently any information about singularities in the signal itself. The regrouping proposed by Littlewood and Paley is based on the concept of dyadic blocks, which can be used to create an orthonormal basis similar to what was later defined as a wavelet basis. The major difference between the Haar basis and the Littlewood-Paley basis is that whereas in the former, the scaling is performed in the time basis, in the latter the scaling is set on the Fourier transform coefficients. For this reason, the Littlewood-Paley basis is considered the dual basis of the Haar function.

While all these developments were taking place in the field of mathematics, Lusin, a physicist, was involved in the analysis and synthesis of functions using "atoms" or "basis elements." Any signal in a given space can be constructed by the combination of these "atoms." Lusin effectively initiated the field of "Harmonic Analysis," even thought this term was coined much later by G. Weiss and R.R. Coifmann in the 1980s. In 1938, Marcinkiewicz showed that the Haar functions are undoubtedly the simplest atomic decomposition. The theory of the harmonic analysis was extended by Calderon in 1964, in what is now known as the Calderon's Identity. Grossman and Morlet (a quantum physicist and an engineer) rediscovered

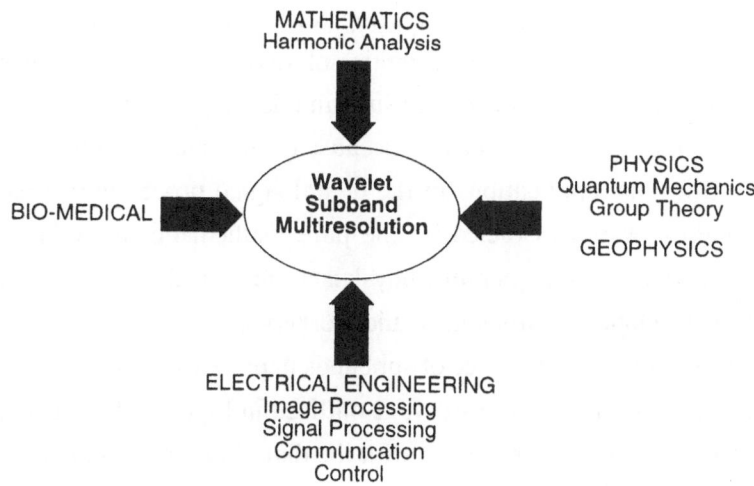

Figure 1.1.3 Different areas of applications of the wavelet, subband, and multiresolution signal processing.

this identity in 1980, 20 years after the original work of Calderon. However, they gave it a different interpretation, by relating it to the "coherent states" of quantum mechanics.

Morlet, a geophysical engineer with the French oil company Elf Acquitaine, came up with the idea of wavelets as an alternative to short-time Fourier transforms. He was interested in analyzing the seismographs related to oil exploration and found that variable-time windows, rather than the fixed window used by Gabor, produced similar results. Morlet is the geophysicist who coined the french word "ondelettes," later translated into the English word "wavelet." Morlet's work was put on a thorough mathematical foundation by Grossman and Meyer who also recognized the connection between wavelets and approximation theory [Gro84]. At this point, many people contributed significantly. Lemarie, Stromberg, Battle, and many others created new basis functions for their applications. But the major impetus to the development and popularity of wavelet theory to the scientific community, in general, came from the works of Daubechies [Dau90] and Mallat [Mal98].

Daubechies introduced the concept of compactly supported wavelets and theory of frames. She also saw the connection between the wavelet theory and the theory of subband decomposition which was independently being pursued by the digital signal processing community of electrical engineers. Mallat introduced the concept of multiresolution, which is intimately related to multirate digital filters used for subband decomposition. At this stage, there was an explosion of interest because of the connection between pure mathematics and applications in the digital signal processing community. This interest was also fueled by the pure mathematicians working with approximation theory, quantum physicists, numerical analysts, computer graphics developers, statisticians, and workers in many other fields.

Because of the importance of subbands in practical applications, it is of interest to show also their history, as outlined in Figure 1.1.2. The history of the subband transform started with the digital processing of speech and audio signals. In the 1970s, speech and music coding was initiated by Croiser et al, and the concept of polyphase filters was introduced by Bellanger et al. The Quadrature Mirror Filter (QMF) was introduced by

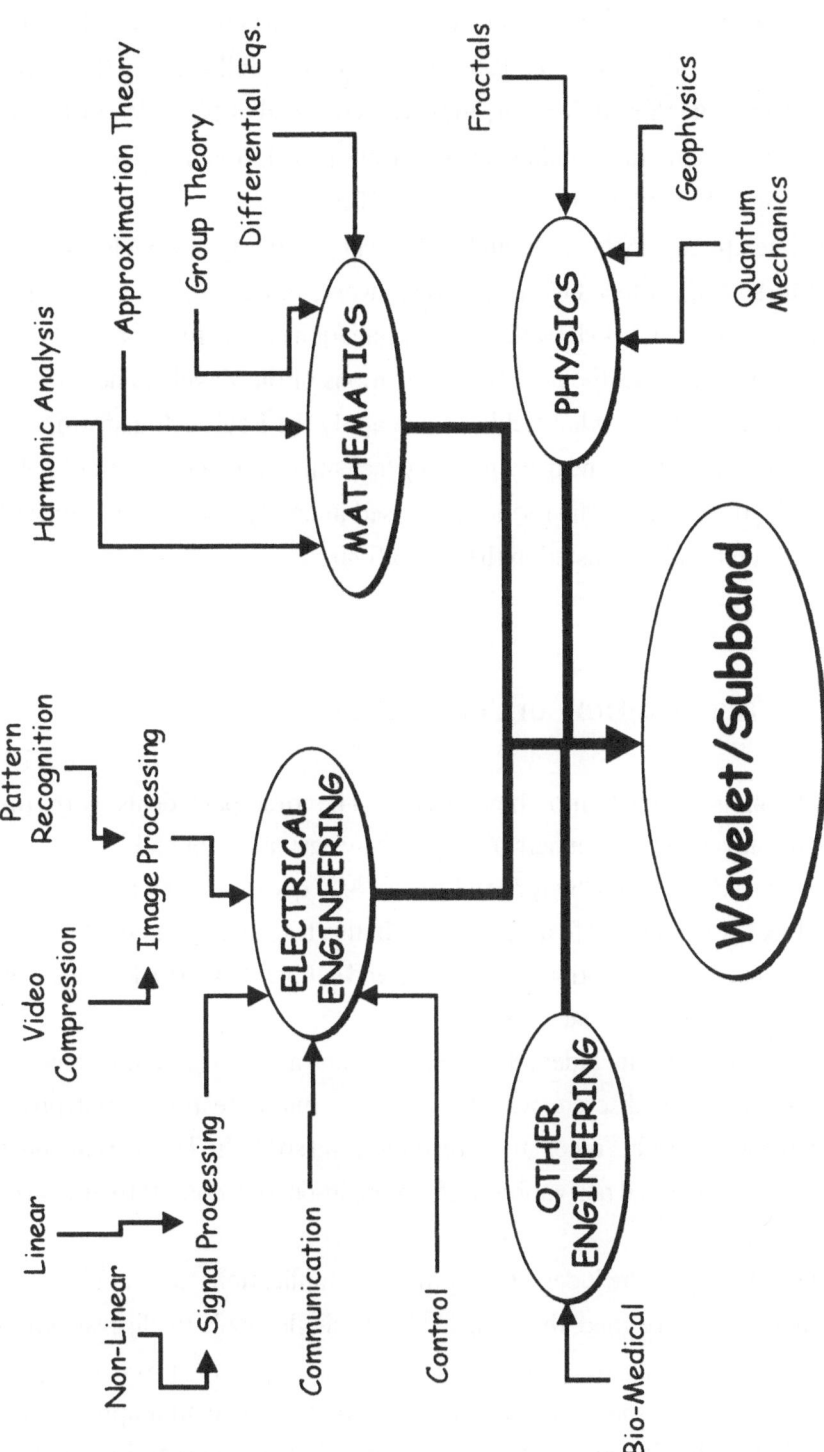

Figure 1.1.4 Disciplines of applications of wavelet/subband signal processing.

Esteban and Galand [Est77] and Perfect Reconstruction (PR) filters by Smith [Smi87], Barnwell, and Ramstead. This was followed by the works of Vaidynathan [Vai90], Malvar, Vetterli, Wood and O'Neil [Wor96]. The work of Burt and Adelson and development of the Laplacian pyramid was pioneering in the image compression field [Bur83].

Contemporary applications and the present popularity of wavelets started with Morlet and exploded with the seminal works of Daubechies and Mallat. Figures 1.1.3 and 1.1.4 show different applications in the form of charts. In general, to summarize the possible applications of the wavelet, one asks the following question: In what field can one apply the Fourier transform? Of course the answer is nearly in every scientific and engineering field. Similarly, the wavelet transform can be used in nearly every field. Some of the important ones are listed in these two figures.

1.2   Organization of This Book

This book is divided into three parts. The first part deals with the fundamentals and basic understanding of wavelets and subbands. This part tries to answer the usual "why?" and "what for?" questions which arise when the subject is introduced to a newcomer. It also tries to explain the linkages between discrete and continuous wavelets and subbands and to describe why certain applications are successful.

The second part includes advanced topics and a more rigorous discussion of continuous and discrete wavelets, frames, multirate filters, polyphase decomposition, perfect reconstruction filters, and so forth. This part includes detailed mathematical derivations and a more in-depth technical treatment of the subject matter.

The third part includes some practical applications of wavelets and subbands. It is beyond the scope of any single book to discuss every application. In this book, the major emphasis will be given to signal processing of ultrasonic signals [Abb94] and communication applications [Med95a]. This includes image processing, image compression, pattern

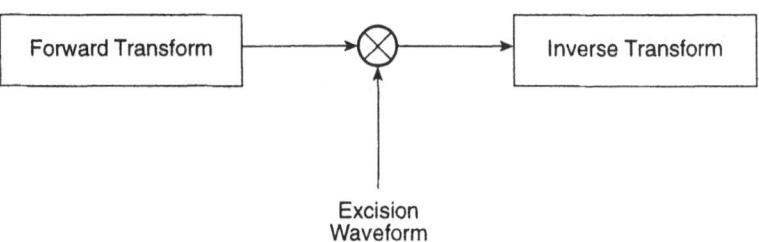

Figure 1.2.1 Block diagram of a transform domain excision system.

recognition, and signal-to-noise improvement [DeC95]. The communication application concentrates on spread spectrum systems, which have applications to wireless communication, digital multitone, code division multiple access, and excision. Implementations using digital VLSI circuits as well as optical signal processing are presented [DeC96].

It should be mentioned at this point that the wavelet transform is a particular example of a very general transform domain signal processing methodology. When a signal is transformed or mapped to a different "space" and then processed, the signal processing is said to have been done in the transform domain or, in other words, that one is using transform domain processing. Note that this mapping should be unique and unambiguous and that an inverse mapping or transformation, which can return the signal to the time domain, should exist. The most widely used continuous-time transform is the Fourier transform, but there are many others of importance such as the Fresnel, Hartley, Mellin, and Hilbert transforms, to name but a few. In communications and radar applications, for example, particularly ones using spread spectrum techniques, transform domain processing can be utilized to suppress undesired interference and, consequently, improve performance. Here, the basic idea is to choose a transform such that the jammer or the undesired signal is nearly an impulse function in the transform domain, and the desired signal is transformed to a waveform that is very "flat" or "orthogonal," with respect to the transformed interference. A simple exciser, that sets the portions of the transform which are jammed to zero, can then remove the interferer without removing a significant amount of desired signal. An inverse transform then produces the nearly interference-free

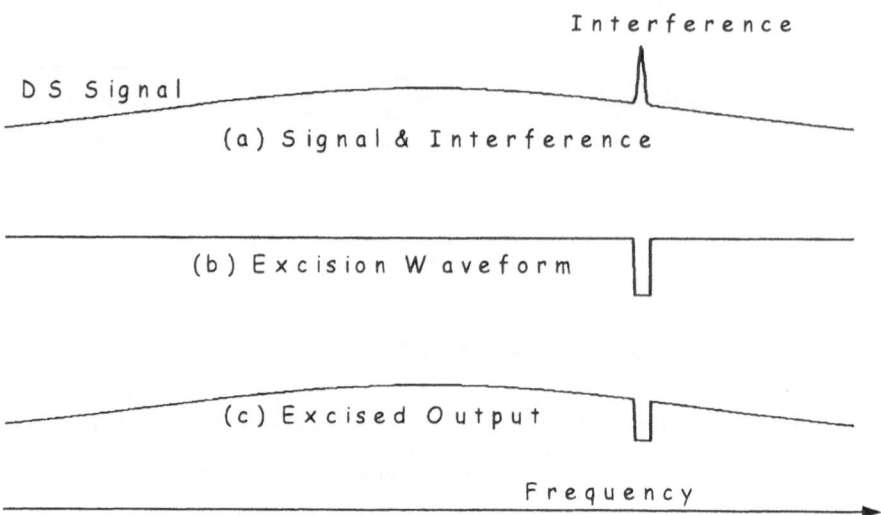

Figure 1.2.2 An illustration of the excision process.

desired signal. Figure 1.2.1 is a block diagram of a transform domain excision system. The excision waveform usually takes only the values of zero and unity, resulting in the complete removal of portions of the transform that are determined to be "jammed." Figure 1.2.2 illustrates the excision process.

The aim of the book is to describe the five transform domain techniques, so that when faced with a problem similar to the one just described, the reader can more easily proceed to a practical solution.

Finally, the book contains four appendices. These are included to avoid the annoying problem for the readers of having to run to the library for a reference to get a simple definition in subjects like linear systems, Fourier transforms, z-transforms, sampling theory, and so forth.

We have assumed that the reader has rudimentary knowledge of the basics of linear systems such as variant and invariant systems, continuous and discrete cases, impulse responses, convolution, correlation, filters including finite impulse response (FIR), infinite impulse response (IIR), matched filters, tapped delay lines, and so forth. It is expected that the reader is also familiar with continuous and discrete Fourier transforms, fast Fourier transforms (FFT), Fourier matrix, sinc functions, time-bandwidth product, uncertainty principle, Nyquist sampling theorem, Q-value of an electronic circuit, and fundamentals of communication and control systems.

At the end of selected chapters, specific references are cited for the reader to find the sources for our material and for further reading.

1.3   References

[Abb94]   A. Abbate, J. Koay, J. Frankel, S.C. Schroeder, and P. Das, Application of Wavelet Transform Signal Processor to Ultrasound, *Proc. of the 1994 IEEE International Ultrasonic Symposium*, Publ. 94CH3468-6, pp. 1147-1152, 1994.

[Bur83]   P.J. Burt and E.H. Andelson, The Laplacian pyramid as a compact image code, *IEEE Trans. Communications* vol. COM-31, pp. 532-540, 1983.

[Dau90]   I. Daubechies, The wavelet transform, time-frequency localization and signal analysis, *IEEE Trans. Inform. Theory*, vol. 36, pp. 961-1005, 1990.

[DeC95]   C. DeCusatis, J. Koay, D.M. Litynski, and P. Das, The wavelet transform: fundamentals, applications, & implementation using acousto-optic correlators, *SPIE Proc.* vol. 2643, pp. 17-37,1995.

[DeC96]   C. DeCusatis, A. Abbate, and P. Das, Wavelet Transform Based Image Processing using Acousto-Optics Correlators, Proc. of 1996 SPIE Conf. on Wavelet Applications, vol. 2762, pp. 302-313, 1996.

[Est77]   D. Estaban and C. Galand, Application of quadrature mirror filters to split band voice coding schemes, *Proc. International Conference on Acoutsics, Speech and Signal Processing ICASSP*, pp. 191-195, 1977.

[Gro84]   A. Grossman and J. Morlet, Decomposition of Hardy functions into square integrable wavelets of constant shape, *SIAM J. Math. Anal.*, vol. 15, no. 4, pp. 723-736, 1984.

[Mal89]   S. G. Mallat, A theory for multiresolution signal decomposition: the wavelet representation,. *IEEE Trans. on Pattern Analys. and Machine Intell.*, vol. 11m n. 7, pp. 674-693, 1989.

[Med95a]   M.J. Medley, G.J. Saulnier, and P.K. Das, The application of wavelet-domain adaptive filtering to spread spectrum communications, *SPIE Proceedings on Wavelet Applications for Dual-Use*, vol. 2491, pp. 233-247, 1995.

[Mey93]   Y. Meyer, *Wavelets. Algorithms and Applications*, translated by R.D. Ryan, SIAM, Philadelphia, 1993.

[Sch96a]   P. Schröder, Wavelets in Computer Graphics, *Proc. of IEEE*, vol. 84, n. 4, pp. 615-625, 1996.

[Smi84]   M.J. Smith and T.P. Barnwell III, A procedure for designing exact reconstruction filter banks for tree structured subband coders, *Proc. IEEE Intl. Conf. ASSP*, pp. 27.1.1-27.1.4, 1984.

[Vai90]   P.P. Vaidyanathan, Multirate digital filters, filterbanks, polyphase networks and applications: A tutorial, *Proc. IEEE* vol. 78, pp. 56-93, 1990.

[Vet86]   M. Vetterli, Filter banks allowing perfect reconstruction, *Signal Process.* vol. 10, no. 3, pp. 219-244, 1986.

[Wor96]   G.W. Wornell, "Emerging applications of mutirate signal processing and wavelets in digital communications," *Proc. IEEE*, vol. 84, pp. 586-603, 1996.

Part I

Fundamentals

Chapter 2

Wavelet Fundamentals

2.1   Introduction

Signal processing is based on transforming a signal in a manner that it is more useful to the application at hand [Pro88]. For example, if we are interested at reducing the noise in a signal, the best representation is the one in which the signal and noise are easily separated.* The various signal processing techniques described in this book are pictured in Figure 2.1.1.

The two most common representations for a one-dimensional signal are the temporal representation (i.e., the time signal), and its spectral dual (i.e., the Fourier transform). Unfortunately, these two representations are orthogonal to each other, meaning that it is not easy to extract frequency information from the time signal and vice versa. For this reason, the top two representations in Figure 2.1.1 belong to the two opposite corners of the time-frequency resolution plane, as the high accuracy in one domain is traded off for a complete uncertainty in the other. We can envision other signal representations in which this trade-off is not so extreme. These are the signal representations depicted in the middle column of Figure 2.1.1. Different choices of the time-frequency resolution will result in different signal

---

* In the book, we will use the term *signal* both for representing functions of one variable (i.e., audio signals), and of two or more variables, (i.e., video images). In many cases the concepts presented are equivalent in one or multiple dimensions. For this reason, the term *signal* refers to a generalized function of *N* dimensions.

.

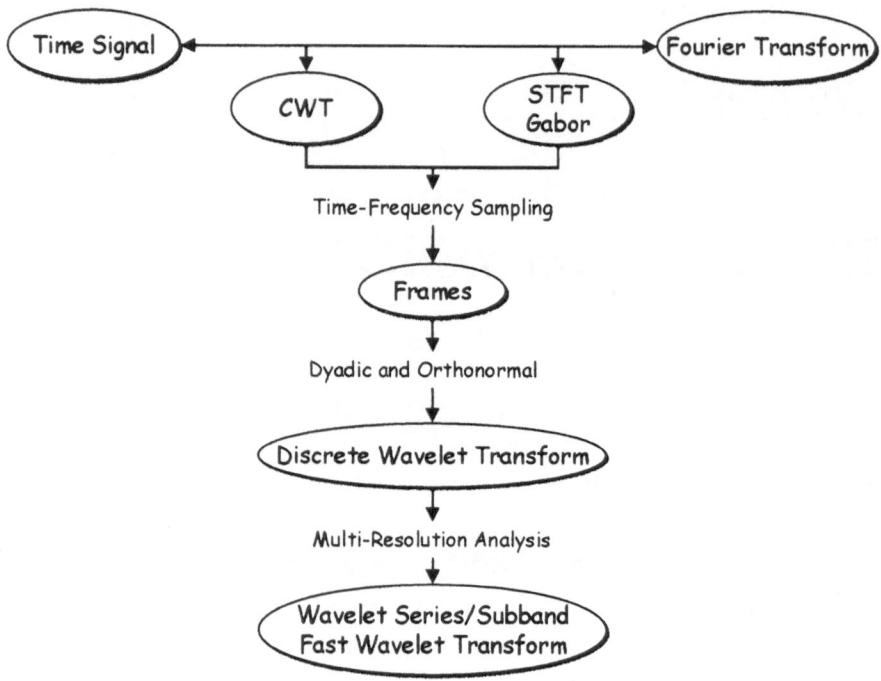

Figure 2.1.1. Signal processing tree representing the different techniques and signal representations described in the book.

representations.   The choice of the proper signal processing technique is based on the signal that needs to be analyzed.   We will see shortly that both the short-time Fourier transform (STFT) and the continuous wavelet transform (CWT) are used to represent a signal with finite resolution in both domains, time and frequency [Chu92].

By adding constraints to the transformation used, we will proceed to define a different signal processing technique, be it frames, the discrete wavelet transform (DWT) or the wavelet series and the subband transform [Dau92, Vet95]. This will result in more compact representations of the signal suited to a specific application, at the cost of reduced versatility. The key principles used to move from one representation to another are shown as labels in the links in Figure 2.1.1. By introducing the concept of time-frequency sampling we are able to introduce the frames, whereas using the concept of orthogonal transformation, we will be able to define a discrete wavelet transform and ultimately the fast wavelet transform.

This evolution in the concept of the wavelet transform from the continuous domain to the dyadic multiresolution and fast transformation can be compared to the different layers of an onion, as shown in Figure 2.1.2. The CWT can be considered the most generalized representation of the wavelet transform. The price to pay is the high redundancy of the transform. By using the sampling theorem in time and frequency, we can reduce the redundancy by using frames, which form a subset of the CWT. The inner layers of the onion thus correspond to more strictly defined and thus less general transformations that have stronger constrains on the basis functions that can be used, but also less redundancy in their output. In the following, the logical connection between Fourier, wavelet and subband will be explained using examples [Abb95, Abb95a, Abb96, DeC97, DeC97b].

The aim of this book is to familiarize the reader with the basic concepts of the signal processing techniques shown in Figure 2.1.1. It is expected that by the end of this chapter, the reader will have some understanding of the following topics, including their implementation, numerical computation, and practical applications:

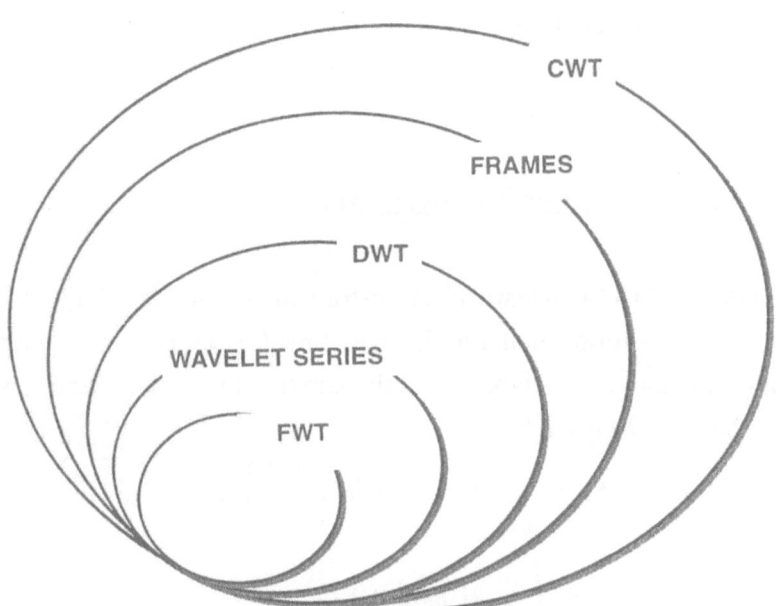

Figure 2.1.2 Wavelet transform signal processing onion.

- Wavelet Transform:
  - → continuous case ($t \rightarrow$ continuous)
  - → discrete scaling parameter a (with $t \rightarrow$ continuous)
  - → discrete scaling parameter a (with $t \rightarrow$ discrete)
  - → admissibility condition for mother wavelets
  - → tiling in the time-frequency plane and time-frequency resolution
  - → multiresolution analysis
  - → wavelet frames
  - → discrete wavelet transform (DWT)
  - → wavelet series

- Subband Transform
  - → discrete "t" and discrete "f"
  - → multirate filters
  - → polyphase decomposition
  - → analysis and synthesis filters
  - → quadrature mirror filters (QMF)
  - → perfect reconstruction (PR) filters
  - → iterated filters for subbands
  - → filter banks for subbands

2.2   Why Wavelet Transforms?

It is traditional to introduce wavelet transform in the following way [Str89]. The wavelet transform, similar to the short-time Fourier transform, maps a function, $s(t)$, into a two-dimensional domain (the time-scale plane) and is denoted by $W_s(a,b)$ given by

$$
W_S(a,b) = \frac{1}{\sqrt{a}} \int_{-\infty}^{+\infty} s(t)\, h^* \left(\frac{t-b}{a} \right)\, dt
$$

$$
= \int_{-\infty}^{+\infty} s(t)\, h_{ab}^*(t)\, dt
$$

(2.2.1)

where $h(t)$ is in general called the mother wavelet, and the basis functions of the transform, called daughter wavelets, are given by:

$$h_{ab}(t) = \frac{1}{\sqrt{a}} h\left(\frac{t-b}{a}\right) . \qquad (2.2.2)$$

Equation 2.2.1 is also known as the expansion formula the forward transform or analysis; $h_{ab}(t)$ is a set of basis functions obtained from the mother wavelet $h(t)$ by compression or dilation using scaling parameter a and temporal translation using shift parameter b. It is to be noted that

$$h(t) = h_{10}(t) . \qquad (2.2.3)$$

The scaling parameter a is positive and varies from 0 to ∞. For $a < 1$, the transform performs compression of the signal, and for $a > 1$, the transform performs dilation of the signal. The signal $s(t)$ can be recovered from the wavelet coefficients $W_s(a,b)$ by the inverse wavelet transform given by:

$$s(t) = \frac{1}{c} \int_{-\infty}^{\infty} \int_{0}^{\infty} W_s(a,b) \, h\left(\frac{t-b}{a}\right) \frac{da}{a^2} \, db \qquad (2.2.4)$$

provided that the constant c is

$$c = \int_{-\infty}^{\infty} \frac{|H(\omega)|^2}{\omega} \, d\omega < \infty . \qquad (2.2.5)$$

Equation 2.2.4 is also referred to as the reconstruction formula, inverse transform, or synthesis, and Equation 2.2.5 is generally known as the admissibility condition [Str94].

As mentioned earlier it is traditional to introduce wavelet transforms in the manner shown above. Actually, in many classes and tutorial talks, we have made this introduction in a few minutes using a couple of viewgraphs. A general silence in the audience is the result of this introduction. You can also feel the thoughts of your audience as follows:

- Why is Equation 2.2.1 called a wavelet transform? What does it mean?

- What is its connection to waves?
- Why do you want to utilize Equations 2.2.1 and 2.2.4?
- Why is the result of Equation 2.2.1 a complex 2-D transform rather than other 1-D transforms like Fourier?
- Why does $da\,/a^2$ appear in the admissibility condition?
- What is the admissibility condition?

Actually, those were the authors' own thoughts when they first heard a talk on wavelets. We were lost from then on for the rest of the talk. The objective of this chapter is to answer those questions and many others without getting into mathematical rigor. A more detailed analysis will be presented in the successive chapters.

2.3   Fourier Transform as a Wave Transform

The word "wavelet" literally means " small wave," so it is natural to talk about waves first before we discuss wavelets [Chu95a]. Consider the Fourier transform $S(\omega)$ of the signal $s(t)$ given by:

$$S(\omega) = \int_{-\infty}^{+\infty} s(t)e^{-j\omega t}\,dt \qquad (2.3.1)$$

and its inverse transform

$$s(t) = \frac{1}{2\pi} \int_{-\infty}^{+\infty} S(\omega)e^{j\omega t}\,d\omega \qquad (2.3.2)$$

where ω is the angular frequency and is equal to $2\pi f$. ** Equation 2.3.1 is the traditional definition of a Fourier transform. However, we can choose to utilize a different notation using the concept of time scaling, $a = 1/\omega$. We can also introduce the concept of a wave function as the basis function of a transformation

---

** Note that f is the frequency; there is some possibility of confusion between frequency, f, and a function of time, $f(t)$; the difference should be obvious from the context.

$$h(t) = e^{jt} \qquad (2.3.3)$$

and

$$h\left(\frac{t}{a}\right) = h_a(t) = e^{jt/a} \qquad (2.3.4)$$

Equation 2.3.1 can now be written as the wave transform:

$$W(a) = \int_{-\infty}^{+\infty} s(t)e^{-jt/a} \, dt = \int_{-\infty}^{+\infty} s(t) \, h_a^*(t) \, dt \,. \qquad (2.3.5)$$

As the basis functions are all created by scaling in time of the same mother wave $h(t)$, they will be called daughter waves. A mother wave and two daughter waves are shown in Figure 2.3.1. The functions are periodic waves that extend over the entire time axis from $-\infty$ to $+\infty$. It is important to note that the basis functions are like waves, each having a different frequency obtained by changing the scale of the horizontal axis. The top curve in Figure 2.3.1 represents the mother wave e^{jt}. If $a > 1$, as in the curve in the middle of the figure, the function is dilated in time, resulting in a lower frequency wave. If $a < 1$, as in the bottom curve, the wave is compressed in time and it has a higher frequency oscillation.

As the waves are orthonormal and extend over the entire frequency space, the inverse transformation is obtained as

$$s(t) = \int_{-\infty}^{+\infty} W(a)e^{jt/a} \frac{da}{a^2} = \int_{-\infty}^{+\infty} W(a) \, h_a(t) \, \frac{da}{a^2} \,. \qquad (2.3.6)$$

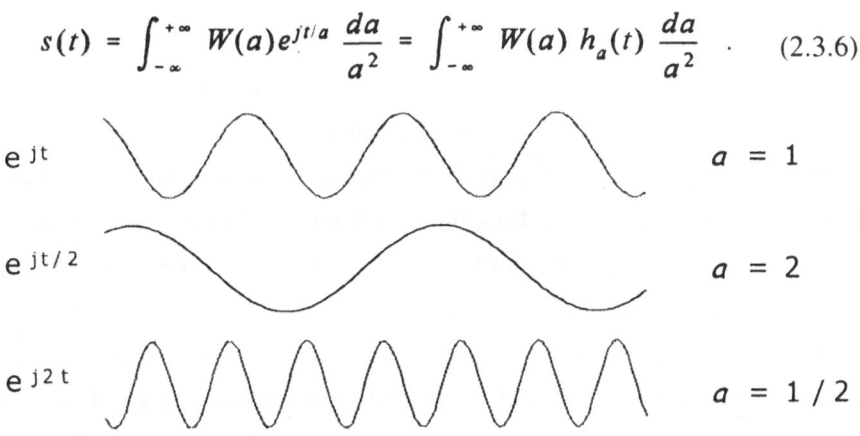

e^{jt} $a = 1$

$e^{jt/2}$ $a = 2$

e^{j2t} $a = 1/2$

Figure 2.3.1 Mother and daughter waves (Fourier bases) as a function of time for $a = 1$, 2, and ½. For $a > 1$, the wave is dilated in time resulting in a lower frequency wave. For $a < 1$, the wave is compressed resulting in the higher frequency wave.

In order to keep consistency with Equation 2.3.2, the limit of integrations of Equation 2.3.6 are set between $-\infty$ to $+\infty$. In reality most of the applications are limited to real signals, hence it is common practice to consider only positive values of the scaling parameter a. Thanks to the relationship between the Fourier transform components at the positive and negative frequencies, no information is lost if considering only positive frequencies.

The term da/a^2 represents the differential change in frequency and is obtained from differentiating the relationship between frequency and scale:

$$d\omega = -\frac{da}{a^2} .$$ (2.3.7)

The Fourier transform is the most used signal transformation in the analysis of signals, even if it is not the optimal for many applications. As an example, let us look at the Fourier transform $S(\omega)$ of the rectangular function $s(t) = u_T(t)$ defined as

$$s(t) = u_T(t) = \begin{cases} 1, & for \quad |t| < T/2 \\ 0, & otherwise \end{cases}$$ (2.3.8)

given by

$$S(\omega) = T\frac{\sin(\omega T/2)}{(\omega T/2)} = T\ sinc\left(\frac{\omega T}{2\pi}\right) .$$ (2.3.9)

The Fourier transform in Equation 2.3.9 is defined (i.e., it has no zero values) over the entire frequency axis. It follows that to define the signal $s(t)$ in the frequency domain, the complete set of values of $S(\omega)$ must be used. The rectangular function of Equation 2.3.8 and its Fourier transform are plotted in Figure 2.3.2. It can be easily seen that the signal $s(t)$ is well localized in time, but it is not well represented in frequency by its Fourier transform. This result is not surprising, as we are trying to represent the signal $s(t)$ of Figure 2.3.2 using waves that exist for every value of the time axis. To properly represent $s(t)$, we need to use enough waves such that their sum is zero for values of t larger than T and smaller than $-T$ (i.e., $|t| > T$).

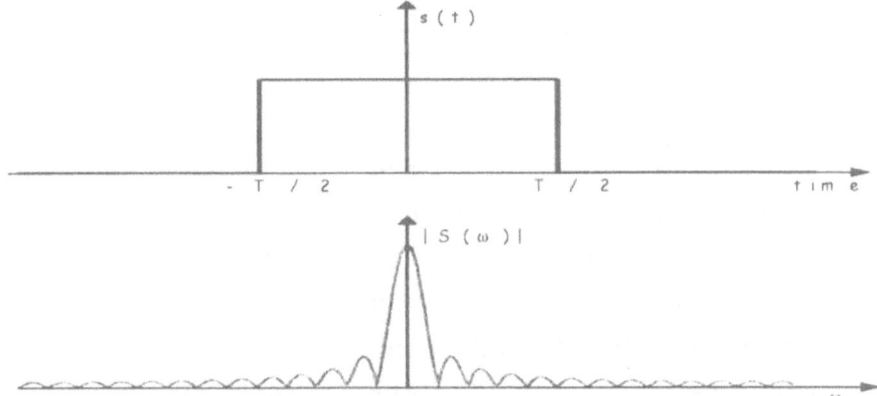

Figure 2.3.2 Rectangular function and its Fourier transform.

For time localized signals, like the rectangular function $s(t)$ shown in Figure 2.3.2, a set of basis functions such as the waves is not convenient, although mathematically correct. A more compact result can be obtained if also the basis functions of the transformation were localized. Thus, we need wavelets, not waves.

2.4   Wavelet Transform

Wavelets are localized waves and they extend not from $-\infty$ to $+\infty$ but only for a finite time duration, as shown in Figure 2.4.1. We can thus think of using $h(t)$ and its scaled daughter functions as the basis for a new transform. Unfortunately, if $h(t)$ is centered around $t = 0$, with extension between $-T$ and $+T$, no matter how many daughter wavelets we use, it will not be possible to properly represent any point at $t > T$ of a signal $s(t)$. Please note that the wave transform did not have this problem as the wave function was defined for every value of t. For the case of using a localized wave or wavelet, we must be able to shift the center location of the function. In other words, we must include a shift parameter, b, and the daughter wavelets should be defined as

$$h_{ab}(t) = \frac{1}{\sqrt{a}} \, h\!\left(\frac{t-b}{a}\right) . \tag{2.4.1}$$

The reason for choosing the factor $1/\sqrt{a}$ in the above equation is to keep the energy of the daughter wavelets constant.   Note that without this normalization factor, for different a values the wavelets dilate or compress and their total energy changes.   Again, this was not the case of the wave functions, as they extended over the entire t-axis.

Thus, we see that the wavelet transform has to be a two-dimensional transformation with the dimensions being a, the scale or inverse frequency parameter, and b, the shift parameter. Let us restate this again for emphasis: since waves extend over the entire space, they do not need any shift parameter.   Thus, a Fourier transform maps 1-D time signals to 1-D frequency signals, whereas the wavelet transform maps 1-D time signals to 2-D scale (frequency) and shift parameter signals.

There are other transforms which are very useful and actually can be more optimum for a particular case than wavelets. For example, it is well known that the Karhunen-Loeve (K-L) transform is optimum for

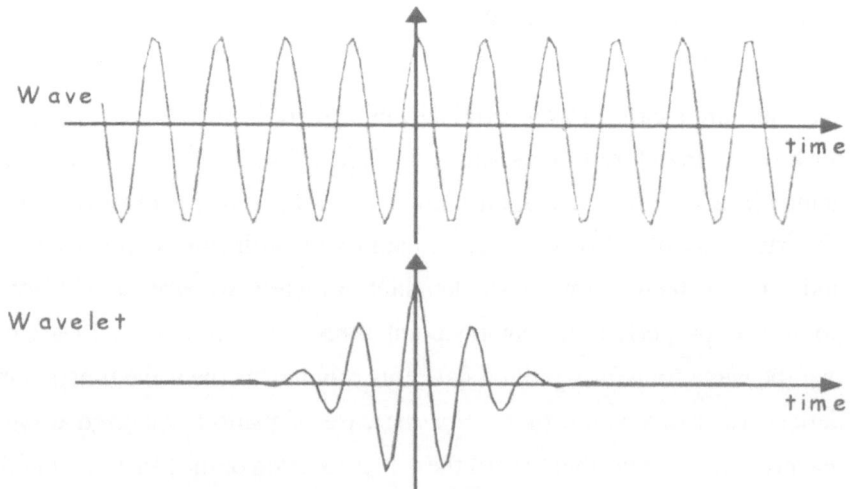

Figure 2.4.1 Difference between a wave function (top curve) as defined in Equation 2.3.3 and a wavelet function $h(t)$ (bottom curve), which has non zero values only for a finite time window.

decorrelating an input signal from other signals or noise and repacking the signal energy. However, it is often not practical to implement the K-L transform due to its computational complexity.

So, what is the important point behind the significance of wavelet transforms? The most important point is that the basis functions are obtained by scaling and shifting one particular function or by the manipulation of a single function. In that sense, the wavelet transform is similar to a Fourier transform, whose basis functions are obtained by manipulating a periodic function. Thus for periodic functions, Fourier analysis is ideal. However, with wavelet transforms, we are not restricted to only the periodic function, but any function (provided it is admissible; that is, it satisfies Equation 2.2.5). In many cases of signal processing, one can choose the signal itself or a theoretical model as the mother wavelet. The advantage of doing this is that only a few wavelet transform coefficients are then required to represent the signal, whereas the noise, interference, distortion, etc. need more coefficients. This forms the basis for interference rejection or excision in communication systems as well as for data and image compression.

As mentioned earlier, the Fourier transform (or other transforms) maps a 1-D signal into a 1-D transform domain (i.e., $t \rightarrow \omega$) whereas the wavelet transform maps a 1-D signal to a 2-D transform domain (i.e., $t \rightarrow a,b$). A wavelet transform thus has a highly redundant number of basis functions. Naturally, the question arises, can one make a less redundant set, or an orthogonal or orthonormal set of basis functions in the wavelet transform? It turns out that under certain conditions, one can do this; the details will be discussed later in the section on wavelet frames.

The need to make a set of basis functions which are localized came in the early days of signal processing. There were different attempts to tackle this problem and two well-known solutions are the so-called short time Fourier transform (STFT) and the Gabor transform. Before we get into these transforms and the discussion of admissibility conditions, it is highly fruitful to discuss another great advantage of the wavelet transform from the point of view of implementation and numerical computation; namely the connection between wavelet transforms and filter banks.

2.5   Connection Between Wavelets and Filters

By definition, Equation 2.2.1 describes the wavelet transform according to the following relationship:

$$W_S(a,b) = \frac{1}{\sqrt{a}} \int_{-\infty}^{+\infty} s(t) \, h^* \left(\frac{t-b}{a} \right) dt \ . \qquad (2.5.1)$$

For the convenience of readers familiar with linear systems notation, let us rewrite the above equation using the variable t instead of b, and the variable τ instead of t

$$W_S(a,t) = \frac{1}{\sqrt{a}} \int_{-\infty}^{+\infty} s(\tau) \, h^* \left(\frac{\tau-t}{a} \right) d\tau \ . \qquad (2.5.2)$$

As is well known from linear systems theory, Equation 2.5.2 can be rewritten as

$$W_S(a,t) = s(t) \otimes h_a^*(t) = s(t) * h_a^*(-t) \qquad (2.5.3)$$

where the symbol $\otimes$ represents the correlation operation, and the symbol $*$ the convolution operator.

The function $h_a(t)$ is given by

$$h_a(t) = \frac{1}{\sqrt{a}} \, h \left(\frac{t}{a} \right) \ . \qquad (2.5.4)$$

The wavelet transform of a signal is nothing but the correlation between the signal and the function $h_a(t)$. The wavelet transform of a signal $f(t)$ can be obtained by applying the signal as the input to a linear system whose impulse response is given by $h_a(-t)$ as shown in Figure 2.5.1. The figure also shows Equation 2.5.3 in the Fourier domain which is given by

$$\mathcal{F}\{W_S(a,t)\} = W_S(a,\omega) = S(\omega) \, H_a(\omega) \ . \qquad (2.5.5)$$

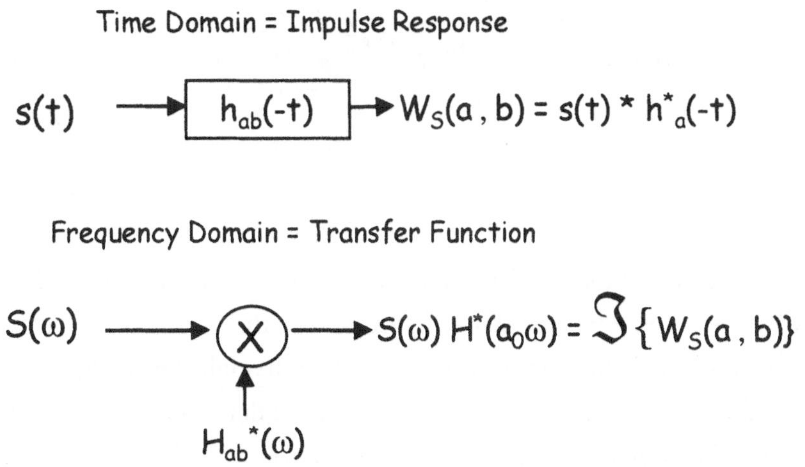

Time Domain = Impulse Response

$$s(t) \longrightarrow \boxed{h_{ab}(-t)} \longrightarrow W_S(a,b) = s(t) * h_a^*(-t)$$

Frequency Domain = Transfer Function

$$S(\omega) \longrightarrow \otimes \longrightarrow S(\omega)\, H^*(a_0\omega) = \mathcal{F}\{W_S(a,b)\}$$
$$\uparrow$$
$$H_{ab}^*(\omega)$$

Figure 2.5.1 Filter bank representation of the wavelet transform of a signal $s(t)$. (a) Time domain representation using a convolver and the impulse response of the filter; (b) same representation in the frequency domain.

where the operator $\mathcal{F}\{\}$ represents the Fourier transform, hence $W_S(a,\omega)$, $S(\omega)$ and $H_a(\omega)$ are the Fourier transforms of $W_S(a,t)$, $f(t)$ and $h_a(t)$ respectively. $S(\omega)$ gives us the frequency components of the signal. $H_a(\omega)$ gives us the frequency response of the linear system or filter whose impulse response is given by $h_a(t)$. $H_a(\omega)$ is given by:

$$H_a(\omega) = \mathcal{F}\{h_a(t)\} = \sqrt{a}\, H(a\omega) \tag{2.5.6}$$

where $H(\omega)$ is the Fourier transform of the mother wavelet.

The wavelet transform is thus the response of the filter bank constructed by the filters $H_a(\omega)$ to the signal $s(t)$. Let us utilize as an example a filter designed by Morlet as a Gaussian modulated tonerbust of center frequency $\omega_0 = 2\pi f_0$ [Gro84]:

$$h(t) = \pi^{-1/4}\, e^{-j\omega_0 t}\, e^{-t^2/2}$$
$$\tag{2.5.7}$$
$$H(\omega) = \sqrt{2}\, \pi^{-1/4}\, e^{-(\omega-\omega_0)^2/2}.$$

The scaled version of the wavelet is thus given by:

$$h_a(t) = \frac{1}{\sqrt{a}}\, \pi^{-1/4}\, e^{-j\omega_0 \frac{t}{a}}\, e^{-\frac{t^2}{2a^2}}$$

(2.5.8)

$$H_a(\omega) = a\,\sqrt{2}\,\pi^{-1/4}\, e^{-\frac{(\omega-\omega_0)^2}{2a^2}}.$$

Figure 2.5.2 shows the Morlet mother and daughter wavelets for different values of a. The left side of the figure shows the wavelets in the time domain, and their Fourier spectra are plotted on the right. From the signals displayed in the figure, we can see that the time duration of the wavelets varies as a function of the scaling parameter a.

If ΔT represents the duration of the mother wavelet, then the daughter wavelet will have a duration $\Delta_{ta} = a\,\Delta T$, (i.e., for $a < 1$ the wavelet is compressed and for $a > 1$, the wavelet is dilated in time). The center frequency and bandwidth for the daughter wavelet filters are given by

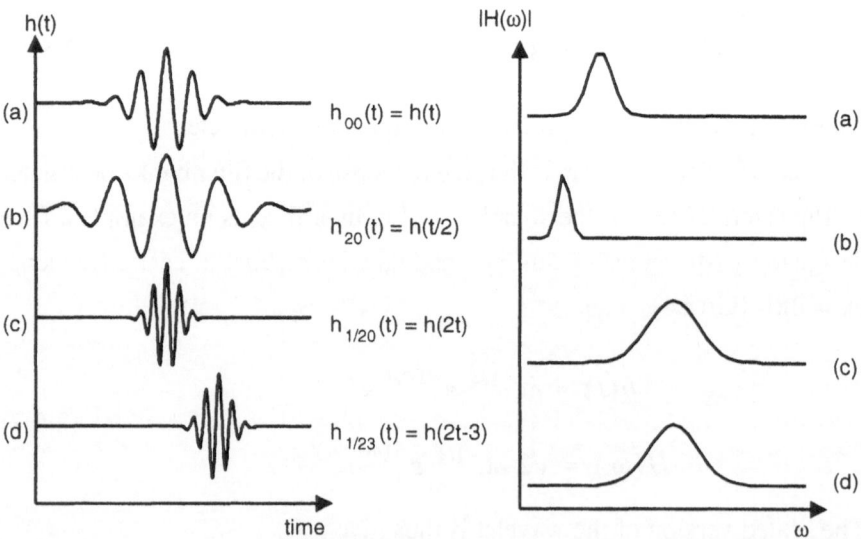

Figure 2.5.2. Scaled and shifted versions of the wavelet and the correspondent Fourier spectra.

$$\omega_a = \frac{\omega_0}{a} = 2\pi f_a = 2\pi \frac{f_0}{a}$$

$$\Delta f_a = \frac{\Delta f_0}{a}$$

(2.5.9)

where $\omega_0 = 2\pi f_0$ is the center frequency of the mother wavelet (Equation 2.5.7) and Δf_0 is its bandwidth. We also note that the Q of the filter, i.e., the ratio between the center frequency and its bandwidth is constant for any value of the scaling a:

$$Q = \frac{f_a}{\Delta f_a} = \frac{f_0/a}{\Delta f_0/a} = \frac{f_0}{\Delta f_0} = constant$$

(2.5.10)

The importance of Equation 2.5.10 will be discussed in the next section.

To obtain the wavelet coefficients, we thus process the signal using a filter bank whose frequency responses are given by $H_a(\omega)$ as shown in the frequency domain in Figure 2.5.3. The time-domain implementation using convolvers is shown in Figure 2.5.4.

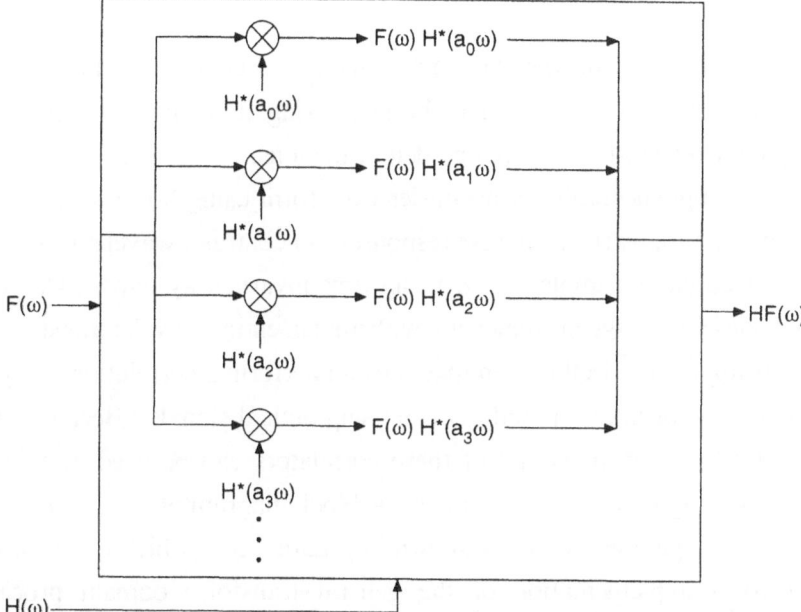

Figure 2.5.3 Filter bank representation of the Continuous Wavelet Transform

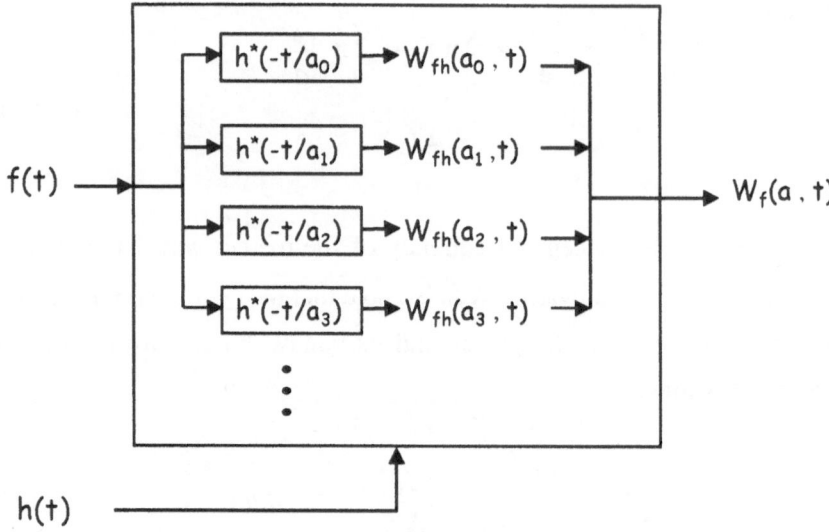

Figure 2.5.4 Time domain filter bank implementation of the continuous wavelet transform.

Although it is trivial, it is instructive to point out that the Fourier transform equation in the frequency domain is given by

$$S(\omega) = S(\omega') \, \delta(\omega - \omega') \tag{2.5.11}$$

where the δ represents the Dirac delta function. Equation 2.5.11 states that the Fourier transform is also obtained using a bank of filters whose frequency responses are given by delta functions.

The implementation of a Fourier transform using linear systems thus involves a time-variant impulse response. In contrast, wavelet transforms as defined can be implemented using time-invariant systems. This is an enormous advantage in connection with implementing wavelet transformers. For example, it is well known that one can perform a correlation of signals in real time or in the spatial domain using optical signals. Because of the time-invariant property, all of these correlators can be used directly for implementing wavelet transformers. A block diagram of a typical wavelet transform signal processor is shown in Figure 2.5.5, which is the wavelet transform implementation of the general transform domain processor discussed in Section 1.2 [Vet90a].

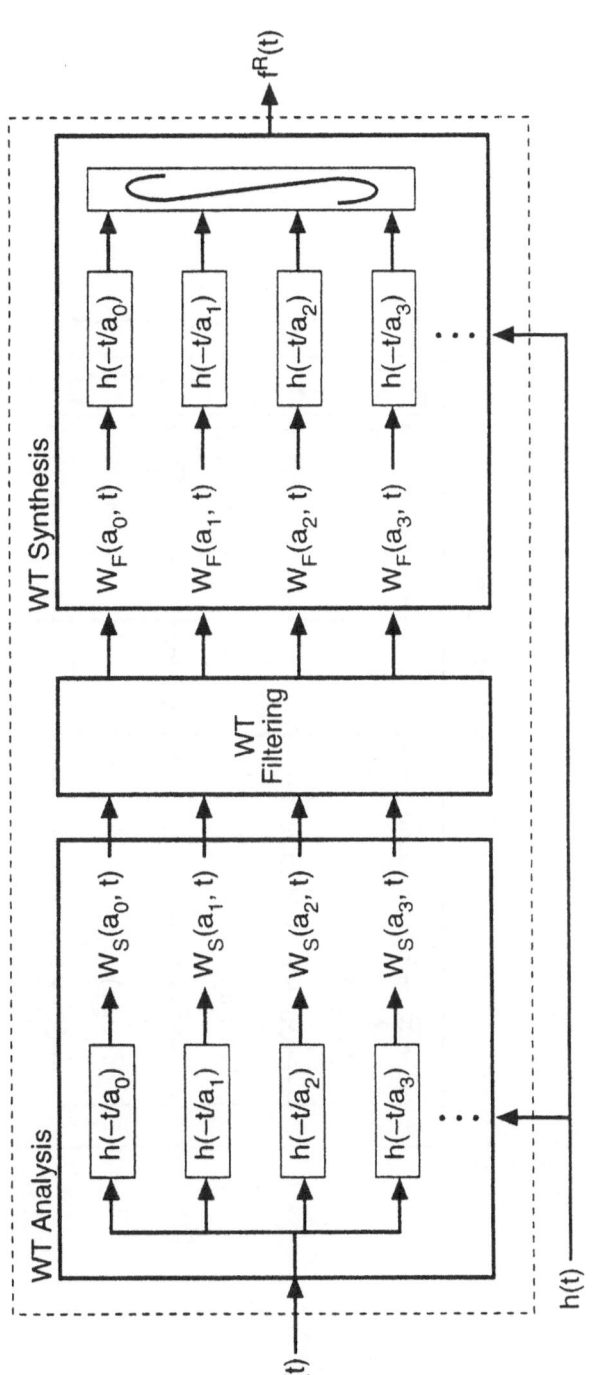

Figure 2.5.5 Block diagram of a wavelet transform signal processor. The signal is filtered in the wavelet domain and the inverse wavelet transform is used to reconstruct the processed output in the time domain.

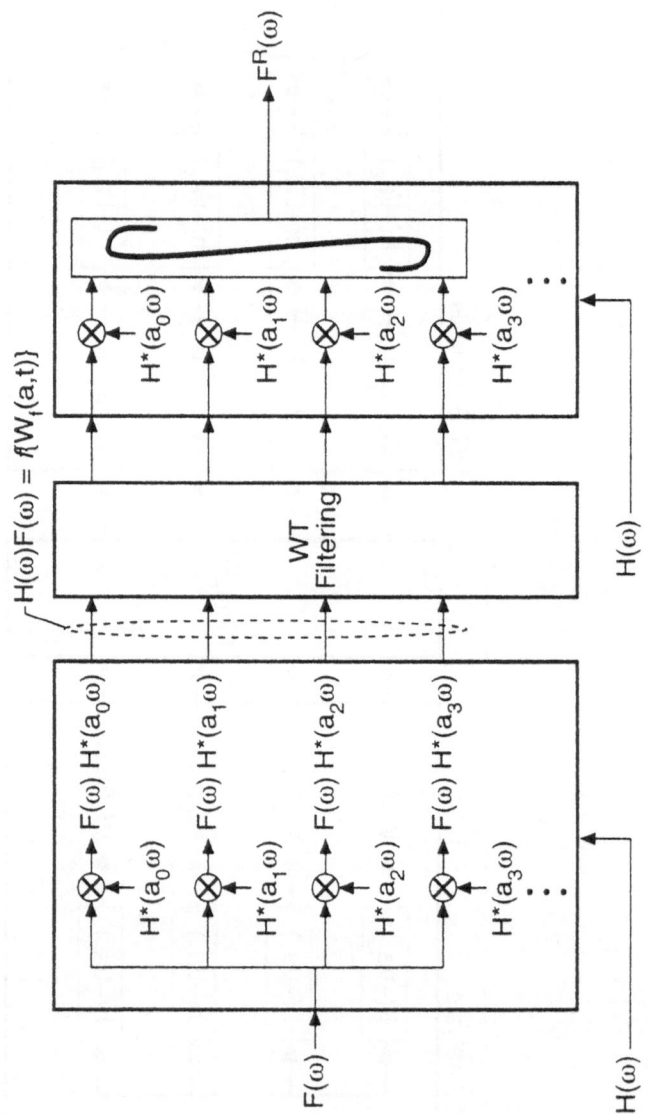

Figure 2.5.6 Frequency domain implementation of the wavelet transform signal processor.

As mentioned before, a wavelet transform can be performed either in the time domain or in the frequency domain using filter banks. The frequency domain implementation is shown in Figure 2.5.6. In this figure $S(\omega)$, $H(\omega)$, and $W_f(a)$ represent the Fourier transform of $f(t)$, $h(t)$, and $w_f(a)$, respectively. Depending on the application, the block labeled *WT filtering* multiplies the wavelet transform coefficients with another function (i.e., this can perform excision of jammers in a spread spectrum communication

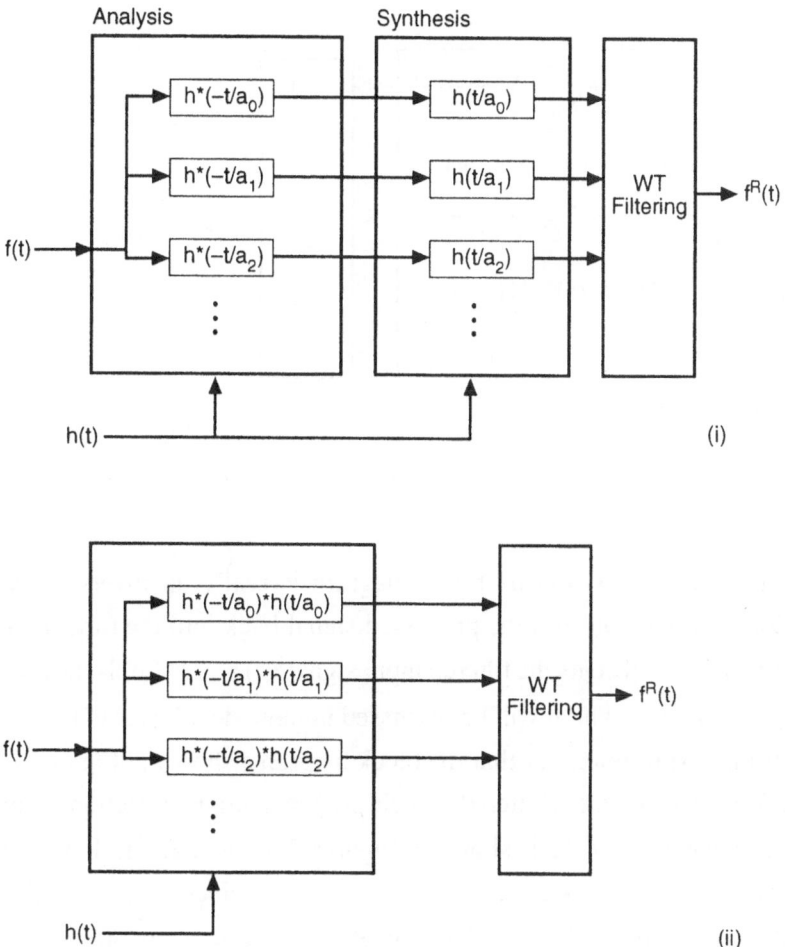

Figure 2.5.7 A different design of the wavelet transform signal processor in which the processing is performed at a later stage. Time domain representation.

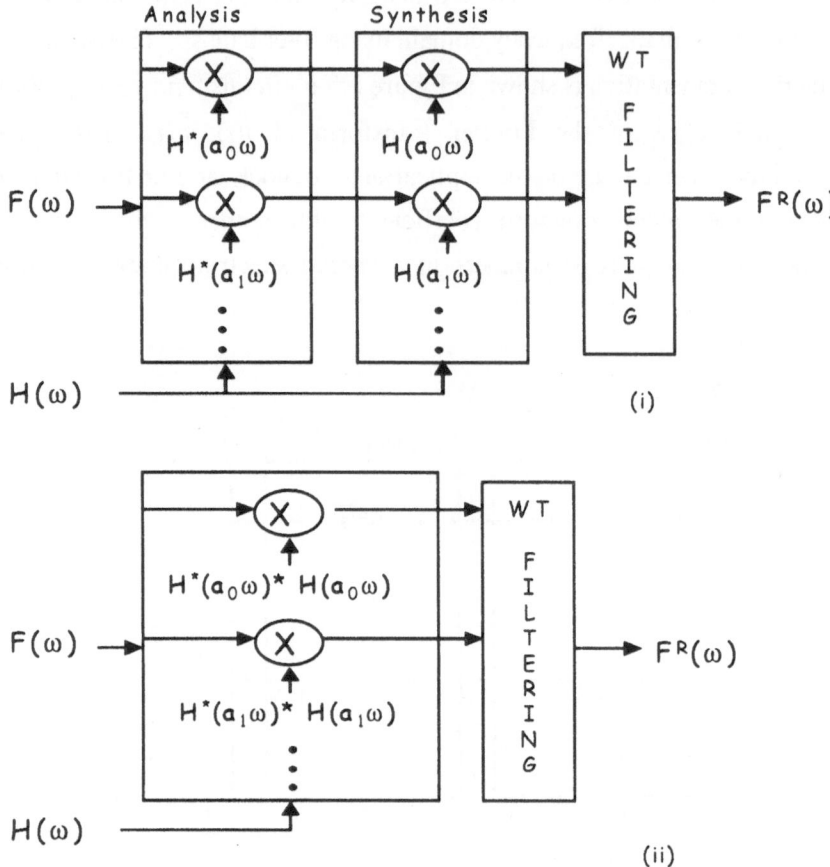

Figure 2.5.8 Frequency domain representation of the alternative design of the wavelet transform signal processor.

system). The *WT synthesis* block diagram is really an inverse wavelet transform, which converts the processed signal back into the time domain. Note that $f^R(t)$ represents the filtered input signal $f(t)$ and $F^R(\omega)$ is the Fourier transform of $f^R(t)$. These will be discussed in more detail later in this book. It is of interest to point out that the block diagrams shown in Figures 2.5.5 and 2.5.6 can be modified such that their practical implementation becomes easier. A particular case is shown in Figures 2.5.7 and 2.5.8. In this case, the filtering is performed after the *WT synthesis* block diagram. This is possible when using the wavelet transform because of its structure. In Figure 2.5.7, the synthesis and analysis blocks are combined. For this case, each filter has a combined impulse response given by $h^*(-t/a)\,h(t/a)$. Note that a similar simplification is not possible for the case of Fourier transforms. Figure 2.5.8 shows the frequency domain implementation of the modified structure. Again, the synthesis and analysis blocks can be combined for simplification. For this case the filters have combined frequency responses given by $|H(a\,\omega)|^2$.

It is convenient to discuss the admissibility condition at this point. For the wavelet transform to be a valid mathematical transform, an inverse wavelet transform should also exist. In other words, we should be able to map back to the signal $s(t)$ in the time domain from the 2-D wavelet coefficients. The inverse transform is given by Equation 2.2.4 and is repeated here for convenience:

$$s(t) = \frac{1}{c} \int_{-\infty}^{\infty} \int_{0}^{\infty} W_S(a,b)\, h\left(\frac{t-b}{a}\right) \frac{da}{a^2}\, db \qquad (2.5.12)$$

where C is an unknown constant. If we choose $s(t) = \delta(t)$ or $S(\omega) = 1$, we know that the output of the inverse transform should also be a delta function.

This situation is shown in the frequency domain in Figure 2.5.9 with $F(\omega) = S(\omega)$. It is obvious that the following equation should be satisfied:

$$S(\omega) \equiv 1 = \frac{1}{C} \int_{-\infty}^{+\infty} |H_a(\omega)|^2\, \frac{da}{a^2} \quad . \qquad (2.5.13)$$

Note that $H_a(\omega) = \sqrt{a}\, H(a\omega)$ hence

$$1 = \frac{1}{C} \int_{-\infty}^{+\infty} \frac{|H(\omega)|^2}{\omega}\, d\omega$$

$$C = \int_{-\infty}^{+\infty} \frac{|H(\omega)|^2}{\omega}\, d\omega < +\infty \quad . \qquad (2.5.14)$$

The constant C has to be less than infinity for the wavelet transform to be valid. Thus, we see that any mother wavelet can be chosen or admissible as long as $C < \infty$. In this way, we obtain the so-called admissibility condition. In lieu of deriving the condition using Figure 2.5.9, one could also directly

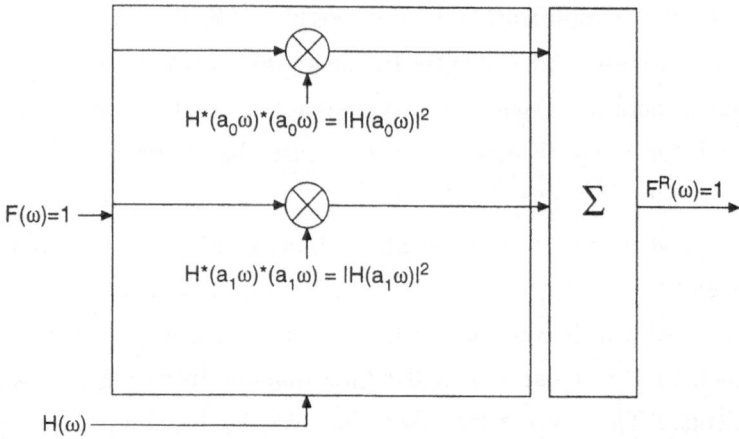

Figure 2.5.9 Filter bank representation of the wavelet transform admissibility condition.

take the Fourier transform of Equation 2.5.14 and obtain the admissibility condition.

Let us point out again the versatility of the wavelet transform. Any function can be chosen as a mother wavelet as long as the admissibility condition is satisfied. Thus, we are not confined to a single set of basis functions as is the case in a Fourier transform. This ability to choose the mother wavelet depending on the particular problem is a significant advantage. Actually, one can employ the usual optimization techniques (such as neural nets, etc.) to obtain the optimum mother wavelet for a particular problem by minimizing a properly chosen cost function. This is possible since the wavelet transform is both linear and time invariant; this means that one can choose a linear combination of admissible wavelet functions [Gro87].

From Equation (2.5.14), we note that $H_0(\omega)$ should tend to 0 as $\omega \to 0$; that is

$$H_0(\omega)\Big|_{\omega \to 0} = 0 \ . \tag{2.5.15}$$

It follows that

$$H_0(\omega) = \int_{-\infty}^{+\infty} h_0(t) \, e^{-j\omega t} \, dt$$

(2.5.16)

$$H_0(\omega) \bigg|_{\omega - 0} = \int_{-\infty}^{+\infty} h_0(t) \, dt = 0 \ .$$

Although Equation (2.5.16) is a weaker condition than the admissible condition, for most practical purposes the condition that the total area of the pulse representing the mother wavelet is zero will be sufficient. We can make the following observations regarding Equation (2.5.16): First, it is a weaker condition than the admissability condition since we do not discuss how $H_0(\omega)$ approaches zero as ω approaches zero. Second, $h_0(t)$ can be any localized function as long as it does not have dc components. Third, $H_0(\omega)$ must be a high-pass filter or a bandpass filter. The most popular wavelets are Morlet (shown previously in Figure 2.5.2), Haar, and Mexican Hat (these will be discussed further in later chapters).

2.6 Time-Frequency Analysis: Short-Time Fourier Transform, Gabor Transform, and Tiling in the Time-Frequency Plane

In this section, we discuss the difference among short-time Fourier transforms, Gabor transforms, and wavelet transforms. Actually, all three transforms are used for the time-frequency analysis of signals which will be discussed rigorously in forthcoming chapters. As we will see, this will lead us to the concept of tiling in the time-frequency plane, which basically means that we want to cover the whole space of time-frequency using different blocks, or tiles, so that we can do some useful signal processing. Historically, musicians were the first to realize the importance of the time-frequency plane. Musical scores involve both time and frequency. The sheet music score tells us the frequency of the music to be played at certain times. So, in a sense, a musical score uses what is known as the short-time Fourier transform (i.e., transformation of blocks from a signal rather than the whole signal at one time) [Coh89].

If the wavelet function $h_a(t)$ is a passband filter of center frequency ω_a and bandwidth Δf_a, then the wavelet transform value $W_s(a_0,b_0)$ contains information of the signal $s(t)$ during the time interval $(b_0 - \Delta T_a/2)$ to $(b_0 + \Delta T_a/2)$, where ΔT_a is the duration in time of $h_a(t)$. Also $W_s(a_0,b_0)$ is limited to the frequencies contained between $(\omega_{a0} - \Delta f_{a0}/2)$ and $(\omega_{a0} + \Delta f_{a0}/2)$. As we vary the shift parameter b_0, we can cover the complete time axis, and by changing the scaling parameter a_0 we can sweep through the frequency axis. Hence, the $W_s(a_0,b_0)$ represents a tile in the time-frequency plane of dimensions $(\Delta T_a , \Delta f_a)$ which vary for different values of a. For large a, ΔT_a is large but Δf_a is small, the opposite is for very small $a < 1$. The time-frequency plane is thus covered by tiles of different dimensions, as shown in Figure 2.6.1a.

A question arises: Could we choose to utilize a filter $h(t)$ in such a way that the frequency axis is covered using frequency shifting instead of scaling? The answer is yes. In reality the choice of shifting was proposed by Gabor in 1946, before the idea of wavelets really took off. Looking at the Morlet filter $h(t)$ of Equation 2.5.7, we can easily construct a shifted version of $h(t)$ as follows:

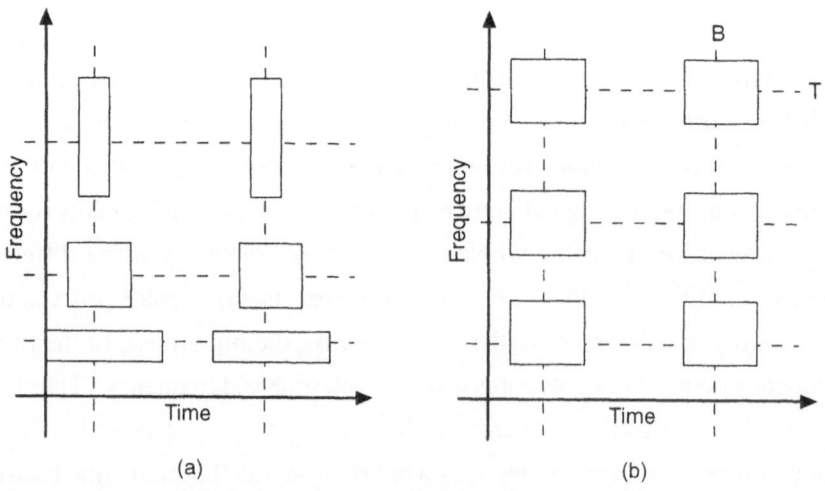

(a)   (b)

Figure 2.6.1 Tiling in the time-frequency plane by: (a) wavelets and (b) short-time Fourier transform.

$$h_{\omega_1}(t) = h(t)e^{j\omega_1 t} = \pi^{-1/4} e^{-j(\omega_0 - \omega_1)t} e^{-t^2/2}$$

$$H_{\omega_1}(\omega) = H(\omega - \omega_1) = \pi^{-1/4} e^{-j(\omega - \omega_0 + \omega_1)^2/2}$$

(2.6.1)

The functions described in Equation 2.6.1 are shown in Figure 2.6.2. Performing the same analysis as for the wavelets, we see that the center frequency of the filter associated with the function of Equation 2.6.1 has:

- constant duration $\Delta T_{\ast_1} = \Delta T_0$
- constant bandwidth $\Delta f_{\ast_1} = \Delta f_0$

hence, the time-frequency plane is covered by tiles of constant dimensions, as shown in Figure 2.6.1b [Coh93]. The functions shown in Equation 2.6.1 utilize a Gaussian window to limit their duration in time. Using these functions, we have the so-called Gabor transform:

$$S(\tau, \omega) = \int_{-\infty}^{+\infty} s(t) \, h_{\omega_1}(t - \tau) \, dt \quad .$$

(2.6.2)

If $h(t)$ is different from the Gaussian window, then $S(\tau, \omega)$ is called the short-time Fourier transform (STFT). An example using rectangular windows is shown in Figure 2.6.3. In the figure, it can be seen that the improved temporal resolution due to the finite duration of the rectangular

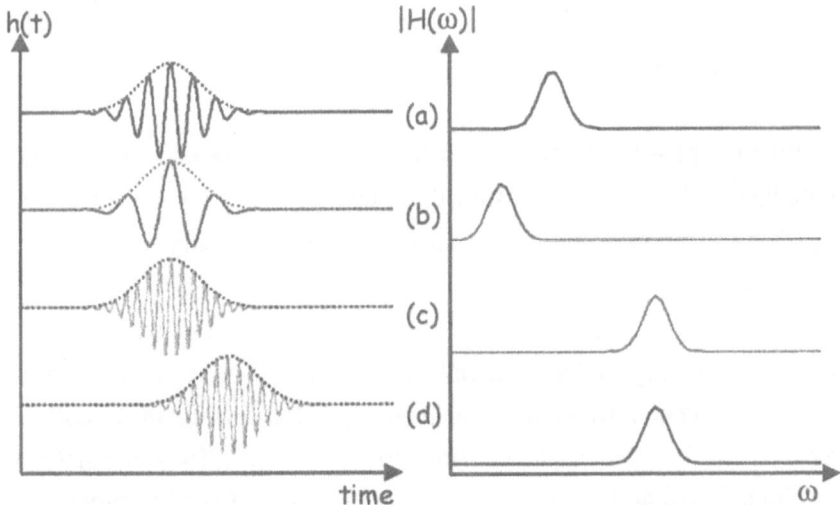

Figure 2.6.2 STFT bases and their correspondent Fourier spectra. A Gaussian windows was used.

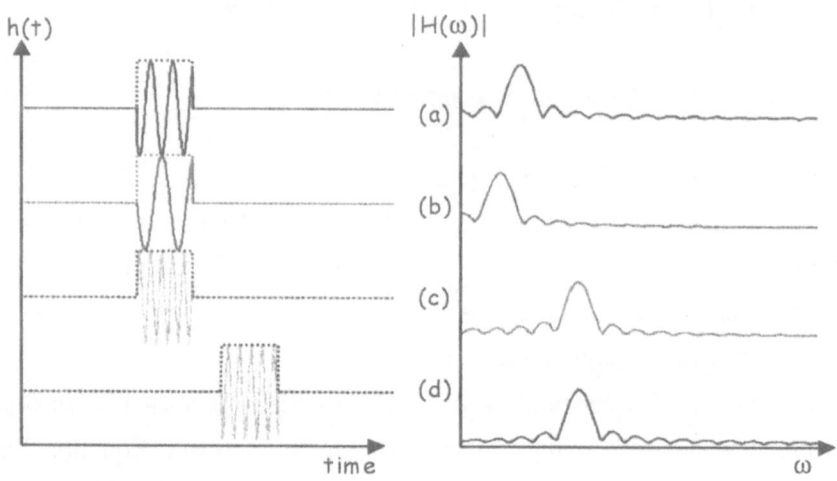

Figure 2.6.3 STFT bases and their correspondent Fourier spectra.   A rectangular windows was used.

function is obtained at the cost of a reduced resolution in frequency.

A note regarding the tiling of the time-frequency plane. As we have presented, the tile represents the area over which the information is smeared by the transformation. Thus, the smaller we make the tile, the more accurate is the information we can obtain. Unfortunately, it can be shown that

$$\Delta T \, \Delta \omega \; \geq \; 2\pi \qquad\qquad (2.6.3)$$

The above equation is known as the uncertainty principle.

Before we continue with our discussion on wavelets and subbands we should note the following points:

(i) What happens to wavelet transforms when the space is not infinite but is finite instead? Does one get something like a Fourier series for the case of wavelets? The answer is yes, and it involves the theory of frames to be discussed later.

(ii) It is customary to define in circuit theory the Q factor of a circuit to be the ratio of center frequency and bandwidth for a resonant circuit. For example, to design a clock for a computer, one looks for a resonator with very high Q so that the frequency does not deviate far from its nominal value. Note that a similar concept is used in Equation 2.4.15.

(iii) The constant Q tiles for wavelets are optimum if one considers the Nyquist sampling theorem. For low frequency, one needs fewer samples, whereas at high frequency, one needs more samples. This is also related to what is known as the multiresolution property of wavelet transforms. Multiresolution is the ability to change resolution as needed, similar to using a microscope with different magnification levels. If one needs to see the gross structure, one uses low magnification; to see very small details, a higher magnification is used. As is well known, for low-magnification one has a large field of view; however, for a higher magnification, the field of view is extremely small.

This multiresolution property of wavelet transforms is probably the most important for different applications. For example, using multiresolution one can perform what is known as progressive pattern recognition, a type of object recognition algorithm which is very similar to the way our human vision system works. For example, we need very low resolution to distinguish between, say, an elephant and a monkey. However, if we look carefully at the details of the elephant's eye, we localize our field of view and increase our resolution to look at the details, which can help us identify what type of elephant we are observing [Abb97a, Mal89].

(iv) It is instructive to discuss the time-frequency analysis properties if one insists on using Fourier transforms. This is shown in Figure 2.6.4. Here, the

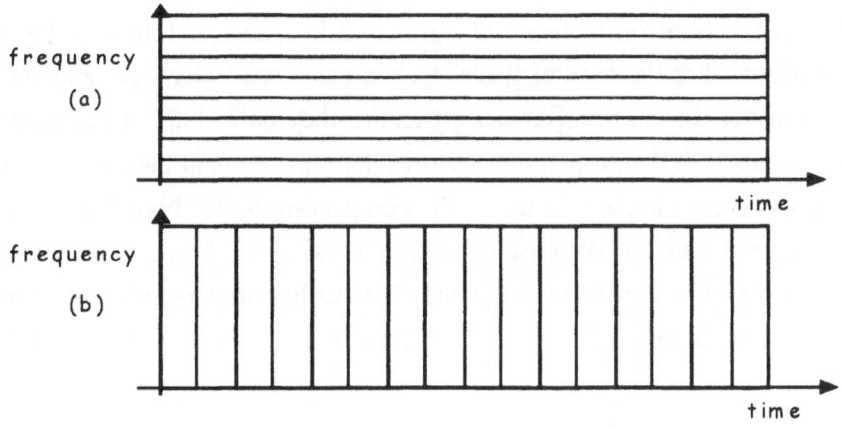

Figure 2.6.4 Tiling of the time-frequency plane using the sampling theorem and the Fourier transform.

tiles are of zero width and one needs an infinite number since there is no localization at all. Figure (2.6.4b) shows the time domain case where the Dirac delta functions are used as basis. What does this mean? This means that there can be no concept of time in the frequency domain and thus no concept of a frequency changing with time. Mathematically this can be seen from Equation 2.6.3. If $\Delta f = 0$, then ΔT must be ∞.

(v) Naturally, the question arises: What is the best way to tile the time-frequency plane so that it is useful for radar or sonar or other applications? The answer to this question will vary with the application. However, the aim of this book is to furnish the reader with the tools to answer the question properly.

(vi) We have discussed the concept of tiling, or trying to cover the whole time-frequency plane. Of course, one can use a highly redundant set of basis functions which becomes like using more tiles than absolutely necessary or one can choose just the minimum necessary. These naive questions lead to the theory of frames and orthonormal sets, which will be discussed more rigorously later.

(vii) Let us point out at this stage the fundamental difference between a wavelet transform and a STFT transform. Note that STFT transforms include the Gabor transform as a special case with a Gaussian window. Both of the transforms map 1-D signal to 2-D spaces and thus are highly redundant. For the case of wavelets, the basis functions scale and shift, whereas for the case of Gabor transforms, they only shift. Thus, Gabor transforms lack the property of multiresolution. As mentioned earlier, the theory of frames applies for the case of compact support at bounded regions; the same is true for the case of Gabor transforms. Some authors have modified Gabor transforms to include the scaling property and called it a Gabor wavelet; the definition and nomenclature becomes quite fuzzy at this stage.

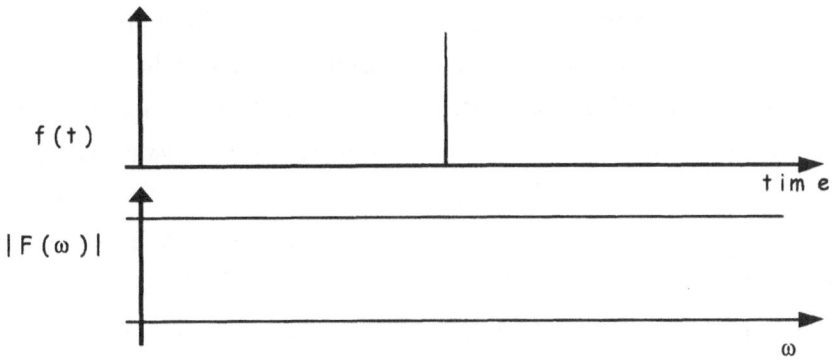

Figure 2.7.1. Dirac pulse in time (top curve) and the magnitude of its Fourier transform (bottom).

2.7   Examples of Wavelets

It is highly instructive to consider some examples to demonstrate the concepts we have developed so far. Figure 2.7.1 shows the delta function in both the time and frequency domains. The wavelet transform of a delta function using Morlet mother wavelets is shown in Figure 2.7.2a. The multiresolution or microscopy property is obvious [Gro87]. By making the scaling parameter

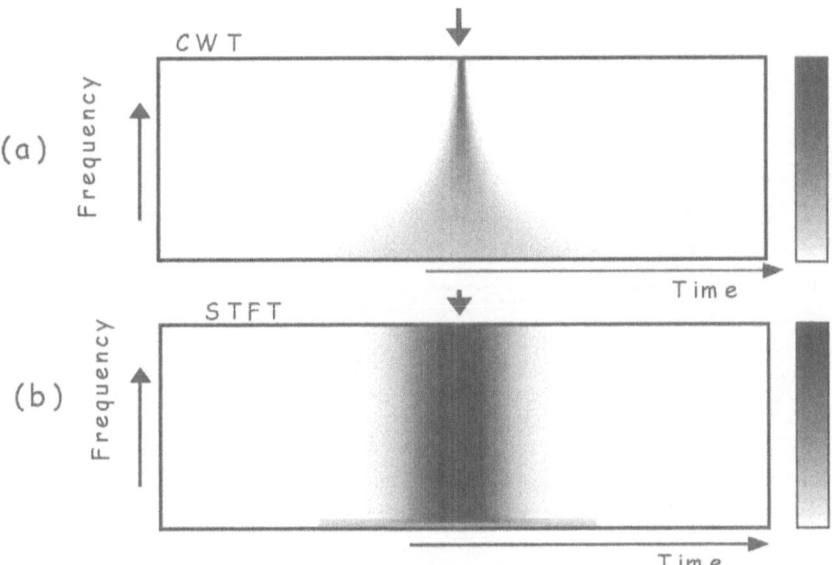

Figure 2.7.2 (a) Continuous wavelet transform and (b) short-time Fourier transform of the Dirac pulse in Figure 2.7.1.

a as small as possible, one can find the location of the impulse with an arbitrary degree of accuracy. For comparison, the STFT of the delta function is shown in Figure 2.7.2b. It is clear that the time resolution does not vary as a function of the frequency. Figure 2.7.3 shows the wavelet transform of a step function.   For the region in time far from the step, the signal $s(t)$ is constant. As the area under $h_{ab}(t)$ is equal to zero, the CWT has nonzero values only in the proximity of the step. For $a \ll 1$, the daughter wavelet is very compressed and thus will have only nonzero value just near the step. However, for $a > 1$, daughter wavelet are wider in time; for this case, the wavelet transform coefficients will have nonzero values away from the step. Thus, depending on the value of a, one can zoom into a singularity or obtain a global view; this is the microscopic or zooming ability of wavelets mentioned earlier. If we want to represent a fast transient as a step function, we only need few values of the wavelet to properly identify the transient. Conversely, you need a large number of frequency components to properly represent the same step.

The CWT and the STFT of a single and double tone bursts are shown in Figures 2.7.4 and 2.7.5. In these figures, the CWT is given in part (b) and the

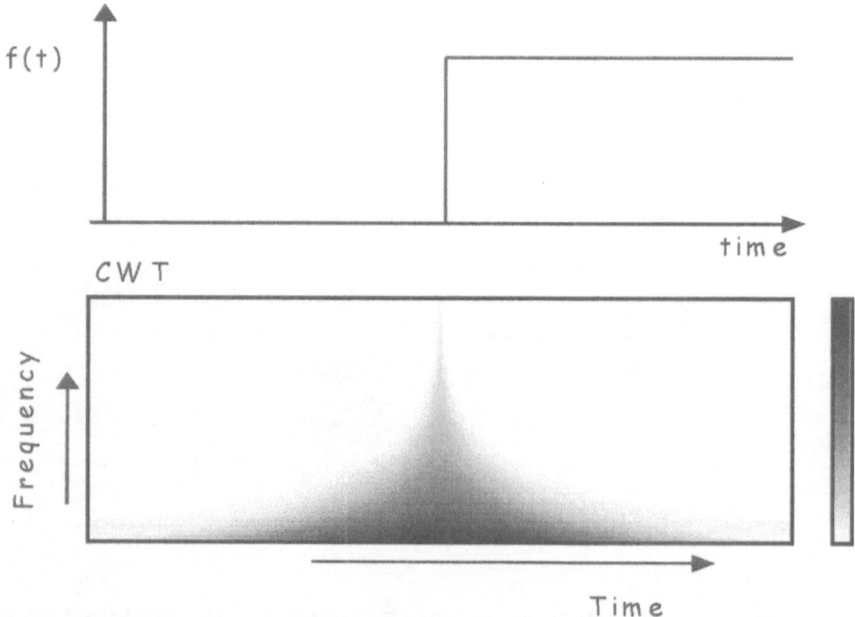

Figure 2.7.3 Step function and its CWT.

STFT is given in part (c). Finally, Figure 2.7.6 shows the more complicated case of a linear up-chirp.   From these figures we can clearly see the differences between Fourier and wavelet analysis as previously discussed.

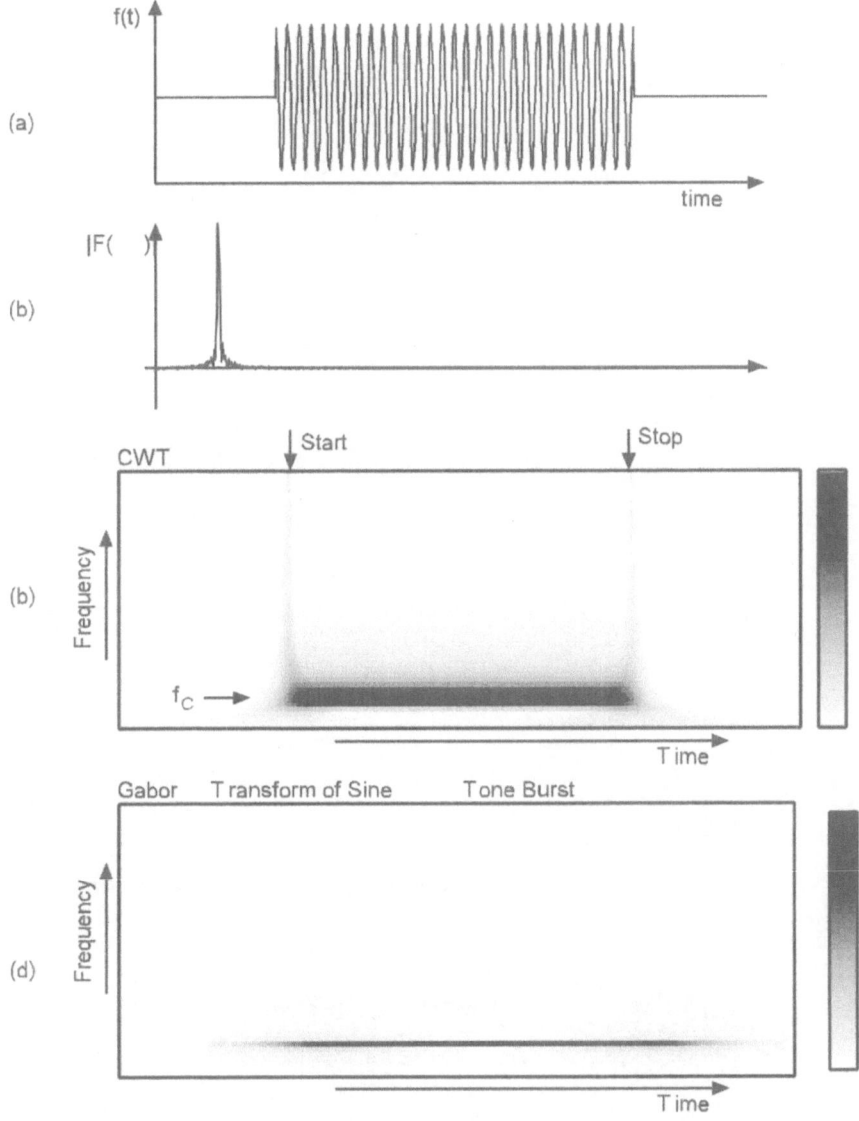

Figure 2.7.4 Analysis of a single frequency tone burst: (a) temporal representation; (b) relative Fourier spectra; (c) continuous wavelet transform and (d) Gabor transform.

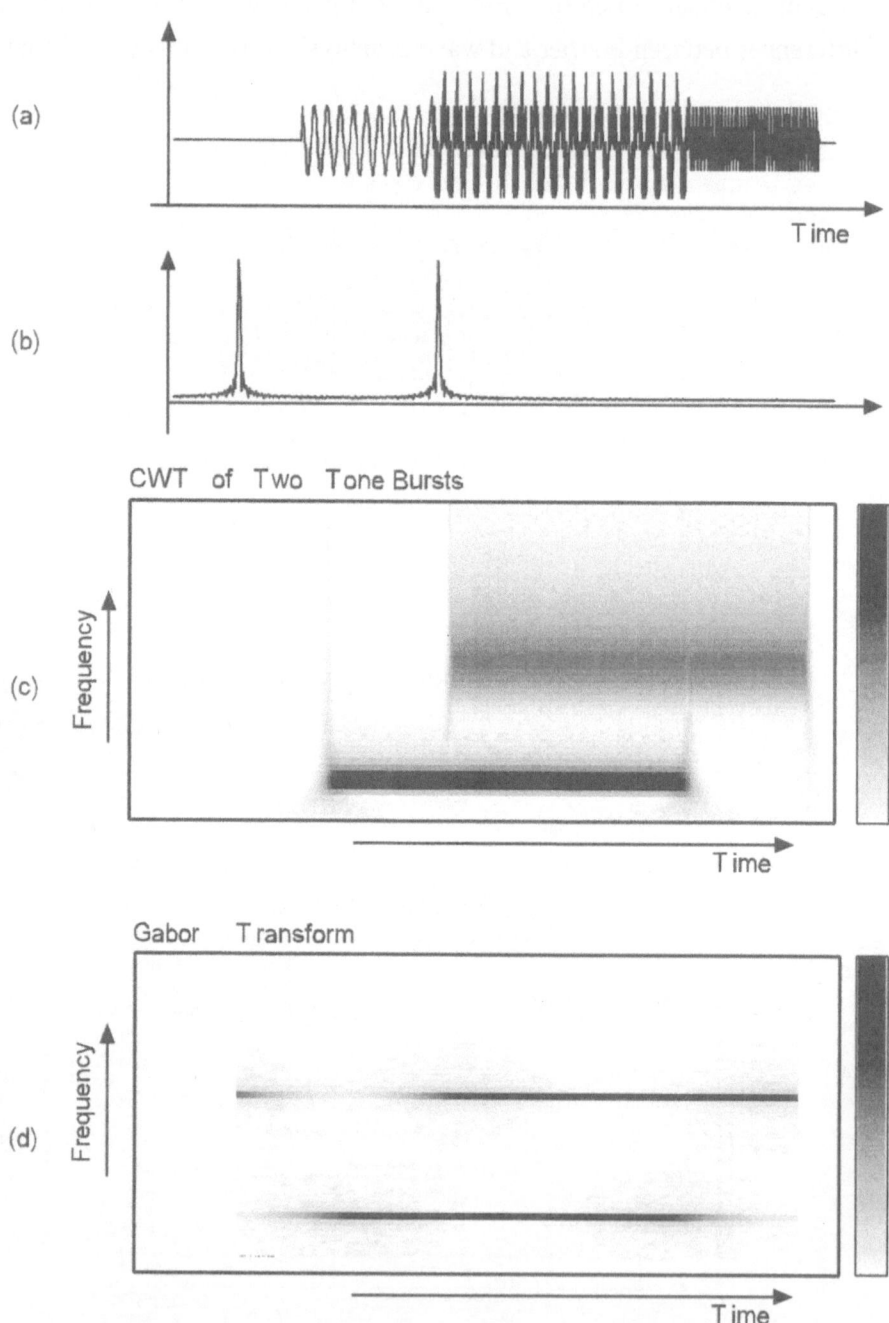

Figure 2.7.5 Analysis of a single-frequency two tone burst signal: (a) temporal representation; (b) relative Fourier spectra; (c) continuous wavelet transform, and (d) Gabor transform.

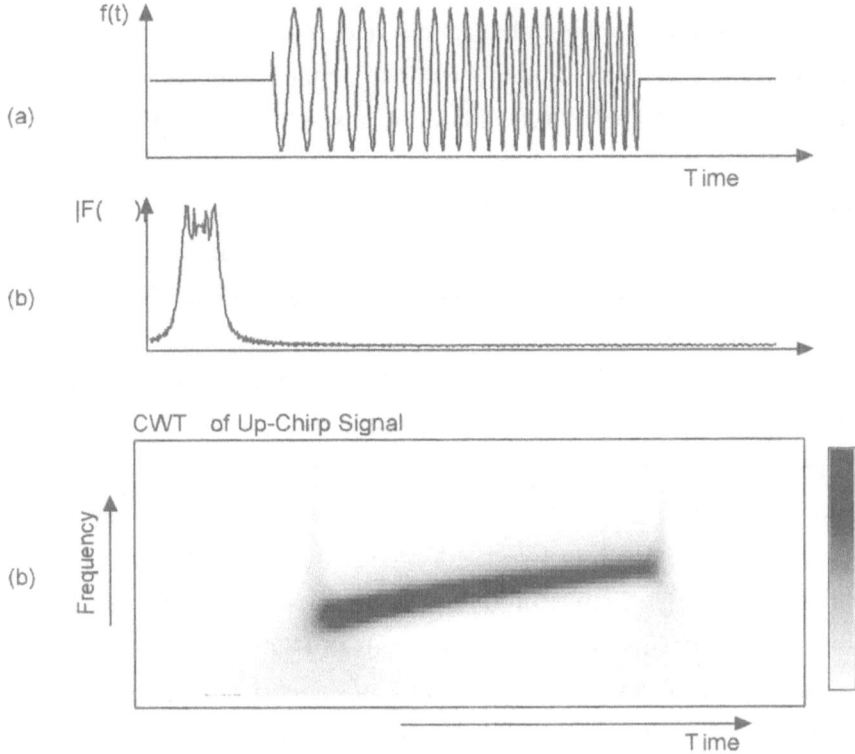

Figure 2.7.6 Analysis of a up-chirp signal. (a) temporal representation; (b) relative Fourier spectra; (c) continuous wavelet transform, and (d) Gabor transform.

2.8   From the Continuous to the Discrete Case

So far, we have considered the time variable, t, to be continuous and the variables a, b and ω to be continuous as well. While considering time-frequency analysis, we mentioned the STFT when ω became discrete and t, although continuous, was bounded. Before we discuss the situation for wavelet transforms and how they change when we go from continuous to discrete variables, it is highly instructive to consider the analogous case for Fourier transforms. A Fourier transform is the case where both time and frequency (t and ω) are continuous. If time is bounded but still continuous and ω is discrete, we obtain a Fourier series. We can also have the complementary case, when frequency is bounded and time is discrete. The next case is where both t and ω are discrete; of course, when both variables are discrete, Nyquist's sampling theorem applies. This is the widely used discrete Fourier transform (DFT). As it is well known, the DFT can be also discussed in the framework of matrix theory, where one deals with a DFT matrix of size $N \times N$, where N is any integer. A special case of this is a fast Fourier transform (FFT) with $N = 2^m$ where m is an integer; this is also known as the dyadic property. The FFT, because of its dyadic nature, is highly computationally efficient.

In the theory of linear systems, one also deals with continuous and discrete times. In the first case, we talk about analog signal processing, whereas the latter is called discrete-time signal processing. Analog signal processing involves the Fourier transform, whereas the so-called z-transform plays an important role for the discrete case. Digital signal processing involves not only discrete time but also values which are quantized.

A Fourier transform maps 1-D signals to 1-D frequency space, whereas a wavelet transform maps 1-D signals to a 2-D space (the dimensions being given by the scale and shift parameters a and b). Thus, when we start to discretize the variables, the situation becomes quite complex [Coh96]. For this case, t can be continuous, bounded, and discrete as before. However, we can have a situation where the signal itself is continuous with continuous t and f but the basis functions have bounded t and discrete a. This is the case which people generally refer to as a discrete wavelet transform (DWT) which is rather unfortunate because people tend to mistakenly associate this with the DFT, where t is discrete and not continuous. Thus, one cannot use matrix theory with discrete t values to discuss DWT as the variable t is still continuous. Equivalent to the DFT, one has the case where t and a are both discrete. This is the case generally known as subbands and was developed independently starting from the concepts of digital signal processing. In wavelet theory, we will call this a wavelet series. Just as we modified the DFT to obtain the FFT, if one restricts subbands to the dyadic case; one obtains high computational efficiency like the FFT case, except using polyphase decomposition.

We will begin to describe the DWT in the next section. However, it is

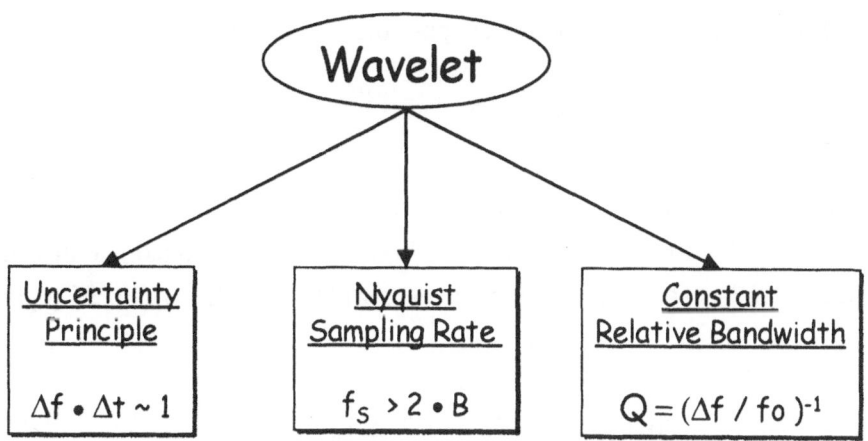

Figure 2.8.1 Link between the wavelet transform and the three fundamental interdependent forces.

instructive to discuss the interplay of three fundamental interdependent "forces" when one goes from continuous analysis to the discrete or digital case. As shown in Figure 2.8.1, these three forces are the uncertainty principle, the Nyquist sampling theorem (and its consequence, aliasing if the signal is under sampled), and the Q of the basis functions (or the wavelet "tile" parameters in the time-frequency plane). In the next section, we discuss localized, bounded, mother and daughter wavelets leading to discrete a and b. It turns out that this discretization can be optimal in the sense that the smallest number of wavelet basis functions are needed. However, for many reasons, such as robustness in signal processing and better performance, one includes many more basis functions than absolutely necessary. This case leads to the theory of frames.

2.9   Frames

The reconstruction formula, given by Equation 2.2.4, is a double integral and involves an overcomplete set of basis functions. It is highly redundant. We would like to perform the expansion and reconstruction with a less redundant set of basis functions. This means that we want to use the discrete case and as low a number of a's and b's as possible. Let us choose

$$a = a_0^{m} \qquad (2.9.1)$$

where a_0 is a constant (not equal to 1) and m is any positive integer ($m \in Z$). We know that the mother wave is localized. Let it be bounded within a time interval ΔT_0. Then the daughter wavelets will be bounded within ΔT_a, given by

$$\Delta T_a = a \, \Delta T_0 = a_0^{m} \, \Delta T_0 \quad . \qquad (2.9.2)$$

From the previous discussion on tiling of the time-frequency plane, we note that to cover the whole time axis, we do not need continuous b's, but rather we can use a discrete number of steps:

$$b = n \, b_0 \, a_0^{m} \qquad (2.9.3)$$

where n is any integer ($m \in Z$) and b_0 is a positive constant. The daughter wavelets can thus be written as

$$h_{n,m}(t) = a_0^{-m/2} \, h(a_0^{-m}t - nb_0) \quad . \qquad (2.9.4)$$

In general, we utilize a dyadic scaling ($a_0 = 2$). Also, we normalize everything to $b_0 = 1$, as commonly done in signal processing. In this case, Equation 2.9.4 becomes

$$h_{n,m}(t) = 2^{-m/2} \, h(2^{-m}t - n) \quad . \qquad (2.9.5)$$

It is now common to ask : What are the conditions that must be imposed on a_0 and b_0 in order to obtain a stable reconstruction? This question is answered by the theory of frames, as we shall now discuss. Let us write the signal $s(t)$ as

$$s(t) = \sum_{n=-\infty}^{n=+\infty} \sum_{m=-\infty}^{m=+\infty} W_{nm} \, h_{nm}(t) = \sum_{n \in Z} \sum_{m \in Z} W_{nm} \, h_{nm}(t) \quad . \qquad (2.9.6)$$

A family of functions $h_{mn}(t)$ in Hilbert space is called a frame if there exists some A and B satisfying the inequality $0 < A \le B$ such that for all signals s in H

$$A \, \|s\|^2 \le \sum \sum |W_{mn}|^2 \le B \, \|s\|^2 \qquad (2.9.7)$$

where

$$\| s \|^2 = \int_{-\infty}^{+\infty} | s(t) |^2 \, dt \qquad (2.9.8)$$

and A and B are called frame bounds. A detailed discussion of Equation 2.9.7 will be given in a later chapter; for now, we will only discuss the results [Dau90].

If $A = B$, it is called a tight frame. The ratio A/B is called the redundancy ratio, which is equal to 1 for a tight frame. For a tight frame, if one normalizes the functions such that $A = B = 1$ and $\|h_{nm}\|^2 = 1$, then that set of h_{nm} forms a set of orthonormal basis functions.

We note that $\| s \|^2$ is the total energy of the signal. $\| W_{nm} \|^2$ is related to the total energy of the signal in the transform domain and is equal to the energy of the signal for the orthonormal case. The physical meaning of the condition $\Sigma\Sigma \, \|W_{nm}\|^2 \le B\|s\|^2$ is that $\Sigma\Sigma\|W_{nm}\|^2$ must be less than infinity for a proper representation of the signal, or in other words, the energy must be finite.   The other condition $A\|s\|^2 \le \Sigma\Sigma \, (W_{nm})^2$ means that the expansion energy cannot go to zero (i.e., $0 < A$).

One can easily show that by substituting $h_{nm}(t)$ given by Equation 2.9.4 in Equation (2.9.7), one obtains [Dau92]

$$\frac{b_0 \ln a_0 \, A}{2\pi} \le \int_0^\infty \frac{|H(\omega)|^2}{\omega} \, d\omega \le \frac{b_0 \ln a_0 \, B}{2\pi} \; . \qquad (2.9.9)$$

For tight frames

$$A = B = \frac{2\pi}{b_0 \ln a_0} \int_0^\infty \frac{|H(\omega)|^2}{\omega} \, d\omega \; . \qquad (2.9.10)$$

For the orthonormal case $A = B = 1$ or

$$\int_0^\infty \frac{|H(\omega)|^2}{\omega} \, d\omega \; = \; \frac{\ln 2}{2\pi} \tag{2.9.11}$$

and for the dyadic case $a_0 = 2$ and $b_0 = 1$.

As mentioned earlier, the dyadic case is also generally referred to as the discrete wavelet transform or DWT [Dau89]. Because of its importance we will summarize the results for this case. For a mother wavelet $h_0(t)$, the daughter wavelets are given by

$$h_{nm}(t) \; = \; 2^{-m/2} \, h(2^{-m}t - n) \quad . \tag{2.9.12}$$

In the Fourier domain, it is proportional to

$$H_{nm}(\omega) \; = \; \sqrt{2^m} \, H(2^{-m}\omega) \, e^{-j\omega n} \quad . \tag{2.9.13}$$

Also note that because the unmodulated or dc term in the series expansion must be zero, $H_{nm}(\omega)$ must represent a bandpass filter. Let us consider a signal space in the frequency domain which goes from 0 to f_{max}. We immediately realize that $H_{nm}(\omega)$ must span the whole frequency space. We also realize that as $H_{nm}(\omega)$ have response functions which are basically bandpass filters, we need very large values of m to cover the low- frequency part of this space. It becomes obvious that if we choose two filters whose frequency responses are low-pass and bandpass, respectively, and then scale them, we can span the required frequency space without resorting to large values of m. This is the preferred solution, and it leads us to what is known as multiresolution analysis, which will be discussed further in Section 2.11. The time domain response of the low-pass and high-pass filters are represented by scaling functions and wavelet functions, respectively [Vet87].

We mentioned that it is possible to form an orthonormal set of basis functions, $h_{nm}(t)$. To be orthonormal we must have

$$\int h_{nm}(t) \, h_{qp}(t) \, dt \; = \; \delta_{m-p}\delta_{n-q} \quad . \tag{2.9.14}$$

The steps leading to the design of an orthonormal set will be discussed after we consider the discrete time case or subbands. Not only did subbands precede the development of the DWT historically, but it is also somewhat easier to understand the DWT if one is familiar with subbands.

2.10   Subbands

Let us consider the wavelet transform when the time axis is not continuous but discrete. The sampling rate, f_s, is determined by the Nyquist sampling theorem, which states that for a signal with highest frequency f_{max}, the sampling frequency must be $f_s > 2f_{max}$, or the sampling interval must be

$$T_S = \frac{1}{f_S} < \frac{1}{2f_{max}} \ . \tag{2.10.1}$$

Thus, for this case, $s(t)$ becomes $s(nT_S) = s(n)$ since in general one normalizes the time variable so that $T_S = 1$. Hence we are analyzing a band-limited signal with $\omega_{max} = 2\pi f_{max} = \pi f_S = \pi$, as shown in Figure 2.10.1. A typical discrete-time linear system is shown in Figure 2.10.2. Similar to the impulse response in the continuous case, we have a delta response denoted by $h(n)$. The convolution integral is replaced by a discrete convolution given by

$$g(n) = f(n) * h(n) = \sum_m f(m) \, h(n-m) \tag{2.10.2}$$

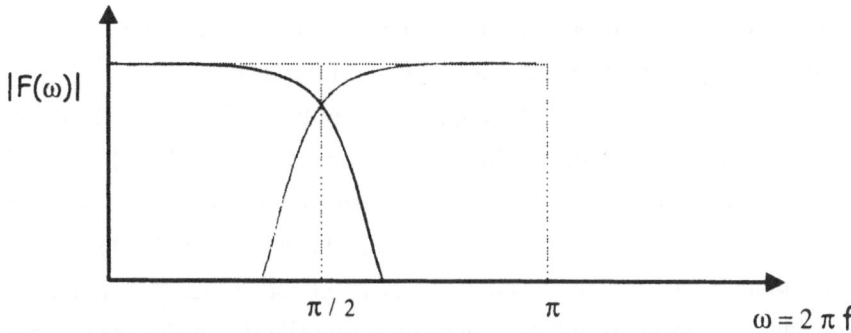

Figure 2.10.1 Frequency space of a discrete-time signal with sampling rate $T_s=1$ leading to $\omega_{max} = \pi$.

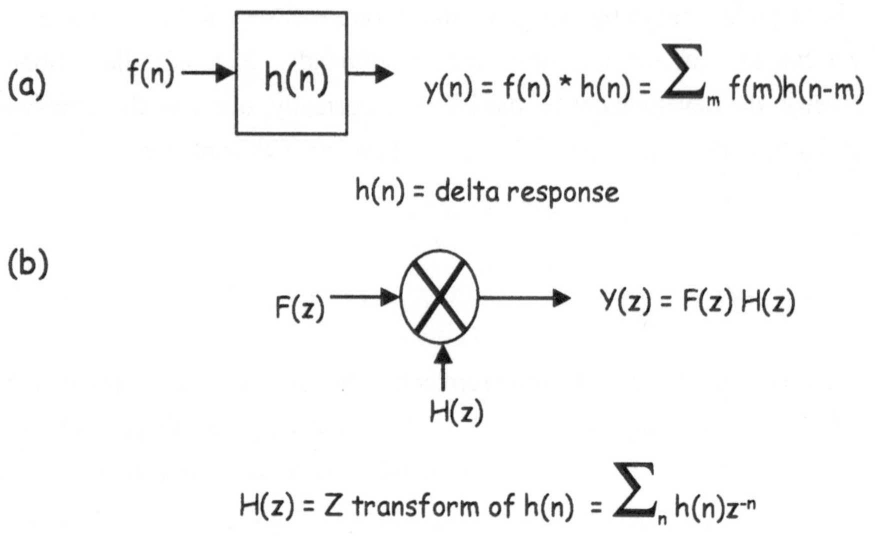

(a) $f(n) \longrightarrow$ | h(n) | $\longrightarrow$ $y(n) = f(n) * h(n) = \sum_m f(m)h(n-m)$

h(n) = delta response

(b)

$F(z) \longrightarrow \bigotimes \longrightarrow$ $Y(z) = F(z) H(z)$

$H(z)$

$H(z) = Z$ transform of $h(n) = \sum_n h(n)z^{-n}$

Figure 2.10.2 Discrete-time linear system (a) in the time domain and (b) in the z-domain.

In order to obtain a representation similar to a Fourier transform, for the discrete-time case, we will utilize the z-transform, and substitute $z = e^{j\omega}$. For example, Equation 2.10.2 is quickly represented in the z-domain as

$$G(z) = F(z) H(z)$$

$$G(z)\big|_{z=e^{j\omega}} = G(e^{j\omega}) = F(e^{j\omega}) H(e^{j\omega}) \qquad (2.10.3)$$

$$G(\omega) = F(\omega) H(\omega)$$

where $G(z)$, $F(z)$, and $H(z)$ are the z-transform of $g(n)$, $f(n)$, and $h(n)$, respectively. For further details on the design and analysis of discrete or digital filters, the readers are referred to the references.

Now, let us introduce the concept of subbands [Aka99]. Consider a signal space in the frequency domain as shown in Figure 2.10.1 going from 0 to π. This is the case if we have normalized the sampling frequency to be 1. Thus, according to the Nyquist sampling theorem, to avoid aliasing error, the signal's angular or radian frequency cannot exceed π. Let us also consider the classic problem of designing two filters, one low-pass and the other one high-pass, such that we split the signal space exactly in half as

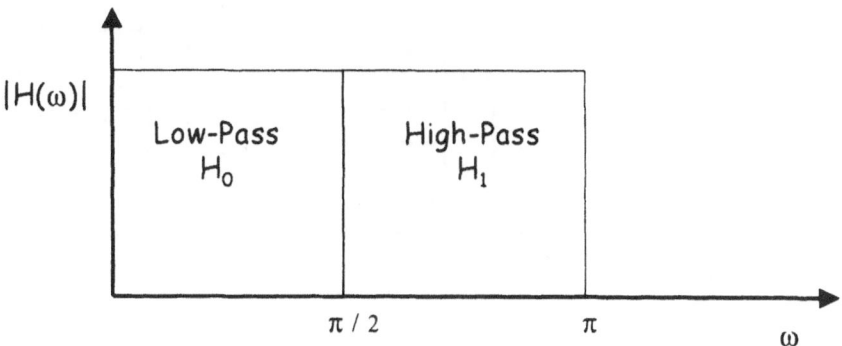

Figure 2.10.3 Splitting of the signal space using and ideal brickwall low-pass (H_0) and high-pass (H_1) filters.

shown in Figure 2.10.3. The low-pass region extends from 0 to $\pi/2$ and the high-pass region extends from $\pi/2$ to π. There are many problems which arise for which it is highly advantageous to split the frequency band. Of course, once we design the filters to split a frequency band, we can split it again and again, forming subbands. Thus, in subband theory, we design a set of two prototype filters, such that repeated application of these filters divides the signal frequency band into equal parts. This is also referred to as a filter bank. There are different ways to split up the frequency space; some examples are shown in Figures 2.10.4 and 2.10.5.

At this stage, the reader might wonder, What is the connection between subbands and the wavelet transform we have discussed so far? In a dyadic wavelet transform, we design a mother wavelet from which we generate a set of daughter wavelets which forms a complete set. In other words, remembering the connection between the wavelet transform and the filter bank, we see that for this case we have designed one filter, and then using the scaling property, we have constructed a filter bank. All the filters in the filter band are identical in properties except for their scale. In a subband implementation, we initially design two filters which are then scaled to form a filter bank or subband. We will show that for most applications, it is convenient to construct the wavelet transforms using two filters. Their corresponding impulse responses are then used to obtain a scaling function (low-pass filter) and a wavelet function (high-pass filter). The difference at this point between subband and wavelet transform becomes very small as, in

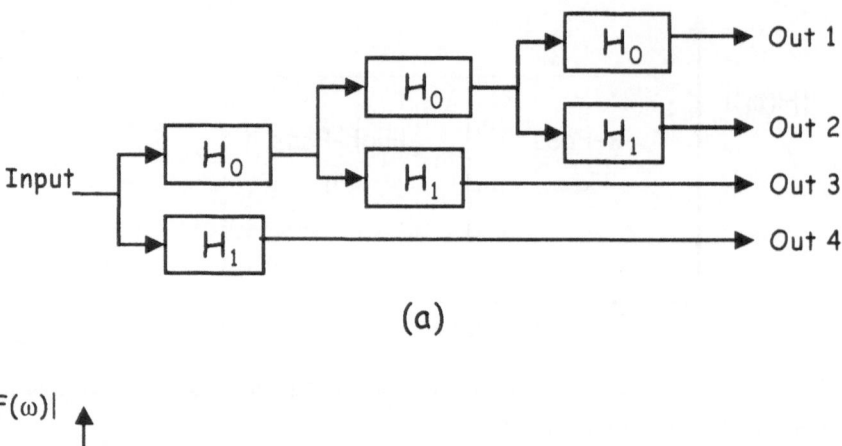

(a)

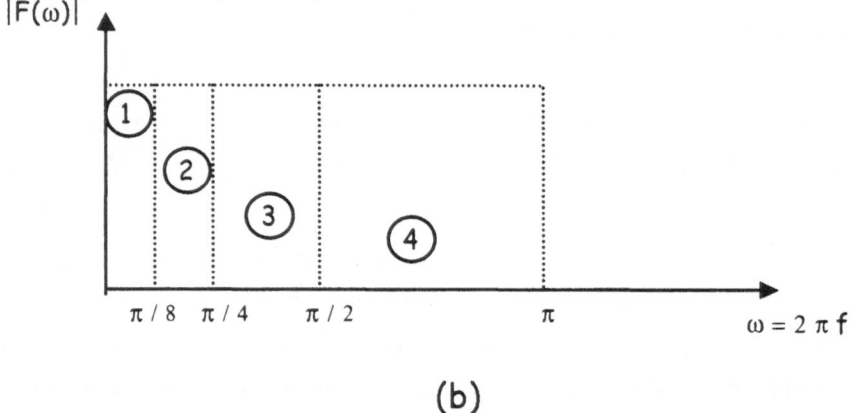

(b)

Figure 2.10.4 (a) The filter bank for splitting the signal bandwidth in the low frequency part only. (b) Frequency domain representation of the non-uniform subband.

general, subbands refer to the discrete-time signal analysis application, whereas wavelet transforms refer to the continuous-time case [Aka99].

Now, consider the design of the subband filters. The obvious choice in the frequency domain is to design two brick wall filters whose frequency responses are rectangular as shown in Figure 2.10.3. Note that this definition of filters was obtained in the frequency domain; to actually implement them, we need to go back into the time domain. The delta response of the brick wall filter is the well-known sinc function. Note that just as for the continuous-time case, the response extends to infinity and an exact implementation is therefore impossible. A delay line implementation of an N-tap filter is shown in Figure 2.10.6. The figure also shows the fundamental elements of digital signal processing. To obtain a perfect response based on the filters in Figure 2.10.3, the number N of coefficients should go to infinity. The simplest

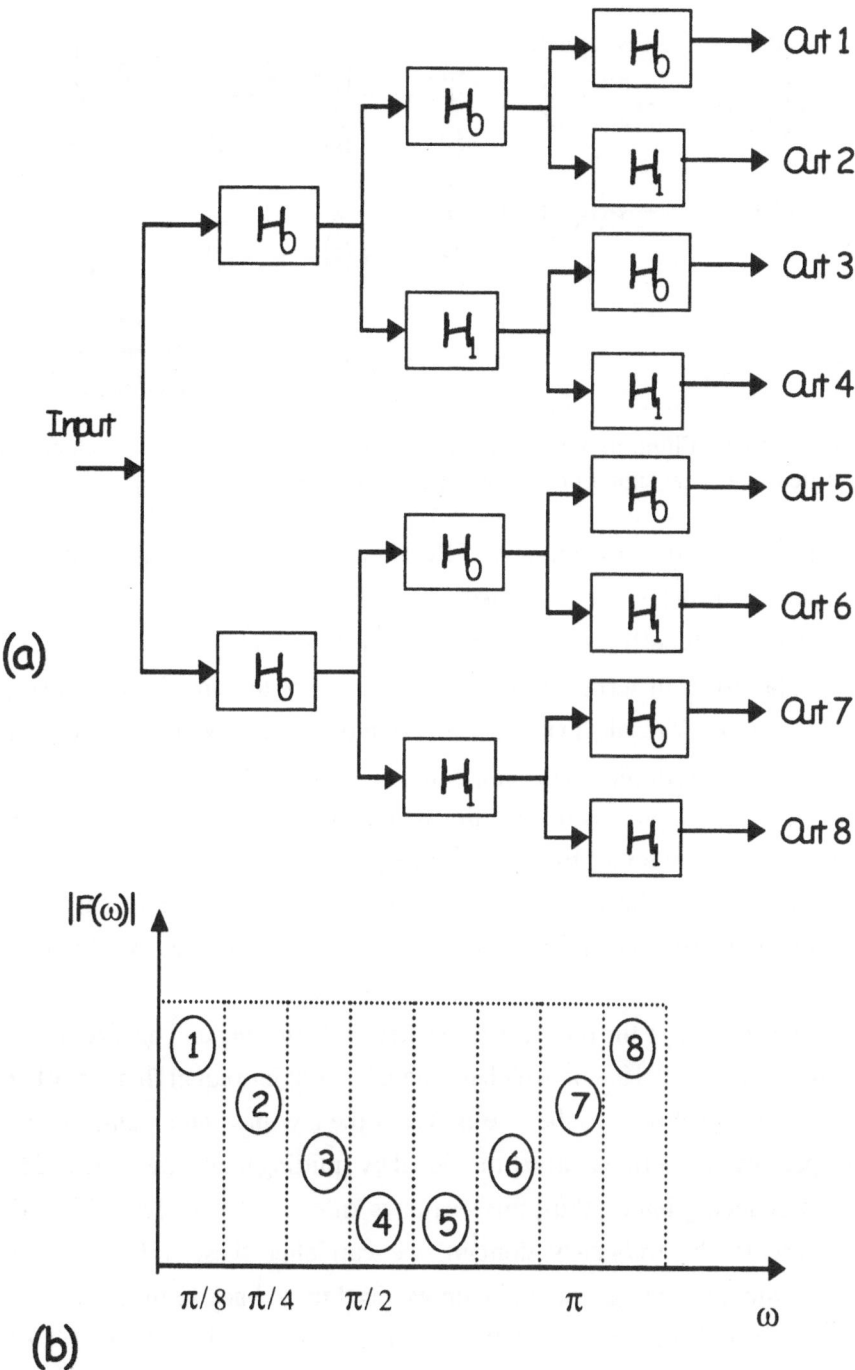

Figure 2.10.5 (a) The filter bank for uniform splitting of the signal bandwidth into eight equal parts; (b) frequency domain representation of the subbands.

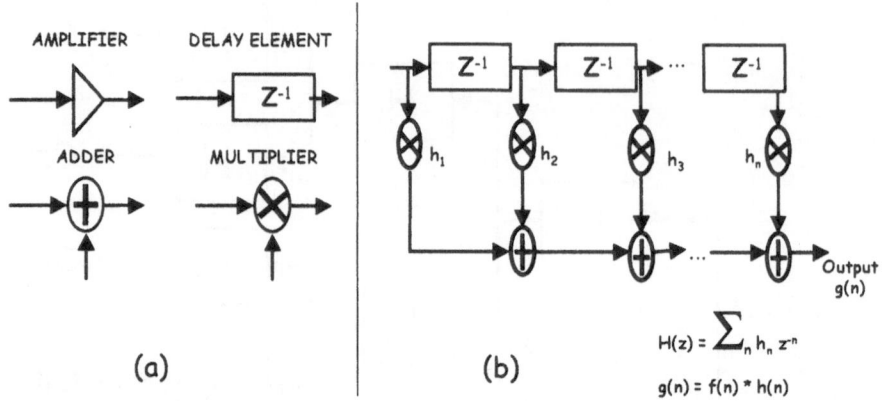

Figure 2.10.6 Finite impulse response (FIR) implementation of (a) elements of single rate digital signal processing and (b) N-tap FIR filter.

practical solution to obtain a FIR (finite impulse response) filter is to terminate the delay line coefficients beyond a certain value N. This assumes that truncated coefficients are very small in amplitude, with respect to the first N. The effect of terminating the coefficients of the impulse response to a finite number N results in the modification of the brickwall response with the introduction of ripples in the stopband, as shown in Figure 2.10.7. As it can be seen, some high- frequency information of the signal still leaks in the low-pass filter as we notice that the filter response for the low-pass case extends beyond $\pi/2$ and introduces ripples in the pass band. This means that this design will introduce aliasing error and distortion, which we would prefer to avoid.

To appreciate the problem associated with aliasing and distortion errors, consider Figure 2.10.8, in which the signal is split into equal frequency bands. $H_0(z)$ and $H_1(z)$ refer to the z-response of the low-pass and high-pass filters, respectively. We first analyze the signal by splitting it into low-pass and high-pass frequency bands, thus, this section is known as the analysis filter. Once we are in the frequency domain, we can alter these values for signal processing, encoding, etc. and then we need to go back to the time domain. This last part is done through the so-called synthesis filter. Synthesis filters are like the reconstruction formula or the inverse transform. If no alteration of the coefficients in the frequency domain is done, then the input signal $s(n)$ must be identical to the output $\hat{s}(n)$. Also note that we have used G_0 and G_1

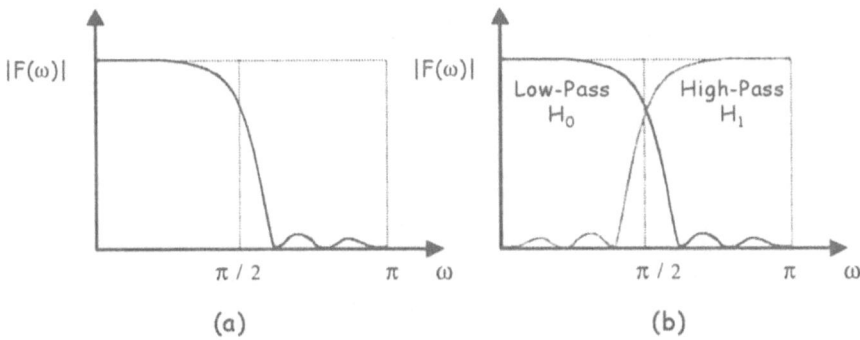

Figure 2.10.7 (a) Typical N-tap FIR low-pass filter response. (b) Typical N-tap high-pass and low-pass filter responses showing distortion and aliasing.

for the low-pass and high-pass synthesis filters, respectively. However, if H_0 and H_1 are brick wall filters, then G_0 and G_1 are also brick wall filters. Thus, for this case $G_0 = H_0$ and $G_1 = H_1$. A point of historical interest is the fact that people initially used H_0 and H_1 for both analysis and synthesis filters. The concept of using different filters to cancel out the distortion and aliasing errors was introduced later.

Let us introduce the concept of multirate filters using Figure 2.10.8. We note that the bandwidth at the output of H_0 is only $\pi/2$, although the output has a sampling frequency corresponding to π. So, every other sample in the output is superfluous and carries no information; we might as well discard these samples. This is shown by the block diagram with a downward arrow and a 2, indicating that every other sample point is discarded; this is called decimation by 2. Similar operations are performed on the high-pass output. Thus, if we started with N samples, we still get N sample at the output. However, $N/2$ of these correspond to the low-pass version and the other $N/2$ correspond to the high-pass version. Similarly, at the output of the next stage shown in Figure 2.10.8, we have $N/4 \times 4 = N$ samples. This is important; basically, we are taking N samples and mapping using a N by N matrix to the next output. This matrix approach to the problem is highly interesting and useful and will be discussed in more detail as part of Section 2.11.

Since the number of samples in the input and output are the same, this process of subband operation is optimum. However, we throw away half of the calculated values at each stage, this approach is also computationally

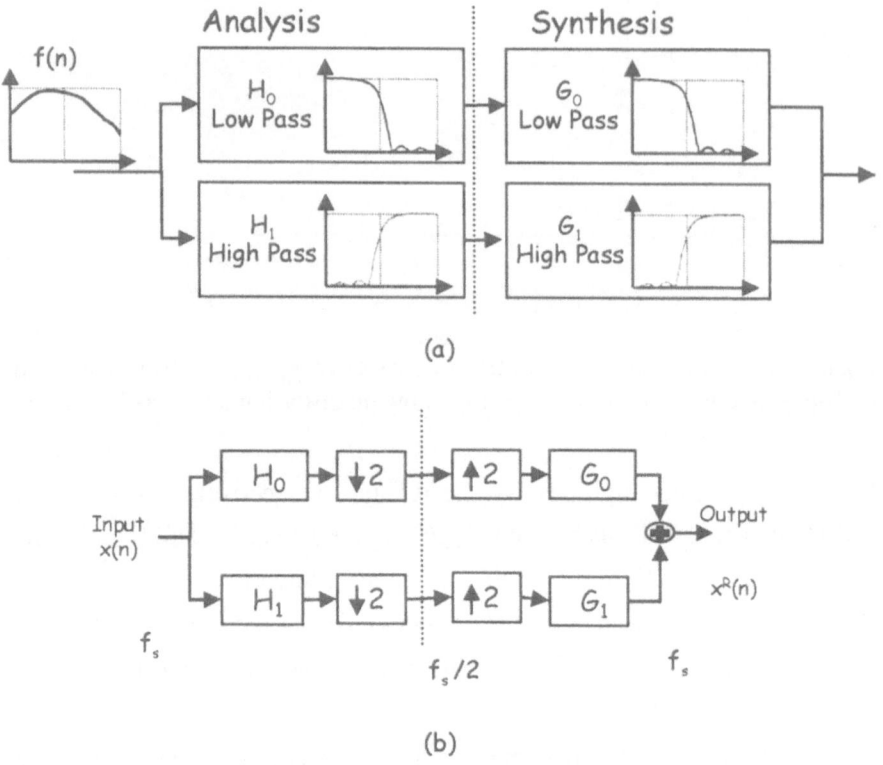

Figure 2.10.8   (a)   Analysis and synthesis filters; (b) with decimation and interpolation.

wasteful. The trick is not to calculate at all of those values which would be thrown away later. We will describe this in more detail later; it is known as a polyphase decomposition. When this is done, the number of operations performed in going from the input to the output is computationally optimum. Actually, the success of subbands in practical applications is due to this computationally efficient algorithm.

Now, let us consider the case when H_0 and H_1 are not ideal; they have ripples and go beyond the $\pi/2$ cutoff frequency as shown in Figure 2.10.7. Distortion error is obvious, as we have ripples in the pass band. However, aliasing error comes if we perform the decimation because the output bandwidths are not limited to $\pi/2$. Initially, the solution for reducing these errors was to use window functions, which reduce ripples in the pass band and extend beyond the cut-off frequency, $\pi/2$. However, one cannot completely eliminate the errors using this technique. The situation is similar to using a Gabor transform (Section 2.6), which does not form a complete set of basis functions.

To solve this problem, the so-called quadrature mirror filter (QMF) was introduced [Vai90]. When using a QMF, one does not use the same filters for analysis and synthesis [Vai93]. The basic idea is to use different filters such that the errors in analysis are minimized by the synthesis. The perfect reconstruction (PR) filters are also QMF, but they completely cancel both the distortion and aliasing error. The situation can be understood by considering an analogous situation, (shown in Figure 2.10.9). If we can map linearly between the time and frequency domains, then both the analysis and synthesis filters are identical and linear. However, if the analysis filter is nonlinear due to practical constraints; then we can correct all the nonlinearity for the whole system by imposing the opposite nonlinearity on the synthesis filter. As shown in Figure 2.10.9, one can choose any kind of nonlinearity and correct it. The case of discontinuous signals will be highly susceptible to quantization error and noise, whereas the continuous, smoother case probably will be more stable. The situation is very similar for the case of PR filter design. The single-stage decimation-interpolation subband decomposition can be extended

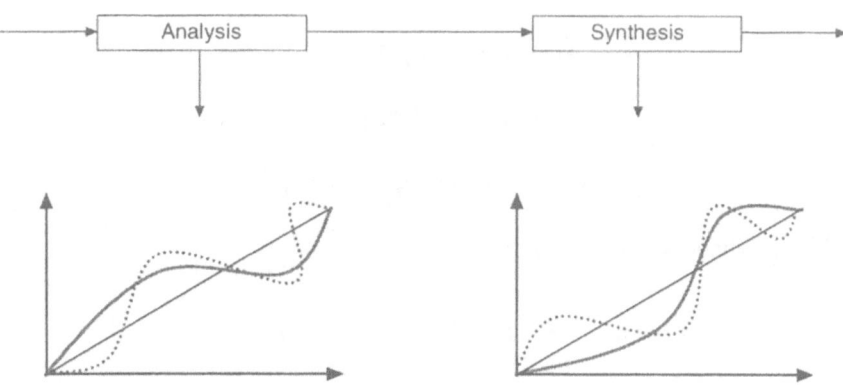

Figure 2.10.9 Linear and nonlinear mapping of the input and output for the analysis and synthesis filters.

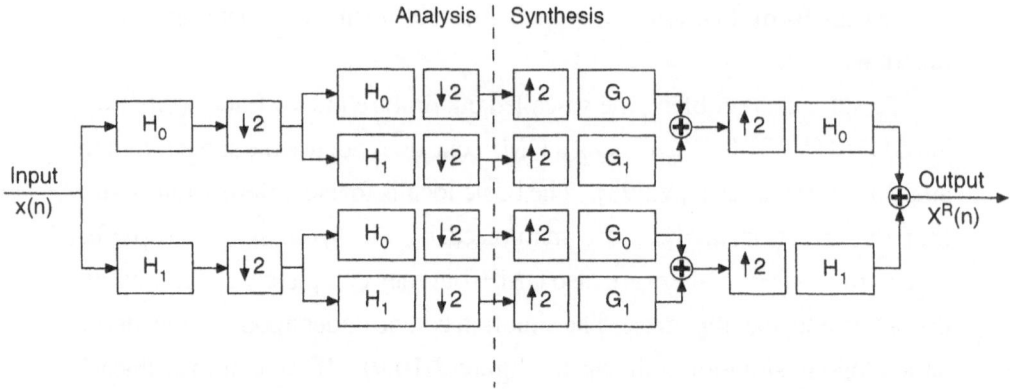

Figure 2.10.10 Two-stage subband filter.

easily, as shown in Figure 2.10.10, where the original input sequence $x(n)$ of length N is mapped into four sequences of $N/4$ samples each. This is very important since basically we are taking N samples and mapping using a $N \times N$ matrix to next output. This matrix approach of the problem is highly interesting and useful and will be discussed in more detail in the next section.

Now, let us consider Figure 2.10.10, which shows a two-stage analysis and synthesis filter (this will be discussed in detail in future sections). The filter responses are given in Figure 2.10.8a and they are mirror images around $\pi/2$; that is the reason for the name Quadrature Mirror Filter. For this case if we consider finite impuse response (FIR) filters extending to N, then the PR condition is satisfied provided that

$$G_0(z) = -H_1(z)$$

$$G_1(z) = -H_0(z) \tag{2.10.4}$$

$$H_1(z) = z^{-(N-1)} H_0(-z^{-1})$$

and in the discrete-time domain

$$h_1(n) = (-1)^{n+1} h_0(N-1-n)$$

$$g_0(n) = h_0(N-1-n) \tag{2.10.5}$$

$$g_1(n) = (-1)^n h_0(n) \;.$$

Note that the above equations fully describe all the filters h_1, g_0, and g_1, once the filter h_0 has been defined. The conditions on h_0 are

$$\sum_{k=0}^{N-1} h(k) \, h(k+2n) = \delta_n$$

$$\left| H_1\left(e^{j\left[\frac{\pi}{2} - \omega\right]}\right) \right| = \left| H_0\left(e^{j\omega}\right) \right| .$$

(2.10.6)

The above equation means that H_1 and H_0 are QMF as shown in Figure 2.10.10. As noted previously, they are called quadrature mirror filters because the magnitudes are a mirror image of each other, with the mirror placed at $\omega = \pi/2$. $H_1(z)$ is in quadrature to $G_0(z)$ and $H_0(z)$ is also in quadrature to $G_1(z)$. As H_0 has N taps, we need to determine N tap weights; however, only $N/2$ equations are specified. Thus, there is quite a bit of arbitrariness in the design. Theoretically, we can choose any other $N/2$ equations to design our filter, which will be PR. However, as discussed in Figure 2.10.9, to make our filters computationally stable we often imposes other conditions. The most well known of these are smoothness conditions, which are given by

$$\left. \frac{d^r H_0 \exp(j\,\omega)}{d\omega^r} \right|_{\omega=\pi} = (-j)^n \sum_n n^r (-1)^n h_0(n) = 0 ,$$

(2.10.7)

$$\text{for} \quad r = 0, \, ... \, , \frac{N}{2} - 1 .$$

Using these expressions, one can solve for a set of equations to obtain tap weights of PR filter weights. As we will discuss later, the same coefficients also come up in connection with other operations, including the DWT, and are generally known as Daubechies wavelets to the wavelet community, after the person who pioneered much of the current work on wavelets.

We have only discussed the dyadic or two-band case. In place of two-band, one can also have m-band filters, where m is an integer. For the m-band case, one can design PR filters as shown in Figure 2.10.11 for the case $m=3$.

There are many variations to the design of PR filters; for example, the cosine-modulated filter bank (also known as a lapped orthogonal transform) is a very useful and important class of subband filters (to be discussed in detail later). Additionally, the subbands do not need to be of uniform bandwidth; they could be nonuniform, too.

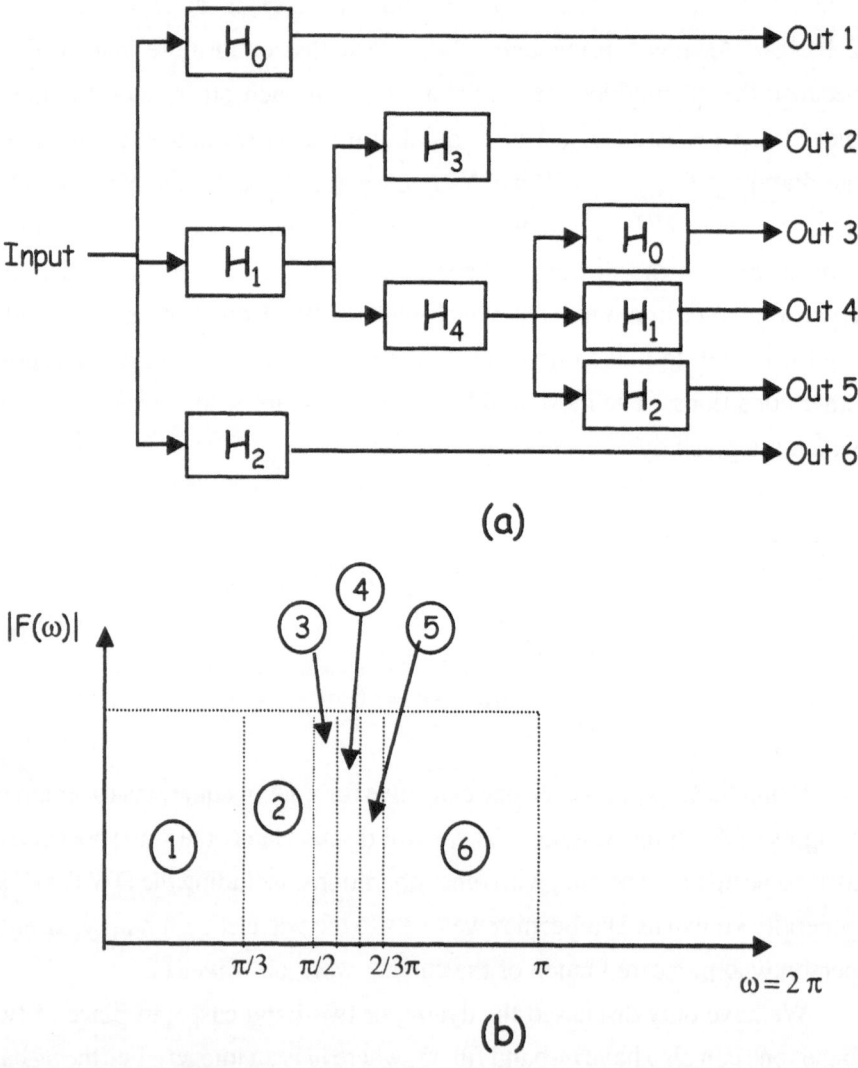

Figure 2.10.11 *M*-Band filter bank for $m = 3$. (a) Filter bank structure; (b) relative subband decomposition.

2.11 Multiresolution Analysis

Multiresolution analysis is similar to subband decomposition; however, for this case, the time variable is continuous rather than discrete [Rio93].   In multiresolution analysis the wavelet function is related to the impulse response of a high-pass filter and, in a similar manner as in the subband case, a low-pass filter is also defined. The impulse response of this filter is the scaling function. Again, in similarity to the subband case, the wavelet and scaling functions are related.

In the previous discussion on subband, the frequency domain was successively split into two halves using a low-pass filter and its mirror image high-pass filter, as shown in Figure 2.10.4.

The high-frequency components of the signal can also be seen as an approximation of the signal itself, whereas the high-pass filtered part is the detail information missing from the signal

$$s(t) = s_{HP}(t) + s_{LP}(t) \qquad (2.11.1)$$

Let us denote a bounded scaling function by $\varphi(t)$; its compression by a factor of 2 is given by $\varphi(2t)$ and its expansion by a factor of 2 is given by $\varphi(t/2)$. As $\varphi(2t)$ is compressed by a factor of 2, it is possible to represent $\varphi(t)$ by shifting components of $\varphi(2t)$, i.e. $\varphi(2t-n)$ where n is an integer. Or we can say that the basis functions for the space 0 to T is given by $\varphi(2t-n)$. We can thus write

$$\varphi(t) = \sqrt{2} \sum_{n} h_0(n) \, \varphi(2t-n)$$

$$\varphi(2t) = \sqrt{2} \sum_{n} h_0(n) \, \varphi(4t-n) \qquad (2.11.2)$$

$$\varphi(t/2) = \sqrt{2} \sum_{n} h_0(n) \, \varphi(t-n)$$

and so on. The above equations are also known as the dilation equations. In this instance, the $h_0(n)$ represent the coefficients of expansion, similar to the tap weights discussed in connection with subband filters. The wavelet function, $\psi(t)$, which is in the detail space, can also be represented using the same scheme:

$$\psi(t) = \sqrt{2} \sum_n h_1(n) \; \varphi(2t-n)$$

$$\psi(2t) = \sqrt{2} \sum_n h_1(n) \; \varphi(4t-n) \qquad (2.11.3)$$

$$\psi(t/2) = \sqrt{2} \sum_n h_1(n) \; \varphi(t-n) \quad .$$

In the Fourier domain, Equation 2.11.2 becomes

$$\varphi(\omega) = \frac{1}{\sqrt{2}} \; \overline{H_0}\left(e^{j\omega/2}\right) \varphi\left(\frac{\omega}{2}\right) \quad . \qquad (2.11.4)$$

In the equation we have used $\varphi(\omega)$ as the Fourier transform of the continuous-time signal $\varphi(t)$; and $H_0(e^{j\omega/2})$ as the Fouries series of the discrete-time sequence $h_0(n)$. If we define a function $M_0(\omega)$ as

$$M_0(\omega/2) = \frac{1}{\sqrt{2}} \; \overline{H_0}\left(e^{j\omega/2}\right), \qquad (2.11.5)$$

then Equation 2.11.4 becomes

$$\varphi(\omega) = M_0\left(\frac{\omega}{2}\right) \varphi\left(\frac{\omega}{2}\right) \quad . \qquad (2.11.6)$$

Utilizing Equation 2.11.6 for $\varphi(\omega/2)$, we obtain

$$\varphi\left(\frac{\omega}{2}\right) = M_0\left(\frac{\omega}{4}\right) \varphi\left(\frac{\omega}{4}\right)$$

$$\varphi(\omega) = M_0\left(\frac{\omega}{2}\right) M_0\left(\frac{\omega}{4}\right) \varphi\left(\frac{\omega}{4}\right) \qquad (2.11.7)$$

and continuing the process, we obtain

$$\varphi(\omega) = \prod_{k=1}^{\infty} M_0\left(\frac{\omega}{2^k}\right) . \qquad (2.11.8)$$

Equation 2.11.8 is an equation which contains an infinite series. Thus, one must be very careful. For an example, at $\omega = 0$ we have

$$\Phi(\omega = 0) = \prod_{k=1}^{\infty} [M_0(0)]^k . \qquad (2.11.9)$$

Thus, $M_0(0) = 1$ is the only stable solution. This leads to $H(0) = 1$ and

$$\varphi(\omega = 0) = \int_{-\infty}^{+\infty} \varphi(t) \, dt = 1 . \qquad (2.11.10)$$

If we consider a FIR filter, then the response is

$$H_0(\omega) = \sum_n h_n \, e^{-jn\omega} \qquad (2.11.11)$$

or

$$H(0) = \sum_n h_n = 1 . \qquad (2.11.12)$$

It is a very important condition. As the function $H_0(e^{j\omega})$ is periodic in 2π, we have

$$H(2\pi) = H(4\pi) = \cdots = H(0) = 1 . \qquad (2.11.13)$$

Furthermore, to obtain a reasonable solution, we must have $\varphi(\omega) \to 0$, as $\omega \to \infty$. This corresponds to imposing

$$H(\pi) = 0 \qquad (2.11.14)$$

and

$$H(\pi) = \sum (-1)^n h_n \quad .$$

$$(2.11.15)$$

Hence, using a FIR filter, we must have

$$\sum (-1)^n h_n = 0 \quad .$$

$$(2.11.16)$$

If one considers that $H_0(\omega)$ is a low-pass filter, it is also obvious that $H(\pi)=0$. So, we obtain, for this special case, the design of scaling and wavelet functions with the following two conditions:

- Sum of tap weights should be equal to 1

$$\sum_n h_n = 1 \quad .$$

- Sum of odd tap weights must equal the sum of even weights for this solution

$$\sum_n (-1)^n h_n = 0 \quad .$$

As an example, we can consider the following cases:
- Lazy Wavelet
  $h_0 = 1; \quad h_n = 0$ for $n \neq 0$
- Haar Wavelet
  $h_0 = h_1 = 1$
- Triangular Wavelet or Hat Wavelet
  $h_{-1} = 1/4; \quad h_0 = 1/2 \quad$ and $\quad h_1 = 1$

These and many others will be discussed in later sections. We will discuss multiresolution in Section 2.13 further, after we introduce matrix formulation in the next section.

2.12   Matrix Formulation

As mentioned in Section 2.10, for the discrete-time system or the subband case, one can formulate the whole problem in terms of matrices. This is often highly convenient and instructive, especially for numerical computation. The signal $f(t)$ becomes $f(n)$ for the discrete case. We can write $f(n)$ as a column vector or a $N \times 1$ matrix as follows:

$$f(n) = f_0, f_1, \cdots , f_N \qquad (2.12.1)$$

or

$$[f] = \begin{bmatrix} f_0 \\ f_1 \\ \cdots \\ f_N \end{bmatrix} = \begin{bmatrix} f_0 & f_1 & \cdots & f_N \end{bmatrix}^T \qquad (2.12.2)$$

Any mapping from one space to another space or the transform operation can be considered as a matrix operation where an $N \times N$ matrix defines the particular transform. For example, the DFT of $[f]$ can be written as

$$[F] = [W]\,[f] \qquad (2.12.3)$$

where

$$[F] = [F_0 \quad F_1 \cdots F_{N-1}]^T \qquad (2.12.4)$$

and

$$[W] = \text{Fourier matrix} = \begin{bmatrix} w^0 & w^0 & \cdots & w^0 \\ w^{-1} & w^{-2} & \cdots & w^{N-1} \\ \cdots & \cdots & \cdots & \cdots \\ w^{-(N-1)} & w^{-2(N-1)} & \cdots & w^{-(N-1)^2} \end{bmatrix} \qquad (2.12.5)$$

and

$$w = \exp(j\, 2\pi\, /\, N\,) . \tag{2.12.6}$$

It is obvious that the columns of the Fourier matrix represent the discrete Fourier basis functions. Of course, there is an inverse DFT given by

$$[f] = [W]^{in} [F] = [W]^{in} [W] [f] . \tag{2.12.7}$$

Thus,

$$[W]^{in} = [W]^{-1} . \tag{2.12.8}$$

This is a common property of all the transform matrices we will be dealing with; they are called unitary matrices. For the case of subbands, and following Figure 2.10.10, we can write, for the analysis filter,

$$[f'] = [W] [f] \tag{2.12.9}$$

where $[W]$ can be written as

$$[W] = \begin{bmatrix} h_{00} & h_{01} & h_{02} & \cdots & & & & \\ \cdots & \cdots & h_{00} & h_{01} & h_{02} & \cdots \\ h_{10} & h_{11} & h_{12} & \cdots & & & \\ \cdots & \cdots & h_{10} & h_{11} & h_{12} & \cdots \end{bmatrix} \tag{2.12.10}$$

where h_{0i} are the coefficients for H_0 and H_{1i} are the tap weights for H_1 (the high-pass filter). Considering low-pass and high-pass filtered signals, we have

$$[f'] = \begin{bmatrix} f_i \\ f_n \end{bmatrix} . \tag{2.12.11}$$

Similarly we can write $[W]$ as

$$[W] = \begin{bmatrix} H_0 \\ H_1 \end{bmatrix} \tag{2.12.12}$$

where

$$[H_0] = \begin{bmatrix} h_{00} & h_{01} & h_{02} & \cdots & \\ \cdots & \cdots & h_{00} & h_{01} & h_{02} & \cdots \end{bmatrix} \qquad (2.12.13)$$

and

$$[H_1] = \begin{bmatrix} h_{10} & h_{11} & h_{12} & \cdots & \\ \cdots & \cdots & h_{10} & h_{11} & h_{12} & \cdots \end{bmatrix}. \qquad (2.12.14)$$

We note that the effect of decimation is that the adjacent row values are shifted by two column positions. Because the system is linear and the convolution theorem applies, the matrices are Toplitz.

Let us consider two cases, where $N = 4$ and $N = 8$. For $N = 4$, we can write

$$[f] = [\, f_0 \; f_1 \; f_2 \; f_3 \,]^T$$
$$[f'] = [\, f'_0 \; f'_1 \; f'_2 \; f'_3 \,]^T \,. \qquad (2.12.15)$$

Assume that $h_{00} = a$, $h_{01} = b$; $h_{10} = c$,; and $h_{11} = d$ ($h_0 = [a \; b]$ and $h_1 = [c \; d]$). Then the above equations simplify to

$$\begin{bmatrix} f_0 \\ f_1 \\ f_2 \\ f_3 \end{bmatrix} = \begin{bmatrix} a & b & 0 & 0 \\ 0 & 0 & a & b \\ c & d & 0 & 0 \\ 0 & 0 & c & d \end{bmatrix} \cdot \begin{bmatrix} f_0 \\ f_1 \\ f_2 \\ f_3 \end{bmatrix} \qquad (2.12.16)$$

provided that we introduce the change of notation $f' = f$. This results in

$$[W] = \begin{bmatrix} a & b & 0 & 0 \\ 0 & 0 & a & b \\ c & d & 0 & 0 \\ 0 & 0 & c & d \end{bmatrix} \qquad (2.12.17)$$

and we note that:

$$[H_0] = \begin{bmatrix} a & b & 0 & 0 \\ 0 & 0 & a & b \end{bmatrix}$$

$$[H_1] = \begin{bmatrix} c & d & 0 & 0 \\ 0 & 0 & c & d \end{bmatrix} .$$

(2.12.18)

The corresponding case for $N = 8$ is left as an exercise for the reader. The system we have considered above is called a block transform. It means that if we have very long signals, we only take blocks of N and process them in turn. In many cases, it is well known that this produces artificial artifacts, especially at the edges of a block; this is obvious for the $N = 8$ case. To avoid this, other transforms are used, including the so-called lapped transform, sliding window, overlap, and other techniques.

2.13 Multiresolution Revisited

It is traditional to define multiresolution analysis from the mathematical stand point as follows. Given a sequence of embedded closed subspaces, i.e.,

$$V_0 \subset V_1 \subset V_2 \subset V_3 \cdots \, , \qquad (2.13.1)$$

the goal of multiresolution is to choose these subspaces in a way that they encompasse the whole function space without any overlap or redundancy. The projection of the function in the subspaces gives finer details as the index j increases or the function is resolved with sampling period $T = (1/2^j) \times T_0$ where T_0 is the sampling period in V_0. The subspaces must satisfy the following conditions:

(i) Completeness (Upward)

$$\bigcup_{m \in Z} V_m = L^2(\mathbb{R}) \qquad (2.13.2)$$

This means that as $m \to \infty$, V_m includes the whole space of square integrable functions.

(ii) Emptiness or downward completeness

$$\bigcap_{m \in Z} V_m = \{\, 0 \,\} \tag{2.13.3}$$

This means that as $m \to - \infty$, V_m defines the subspace with no details and thus the intersection of all these subspaces is the empty space with no information.

(iii) Scale Invariance

$$f(t) \in V_0 \ \leftrightarrow \ f(2^{-m}t) \in V_m \ . \tag{2.13.4}$$

(iv) Shift Invariance Basis

$$f(t) \in V_m \ \leftrightarrow \ f(t-n) \in V_m \ , \ \text{for } n \in Z \tag{2.13.5}$$

If V_j has an orthonormal basis denoted by $\varphi_j(t-n)$ where n is an integer, the orthonormal condition is not necessary. The requirement is that a stable basis (Reisz basis) exists.   Also with the shift invariance property, any non-orthogonal basis can always be orthogonalized.

The subspace V_j has more resolution than the subspace V_{j-1}. Thus, the function $f(t)$ in V_{j-1} has lost some information which can be represented in the detail space W_{j-1}. Thus, one can define W_j space as

$$V_j + W_j = V_{j+1} \ . \tag{2.13.6}$$

$$
\begin{aligned}
V_{j+1} &= W_j + V_j = V_{j-1} + W_{j-1} + W_j \\
&= V_0 + W_0 + W_1 + \cdots + W_j \ .
\end{aligned}
\tag{2.13.7}
$$

We observe that $V_j \perp W_j$ and that $V_j \cap W_j = 0$. For the V_j subspace, one can define the basis function as

$$\Phi_j = \sum_{k=-\infty}^{+\infty} \Phi_{jk}(t) \tag{2.13.8}$$

where

$$\Phi_{jk}(t) = 2^{1/2} \Phi(2^j t - k) \quad . \tag{2.13.9}$$

One thus obtains the function $f(t)$ in the V_j space as:

$$f_j(t) = \sum_{k=-\infty}^{+\infty} a_{jk} \Phi_{jk}(t) \quad . \tag{2.13.10}$$

We see that $f_j(t) \rightarrow f(t)$ as $j \rightarrow \infty$. Similar to ϕ_j, one defines ψ_j, which is the set of basis functions for W_j. Thus,

$$\psi_j = \sum_{k=-\infty}^{+\infty} \psi_{jk}(t) \tag{2.13.11}$$

where

$$\psi_{jk}(t) = 2^{1/2} \psi(2^j t - k) \quad . \tag{2.13.12}$$

Following the arguments presented in Section 2.1, for a novice not well-versed in the mathematical subtleties, the above discussion is too difficult to comprehend. What follows is an introduction of multiresolution for the reader who is not an expert in these fields. First of all, let us explain what the embedded subspaces are with an example. Consider Figure 2.13.1, where we show a function $f(t)$ as a function of t in the top graph. To keep the discussion simple, let us also consider that this function can be made discrete by 64 sampled values using the Nyquist criterion. As $64 = 2^6$, we say that $f(t)$ is completely represented in the space V_6.

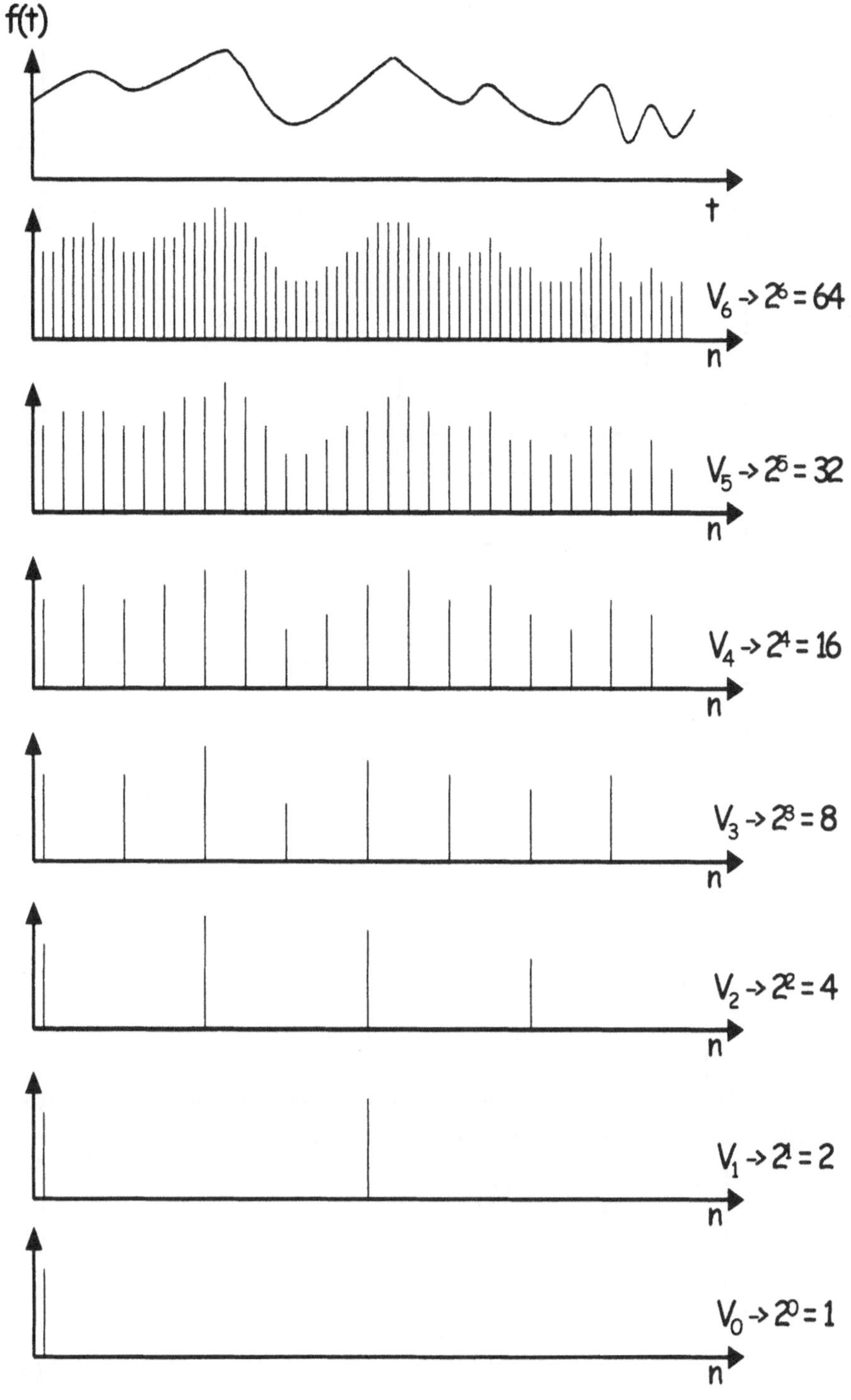

f(t)

$V_6 \rightarrow 2^6 = 64$

$V_5 \rightarrow 2^5 = 32$

$V_4 \rightarrow 2^4 = 16$

$V_3 \rightarrow 2^3 = 8$

$V_2 \rightarrow 2^2 = 4$

$V_1 \rightarrow 2^1 = 2$

$V_0 \rightarrow 2^0 = 1$

Figure 2.13.1 Signal $f(t)$ represented in spaces V_6 to V_0. See text for complete explanation.

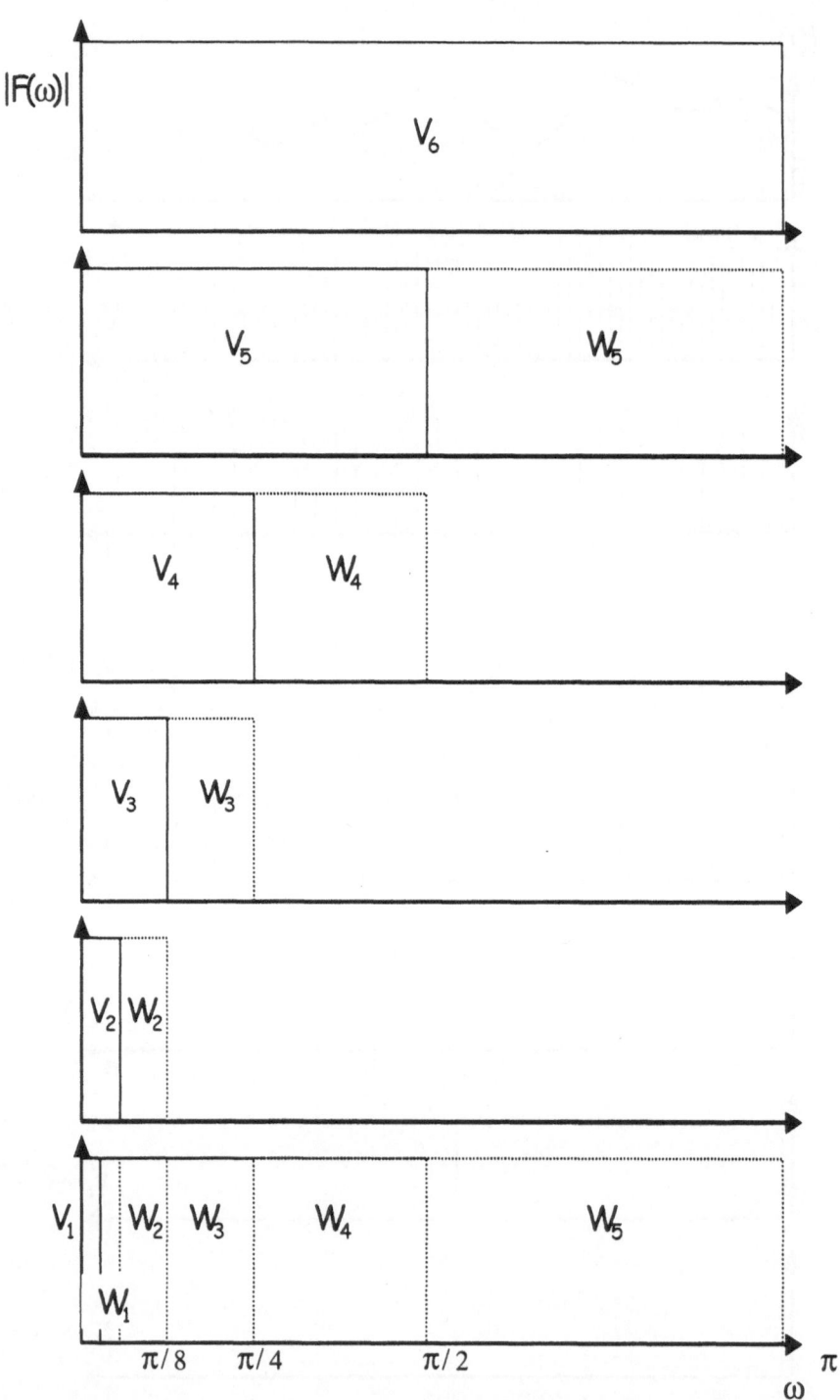

Figure 2.13.2 Representation of the spaces V_6 to V_1 and W_5 to W_1 in the frequency domain.

In the next graph of Figure 2.13.1, we show the same function in V_5 or when there are only $32 = 2^5$ sampled values. In this space, we are sampling every $2T_0$, whereas we sampled every T_0 in V_6. This means that $f(t)$ in V_5 is contained in V_6. That means that $V_5 \subset V_6$. We can define subspaces V_4, V_3, V_2, V_1, and, finally, V_0 using the same approach. Each lower subspace has less information and V_0 is a simple dc value. Thus, we can write

$$V_0 \subset V_1 \subset V_2 \subset V_3 \subset V_4 \subset V_5 \subset V_6 \qquad (2.13.13)$$

and the jth subspace has 2^j sampled values, with sampling period given by $(2^j T_0)$.

In the subspaces with index j lower than 6, it is obvious that the function is represented by a coarser approximation, that is, some derailed information about the function is lost. We can define a set of detail subspaces, denoted by W_j, which will contain this lost information. For this particular example we can write

$$V_6 = (V_5 + W_5) = (V_4 + W_4) + W_5$$

$$V_6 = (V_3 + W_3) + W_4 + W_5 = (V_2 + W_2) + W_3 + W_4 + W_5$$

$$V_6 = (V_1 + W_1) + W_2 + W_3 + W_4 + W_5$$

$$V_6 = (V_0 + W_0) + W_1 + W_2 + W_3 + W_4 + W_5 . \qquad (2.13.14)$$

The meaning of the detail subspace becomes clearer if we consider the spaces in the frequency domain as shown in Figure 2.13.2. The top graph shows the V_6 subspace which contains the full bandwidth of the signal. We split the bandwidth into half: The lower half represents V_5 and the higher one represents W_5. We split V_5 again into two halves, V_4 and W_4, and so on. This also shows that the subband decomposition discussed earlier in Section 2.10 is another manifestation of multiresolution.

It is highly instructive for a beginner to consider a numerical example with only eight samples values, to keep the size of matrices manageable. This is shown in Figure 2.13.3, where we start with V_3. We also use matrix formulation for this example. The signal in V_3 is given by

$$[f_3]^T = [5\ 7\ 6\ 5\ 5\ 6\ 4\ 5] \ . \tag{2.13.15}$$

To go from V_3 to V_2 and W_2, we will use what is known as Haar wavelet in the wavelet community, Chaikin's Algorithm in the computer graphics community, FIR filter with $h_0 = 1$ and $h_1 = 1$ to implement the low-pass perfect reconstruction filter in the subband community, and first-order spline to the numerical computation community. However, it is really common sense or simplest possible approximation, take the average of the two values when making the approximation to go from V_3 to V_2. Thus, $[f_2]$ is given by

$$[f_2]^T = [6\ 5.5\ 5.5\ 4.5] \ . \tag{2.13.16}$$

For detailed coefficients or W_2 space, take the difference between the two and divide the difference by 2. Thus, the detail coefficients $[f_{02}]$ is given by

$$[f_{02}]^T = [-1\ 0.5\ -0.5\ -0.5] \ . \tag{2.13.17}$$

Note that V_3 has eight values. V_2 and W_2 each have 4 giving us a total of 8. This is expected as we have not lost any information, what we have done is transformed the signal into a different space. We continue with the process and go to V_1, W_1 and V_0, W_0. For these we have

$$[f_1]^T = [5.75\ \ 5]$$

$$[f_{01}]^T = [0.25\ \ 0.5]$$

$$[f_0]^T = [5.375] \tag{2.13.18}$$

$$[f_{11}]^T = [0.375] \ .$$

Table 2.13.1 shows all of these values together. Therefore, using the Haar filter coefficients, we can represent the signal using a sequence of average or coarse components and retaining the detail coefficients lost in the approximation. Using this approach we can write

$$[f_3]^T = [5 \quad 7 \quad 6 \quad 5 \quad 5 \quad 6 \quad 4 \quad 5]$$

$$[f_2]^T = [6 \quad 5.5 \quad 5.5 \quad 4.5 \quad -1 \quad 0.5 \quad -0.5 \quad -0.5]$$

$$[f_1]^T = [5.75 \quad 5 \quad 0.25 \quad -1 \quad -1 \quad 0.5 \quad -0.5 \quad -0.5] \tag{2.13.19}$$

$$[f_0]^T = [5.375 \quad 0.375 \quad 0.25 \quad -1 \quad -1 \quad 0.5 \quad -0.5 \quad -0.5]$$

or in tabular format as in Table 2.13.2.

Table 2.13.1 Example of signal components		
Resolution	Average Components	Detail Components
$8 \leftrightarrow V_8$	[5 7 6 5 6 4 5]	
$4 \leftrightarrow V_4$	[6 5.5 5.5 4.5]	[−1 0.5 −0.5 −0.5]
$2 \leftrightarrow V_2$	[5.75 5]	[0.25 0.5]
$1 \leftrightarrow V_1$	[5.375]	[0.375]

Table 2.13.2. Different representation of a signal using Haar basis								
Sample No.	1	2	3	4	5	6	7	8
f_3	5	7	6	5	5	6	4	5
f_2	6	5.5	5.5	4.5	−1	0.5	−0.5	−0.5
f_1	5.75	5	0.25	−1				
f_0	5.37	0.37						

The transformation from one subspace to the next can be elegantly written in the matrix equation [Str96]. For our example, the matrix equation from V_2 to V_3 is given by:

$$[M_3]\,[f_3] = [f_2 \ f_{02}] \qquad (2.13.20)$$

The transformation matrix $[M_3]$ is 8×8. We have added $[f_{02}]$ in the lower half of $[f_2, f_{02}]$ matrix. This is the reason for the $[M_3]$ matrix to have this particular structure: The top four rows have repeated values like (1   1) and (0   0). This is related to the downsampling by 2, which will be discussed further when we consider multirate filters and subbands. $[M_3]$ can also be considered as the combination of two matrices: The top half for the transformation from V_3 to V_2 and the bottom half for V_3 to W_2. Thus,

$$[M_3] = \begin{bmatrix} L_3 \\ H_3 \end{bmatrix} \qquad (2.13.21)$$

or

$$[L_3]\,[f_3] = [f_2]$$
$$[H_3]\,[f_3] = [f_{02}] \quad . \qquad (2.13.22)$$

Using the values in the V_2 and W_2 spaces we are able to recover the original V_3 values using inverse transformation. This is shown in the equation

$$\begin{bmatrix} 1 & 0 & 0 & 0 & 1 & 0 & 0 & 0 \\ 1 & 0 & 0 & 0 & -1 & 0 & 0 & 0 \\ 0 & 1 & 0 & 0 & 0 & 1 & 0 & 0 \\ 0 & 1 & 0 & 0 & 0 & -1 & 0 & 0 \\ 0 & 0 & 1 & 0 & 0 & 0 & 1 & 0 \\ 0 & 0 & 1 & 0 & 0 & 0 & -1 & 0 \\ 0 & 0 & 0 & 1 & 0 & 0 & 0 & 1 \\ 0 & 0 & 0 & 1 & 0 & 0 & 0 & -1 \end{bmatrix} \begin{bmatrix} 6 \\ 5.5 \\ 5.5 \\ 4.5 \\ -1 \\ 0.5 \\ -0.5 \\ -0.5 \end{bmatrix} = \begin{bmatrix} 5 \\ 7 \\ 6 \\ 5 \\ 5 \\ 6 \\ 4 \\ 5 \end{bmatrix} \quad . \qquad (2.13.23)$$

We notice that this matrix and M_3 are transposes of each other. Thus, they form a unitary matrix, a necessary condition for orthogonal transformation. More about this will be discussed shortly. Let us continue with the transformation $V_2 \rightarrow V_1$. This is given by

$$
\frac{1}{2} \cdot
\begin{bmatrix}
1 & 1 & 0 & 0 & 0 & 0 & 0 & 0 \\
0 & 0 & 1 & 1 & 0 & 0 & 0 & 0 \\
1 & -1 & 0 & 0 & 0 & 0 & 0 & 0 \\
0 & 0 & 1 & -1 & 0 & 0 & 0 & 0 \\
0 & 0 & 0 & 0 & 1 & 0 & 0 & 0 \\
0 & 0 & 0 & 0 & 0 & 1 & 0 & 0 \\
0 & 0 & 0 & 0 & 0 & 0 & 1 & 0 \\
0 & 0 & 0 & 0 & 0 & 0 & 0 & 1
\end{bmatrix}
\begin{bmatrix}
6 \\
5.5 \\
5.5 \\
4.5 \\
-1 \\
0.5 \\
-0.5 \\
-0.5
\end{bmatrix}
=
\begin{bmatrix}
5.75 \\
0.5 \\
0.25 \\
0.5 \\
-1 \\
0.5 \\
-0.5 \\
-0.5
\end{bmatrix}
. \quad (2.13.24)
$$

As the bottom four values on the right-hand side of the equation remain the same (they are $[f_{02}]$ values), the important part of the above equation can be simplified to

$$
\frac{1}{2} \cdot
\begin{bmatrix}
1 & 1 & 0 & 0 \\
0 & 0 & 1 & 1 \\
1 & -1 & 0 & 0 \\
0 & 0 & 1 & -1
\end{bmatrix}
\begin{bmatrix}
6 \\
5.5 \\
5.5 \\
4.5
\end{bmatrix}
=
\begin{bmatrix}
5.75 \\
0.5 \\
0.25 \\
0.5
\end{bmatrix}
\quad (2.13.25)
$$

or $[M_2] \, [f_2] = [f_1 \; f_{01}]$.

The Matrix equation for $V_1 \rightarrow V_0$ is given by

$$
\begin{bmatrix}
1 & 1 \\
1 & -1
\end{bmatrix}
\begin{bmatrix}
5.75 \\
5
\end{bmatrix}
=
\begin{bmatrix}
5.375 \\
0.375
\end{bmatrix}.
\quad (2.13.26)
$$

It is also obvious that we can directly come to V_1 and W_1 from V_3 using the following matrix equation

$$\frac{1}{2} \cdot \begin{bmatrix} 1 & 1 & 1 & 1 & 0 & 0 & 0 & 0 \\ 0 & 0 & 0 & 0 & 1 & 1 & 1 & 1 \\ 1 & -1 & 1 & -1 & 0 & 0 & 0 & 0 \\ 0 & 0 & 0 & 0 & 1 & -1 & 1 & -1 \\ 1 & -1 & 0 & 0 & 0 & 0 & 0 & 0 \\ 0 & 0 & 1 & -1 & 0 & 0 & 0 & 0 \\ 0 & 0 & 0 & 0 & 1 & -1 & 0 & 0 \\ 0 & 0 & 0 & 0 & 0 & 0 & 1 & -1 \end{bmatrix} \begin{bmatrix} 5 \\ 7 \\ 6 \\ 5 \\ 5 \\ 6 \\ 4 \\ 5 \end{bmatrix} = \begin{bmatrix} 5.75 \\ 0.5 \\ 0.25 \\ 0.5 \\ -1 \\ 0.5 \\ -0.5 \\ -0.5 \end{bmatrix} . \qquad (2.13.27)$$

It is of interest to look at the transformation matrices only. For an example, the transformation matrices L_3, L_2, L_1 to transform the signal from V_3 to V_2, V_2 to V_1 and V_1 to V_0, respectively, is given by

$$L_3 \; [8 \times 4] \;\rightarrow\; L_2 \; [4 \times 2] \;\rightarrow\; L_1 \; [2 \times 1] \quad . \qquad (2.13.28)$$

We immediately notice that the fundamental matrix really is L_1, from which we can generate L_2, L_3 and so on. The same applies for H_3, H_2, and H_1, matrices:

$$H_3 \; [8 \times 4] \;\rightarrow\; H_2 \; [4 \times 2] \;\rightarrow\; H_1 \; [2 \times 1] \quad . \qquad (2.13.29)$$

Let us define the basis function matrix for V_j as $[\phi_j]$ and W_j as $[\psi_j]$. Note that both $[\phi_j]$ and $[\psi_j]$ have dimensions $2^j \times 1$. For the particular example we are considering, we have the following relationships between V_0 and V_1 subspaces:

$$[\Psi_0] = [1 \; 1] \cdot \begin{bmatrix} \Phi_1^{11} \\ \Phi_1^1 \end{bmatrix} = [P_1] \cdot [\Phi_1]$$

$$[\Phi_0] = [1 \; -1] \cdot \begin{bmatrix} \Phi_1^{10} \\ \Phi_1^1 \end{bmatrix} = [Q_1] \cdot [\Phi_1] \quad . \qquad (2.13.30)$$

We define $[P_j]$ and $[Q_j]$ matrices as $2^{(j+1)} \times 2^j$ dimensional matrices. $[P_j]$ and $[Q_j]$ are the matrices needed to go from subspace $(j-1)$ to j. We can write for the $V_1 \rightarrow V_2$ and $W_1 \rightarrow W_2$ case as follows.

$$\begin{bmatrix} \Phi_1^0 \\ \Phi_1^1 \end{bmatrix} = \begin{bmatrix} 1 & 1 & 0 & 0 \\ 0 & 0 & 1 & 1 \end{bmatrix} \cdot \begin{bmatrix} \Phi_2^0 \\ \Phi_2^1 \\ \Phi_2^2 \\ \Phi_2^3 \end{bmatrix} \qquad (2.13.31a)$$

$$\begin{bmatrix} \Psi_1^0 \\ \Psi_1^1 \end{bmatrix} = \begin{bmatrix} 1 & 1 & 0 & 0 \\ 0 & 0 & 1 & 1 \end{bmatrix} \cdot \begin{bmatrix} \Phi_2^0 \\ \Phi_2^1 \\ \Phi_2^2 \\ \Phi_2^3 \end{bmatrix} \qquad (2.13.31b)$$

or

$$[\Phi_1] = [P_2] \cdot [\Phi_2]$$
$$[\Psi_1] = [Q_2] \cdot [\Phi_2] \quad . \qquad (2.13.32)$$

If we continue the process, we obtain the general matrix equations (also known as refinement equations) given by

$$[\Phi_{j-1}] = [P_j] \cdot [\Phi_j]$$
$$[\Psi_{j-1}] = [Q_j] \cdot [\Phi_j] \quad . \qquad (2.13.33)$$

It is customary to combine the above two equations:

$$\begin{bmatrix} \Phi_{j-1} \\ \Psi_{j-1} \end{bmatrix} = \begin{bmatrix} P_j \\ Q_j \end{bmatrix} \cdot [\Phi_j] \quad . \tag{2.13.34}$$

It is instructive for the readers to go back to the Equations 2.13.21 - 2.13.27 in our example and identify the different matrices defined here. Substituting for $[\phi_{j-1}]$ and $[\psi_{j-1}]$ we obtain the synthesis equation

$$[C_j] = [C_{j-1}] \cdot [P_j] + [d_{j-1}] \cdot [Q_j] \quad . \tag{2.13.35}$$

We can also go from V_{j-1} and W_{j-1} space to V_j. Thus we can write the refine

$$\begin{bmatrix} \Phi_{j-1} \\ \Psi_{j-1} \end{bmatrix} \cdot \begin{bmatrix} A_j & B_j \end{bmatrix} = [\Phi_j] \quad . \tag{2.13.36}$$

Thus we have the analysis equations

$$\begin{aligned} C_{j-1} &= A_j \cdot C_j \\ d_{j-1} &= B_j \cdot C_j \quad . \end{aligned} \tag{2.13.37}$$

Equating Equations 2.13.36 and 2.13.37, we obtain

$$\begin{bmatrix} A_j & B_j \end{bmatrix} = \begin{bmatrix} P_j \\ Q_j \end{bmatrix}^{-1} \quad . \tag{2.13.38}$$

Thus, both $[A_j : B_j]$ and $\begin{bmatrix} P_j \\ Q_i \end{bmatrix}^{-1}$ must be invertible.

To satisfy Equation 2.13.38, one can make further simplifications to obtain three cases.

(A) Orthogonal or orthonormal

(B) Semiorthogonal

(C) Biorthogonal with dual base.

The orthogonal case is given by

$$\begin{bmatrix} P_j \\ Q_j \end{bmatrix}^T = \begin{bmatrix} P_j \\ Q_j \end{bmatrix}^{-1} . \tag{2.13.39}$$

Thus,

$$[A_j] = [P_j]^T$$

$$[B_j] = [Q_j]^T . \tag{2.13.40}$$

The QMF filters, Daubechies filters, are orthogonal filters. For this case one can easily prove that

$$\langle \Phi_j^k \mid \Phi_j^l \rangle = \delta_{kl}$$

$$\langle \Psi_j^k \mid \Psi_j^l \rangle = \delta_{kl} \tag{2.13.41}$$

$$\langle \Phi_j^k \mid \Psi_j^l \rangle = 0 .$$

We will discuss the other two cases in detail during later chapters.

The summary of conditions defining orthogonal, semiorthogonal, and biorthogonal cases are given in Table 2.13.3. Table 2.13.4 gives the summary of refinement, analysis, synthesis, and invertibility conditions. At this point, we will point out some important comments:

(A) So far, we have discussed multiresolution from the matrix point of view. The filter or subband point of view was discussed in Section 2.11. One can also use differential equations for the same purpose. For this case, the bases in the V_j space become scaling function $\phi_j(t)$ and the wavelet function $\psi_j(t)$ in the W_j space. This will be further discussed in Chapter 5.

For the simple case of Haar wavelets or basis functions we note that corresponding to Equation 2.13.30, we have for the continuous case,

$$\Phi_0(t) = \Phi_1(2t) + \Phi_1(2t-1)$$

$$\Phi_1(t) = \Phi_2(2t) + \Phi_2(2t-1) \ .$$

(2.13.42)

Property	Basis Function Constraints	Matrix Constraints
Table 2.13.3. Summary of conditions defining orthogonal, semiorthogonal, and biorthogonal wavelets.		
Orthogonality	$[\langle \Phi^j \mid \Phi^j \rangle] = I$ $[\langle \Psi^j \mid \Psi^j \rangle] = I$ $[\langle \Phi^j \mid \Psi^j \rangle] = 0$	$[\,P^j \mid Q^j\,]$ invertible & orthogonal
Semiorthogonality	$[\langle \Phi^j \mid \Psi^j \rangle] = 0$	$[\,P^j \mid Q^j\,]$ invertible $(P^j)^T [\langle \Phi^j \mid \Phi^j \rangle] Q^j = 0$
Biorthogonality	$[\langle \Phi^j \mid \Phi^j \rangle] = I$ $[\langle \Psi^j \mid \Psi^j \rangle] = I$ $[\langle \Phi^j \mid \Psi^j \rangle] = 0$ $[\langle \Psi^j \mid \underline{\Phi} \rangle] = 0$	$[\,P^j \mid Q^j\,]$ invertible

(B) In the numerical example, we considered one of the simplest cases, the Haar wavelet. It is obvious that one can consider more complex cases. As discussed in Section 2.11, the tap weights h_n must obey Equations 2.11.19 and 2.11.25. Thus we have:

(i) Lazy wavelet with $h = h_0 = 1$.

(ii) Haar wavelet with $h = [h_{-1} \ \ h_0 \ \ h_1] = [.25 \ \ .5 \ \ .25]$.

(iii) Daubechies 2-tap wavelet with $h = [h_0 \ \ h_1] = \frac{1}{2} \cdot [\ 1+\sqrt{3} \ \ \ \ 1-\sqrt{3}]$.

(C) In the numerical example, we have considered a very simple case of two-tap filters. It is obvious that one can have more taps for the FIR case; not only that, one can have other solutions possible (i.e., using infinite impulse response IIR filters). Naturally the question arises as to how we design these filters, or tap coefficients for different cases.

It is also obvious that because there are different approaches to the design, different mathematical tools are used: matrix and z-transform for the discrete or subband case and Fourier transform for the continuous case. Although often for the same problem, many times all three approaches can be used, producing, of course, identical results. However, different approaches give different insights. Not only that, depending on the readers background, the reader might feel quite comfortable in one particular approach but might need supplemental reading for the other approaches. Electrical engineering background readers tend to favor z-transform, whereas computer scientists favor the matrix approach and mathematicians the Fourier transform.

Table 2.13.4 Summary of refinement, analysis, synthesis and invertibility equations

Process	Definition	Block-Matrix Form
Refinement	$\Phi^{j-1}(x) = \Phi^{j}(x) \cdot P^{j}$ $\Psi^{j-1}(x) = I(x) \cdot Q^{j}$	$[\ \Phi^{j-1}\ \ \Psi^{j-1}\ ] =$ $= \Phi^{j} \cdot [P^{j}\ \ Q^{j}]$
Analysis	$c^{j-1} = A^{j} \cdot c^{j}$ $d^{j-1} = B^{j} \cdot c^{j}$	$\begin{bmatrix} c^{j-1} \\ d^{j-1} \end{bmatrix} = \begin{bmatrix} A^{j} \\ B^{j} \end{bmatrix} \cdot c^{j}$
Synthesis	$c^{j} = P^{j} \cdot c^{j-1} + Q^{j} \cdot d^{j-1}$	$c^{j} = \begin{bmatrix} P^{j} \\ Q^{j} \end{bmatrix} \cdot \begin{bmatrix} c^{j-1} \\ d^{j-1} \end{bmatrix}$
Invertibility	$A^{j} \cdot P^{j} = B^{j} \cdot Q^{j} = I$ $A^{j} \cdot Q^{j} = B^{j} \cdot P^{j} = 0$ $P^{j} \cdot A^{j} + Q^{j} \cdot B^{j} = I$	$\begin{bmatrix} A^{j} \\ B^{j} \end{bmatrix} = \begin{bmatrix} P^{j} \\ Q^{j} \end{bmatrix}^{-1}$

(D) Once we find the tap coefficients which are acceptable, one needs to find the scaling function. We should mention again that when we go from the discrete case to the subband case, we should be able to subdivide for the limit $j \to \infty$. Note that in the numerical case we considered with 64 or 2^6, $j = 6$. However, for the continuous case there is no limit to this subdivision. Thus, one must consider the case $j \to \infty$. For this case, one must obtain a finite solution - this leads to different conditions discussed earlier. Once these conditions are satisfied, the next question is how to obtain $\phi(t)$ and $\psi(t)$.

(E) As we go from the finite discrete case to the continuous case (i.e., the precision tending to infinity), we will see in Part III that the conditions to be satisfied are

$$\sum_i h_i^2 = 1$$

$$h_0 h_2 + h_1 h_3 + \cdots = 0 \qquad\qquad (2.13.43)$$

$$h_0 h_4 + h_1 h_5 + \cdots = 0 \ .$$

(i) For the matrix case, the matrix with basis functions (HH^T) must have at least one eigenvalue 1 and all others less than 1. In the case discussed before, the Haar case, we have

$$H = \begin{bmatrix} 1 & 1 \\ 1 & -1 \end{bmatrix}$$

and the eigenvalues are 1 and 1/2. For the spline case, we will see that the eigenvalues will be given by $1, \dfrac{1}{2}, \ldots, \dfrac{1}{2^n}$ for the nth order spline.

(ii) In the z-domain, the filter response must have response given by $(1+z^{-1})^n$ multiplied by other factors. Actually, we will see that if we only use $(1+z^{-1})^n$ where n is an integer, we get nth-order splines. And if we use a function $(1+z^{-1})^n Q(z)$, where $Q(z)$ is such that the even powers of z are canceled, we obtain the orthogonal Daubechies wavelets. Spline wavelets are semiorthogonal, whereas Daubechies wavelets are orthogonal. We also note that spline semiorthogonal wavelets can be orthogonalized.

(iii) In the frequency domain, it must be a half-band filter. As we are considering two channel filters, when we downsample by 2, we introduce aliasing. To eliminate aliasing error, the filter must have $H(0) = 1$ and $H(\pi) = 0$. For the orthonormal case, it also must be QMF.

(F) In D, we discussed the conditions for obtaining the scaling function. Once we know the scaling function, the next problem is how to determine the wavelet function. This determination is very interesting for the case of the spline and will be discussed in detail in Part II.

(G) As mentioned earlier for the spline case or Daubechies case of order n, we have n eigenvalues, or nth-order zero on $z=1$ or the condition of nth-order zero at $\omega = 0$. This also relates to what is known as the regularity condition. This determines how good the low-pass filter is. The regularity condition can also be related to the Taylor series expansion of a function and how many derivatives are zero. This is discussed in detail in Part II.

(H) The numerical example we chose is what is known as the two-channel filter. As we will discuss later, one can have an m-channel filter also, where m is an integer. However, a two-channel filter is so important for practical purposes; we will consider it again and again in different parts of this book. Here we want to point one particular aspect of the two-channel filter, for critical sampling, after we process the samples into low-pass and high-pass we need to downsample it by 2 in the analysis filter and upsample by 2 in the synthesis section. From the matrix point of view, the downsampling by 2 means the matrices with the tap coefficients of the filter are as follows

$$
H = \begin{bmatrix} h_1 & h_2 & h_3 & h_4 & \cdots \\ 0 & 0 & h_1 & h_2 & \cdots \end{bmatrix} .
$$

Ordinarily, for a linear system without downsampling we should have obtained

$$H = \begin{bmatrix} h_1 & h_2 & h_3 & h_4 & \cdots \\ 0 & h_1 & h_2 & h_3 & \cdots \end{bmatrix} .$$

This shift by two zeros rather than one zero in the successive rows is the effect of downsampling. It is of interest to consider this matrix further for the orthogonal case. As mentioned before we must have $HH^T = I$ or the matrix must be unitary. This condition results in the same equations discussed earlier in connection with subband filters. The only difference is that they were obtained differently using matrix methods. Using the z-transform analysis, the downsampling by 2 imposes that the transfer function $H(z)$ between input and output to have the property that $H(z)H(z^{-1})$ must have only even powers of z.

2.14   Two-Dimensional Case

So far, we have discussed 1-D signals only and, in general, we were concerned with time signals. However, in place of $f(t)$ we can also consider $f(x)$, where x can be any variable, (e.g. space). The concept of wavelets can be easily extended to 2-D signals and multidimensional signals. In particular, 2-D signals are very important since they include processing of images, which can be represented by signals of the form $f(x,y)$.

As expected, the case of 2-D processing is more complex. For example, the simplest case is where the x and y components are separable, meaning

$$f(x,y) = f_x(x) \, f_y(y) \quad . \tag{2.14.1}$$

The case of $f(x,y) = f(r,\sigma)$ where $r = \sqrt{x^2 + y^2}$ and $\tan\sigma = (x \, / \, y)$, is more complex.

Note that even in the complex cases, one can perform wavelet transforms or subband decompositions. The fundamental concept is to choose a particular mother function such that by scaling and shifting the mother function, one can represent any other desired function; that is, the mother function shifted and scaled forms a set of complete basis functions. For the subband case, one chooses the same filter response which uses multirate to achieve the decomposition. For the continuous case, we can easily define the wavelet transform to be

$$W_f\,(\overline{a},\overline{b})\ =\ \int_{-\infty}^{+\infty} \int_{-\infty}^{+\infty} f(x,y)\ h^*_{\overline{a},\overline{b}}\,(x,y)\ dx\ dy \qquad (2.1.4.2)$$

where now the scale parameter a is a vector $\overline{a}\ =\ (a_x\ a_y)$. Similarly, the shift parameter b has become a vector, $\overline{b}\ =\ (b_x\ b_y)$. Thus, a 2-D signal produces a four-dimensional or 4-D transform. However, in many practical situations, one can make the simplification $a_x = a_y = a$. With this simplification, one deals with three-dimensions (3-D) rather than 4-D. For this case, Equation 2.14.2 can be written as

$$W_f\,(a,\overline{b})\ =\ \int_{-\infty}^{+\infty} \int_{-\infty}^{+\infty} f(x,y)\ h^*_{a,\overline{b}}\,(x,y)\ dx\ dy \qquad (2.1.4.3)$$

where the daughter wavelets are given by

$$h^*_{a,\overline{b}}\,(x,y)\ =\ \frac{1}{a}\ h\!\left(\frac{x-b_x}{a}\ ,\ \frac{y-b_y}{a}\right) \qquad (2.14.4)$$

In the frequency domain, Equation 2.14.4 becomes

$$\mathscr{F}\,[\ W_f\,(a,\overline{b}\ )\ ]\ =\ H^*(af_x,af_y)\ F(f_x,f_y)$$

$$W_f\,(a,\overline{b})\ =\ a\int_{-\infty}^{+\infty} \int_{-\infty}^{+\infty} H^*(af_x,af_y)F(f_x,f_y) \qquad (2.14.5)$$

$$e^{\,j2\pi\,(f_x\,b_x+f_y b_y)}\,df_x df_y$$

2.15   DWT and Subband Example

As a practical example which will help motivate our study of the DWT, consider Figure 2.15.1, which shows a 2-D signal of some practical interest, a palm print or handprint. We see that in the 2-D Fourier transform of this signal, shown in Figure 2.15.2, we have completely lost any recognition that it is a hand, although all the information contained in the original image is still there. Figure 2.15.3 shows the 2-D DWT of the palm print. Even for resolution of only 8 × 8 pixels in the top leftmost of the DWT, we can still

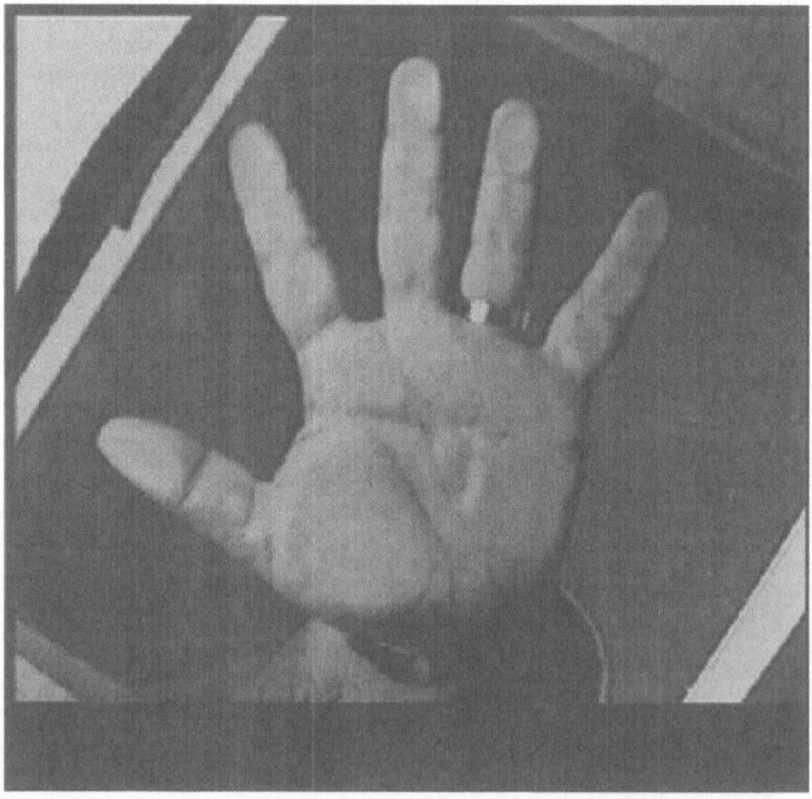

Figure 2.15.1 Image of an handprint as scanned.

2D-Fourier Transform [log scale]

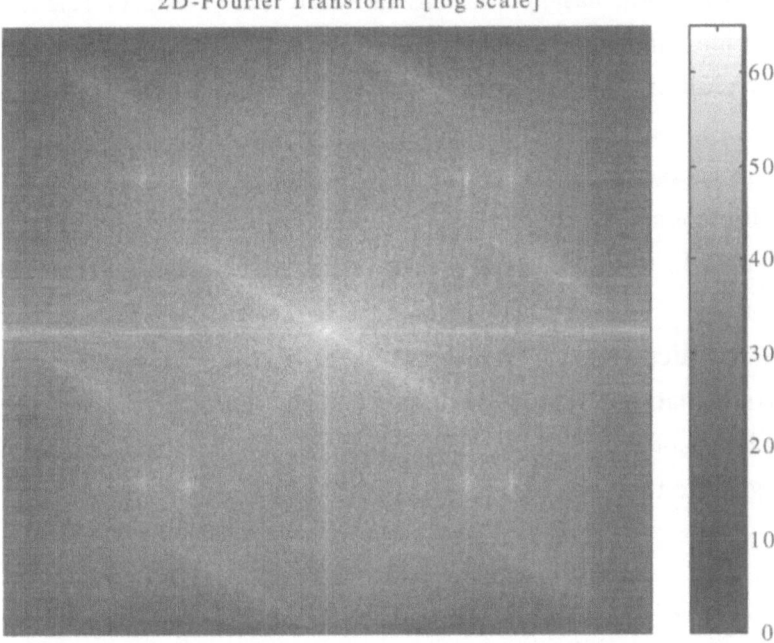

Figure 2.15.2 Two-dimensional Fourier transform of the handprint of Figure 2.15.1.

recognize the hand, although we have lost the resolution or sharpness of the image. Since this is the low-frequency component, we are actually seeing the local average of the pixels, also known as blurring. The high-frequency component shows the edges of the hand more clearly, as it involves differentiation, whereas low-frequency components involve integration. This is the basis for an algorithm known as progressive pattern recognition; we will discuss this method later in the book, in connection with more advanced signal processing methods [DeC97b].

All of the possible applications of wavelet transforms and subbands are too extensive to be covered in this book. The objective of our book is to give the reader a feeling for the applications and to cover in somewhat more detail the applications in the field of signal processing, image processing, and communication. It is to be mentioned that one might compare possible applications of wavelet transforms with that of Fourier transforms. The

Fourier transform has applications in nearly every field of science and engineering. Similarly, wavelet transforms already have found applications in nearly all branches of science and engineering, including mathematics, physics, electrical engineering, geophysics, bioengineering, computer vision, and others as outlined in the following. For details of many applications, refer to the Bibliography in the book or the technical proceedings of the Society of Photo-optic Instrumentation Engineers (SPIE), which are an excellent starting point.

Some Applications of Wavelets and Subbands

Mathematics . Statistics, numerical computation, approximation theory, solution of partial differential equations, inverting transform domain signals to the time domain

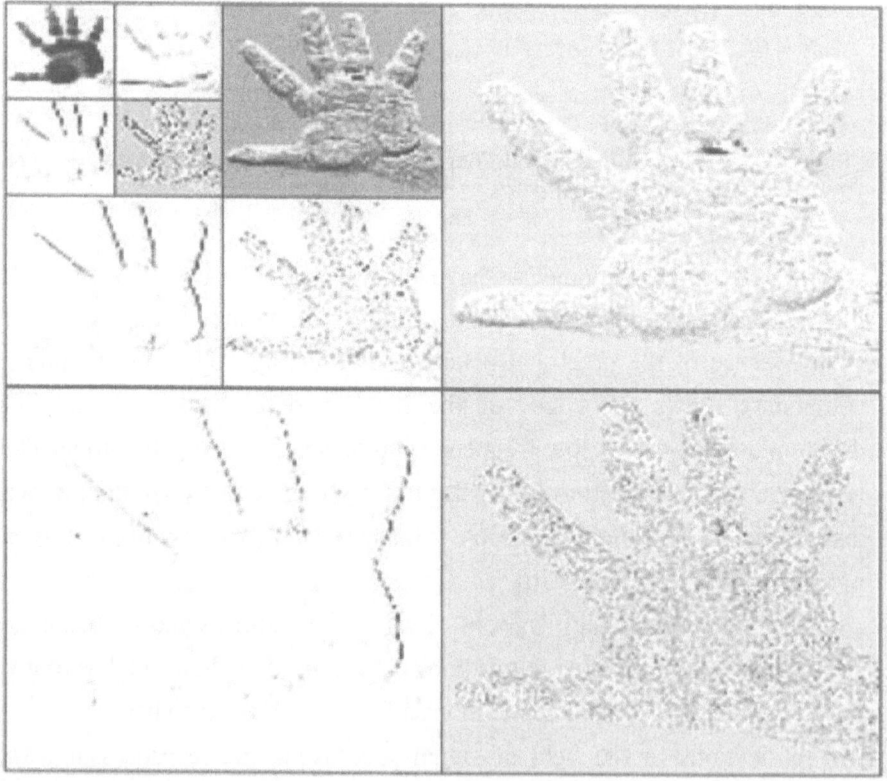

Figure 2.15.3 Two-dimensional wavelet transform of the handprint image in Figure 2.15.1.

Physics. Solutions of the wave equation., Green's function, solitons, fractals, chaos theory, quantum mechanics, photons, coherent states

Electrical Engineering. Signal processing, image processing, communication, control and systems theory, pattern recognition, radar, sonar, image compression, storage and transmission, nonlinear signal processing, progressive pattern recognition

Materials Science. Nondestructive testing, echo location, material dispersion

Bioengineering. Analysis of EKG, EEG, and ECG signals

Geophysics. Analysis of seismograph data, for example as used in oil exploration

Computer Vision. Computer graphics, morphing, 3D rendering

Communication Applications. Spread spectrum and pseudonoise codes, wireless communication, code division multiple access (CDMA), excision and denoising, discrete multitone (DMT), optical frequency division multiplexing (OFDM), asynchronous digital subscriber line (ADSL), and the so-called "last mile problem," transmultiplexers

2.16 Implementations

Wavelet transform implementation simply means realizing the correlation between a signal and a set of mother and daughter wavelets. There are many, many ways to implement correlators: they can be real time or non-real-time, analog or digital. However, one needs a bank of correlators or matched filters, as the wavelet transform maps 1-D signals into 2-D, or maps 2-D signals to 4-D signals in general, or, in the simplified case, to 3-D signals. Because of this multidimensional mapping, the optical implementation of these transforms has certain inherent advantages; an optical system is inherently 2-D, which makes it well suited for this type of operation.

We will devote an entire chapter to the implementation of wavelets and subbands later in this book; for now, we list some examples. For the digital case, dedicated semiconductor chips are available, including application-specific integrated circuits (ASICs) and dedicated digital signal processing

(DSP) chips, which realize wavelet or subband transforms in hardware. Alternately, many computer models and simulations are available, including Java and Matlab implementations.

For an example of an all-optical implementation, see [Sho97]. Acousto-optic (AO) devices can be used for designing convolvers and correlators in many ways; most of these can be adapted to wavelet and subband implementation [Das93, DeC96].

2.17 Summary and Conclusions

Transform domain processing of a signal involves mapping it from the signal space to the transform space using a set of basis functions. For wavelet transforms or subbands, one chooses a particular function as the mother wavelet, and by scaling and shifting, defines a family of daughter wavelets which form a complete set of basis functions. The important point is to use the same function and repeat it with different scale and shift parameters. The situation has been likened to the theory of natural evolution: By repeating a particular DNA sequence or an individual type of cell, we form a living being or an organism with individual characteristics.

Of course, not any function can be defined to be a mother wavelet. It must satisfy the admissibility condition. In many cases, the mother wavelet and daughter wavelets can be made orthogonal, and upon normalization, they can form an orthonormal set. For implementation of wavelet transforms, one notes that, by definition, such transforms are a correlation in the time-domain or multiplication in the frequency domain of a signal and the basis functions; thus, a band of matched filters or bandpass filters can be used.

Since a wavelet transform maps a 1-D signal to a 2-D transform domain, its representation is highly redundant. This redundancy can be reduced by using frames, which are applicable for compactly bounded wavelets. Frame theory leads to the orthonormal case in the limit, which, in turn, leads to the discrete wavelet transform or DWT.

Multiresoluton analysis is a natural consequence of choosing a set of basis functions which originate from a single mother wavelet. For this case, one defines a scaling function in association with the wavelet function. This elegant decomposition of the signal in the meshed spaces of increasing frequency resolution without losing any part of the signal (the errors are contained in the detail space) is a very important property and leads to many practical applications.

The signal to be analyzed can be either continuous, discrete, or digital. For the latter two cases, one talks about subband decomposition, as this was developed independently by the digital signal processing community. The subband decomposition involves multirate filtering, polyphase decomposition, QMF and PR filters, the unitary transform, and other topics.

These ideas can also be extended to more than just 1-D signals. In particular, 2-D signals or images are discussed because of the importance of image processing, pattern recognition, and video compression. It is possible to define multidimensional wavelets for more than 2-D signals, although they are quite complex. A particular 4-D case is the wave which is a function of 3-space coordinates and time and constitutes a solution of the wave equation. Wavelet transform or subband decomposition can also be formulated in a matrix framework; this formulation is extremely instructive and useful for numerical computation.

Any real-time correlator can be considered as a wavelet transformer. Thus, many optical correlators can be easily adapted to perform wavelet transforms. Digital integrated circuit (IC) chips have also been introduced which perform dedicated transforms for video compression and application in spread spectrum communication, among other areas.

2.18 References

[Abb95]   A. Abbate, M. Doxbeck, and P. Das, Applications of wavelet transform in signal processing, *Proc. of the International Conference on Signal Processing Applications and Technology*, pp. 652-655, 1995.

[Abb95a]   A. Abbate, Wavelet transform applied to ultrasonics, *US Army Tech. Rep. ARCCB-TR-95013*, 1995.

[Abb96]   A. Abbate, J. Frankel, R.W. Reed, and P. Das, Ultrasonic gauging and wavelet image processing for wear and erosion mapping, *Proc. of the 1996 QNDE*, vol. 18, 1996.

[Abb97a]   A. Abbate, J. Frankel, and P. Das, Application of wavelet image processing for ultrasonic gaging, *Proc. 1997 SPIE Conference on Wavelets*, 1997.

[Aka94]   A. N. Akansu and R.A. Haddad, *Multiresolution Signal Decomposition Transforms, Subbands, Wavelets*, Academic Press, New York, 1994.

[Aka99]   A.N. Akansu and M.J. Medley Eds., *Wavelet, Subband and Block Transforms in Communications and Multimedia*, Kluwer, Boston, 1999.

[Chu92]   C.K. Chui, *An Introduction to Wavelets*, Academic Press, New York, 1992.

[Chu95a]   C.K. Chui, *Wavelets: Theory, Algorithms, and Applications*, Academic Press, New York, 1995.

[Coh89]   L. Cohen, Time-frequency distributions - A review, *Proc. IEEE*, vol. 77 no. 7, pp. 941-981, 1989.

[Coh93]   L. Cohen, The scale representation, *IEEE Trans. Signal Process.*, vol. 41, pp. 3275-3292, 1993.

[Coh96]   A. Cohen and J. Kovačević, Wavelets: The mathematical background, *Proc. IEEE*, vol. 84, no. 4, pp. 514-522, 1996.

[Das93]   P. Das and C. DeCusatis, A review of acousto-optic image correlators, *Proc. SPIE 5th Annual School Seminar on Acousto-optics and Applications*, vol. 1844, pp. 33-48, 1993.

[Dau89]   I. Daubeches and J. Lagarias, Two scale differential equations, 11 local regularity, infinite products of matricies, and fractals, *AT&T Bell Labs Tech. Report*, 1989.

[Dau90]   I. Daubechies, The wavelet transform, time-frequency localization and signal analysis, *IEEE Trans. Inform. Theory*, vol. IT-36, pp. 961-1005, 1990.

[Dau91]   I. Daubechies, The wavelet transform: a method for time-frequency localization, in *Advances in Spectrum Analysis and Array Processing*, vol. 1, edited by S. Haykins, Prentice-Hall, Englewood Cliffs, NJ, pp. 366-417, 1991.

[Dau92]   I. Daubechies, *Ten Lectures on Wavelets*, SIAM, Philadelphia, 1992.

[DeC94]   C. DeCusatis, P. Das, and J. Koay, Perfect reconstruction wavelets using acousto-optic correlators, *Proc. OSA Annual Meeting*, 1994.

[DeC96]   C. DeCusatis, A. Abbate, and P. Das, Wavelet transform based image processing using acousto-optics correlators, *Proc. of 1996 SPIE Conf. on Wavelet Applications*, vol. 2762, pp. 302-313, 1996.

[DeC97]   C. DeCusatis, A. Abbate, D.M. Litynski, and P. Das, Wavelet image processing for optical pattern recognition and feature extraction, *SPIE Proc.*, vol. 3110, pp. 804-815, 1997.

[DeC97b] C. DeCusatis, A. Abbate, and P. Das, Progressive pattern recognition using the wavelet transform, *Int. J. Optoelectron..*, vol. 11, p. 425-432, 1997.

[Dud84]   D.E. Dudgeon and R.M. Mersereau, *Multidimensional Digital Signal Processing*, Prentice-Hall, Englewoods Cliffs, NJ, 1984.

[Gou84]   P. Goupillaud, A. Grossmann and J. Morlet, Cycle-octave and related transforms in seismic signal analysis, *Geoexploration*, vol. 23, pp. 85-102, 1984.

[Gro84]   A. Grossman and J. Morlet, Decomposition of Hardy functions into square integrable wavelets of constant shape, *SIAM J. Math. Anal.*, vol. 15, no. 4, pp. 723-736, 1984.

[Gro87]   A. Grossmann, M. Holschneider, R. Kronland-Martinet, and J. Morlet, Detection of abrupt changes in sound signals with the help of wavelet transforms, in *Inverse Problems*, Academic Press, New York, pp. 289-306, 1987.

[Mal89]   S. G. Mallat, A theory for multiresolution signal decomposition: the wavelet representation, *IEEE Trans. on Pattern Analys. and Machine Intell.*, vol. 11, no. 7, pp. 674-693, 1989.

[Pro88]   J.G. Proakis and D.G. Manolakis, *Introduction to Digital Signal Processing*, Macmillan, New York, 1988.

[Rab78]   L.R. Rabiner and R.W. Schafer, *Digital Signal Processing of Speech Signals*, Prentice-Hall, Englewwod Cliffs, NJ, 1978.

[Rio93]   O. Rioul, A discrete-time multiresolution theory, *IEEE Trans. Signal Process.*, vol. 41 no. 8, pp. 2591-1606, 1993.

[Sho97]   B.L. Shoop, A.H. Sayles, G.P. Dudevoir, D.A. Hall, D.M. Litynski, and P. K. Das, Smart pixel based wavelet transformation for wideband radar and sonar signal processing, *SPIE Proc.*, vol. 3078, pp. 415-423, 1997.

[Str89]   G. Strang, Wavelets and dilation equation: A brief introduction, *SIAM*

Rev., vol. 31, pp. 614-627, 1989.

[Str94]   G. Strang, Wavelets, *American Scientist*, vol. 82, pp. 250-255, 1994.

[Str96]   G. Strang and T. Nguyen, *Wavelets and Filter Banks*, Wellesley-Cambridge Press, Cambridge, 1996.

[Vai90]   P.P. Vaidyanathan, Multirate digital filters, filterbanks, polyphase networks and applications: A tutorial, *Proc. IEEE*, vol. 78, pp. 56-93, 1990.

[Vai93]   P.P. Vaidyanathan, *Multirate Systems and Filter Banks*, Prentice-Hall, Englewwod Cliffs, NJ, 1993.

[Vet87]   M. Vetterli, A theory of multirate filter banks, *IEEE Trans. Acoust. Speech Signal Process.*, vol. 35, pp. 356-372, 1987.

[Vet90a]   M. Vetterli and C. Herley, Wavelets and filter banks: relationships and new results, *Proc. ICASSP (Int. conf on acoustics, speech, and signal processing)*, vol. 3, pp. 1723-1726, 1990 .

[Vet95]   M. Vetterli and J. Kovačević, *Wavelets and Subband Coding*, Prentice-Hall, Englewood Cliffs, NJ, 1995.

Part II

Wavelets and Subbands

Chapter 3

Time-Frequency Analysis of Signals

3.1   Introduction

3.1.1   Fundamentals of Signal Analysis

In this chapter, the fundamentals of time-frequency analysis of transient signals will be introduced [Coh95, Dau90].   If we look up the term "analysis" in Webster's dictionary, it is defined as "a separating or breaking up of any whole into parts so as to find out their nature, proportion, function, relationship, etc." This is a good description of signal analysis; our intent is to decompose a transient signal into its fundamental components, which are then utilized to obtain information about the original signal more easily [Pap77].

The most obvious property of a transient signal is its variation with respect to time; however, many times, the temporal variation of a signal alone is not enough to understand the physical processes involved. For this reason, alternate forms of representation have been developed in order to decompose the signal into components better suited to our needs. One of such representations is the spectral representation, in which the amplitude of the signal is given as a function of a linear combination of pure harmonic functions. This representation was introduced in 1807 by Fourier and has come to be known as Fourier analysis. We will study both approaches and it will be shown that they may not be enough to separate different components of the signal that overlap both in time and frequency. For this

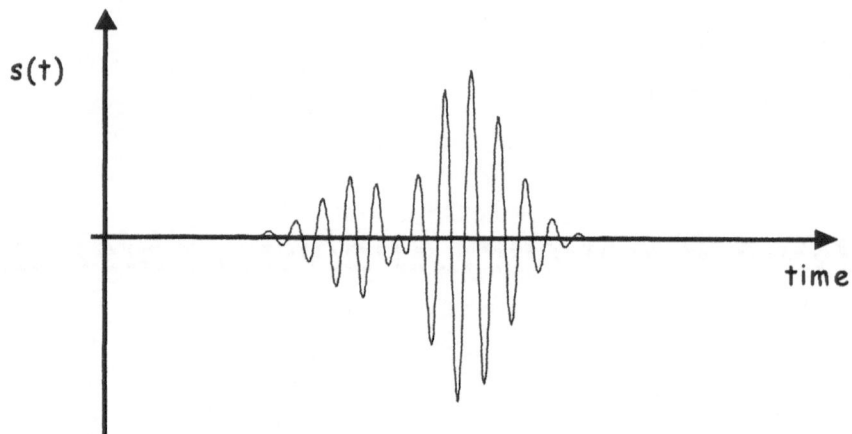

Figure 3.1.1   Temporal representation of a signal $s(t)$. The horizontal axis represents the time elapsed and the vertical axis the amplitude of the signal. A real signal is assumed.

reason different types of representation were created. Among these, the short-time Fourier transform (STFT) and the continuous wavelet transform (CWT) are described in detail in this chapter. The concept of time and frequency resolution and their connection through the uncertainty principle will be discussed. This principle is fundamental in understanding the joint time-frequency analysis properties of wavelets and subbands [Coh96].

Let us begin with a description of purely temporal analysis. Fundamental physical quantities such as electromagnetic fields, pressure, and voltage change their value in time. Their temporal variation is usually defined as a time waveform or more simply as a *signal*. A signal $s(t)$ can be expressed by many different functional forms, or also be random or stochastic in value [Pap84]. A typical representation of a signal $s(t)$ is given in Figure 3.1.1; the amplitude of the signal, assumed in this case of real value, is plotted as a function of time. Fortunately, most of the more complicated signals can be constructed using a combination of simpler signals, such as those defined by a sinusoidal function.

A pure harmonic signal can be represented by the cosine or sine function:

$$s(t) = A \cos(\omega t) \tag{3.1.1}$$

where the signal has a constant amplitude, A, modulated by a sinusoidal function of frequency ω. This signal is fully represented by the two parameters $\{A, \omega\}$. A typical example of functions of the type described by Equation 3.1.1 is given in Figure 3.1.2, with different values of ω. Generalizing this concept leads us to define the signal $s(t)$ as:

$$s(t) = a(t) \cos[\theta(t)] \tag{3.1.2}$$

where both $a(t)$ and $\theta(t)$ are now functions of time. Again this signal can be defined by the pair of functions $\{a(t), \theta(t)\}$. In Figure 3.1.3 are plotted $a(t)$ and $\theta(t)$ which result in the signal $s(t)$ in Figure 3.1.1. Please note that $\cos(2n\pi+t)=\cos(t)$, with n an integer. Hence a signal $s(t)$ can be represented by an infinite number of functions $\theta(t)$, which satisfy the previous equation. We will here introduce other signal representations that eliminate this ambiguity.

Another possible representation of a signal is using the complex domain, in which the signal is decomposed into real and imaginary components, or

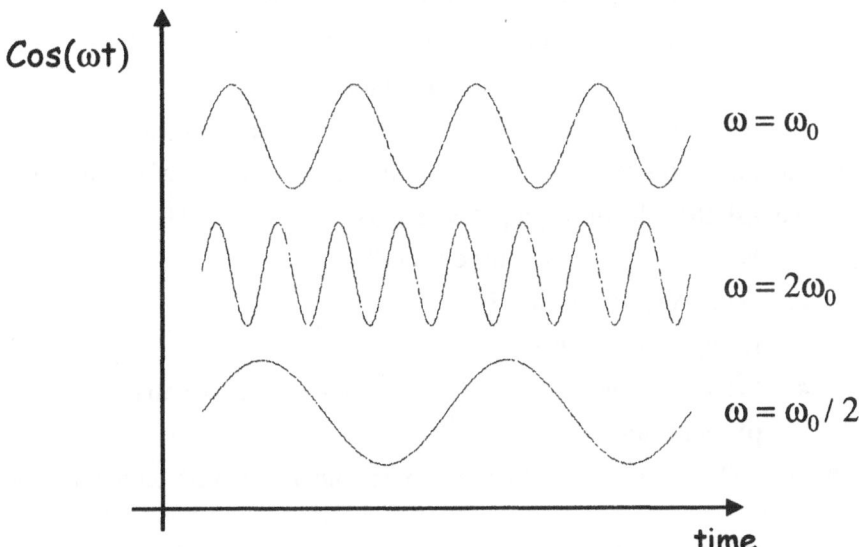

Figure 3.1.2 Plot of basis functions $\cos(\omega t)$ with different values of angular frequency ω.

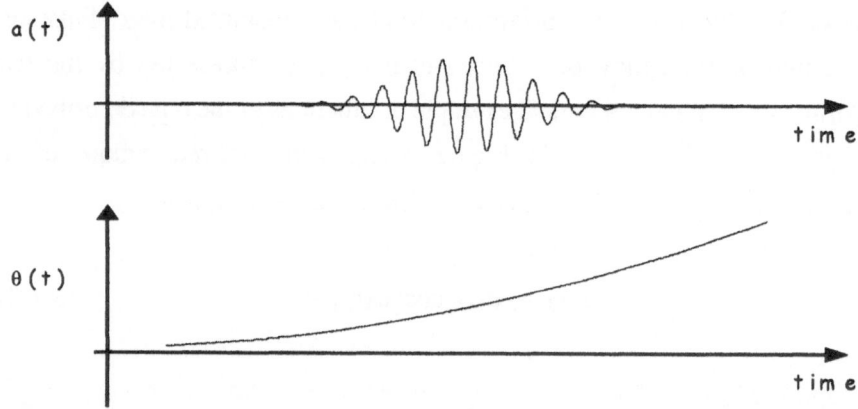

Figure 3.1.3   Signal amplitude a(t) versus time (top curve) and relative phase θ(t) versus time (bottom curve). The signal s(t) calculated using Equation 3.1.2 is plotted in Figure 3.1.1.

into amplitude and phase, this time using a complex exponential. The signal can thus be represented as

$$s(t) \; = \; s_R(t) \; + \; j \; s_I(t) \; = \; A(t) \; e^{j\theta(t)} \tag{3.1.3}$$

where $j = \sqrt{(-1)}$, and $s_R(t)$ represents the real part of the signal and $s_I(t)$ its imaginary counterpart. Note that if $s(t)$ is a real signal, then $s(t) = s_R(t)$ and $s_I(t)=0$. Both $A(t)$ and $\theta(t)$ can be functions with complex values, but it is common practice to have $A(t)$ of real values.

Assuming that the signal represents the voltage consumption on a electrical line, and we are interested at analyzing the effect of power surges, we may ask the following simple questions:

1. When did the power surge occur?
2. What was its duration?
3. How often it occurs?
4. What is its energy, i.e., what is the work necessary to reproduce the phenomenon?

These are basic questions that a proper signal analysis technique must answer.

Let us start by defining the energy of a signal. The *energy* of a signal is a basic concept, because it represents the amount of work necessary to reproduce the signal $s(t)$. Hence, how do we define its energy? There are many different ways, depending on the type of signal being considered. For example, in electromagnetic theory, the absolute square of the electric and magnetic fields are proportional to the energy density (*Poynting's theorem*). In acoustics, the energy is represented by the square of the sound pressure. If $s(t)$ represents the voltage across a resistor R, the current necessary to generate such voltage is given by the following ratio:

$$i(t) = \frac{s(t)}{R} \tag{3.1.4}$$

and the electromagnetic energy needed to generate this signal is:

$$P(t) = \frac{dE(t)}{dt} = i(t)\, s(t) = \frac{|s(t)|^2}{R} \tag{3.1.5}$$

The power intensity $P(t)$ (i.e., the energy per unit time), is proportional to the square of the signal $s(t)$. Therefore, it seems natural to define the energy of a signal as the square of its absolute instantaneous value:

$|s(t)|^2$ = Energy of the signal $s(t)$ per unit time at instant t

   (also known as the *energy density* or *instantaneous power*)

We can also define the fractional energy of a signal as:

$|s(t)|^2 \Delta t$ = Fractional energy in time interval Δt around t

The *total energy*, E, of the signal $s(t)$ is thus given by the integral of its fractional energy for the duration of the signal:

$$E = \int_{-\infty}^{+\infty} |s(t)|^2\, dt \quad . \tag{3.1.6}$$

Also, it is important to know how the signal is concentrated in time, (i.e., is if it lasts a minute, a day, or a year), and at what time the event represented by the signal occurred. For this reason we define an average (mean) time $<t>$, and its range or standard deviation σ_t as

$$<t> = \frac{1}{E} \int_{-\infty}^{+\infty} t \, |s(t)|^2 \, dt \qquad (3.1.7)$$

$$\sigma_t^2 = \frac{1}{E} \int_{-\infty}^{+\infty} (t - <t>)^2 \, |s(t)|^2 \, dt = <t^2> - <t>^2 \ . \qquad (3.1.8)$$

These two quantities are very useful in defining the temporal duration of the signal $s(t)$: $<t>$ and σ_t tell us the duration in time of the signal, since most of the signal energy is localized in a time interval $2\sigma_t$ around $<t>$. In many cases, the duration $2\sigma_t$ is also represented with the symbol T.

As an example, let us consider a signal with a Gaussian envelope:

$$s(t) = (\alpha/\pi)^{1/4} \, e^{-\alpha(t - t_o)^2/2} \, \sin(2\pi f t) \qquad (3.1.9)$$

where $\alpha = 0.0003$, $t_0 = 500$, and $f = 1/20$.

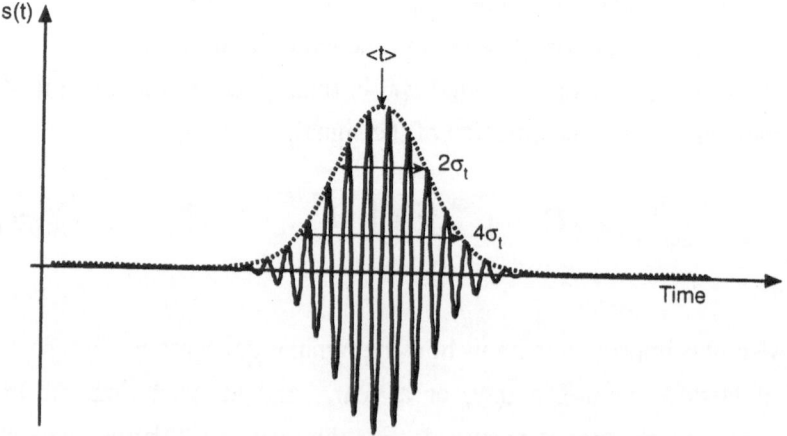

Figure 3.1.4 Signal amplitude versus time, with indications of the mean time $<t>$ and the signal duration $T = \sigma_t$.

For this example, we may calculate

$$\langle t \rangle = t_0$$
$$\sigma_t^2 = 1/2\alpha = 1667$$
$$\sigma_t = 40.82$$

A plot of this expression is shown in the Figure 3.1.4; we can clearly see that most of the signal's energy is contained within twice the standard deviation.[*] For non-Gaussian signals, this approximation may not hold as strongly, but it remains a good rule of thumb.

Another example is the rectangular signal (i.e., a signal $s(t)$ with constant amplitude from time t_1 to time t_2 and zero otherwise):

$$s_1(t) = \begin{cases} A, & t_1 \le t \le t_2 \\ 0, & \text{otherwise} \end{cases} \tag{3.1.10}$$

We can calculate the total energy of the signal E as

$$E = \int_{-\infty}^{+\infty} |s(t)|^2 \, dt = \int_{t_1}^{t_2} A^2 \, dt = A^2 (t_2 - t_1) \tag{3.1.11}$$

and the mean time is, as expected, the average between t_2 and t_1:

$$\langle t \rangle = \frac{1}{E} \int_{t_1}^{t_2} t A^2 \, dt = \frac{A^2}{E} \left[\frac{t^2}{2} \right]_{t_1}^{t_2} = \frac{\frac{A^2}{2}\left(t_2^2 - t_1^2\right)}{A^2 (t_2 - t_1)} \tag{3.1.12}$$

$$= \frac{(t_2 - t_1)(t_2 + t_1)}{2(t_2 - t_1)} = \frac{(t_2 + t_1)}{2}$$

For the duration, we can calculate σ_t from

---

[*] For a Gaussian signal, 67% of its energy is contained within 2σ, around $\langle t \rangle$, and 99% is contained in the interval $\langle t \rangle \pm 3\sigma_t$.

$$\sigma_t = \sqrt{<t^2> - <t>^2} = \frac{(t_2 - t_1)}{2\sqrt{3}} \qquad (3.1.13)$$

It seems counterintuitive to obtain for σ_t a value less than the real duration of the rectangular signal; this is because by means of Equation 3.1.7, we have fitted a Gaussian signal to $s(t)$. It follows that the more the signal resembles a Gaussian shape, σ_t will more accurately estimate the duration of the signal.

Although temporal signal analysis yields important information, many physical processes cannot be easily understood without the aid of frequency analysis. For example, we know from physics that the shape of a propagating wave depends on its frequency (dispersion). In chemical analysis of compounds, we need to analyze the absorption or emission spectrum, or frequencies of light, to determine the chemical structure. Frequency or Fourier analysis is also a very useful tool for the solution of differential equations. In frequency analysis, the signal $s(t)$ is expanded in terms of sinusoidal functions of different frequencies:

$$s(t) = \frac{1}{\sqrt{2\pi}} \int_{-\infty}^{+\infty} S(\omega)e^{j\omega t} \, dt \qquad (3.1.14)$$

The amplitude $S(\omega)$ of each term in the series expansion can be utilized to characterize the signal. These terms are calculated using the *Fourier transform* of $s(t)$:

$$S(\omega) = \frac{1}{\sqrt{2\pi}} \int_{-\infty}^{+\infty} s(t)e^{-j\omega t} \, dt \qquad (3.1.15)$$

Both the Fourier transform $S(\omega)$ and the temporal waveform $s(t)$ represent the same signal. The Fourier transform and its inverse (Equations 3.1.14 and 3.1.15) show us the relationship between the two representations:

$$s(t) = a(t) \, e^{j\varphi(t)}$$

$$S(\omega) = A(\omega) \, e^{j\Psi(\omega)}$$

(3.1.16)

The amplitude of the Fourier transform of the signal in Figure 3.1.4 is plotted in Figure 3.1.5 as a function of the angular frequency ω. Please note that the Fourier transform $S(\omega)$ is, in general, a complex signal, especially if the signal $s(t)$ is a real function. For a complete list of properties of the Fourier transform see the Appendix and related references.

In analogy with the temporal representation, we can define the energy of the signal per unit frequency at a frequency ω as the modulus square of $|S(\omega)|^2$:

$|S(\omega)|^2$ = Energy of the signal $s(t)$ per unit frequency at frequency ω

(Energy density or instantaneous power).

$|S(\omega)|^2 \cdot \Delta\omega$ = Fractional energy in time interval $\Delta\omega$ around ω.

The *total energy*, E, of the signal $s(t)$ is thus given by the integral of its fractional energy over its frequency range:

$$E = \int_{-\infty}^{+\infty} |S(\omega)|^2 \, d\omega = \int_{-\infty}^{+\infty} |s(t)|^2 \, dt$$

(3.1.17)

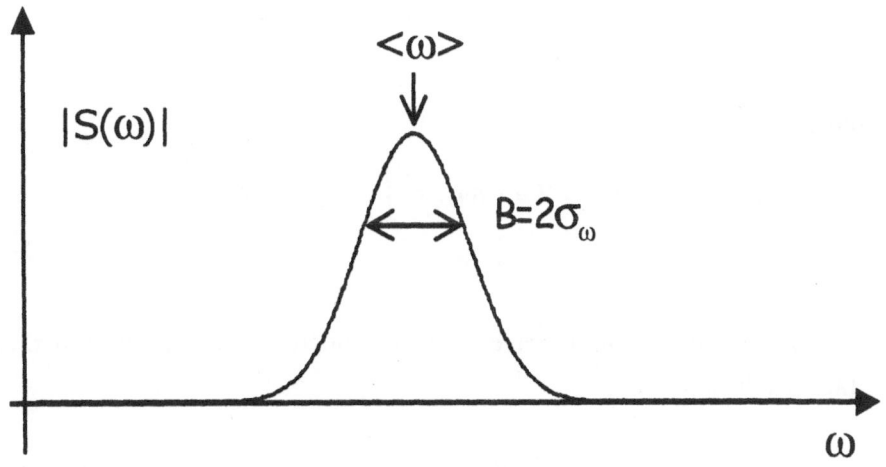

Figure 3.1.5   Signal amplitude in frequency domain, with mean frequency $<\omega>$ and signal bandwidth $B = \sigma_\omega$.

The equivalency between the two relationships is a result of the fact that both are representations of the same signal. This last relationship is commonly called *Parceval's theorem* or *Rayleigh's theorem*. The proof of this theorem is fairly straightforward; beginning with the definition of signal energy, we substitute the Fourier transforms of the signal $s(t)$ to obtain the following:

$$E \doteq \int |s(t)|^2 \, dt = \int_{-\infty}^{+\infty} s(t) \, s^*(t) \, dt \qquad (3.1.18)$$

$$E = \int_t \left| \frac{1}{\sqrt{2\pi}} \int_\omega S(\omega) e^{j\omega t} \, d\omega \right| \left| \frac{1}{\sqrt{2\pi}} \int_{\omega_1} S^*(\omega_1) \, e^{j\omega_1 t} d\omega_1 \right| dt \qquad (3.1.19)$$

$$E = \frac{1}{2\pi} \int_\omega \int_{\omega_1} S(\omega) S^*(\omega_1) \, d\omega \, d\omega_1 \int_t e^{j\omega t} \, e^{j\omega_1 t} dt \qquad (3.1.20)$$

but

$$\int_t e^{j\omega t} e^{j\omega_1 t} dt = 2\pi \, \delta(\omega - \omega_1)$$

$$= \int_\omega \int_{\omega_1} S(\omega) S^*(\omega_1) \, d\omega \, d\omega_1 \, \delta(\omega - \omega_1) \qquad (3.1.21)$$

$$= \int_\omega S(\omega) \, d\omega \int_{\omega_1} S^*(\omega_1) \, d\omega_1 \, \delta(\omega - \omega_1)$$

thus,

$$E = \int_\omega S(\omega) S^*(\omega) \, d\omega = \int_\omega |S(\omega)|^2 \, d\omega \quad . \qquad (3.1.22)$$

We can thus define an average (mean) frequency $\langle \omega \rangle$, and its range or standard deviation σ_ω as

$$\langle \omega \rangle = \frac{1}{E} \int_{-\infty}^{+\infty} \omega |S(\omega)|^2 \, d\omega \qquad (3.1.23)$$

$$\sigma^2_\omega = \frac{1}{E} \int_{-\infty}^{+\infty} (\omega - <\omega>)^2 \, |S(\omega)|^2 \, d\omega \quad . \tag{3.1.24}$$

These two quantities are very useful in defining the frequency range of the signal $s(t)$. The square root of the standard deviation of the frequency is commonly called the *frequency bandwidth* of the signal

$$B = 2\sigma_\omega \quad . \tag{3.1.25}$$

In this expression, $\langle\omega\rangle$ and σ_ω tell us that the signal is localized in a frequency range $2\sigma_\omega$ around $\langle\omega\rangle$. As an example, consider once again a signal with a Gaussian envelope given by

$$s(t) = \left(\frac{\alpha}{\pi}\right)^{1/4} e^{-\alpha t^2/2} \, e^{(j2\pi ft)} \tag{3.1.26}$$

$$S(\omega) = \left(\frac{\alpha}{\pi}\right)^{1/4} \frac{1}{\sqrt{\alpha}} \, e^{-(\omega - \omega_0)^2/2\alpha} \tag{3.1.27}$$

where

$$
\begin{aligned}
\alpha &= 0.0003 \\
t_0 &= 500 \\
f &= 1/20 \\
\langle\omega\rangle &= 2\pi f = 0.32 \\
\sigma_\omega^2 &= \alpha/2 = 0.00015 \\
\sigma_\omega &= 0.0122
\end{aligned}
$$

The mean frequency $\langle\omega\rangle$ and bandwidth $B = \sigma_\omega$ are indicated in Figure 3.1.5. The mean frequency is also called the center frequency of the signal.

We have shown how to calculate the mean frequency and bandwidth of a signal using its spectrum, but can these properties also be determined from the temporal representation $s(t)$ of the signal?[**]

Using the temporal representation of a signal $s(t)$ of Equation 3.1.3, we repeat for convenience with its Fourier transform:

$$s(t) = s_r(t) + js_i(t) = A(t)e^{j\varphi(t)}$$

$$S(\omega) = B(\omega) e^{j\Phi(\omega)} .$$

(3.1.28)

The total energy of the signal is given by

$$E = \int |s(t)|^2 dt = \int A^2(t) dt .$$

(3.1.29)

The amplitude $A^2(t)$ can thus be utilized to represent the energy density of the signal per unit time and can be used in Equations 3.1.7 and 3.1.8 to estimate the mean time and duration of the signal

$$<t> = \frac{1}{E} \int t A^2(t) dt$$

(3.1.30)

$$\sigma_t^2 = \frac{1}{E} \int_{-\infty}^{+\infty} (t - <t>)^2 A^2(t) dt .$$

(3.1.31)

What about the phase $\varphi(t)$ of the signal? From intuition, we could say that it must yield information about the frequency content of the signal. It can be shown that the instantaneous frequency of the signal is given by the time derivative of the phase $\varphi(t)$:

$$\omega_t = \varphi'(t) = \frac{d\varphi(t)}{dt}$$

(3.1.32)

The concept of instantaneous frequency is linked to the change in signal properties as a function of time. In practice, this is experienced daily as

---

[**] For reasons of space, only results are here given, and the interested reader can look up the references.

changing colors in a light spectrum of changing pitch in a musical score or in a song. The average frequency is given by the definition of the average of a function, using a distribution function:

$$<\omega> = \frac{\int \omega_r A^2(t)\, dt}{\int A^2(t)\, dt} = \frac{1}{E} \int \varphi'(t)\, A^2(t)\, dt \quad . \tag{3.1.33}$$

The bandwidth B of the signal can thus be calculated using the following relationship:

$$B^2 = \frac{1}{E} \int_{-\infty}^{+\infty} \left(\varphi'(t) - <\omega>\right)^2 A^2(t)\, dt +$$

$$+ \frac{1}{E} \int_{-\infty}^{+\infty} \left(\frac{A'(t)}{A(t)}\right)^2 A^2(t)\, dt \quad . \tag{3.1.34}$$

The bandwidth B is given by the sum of two terms, one depending on the phase (FM component), as we would intuitively expect, and a second depending on the derivative of the amplitude $A(t)$ of the signal (AM component). The amplitude-modulated (AM) and frequency-modulated (FM) terms are thus expressed as

$$B^2 = B^2_{AM} + B^2_{FM} \tag{3.1.35}$$

$$B^2_{AM} = \frac{1}{E} \int [A'(t)]^2\, dt \tag{3.1.36}$$

$$B^2_{FM} = \frac{1}{E} \int [\varphi'(t) - <\omega>]^2 A^2(t)\, dt \quad . \tag{3.1.37}$$

What do Equations 3.1.35 through 3.1.37 really mean, and what contributes to the total bandwidth of a signal? The bandwidth of a signal does not represent only the frequencies generated by that signal. For example, if a sound is produced at 1000 Hz and increases in frequencies to 1200 Hz at more or less constant amplitude, the spread in frequency (i.e., the bandwidth) is intuitively 200 Hz. However, we can obtain a similar spread in frequency with a signal of constant frequency 1000 Hz by making the signal of finite time duration or by varying its amplitude. As an example, consider a Gaussian signal $s_1(t)$ and a Chirp signal $s_2(t)$ as follows:

$$s_1(t) = \left(\frac{\alpha}{\pi} \right)^{1/4} e^{-\alpha t^2/2} e^{j\omega_0 t} \tag{3.1.38}$$

$$s_2(t) = \left(\frac{\gamma}{\pi} \right)^{1/4} e^{-\gamma t^2/2 \, + \, j\beta t/2} e^{j\omega_0 t} \tag{3.1.39}$$

where α, β, and γ are constants. The two signals are plotted in Figure 3.1.6a and the magnitude of the Fourier transform is given in Figure 3.1.6b. By looking at the magnitude of the Fourier transform, it seems that the two signals have the same frequency components. This is true, but the major difference between the two signals is the temporal occurrence of such

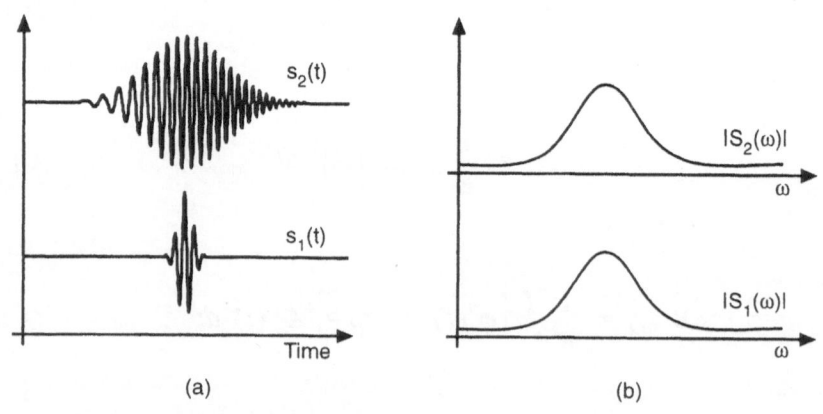

(a)   (b)

Figure 3.1.6   Plot of a Gaussian signal $s_1(t)$ and a chirp signal $s_2(t)$ which have the same bandwidth.

components. In the case of $s_1(t)$, all frequency components are in phase and occur at the same time, whereas for $s_2(t)$ the components exist at different time instances. This is the fundamental limitation of the Fourier transform when analyzing signals. It is not easy to determine the temporal existence of the various frequency components. We will see in the following that other processing techniques are better suited to this task. Nevertheless, it must be pointed out that since the Fourier transform is a lossless transformation, the temporal information is actually embedded in the complex phase of Equation 3.1.29, but it is not easily resolved. The bandwidth equations for these signals are given by

$$B^2_{s_1} = 2\,\alpha \qquad (3.1.40)$$

$$B^2_{s_2} = 2\,\gamma + 2\,\frac{\beta^2}{\gamma} \; . \qquad (3.1.41)$$

It is clear that we can make the two bandwidths the same, even though the frequencies covered by the signals are different. How is this possible? The bandwidth of a signal $s(t)$ is due to the deviations of the instantaneous frequency of the signal from the average frequency and by the fast amplitude changes of the signal. In most applications, the AM term is negligible and thus only the FM term is used in Equation 3.1.32. In a similar manner, we can redefine the temporal averages using the spectrum of the signal:

$$S(\omega) = B(\omega)e^{j\,\phi(\omega)} \qquad (3.1.42)$$

$$E = \int |S(\omega)|^2 \, d\omega \; . \qquad (3.1.43)$$

Using this expression, we obtain for the average temporal variables

$$<t> = -\frac{1}{E} \int \phi'(\omega) \, |S(\omega)|^2 \, d\omega \qquad (3.1.44)$$

$$T^2 = \sigma_t^2 = \frac{1}{E} \int [B'(\omega)]^2 \, d\omega$$

$$+ \frac{1}{E} \int [\phi'(\omega) + <t>]^2 \, B^2(\omega) \, d\omega \quad .$$

(3.1.45)

The average time $<t>$ is obtained by averaging the term $-\phi'(\omega)$ over all frequencies; this is similar to the calculation we have performed to estimate the average frequency $<\omega>$ using the instantaneous frequency ω_i. Please note that the prime sign in this case refers to a derivative versus frequency,

$$\phi'(\omega) = \frac{d\phi(\omega)}{d\omega}.$$

We can thus define an average time for a particular frequency; this is the so-called group delay $\tau_g(\omega)$:

$$t_g(\omega) = -\phi'(\omega) = \frac{d\phi(\omega)}{d\omega} \quad .$$

(3.1.46)

To summarize, the temporal representation of a signal, $s(t)$, tells how the signal varies in time, but it gives no indication about the period or frequency of any component in the signal. The frequency representation of a signal, $S(\omega)$, sometimes known as the spectrum, tells which frequencies are present during the total duration of the signal, but it gives no indication as to when these frequencies existed. The energy of a signal is related to the temporal or frequency variation by

$$E = \int_{-\infty}^{+\infty} |s(t)|^2 \, dt = \int_{-\infty}^{+\infty} |S(\omega)|^2 \, d\omega \quad .$$

(3.1.47)

3.1.2 Uncertainty Principle

We have defined the duration T and the bandwidth B of a signal using Equations 3.1.8 and 3.1.24. These two quantities can be used to estimate the

time and frequency range covered by the signal. We can thus ask ourselves
if it is possible to construct a signal such that its energy is well localized in
time and frequency that is, a signal for which T and B are arbitrarily small?
The answer is NO, as we cannot arbitrarily reduce these two terms. The
Uncertainty Principle states exactly that the product time duration and
bandwidth cannot be arbitrary small:

$$\sigma_t \, \sigma_\omega = \frac{T B}{4} \geq \frac{1}{2} \ .$$

(3.1.48)

In signal analysis the Uncertainty Principle states that a narrow
waveform yields a wide spectrum, and a wide waveform yields a narrow
spectrum, and both time waveform and frequency spectrum cannot be made
arbitrarily small simultaneously. The density of energy in time is given by
$|s(t)|^2$, whereas the density in frequency is $|S(\omega)|^2$. But these two terms are
related through the Fourier transform, hence we should not be surprised to
find out that there is a relationship between the two densities. This
relationship is such that if one density function is narrow in its domain, then
the other is broad [Rab78]. When is the equal sign valid for the Uncertainty
Principle? A more accurate definition of the Uncertainty principle is given
by:

$$\sigma_t^2 \, \sigma_\omega^2 \geq \frac{1}{4} + \text{Cov}_{t\omega}^2$$

(3.1.49)

where $\text{Cov}_{t,\_}$ is the covariance of the signal and is always a positive term.
It can be shown that a Gaussian signal is the most general signal that has a
time-bandwidth product equal to one half.

Example: Chirp Signal
The Chirp signal is expressed by the following expression:

$$s_1(t) = \left(\frac{\gamma}{\pi} \right)^{1/4} e^{-\gamma t^2/2 \, + \, j \, \beta t/2} e^{j\omega_0 t}$$

(3.1.50)

We have established that the time duration and frequency bandwidth are

$$\sigma_t^2 = <t^2> - <t>^2 = \frac{1}{2\gamma}$$

$$\sigma_\omega^2 = \frac{\gamma^2 + \beta^2}{2\gamma} \tag{3.1.51}$$

Hence, the uncertainty inequality is

$$\sigma_t \, \sigma_\omega = \sqrt{\frac{1}{2\gamma}} \sqrt{\frac{\gamma^2 + \beta^2}{2\gamma}} = \frac{1}{2} \sqrt{1 + \frac{\beta^2}{\gamma^2}}$$

$$= \frac{1}{2} \sqrt{1 + 4\mathrm{Cov}_{t\omega}} \tag{3.1.52}$$

$$\sigma_t \, \sigma_\omega \geq \frac{1}{2}$$

Example: Sinusoidal Modulation Signal

A signal whose frequency is modulated by a sinusoidal term is given by the following expression:

$$s_1(t) = \left(\frac{\alpha}{\pi}\right)^{1/4} e^{-\alpha t^2/2 + jm\sin(\omega_m t)} e^{j\omega_0 t} \quad . \tag{3.1.53}$$

Hence, the uncertainty inequality is

$$\sigma_t \, \sigma_\omega = \sqrt{\frac{1}{2\alpha}} \cdot \sqrt{\frac{\alpha}{2} + \frac{\alpha m^2 \omega_m^2}{2} \left(1 - e^{-\omega_m^2/2\alpha}\right)^2} \tag{3.1.54}$$

$$\sigma_t \, \sigma_\omega \geq \frac{1}{2} \sqrt{1 + \frac{m^2 \omega_m^2}{2} \left(1 - e^{-\omega_m^2/2\alpha}\right)^2}$$

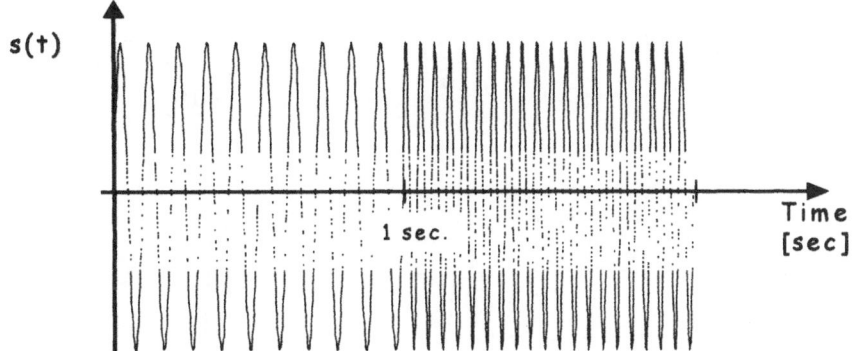

Figure 3.2.1   Signal amplitude in time for a duration of 2 sec. The signal is constituted of two pure sine bursts of duration of 1 sec. each, and not overlapping.

It can be shown that the term within the square root is always larger than unity.

3.2   Windowed Fourier Transform: Short-Time Fourier Transform and Gabor Transform

In the temporal and spectral representation of a signal, it is possible to extract information about transients in either domain. For example, it is easy to notice a surge on a power line by looking at its voltage in time. Conversely, it is easy to detect the 50-Hz electro-magnetic noise by looking at its frequency spectrum. We have also seen that by defining an instantaneous frequency $\omega_i(t)$ and a group delay $\tau_g(\omega)$, we can successfully describe the change in characteristics of a chirp signal. Unfortunately, there are many causes in nature for the spectrum of a signal to change and it is not easy to analyze a signal with multiple frequency components which occur at different and possible overlapping time intervals. As we have seen, Fourier analysis decomposes the signal into a series of harmonic waves; hence, using the frequency information, we can separate these reasons into two large groups. First, we may have a case in which the physical parameters that affect the generation of the signal harmonics change with time. For example, the mechanical oscillation of a string is a function of the length and tension applied to the string. If any of these parameters changes (for example with

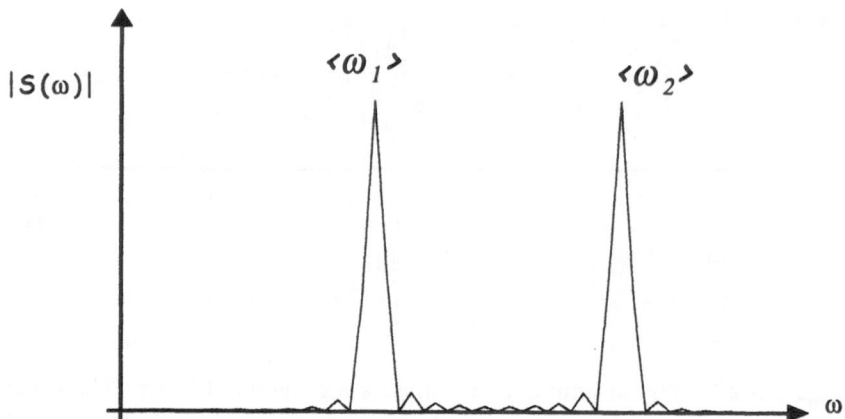

Figure 3.2.2   Fourier transform of the signal in Figure 3.2.1, clearly showing the two mean frequencies $<\omega_1>$ and $<\omega_2>$.

aging) a different beat frequency will result. This concept is fundamental in the generation of musical scores and speech.   The other reason for frequencies to vary with time is that we detect these harmonics at a certain distance from the source. In many materials, the velocity of propagation of the harmonic wave (such as sound or light) is a function of its frequency. The signal is said to disperse, and the dispersion is a function of the distance traveled (or, conversely, the time elapsed). Also, the strength of the various harmonics can change; this is known as frequency-dependent attenuation. There are many practical examples of this situation; for instance, time-frequency analysis can be used as a tool to characterize or classify the sounds of different whale types or of human voices. A similar method is used for fault analysis of car windshields during production [Coh95].

The Fourier transform analyzes a signal over its whole temporal duration; hence, it yields information over the complete spectrum of the signal. Given the signal in Figure 3.2.1, its frequency spectrum, shown in Figure 3.2.2, can be used to determine that two frequencies are present in the signal, but it cannot tell us if these two frequency components do completely overlap in time or not. This is because by performing the Fourier transform, we project the signal in the frequency domain, but lose completely the link with the time domain. The Fourier transform utilizes the information of the signal over all the time (from $-\infty$ to $+\infty$); this means that there can be no concept of time in the frequency domain, and thus no concept of frequency

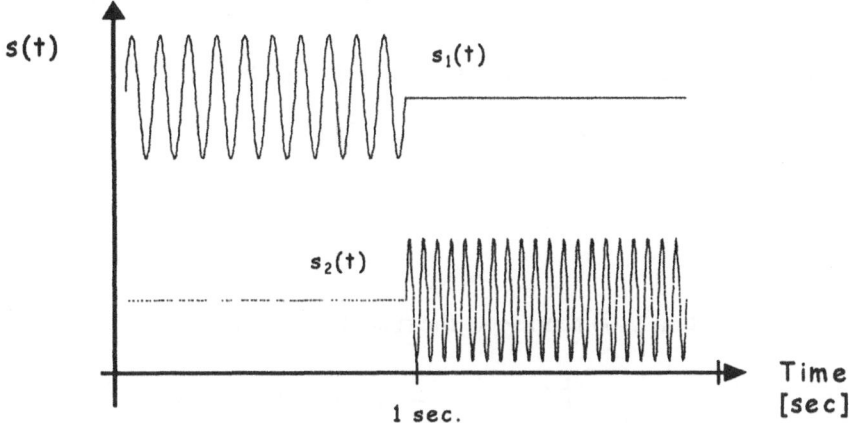

Figure 3.2.3   Signals $s_1(t)$ and $s_2(t)$ defined in Equations 3.2.1 and 3.2.2.

changing with time.   Mathematically, this is equivalent of saying that frequency and time are orthogonal domains. It must be noted that since the Fourier transform is a complete representation of the signal, the temporal information is not lost, but it is not easily accessible for analysis. But if we decide to perform the Fourier transform of two different signals, the first is

$$s_1(t) = \begin{cases} s(t), & 0 \le t \le 1 \text{ sec.} \\ 0, & \text{otherwise} \end{cases} \qquad (3.2.1)$$

and

$$s_2(t) = \begin{cases} s(t), & 1 \le t \le 2 \text{ sec.} \\ 0, & \text{otherwise} \end{cases} \qquad (3.2.2)$$

The two signals are shown in Figure 3.2.3. For clarity of representation, the two signals are shifted in the vertical axis and thus show a bias. The Fourier transform of the two signals is given in Figure 3.2.4. For the same reason, a bias in the vertical axis was added.

From Figure 3.2.4 we can easily conclude that the component at frequency ω_1 was present in the time interval (0, 1 sec) and not in the (1, 2 sec) interval. A similar conclusion can be reached for the component at frequency ω_2. We can thus think of dividing the signal $s(t)$ into smaller time segments, as we have done using $s_1(t)$ and $s_2(t)$, and perform the Fourier transform on these segments, in order to extract the frequencies that exist in that time interval.

Mathematically, we can write the previous equations as

$$s_1(t) = s(t) \, u_1\!\left(t - 1/2\right)$$
$$s_2(t) = s(t) \, u_1\!\left(t - 3/2\right)$$

(3.2.3)

where the function $u_T(t)$ represents the rectangular function:

$$u_T(t - \tau) = \begin{cases} 1 \,, & \tau - \dfrac{T}{2} \leq < t \leq \tau + \dfrac{T}{2} \\ 0 \,, & \text{otherwise} \end{cases}$$

(3.2.4)

More generally, we can define a window function $h(t)$ centered at a time τ to produce a modified signal

$$s_\tau(t) = s(t) \, h(t-\tau) \ .$$

(3.2.5)

The term "window" derives from the fact that we are now looking at the signal only over a small time interval, and we have in fact lost the whole picture of the signal, as if we were looking through a window [Coh89]. The simplest window we can have is a rectangular window, of width b:

$$s_\tau(t) = s(t) \cdot u_T(t - \tau) = \begin{cases} s(t) \,, & \tau - T \leq t \leq \tau + T \\ 0 \,, & \text{otherwise} \end{cases}$$

(3.2.6)

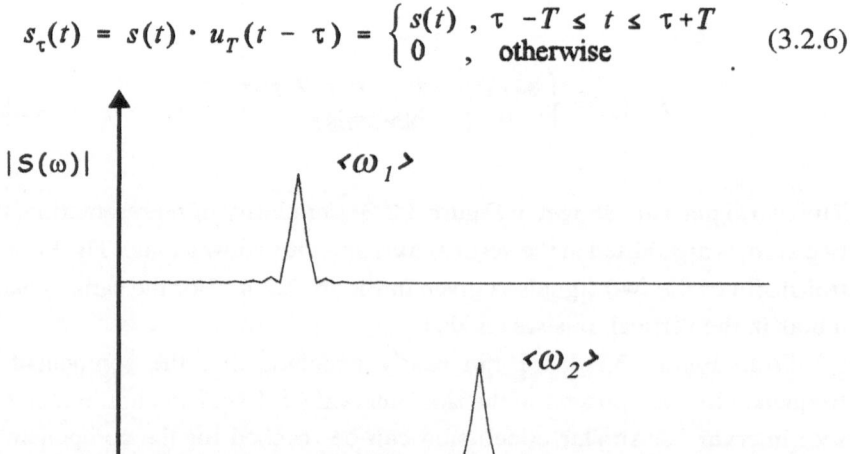

Figure 3.2.4   Fourier transform of the signals s_1 and s_2 in Figure 3.2.3.

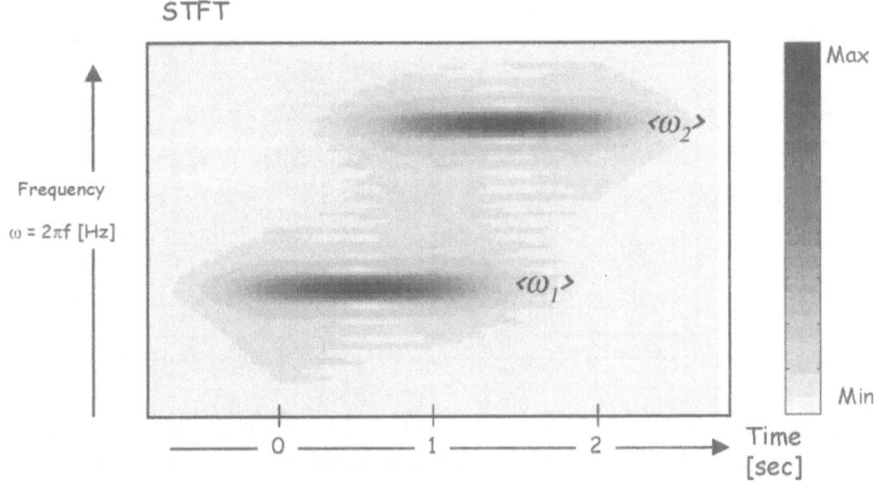

Figure 3.2.5   Short-time Fourier transform of the signal in Figure 3.2.1, clearly showing the two frequencies ω_1 and ω_2 and their duration in time.

The Fourier transform of $s_\tau(t)$ is

$$S(\tau,\omega) = \frac{1}{\sqrt{2\pi}} \int_{-\infty}^{+\infty} s_\tau(t) e^{-j\omega t} \, dt$$

$$= \frac{1}{\sqrt{2\pi}} \int_{-\infty}^{+\infty} s(t) \, h(t - \tau) e^{-j\omega t} \, dt \qquad (3.2.7)$$

This is the so-called *windowed Fourier transform (WFT)* or *short-time Fourier transform (STFT)*, for obvious reasons. It was originally proposed by Gabor in 1946. In that case, Gabor choose, for the window $h(t)$, a Gaussian function [Gab46]. For this reason, in the literature the Gaussian windowed Fourier transform is sometimes referred to as the *Gabor transform*. In the following, we will refer to the windowed Fourier transform, the Short-time Fourier transform (STFT) and Gabor transform without distinction. In the case of the Gabor transform, a Gaussian window is assumed. An image density plot of the STFT of the signal in Figure 3.2.1 is given in Figure 3.2.5. The magnitude of $S(\omega,\tau)$ is plotted as gray-scale intensity, whereas the x axis represents the time delay τ, and the y axis represents the angular frequency ω. From Figure 3.2.5, we can easily determine that two harmonic components are present on the signal. These

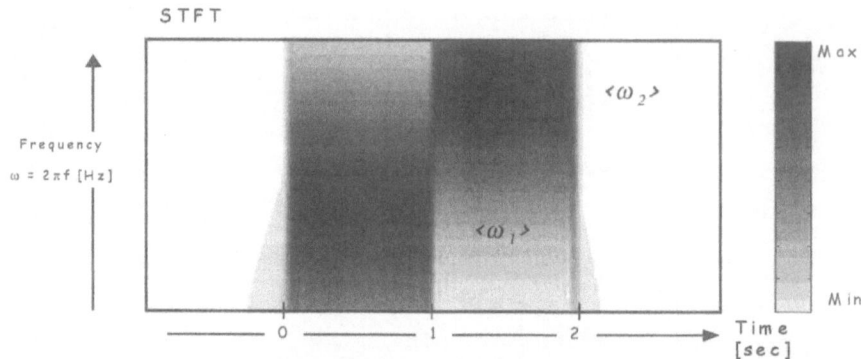

Figure 3.2.6   STFT of the signal in Figure 3.2.1, obtained using a different window h(*t*). The duration of the two components is better determined, but with a loss in resolution in frequency content.

components have a finite duration and different mean time. There seems to be some overlap in time between the components, which is not consistent with Equations 3.2.1 and 3.2.2. The reason for this overlap is given by the size of the window $h(t)$, which is Gaussian in this case. As we will discuss in detail later in the chapter, the time-frequency Uncertainty Principle affects the resolution of our analysis. The smaller the time window, the better temporal resolution can be achieved, at the cost of poorer resolution in frequency. This is easily verified in Figure 3.2.6, where a very narrow window was used. In this case, the duration of the bursts can be easily determined, but it is no longer clear if the two signals $s_1(t)$ and $s_2(t)$ are pure tone bursts or an ensemble of multitones. We will come back to these concepts later in the chapter, since the main difference between the wavelet transform and the windowed Fourier transform is exactly how the Uncertainty Principle is satisfied. It must be noted that a pure tone burst $\sin(\omega t)$ is the equivalent in the frequency domain of a delta function in the time domain, and as such, they are both theoretical extrapolations.

A complementing approach can be taken by windowing the signal in the frequency domain:

$$S_{\Omega}(\omega) \;=\; S(\omega)\, H(\omega - \Omega) \tag{3.2.8}$$

and

$$S_F(\tau,\Omega) = \frac{1}{\sqrt{2\pi}} \int_{-\infty}^{+\infty} S(\omega)\, H(\omega-\Omega)\, e^{j\omega\tau}\, d\omega \quad . \qquad (3.2.9)$$

$S_F(\tau,\Omega)$ is commonly referred to as the short-frequency Fourier transform. If we impose

$$H(\omega) = \frac{1}{\sqrt{2\pi}} \int_{-\infty}^{+\infty} h(t)\, e^{-j\omega t}\, dt \quad ; \qquad (3.2.10)$$

then the two representations are the same except for a phase factor:

$$S(\tau,\omega) = S_F(\tau,\omega)\, e^{-j\omega\tau} \quad . \qquad (3.2.11)$$

If we expect to utilize the new signal representations in a manner similar to the $s(t)$ and $S(\omega)$, the modulus square of these new representations must also be linked to the power density of the signal. This means that we can define a new function:

$$\left| S(\tau,\omega) \right|^2 = \left| S_F(\tau,\omega) \right|^2 = P(\tau,\omega) \qquad (3.2.12)$$

the quantity $P(\tau,\omega)$ is commonly referred to as the *spectrogram*, which represents the energy density per unit time and unit frequency of the signal $s_\tau(t)$. In order to this to be true, its time and frequency integral must equal the total signal energy:

$$E_\tau = \int_{-\infty}^{+\infty} \left| s_\tau(t) \right|^2 dt = \frac{1}{2\pi} \int_{-\infty}^{+\infty} \int_{-\infty}^{+\infty} \left| S(\tau,\omega) \right|^2 d\omega\, d\tau \qquad (3.2.13)$$

Please note that this energy is the total energy of the $s(t)$, (i.e., of the signal $s(t)$ after modulation by the window function).

Example: STFT of a Pure Harmonic Signal

In this case, the signal, shown in Figure 3.2.7, can be expressed by the function

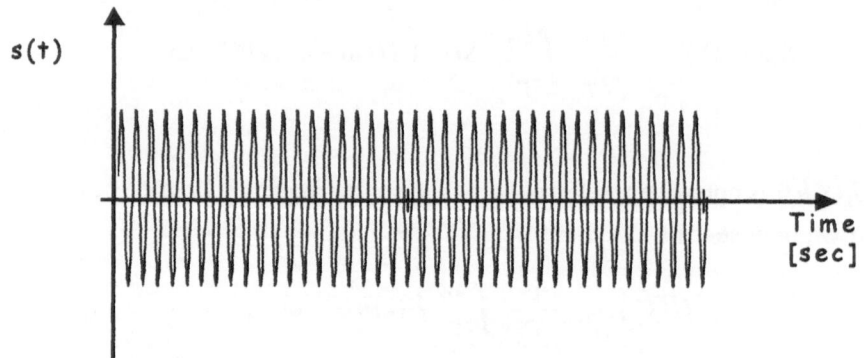

Figure 3.2.7   Plot of the harmonic signal defined in equation 3.2.14.

$$s(t) = e^{-j\omega_o t} \tag{3.2.14}$$

and we will utilize the Gaussian window

$$h(t) = \left(\frac{a}{\pi} \right)^{1/4} e^{-at^2/2} \ . \tag{3.2.15}$$

The correspondent short-time Fourier transform is this given by

$$S(\tau,\omega) = \frac{1}{(a\pi)^{1/4}} e^{\frac{-(\omega - \omega_o)^2}{2a}} e^{-j(\omega - \omega_o) t} \ . \tag{3.2.16}$$

which yields the following spectrogram:

$$P(\tau,\omega) = |S(\tau,\omega)|^2 = \frac{1}{(a\pi)^{1/2}} e^{\frac{-(\omega - \omega_o)^2}{a}} \ . \tag{3.2.17}$$

A plot of the resulting spectrogram is shown in Figure 3.2.8.   A plot of $P(\tau,\omega)$ in 3D is given in Figure 3.2.9, together with the projection on the τ and ω axes for fixed values of the other variable.   As expected, the $P(\tau,\omega)$ is constant in the frequency axis and has a Gaussian distribution in the time axis.   As the signal $s(t)$ to analyze was a pure harmonic function, the width of the Gaussian curve is defined by the window $h(t)$ used.

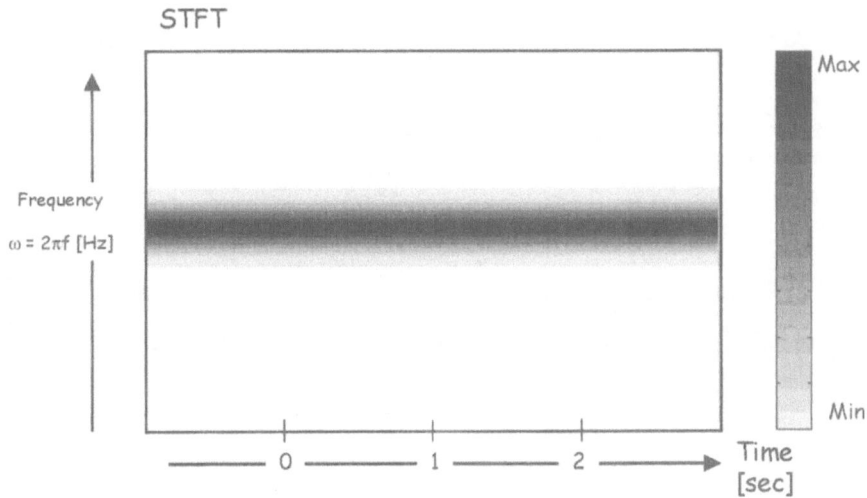

Figure 3.2.8   STFT of the harmonic signal of Equation 3.2.14. The power density plot $P(\tau, \omega)$ is displayed as a pseudo-color intensity gray scale. Darker areas mean higher values.

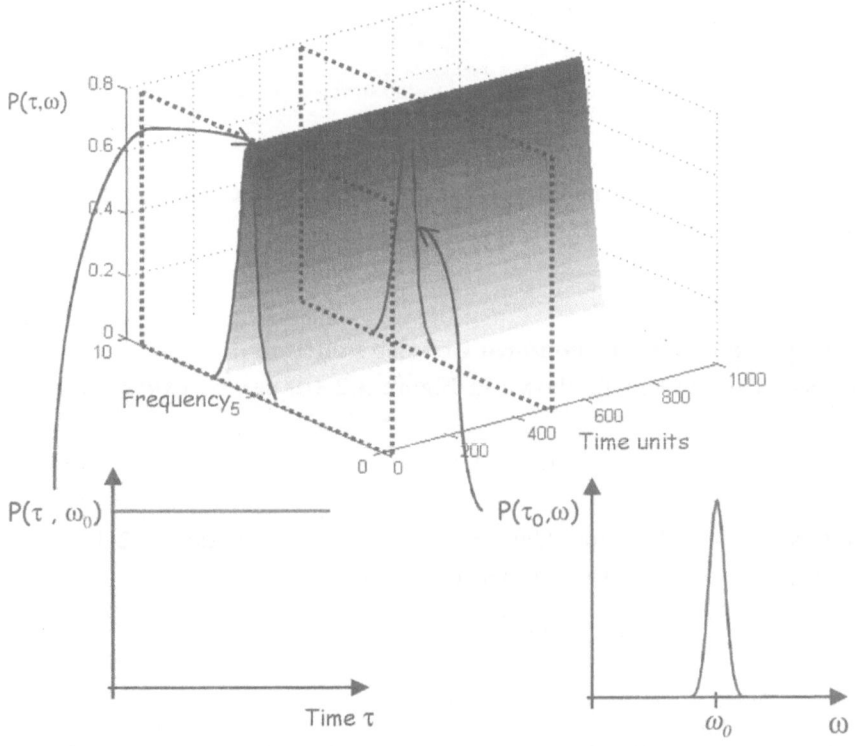

Figure 3.2.9   Three-dimensional surface plot of the $P(\tau, \omega)$ of Equation 3.2.1 and plotted as density image in Figure 3.2.8. with projections in the τ and ω axes.

Figure 3.2.10   Plot of the impulse signal defined in Equation 3.2.18.

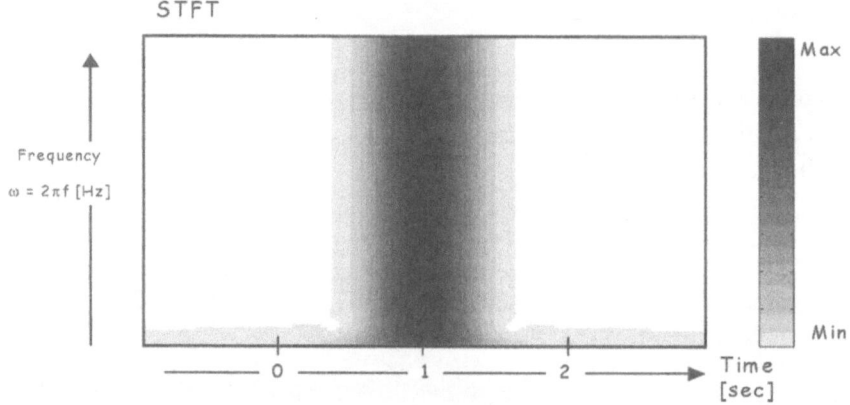

Figure 3.2.11   Short-time Fourier transform of the temporal impulse signal of Equation 3.2.18.

Example: STFT of a Temporal Impulse Function

In this case, the signal, shown in Figure 3.2.10, can be expressed by the function

$$\Delta(t - t_0) \tag{3.2.18}$$

and we will utilize the Gaussian window of Equation 3.2.15.   The correspondent short-time Fourier transform is this given by

$$S(\tau,\omega) = \left(\frac{a}{\pi} \right)^{1/4} e^{-a(t-t_o)^2/2} \, e^{-j\omega t_o} \tag{3.2.19}$$

which yields the following spectrogram:

$$P(\tau,\omega) = |S(\tau,\omega)|^2 = \left(\frac{a}{\pi}\right)^{1/2} e^{-a(t-t_o)^2} . \qquad (3.2.20)$$

A plot of the resulting spectrogram is shown in Figure 3.2.11. In this case, the $P(\tau,\omega)$ is constant on the frequency axis and has a Gaussian distribution in the time axis.

3.2.1   General Properties of the Windowed Fourier Transform

Equation 3.2.20 defines the properties of the signal $s_\tau(t)$, and thus of $s(t)$ and the window $h(t)$. But if the aim of the analysis is to extract information about $s(t)$, can $P(\tau,\omega)$ be used? If we define the signal $s(t)$ and the window $h(t)$ as

$$s(t) = A(t) e^{j\varphi(t)} \qquad (3.2.21)$$

and their Fourier transforms as

$$S(\omega) = B(\omega) e^{j\Psi(\omega)}$$

$$H(\omega) = B_H(\omega) e^{j\Psi_H(\omega)} , \qquad (3.2.23)$$

the total energy of the signal can be calculated as:

$$E_{SP} = \int_{-\infty}^{\infty} \int_{-\infty}^{\infty} P(\tau,\omega) \, d\tau \, d\omega = \int_{-\infty}^{\infty} \int_{-\infty}^{\infty} |S(\tau,\omega)|^2 d\tau \, d\omega \qquad (3.2.24)$$

$$E_{SP} = \int_{-\infty}^{\infty} |s(t)|^2 \, dt \int_{-\infty}^{\infty} |h(t)|^2 \, dt = E_s E_h . \qquad (3.2.25)$$

The total energy of the combined signal E_{SP} is thus given by the product of the energy of the signal E_s and of the window E_h. This result is extremely important, as it states that if we use a window $h(t)$ such that its total energy is 1, $(E_h = 1)$, then the energy of the spectrogram is equal to

$$E_{SP} = E_s \qquad \text{if} \quad E_h = 1 \; . \tag{3.2.26}$$

What about other terms such as the mean time and frequency? The mean time and mean frequency can be calculated by the double integral of the spectrograms expected from a density function:

$$< t >_{SP} = \int_{-\infty}^{+\infty} \int_{-\infty}^{+\infty} t \, |S(\tau,\omega)|^2 \, dt \, d\omega$$

$$< t >_{SP} = <t>_s - <t>_h \tag{3.2.27}$$

$$< \omega >_{SP} = \int_{-\infty}^{+\infty} \int_{-\infty}^{+\infty} \omega \, |S(\tau,\omega)|^2 \, dt \, d\omega$$

$$< \omega >_{SP} = <\omega>_s - <\omega>_h \; . \tag{3.2.28}$$

If we choose a window $h(t)$ whose mean time $< t >_h$ and mean frequency $< \omega >_h$ are zero, then the mean time and frequency as calculated using the spectrogram will coincide with those of the signal:

$$\text{if} \quad <t>_h = <\omega>_h \neq 0 \quad \rightarrow \quad \begin{cases} <t>_{SP} = <t>_s \\ <\omega>_{SP} = <\omega>_h \; . \end{cases} \tag{3.2.29}$$

Unfortunately, it is not as simple for the duration and bandwidth of the spectrogram. It can be shown that:

$$T_{SP}^2 = T_s^2 + T_h^2 \tag{3.2.30}$$

$$B_{SP}^2 = B_s^2 + B_h^2 \tag{3.2.31}$$

and imposing either T_h or B_h equal to zero is not a physical solution.

Depending on the analysis needed we will impose $T_h \ll T_s$, which will result in

$$T^2_{SP} \sim T^2_s$$

$$B^2_{SP} \sim B^2_h \quad , \quad \text{since} \quad B_h \gg B_s \quad .$$

The window is very small and we can properly estimate the duration of the signal which is practically a narrow-band signal. If we impose $T_s \ll T_h$, then we obtain

$$T^2_{SP} \sim T^2_h$$

$$B^2_{SP} \sim B^2_s \quad , \quad \text{since} \quad B_s \gg B_h \quad .$$

But, the more important quantities are the instantaneous frequency and group delay as discussed in the following.

Instantaneous or Local Averages

So far, we have discussed the results of integrating the density function over both domains. We can instead decide to look at local averages, in which one of the two variables, t or ω, is not used in the integration. From Equations 3.2.21 and 3.2.22, we can rewrite the signal $s_\tau(t)$ as

$$s_\tau(t) = A(t) \, A_h(t - \tau) \, e^{j[\varphi(t) + \varphi_h(t - \tau)]} \tag{3.2.32}$$

and we can define a global amplitude $A_\tau(t)$ and phase $\varphi_\tau(t)$ as

$$A_\tau(t) = A(t) \, A_h(t - \tau)$$

$$\varphi_\tau(t) = \varphi(t) + \varphi_h(t - \tau) \tag{3.2.33}$$

hence,

$$s_\tau(t) = A_\tau(t) \, e^{j[\varphi_\tau(t)]} \quad . \tag{3.2.34}$$

The *instantaneous frequency* is defined as the mean frequency calculated from the signal $s(t)$ when the window $h(t)$ is centered at the time τ:

$$\omega_i(\tau) = <\omega>_\tau = \frac{1}{E_\tau} \int_{-\infty}^{+\infty} \varphi'_\tau(t) A^2_\tau(t) \, dt \qquad (3.2.35)$$

where the energy E_τ is calculated as:

$$E_\tau = \int_{-\infty}^{+\infty} A^2_\tau(t) \, dt = \int_{-\infty}^{+\infty} A(t) A_h(t-\tau) \, dt \quad . \qquad (3.2.36)$$

From the above equations we can thus calculate the instantaneous frequency as a function of the signal and window properties:

$$<\omega>_\tau = \frac{1}{E_\tau} \int_{-\infty}^{+\infty} \left\{ \varphi'(t) + \varphi'_h(t-\tau) \right\} \left\{ A^2(t) + A^2_h(t) \right\} dt \quad . \qquad (3.2.37)$$

Using the same approach, we can calculate the relationship between the bandwidths:

$$B^2_\tau = \frac{1}{E_\tau} \int_{-\infty}^{+\infty} \left[\varphi'(t) - <\omega>_\tau \right]^2 A^2_\tau(t) \, dt$$

$$+ \frac{1}{E_\tau} \int_{-\infty}^{+\infty} \left[\frac{dA_\tau(t)}{dt} \right]^2 dt \quad . \qquad (3.2.38)$$

From the above equations, it is clear that the results obtained using the STFT are a combination of the properties of the window $h(t)$ and of the signal $s(t)$. Hence, it is important to fully understand the properties of the window $h(t)$ and to try to obtain results that are window independent. If the window $h(t)$ is approximated by a pulse (i.e., $A_h^2(t) \to \delta(t)$), then the instantaneous frequency is given by

$$<\omega>_\tau \to \frac{1}{E_\tau} \int_{-\infty}^{+\infty} A^2(t) \, \delta(t-\tau) \, \varphi'(t) \, dt = \frac{1}{E_\tau} A^2(\tau) \, \varphi'(\tau) \quad . \qquad (3.2.39)$$

By the same token

$$E_\tau \rightarrow \int_{-\infty}^{+\infty} A^2(t)\, \delta(t-\tau)\, dt = A^2(\tau) \qquad (3.2.40)$$

hence,

$$<\omega>_\tau \rightarrow \frac{A^2(\tau)\, \varphi'(\tau)}{A^2(\tau)} = \varphi'(\tau)\ . \qquad (3.2.41)$$

The above results show that if we make the window very small in order to get increasing time resolution, the limiting value of the estimated instantaneous frequency $<\omega>_\tau$ is the derivative of the phase of the signal (i.e., its instantaneous frequency). Unfortunately, this result comes with a price. The time-bandwidth product is limited by the Uncertainty Principle. As a matter of fact, as we have narrowed the window to a delta function, we have increased its bandwidth (Equation 3.1.48), resulting in $B_\tau^2 \rightarrow \infty$ for $A_h^2(t) \rightarrow \delta(t)$. Hence, we need to perform a trade-off between frequency and time resolution, as dictated by the Uncertainty Principle. We will discuss a possible solution for finding an optimal window in the following, but we must first introduce the dual concepts of mean time and time resolution for a given frequency.

Because of the duality between time and frequency, we can promptly write

$$<\tau>_\omega = \frac{1}{E_\omega} \int_{-\infty}^{+\infty} B^2(\omega')\, B^2{}_H(\omega - \omega') \qquad (3.2.42)$$
$$\cdot \left[\Psi'(\omega') - \Psi'{}_H(\omega'-\omega) \right] d\omega'$$

where the energy E_ω is given by

$$E_\omega = \int_{-\infty}^{+\infty} B^2(\omega')\, B_H^2(\omega - \omega')\, d\omega' \qquad (3.2.43)$$

If the window is narrowed in frequency to a delta function, it can be shown that

$$<\tau>_\omega \rightarrow -\Psi'(\omega), \quad \text{for } H^2(\omega) \rightarrow \delta(\omega) \qquad (3.2.44)$$

but by the same token

$$T^2_\omega = \frac{1}{E_\omega} \int_{-\infty}^{+\infty} \left[\frac{d\,B(\omega') \cdot B_h(\omega' - \omega)}{d\omega'} \right]^2 d\omega'$$

$$+ \frac{1}{E_\omega} \int_{-\infty}^{+\infty} B^2(\omega') B_H^2(\omega' - \omega) \left[\Psi'(\omega') \Psi'_H(\omega' - \omega) \right] \qquad (3.2.45)$$

and thus $T_\omega^2 \rightarrow \infty$ for $B_h^2(\omega) \rightarrow \delta(\omega)$.

3.2.2 Uncertainty Principle for the Windowed Fourier Transform

In many applications, it is important to determine very accurately the time when certain frequencies occurred. Can we make the windows as narrow as we want in order to achieve finer frequency and finer time localization? Unfortunately, the answer is no; short-duration signals (or windows) have large frequency bandwidths; thus, reducing the temporal width of the window may result in a finer time localization, but will also result in a worse frequency localization. This is the so-called *Uncertainty Principle*, first derived by Heisenberg in 1927. It states that given a signal $s(t)$ of duration T and bandwidth B, the time-bandwidth product TB cannot be made arbitrarily small, but instead must be larger than ½, as defined in Equation 3.1.48.

Now, let us consider the implications of the Uncertainty Principle for the short-time Fourier transform. When performing the STFT, we are modifying the original signal $s(t)$ by multiplying it with the window $h(t)$. Hence, the time duration T and the bandwidth B to be considered in the Uncertainty Principle are no longer the ones related to $s(t)$, but must be calculated using the product $s(t)\,h(t-\tau)$. Let us define a function $w_\tau(t)$ as

$$w_\tau(t) = \frac{s(\tau)\, h(\tau - t)}{\sqrt{\int_{-\infty}^{+\infty} |s(\tau)\, h(\tau - t)|^2 \, d\tau}} \; .$$
(3.2.46)

The denominator is used to normalize the function $w_\tau(t)$ such that

$$\|w_\tau\|^2 = \int_{-\infty}^{+\infty} |w_\tau(t)|^2 \, d\tau = 1 \; .$$
(3.2.46a)

If $W(\omega)$ is the Fourier transform of $w_\tau(t)$, the mean time and frequency with the associated standard deviations are

$$<\tau>_t = \int_{-\infty}^{+\infty} \tau \, |w_t(\tau)|^2 \, d\tau$$

$$T_t^2 = \int_{-\infty}^{+\infty} (\tau - <\tau>_t)^2 \, |w_t(\tau)|^2 \, d\tau$$

$$<\omega>_t = \int_{-\infty}^{+\infty} \omega \, |W_\tau(\omega)|^2 \, d\omega$$
(3.2.47)

$$B_t^2 = \int_{-\infty}^{+\infty} (\omega - <\omega>_t)^2 \, |W_\tau(\omega)|^2 \, d\omega \; .$$

Hence, for the case of the STFT, the Uncertainty Principle is related to time-dependent variables, and, in particular, it is a function of the signal and the window functions:

$$B_t \, T_t \le \frac{1}{2} \; .$$
(3.2.48)

Since the window function is so important in determining the bandwidth, naturally the question of how do we choose the optimal window function (if, in fact, an optimal window exists) arises. The choice of an optimal window is based on the particular application. For example, if we are interested in obtaining the finest resolution of the instantaneous frequency $<\omega>_t$, we need to minimize B_t. Since B_t also depends on the signal, clearly the optimal window will depend on the signal. Assuming for the signal $s(t)$ the following relationships:

$$s(t) = A(t)e^{j\phi(t)} \qquad (3.2.49)$$

and assuming that the signal is purely frequency modulated (i.e., that the $A(t)$ represents the low-frequency components of the spectrum of the signal $s(t)$ and the phase $\sum(t)$ the higher components), we can then write for the signal $s(t)$

$$B_s^2 \sim \frac{1}{4\,T_s^2} + T_s^2 \left[\, \phi''(t)\right]^2 \qquad (3.2.50)$$

where T_s and B_s are the time and frequency bandwidths of $s(t)$. If the purpose of signal analysis is to extract information about the temporal variation of the signal frequency spectrum, we must choose a window $h(t)$ such that $T_t = T_s$ and $B_t = B_s$. Assuming that we are interested in obtaining the finest resolution of the instantaneous frequency $<\omega>_t$, we need to minimize B_t in the previous equation. This is achieved with

$$T_t^2 \Big|_{min} \sim T_s^2 \Big|_{min} \sim \frac{1}{2\,|\,\phi''(t)|}$$

$$B_t^2 \Big|_{min} \sim B_s^2 \Big|_{min} \sim |\,\phi''(t)| \qquad . \qquad (3.2.51)$$

$$T_\omega^2 \sim \frac{1}{4\,\varphi''(t)}$$

It follows that the optimal temporal width of the window is:

$$\frac{1}{2\,|\,\phi''(t)|} \qquad . \qquad (3.2.52)$$

The optimal width of the window is thus inversely proportional to the second derivative of the phase (i.e., to the variation of the instantaneous frequency of the signal).

Example: Linear Chirp with Rectangular Envelope Signal

We will present the concepts of time-frequency resolution with two examples, one graphically, by calculating the STFT of a signal using Gaussian windows of varying bandwidth. In the second example, we will utilize Equation 3.2.52 to estimate the optimal window width. The down-chirp signal of Figure 3.2.12 is analyzed using Gaussian windows of three different bandwidths: 0.05, 0.2, and 1 Hertz, respectively. As we have already established, the window with the largest frequency bandwidth (1 Hz) will have the best temporal resolution. The sharp start and stop of the signal burst is clearly identified in the bottom graph of Figure 3.2.13, while the lowest and highest frequencies contained in the burst, can more easily be identified using the top graph of the same figure. From the graphs in Figures 3.2.13a-c, we can conclude that there is an optimal window for the analysis of the signal in Figure 3.2.12. In the case for which both the temporal and frequency resolutions are of importance, the optimal window seems to be the one with bandwidth of 0.2 Hz.

The estimated group delay $t_g(\omega) = \langle t \rangle_\omega$ versus the instantaneous frequency $\langle \omega \rangle_t$ was calculated using the STFT of Figures 3.2.13a-c, and plotted in Figure 3.2.14. For comparison, the theoretical value is reported as a continuous line. If the width of the analyzing window is too large in time (i.e., low frequency bandwidth = 0.05 Hz), the lower frequencies are better estimated, but the analysis completely fails for higher frequencies. The opposite is true if we utilize a window in time which is very small. In the next example, we will obtain similar results analytically.

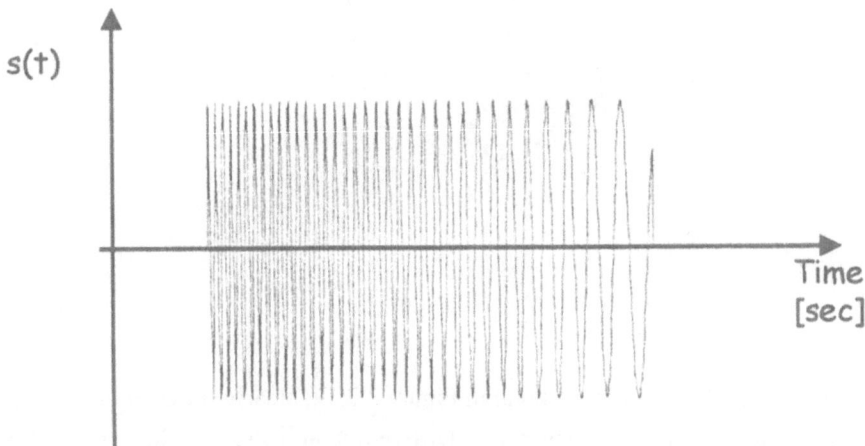

Figure 3.2.12. Linear chirp with a rectangular window.

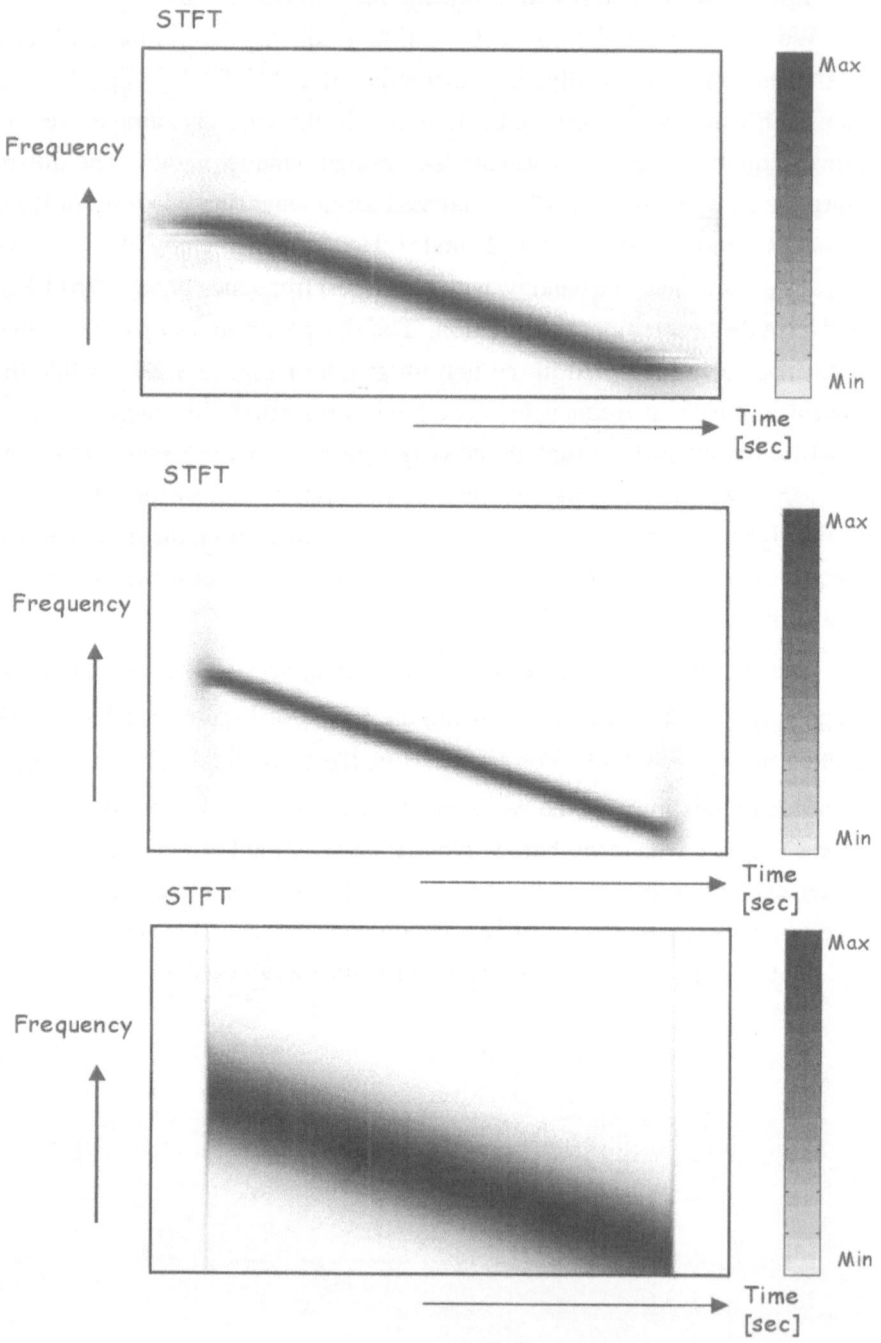

Figure 3.2.13 STFT of the signal in Figure 3.2.12, clearly showing the change in time of the mean frequency $\langle\omega\rangle_t$. Different window bandwidths are used with B_h=0.05, .2 and 1 Hz for top, middle and bottom images, respectively.

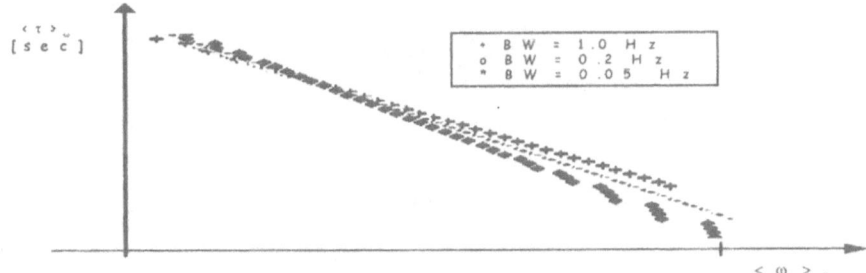

Figure 3.2.14   Plot of the group delay versus the instantaneous frequency, calculated using the STFT of Figures 3.2.13a–3.2.13c. The line represents the theoretical calculated value.

Example: Linear Chirp with Gaussian Envelope Signal and Window

Assuming that the signal $s(t)$ plotted in Figure 3.2.15,

$$s(t) = \left(\frac{\alpha}{\pi} \right)^{1/4} e^{-\alpha t^2/2 + j\beta t^2/2 + j\omega_0 t} \tag{3.2.53}$$

and for the window function

$$h(t) = \left(\frac{a}{\pi} \right) e^{-at^2} \tag{3.2.54}$$

the instantaneous frequency $\langle \omega \rangle_\tau$ is calculated as

$$\langle \omega \rangle_\tau = \frac{a}{\alpha + a} \beta t + \omega_0 \ . \tag{3.2.55}$$

As the window becomes narrow ($a \to \infty$), the estimate of the instantaneous frequency approaches $\beta t + \omega_0$, which is what we would like to measure. Unfortunately, in this limit, the estimate of the group delay $\langle \tau \rangle_\omega$ approaches zero, which is not accurate. As a matter of fact, this is the average time of the signal. The expression for the group delay is

$$\langle \tau \rangle_\omega = \frac{a\beta}{\alpha a^2 + a(\alpha^2 + \beta^2)} \omega \ . \tag{3.2.56}$$

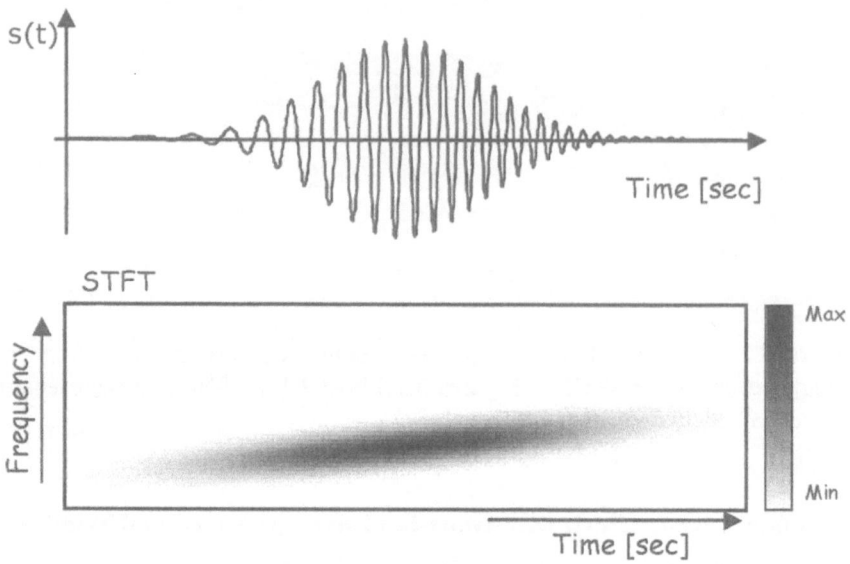

Figure 3.2.15 Chirp signal and correspondent STFT.

Conversely, if we choose $a \to 0$, we obtain $\langle \tau \rangle_\omega \to \beta/(\alpha^2 + \beta^2)\,\omega$, but, on the other hand, we have $<\omega>_\tau \to \omega_0$, which is not the instantaneous frequency, but instead the mean frequency.

The bandwidth and the duration of the signal can be calculated as

$$B^2_\tau = \frac{1}{2} \left| (\alpha + a) + \frac{\beta^2}{\alpha + a} \right| \tag{3.2.57}$$

$$T^2_\omega = \frac{1}{2} \frac{(\alpha + a)^2 + \beta^2}{\alpha\,a^2 + a(\alpha^2 + \beta^2)} \ . \tag{3.2.58}$$

Thus, we can try to minimize B_τ^2 from Equation 3.2.57:

$$\frac{dB^2_\tau}{da} = \frac{1}{2} + \frac{1}{2} \frac{\beta^2}{(\alpha + a)^2} = 0 \tag{3.2.59}$$

whose solution is $a_{min} = |\beta| - \alpha$. The temporal width of the Gaussian window is thus (Equation 3.2.53) $\sigma_\omega^2 = 1/2a$; hence

$$\sigma_{\omega}^{2}\Big|_{optimal} = \frac{1}{2a_{min}} = \frac{1}{2(|\beta| - \alpha)} . \tag{3.2.60}$$

The above equation exactly estimates the optimal width of the window. Returning to Equation (3.2.53), we have

$$\varphi(t) = -\frac{\alpha t^{2}}{2} + j\frac{\beta t^{2}}{2} + j\omega_{0}t \tag{3.2.61}$$

and thus

$$\varphi'(t) = -\alpha t + j\beta t + j\omega_{0} \tag{3.2.62}$$

$$\varphi''(t) = j\beta - \alpha \tag{3.2.63}$$

the estimate of the optimal width is

$$T_{\omega}\Big|_{optimal} \sim \frac{1}{2 \ ine \ \varphi''(t) |} = \frac{1}{2\sqrt{\beta^{2} + \alpha^{2}}} . \tag{3.2.64}$$

If $\beta >> \alpha$, Equations 3.2.60 and 3.2.64 yield the same result.

One more comment regarding the need for understanding the time-frequency resolution of the analysis performed on the signals of interest. As will be shown in Chapter 6, when we present some applications of wavelet analysis, in many cases it is important to determine the group delay versus frequency curve as it defines the system under study. In the case of ultrasonics, the dispersion curve can be traced to the elastic and mechanical properties of materials, and it is used to determine such properties. For the case of thin materials, there are multiple curves which overlap in time and frequency, and it is useful to determine their crossing points.

For example, assuming a signal as the one in Figure 3.2.16, which is composed of two chirp signals, one in which the frequency decreases linearly

s(t)

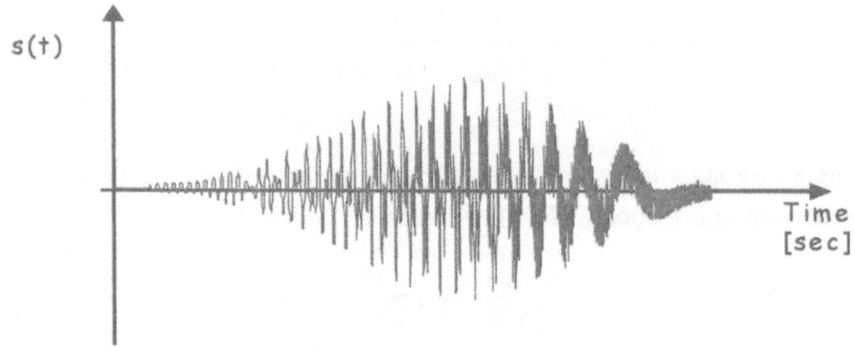

Time
[sec]

Figure 3.2.16   Signal composed by a linear down-chirp and a quadratic up-chirp.

with time and the other in which the frequency increases using a quadratic relationship with time. The theoretical calculation of the instantaneous frequencies versus time are given in Figure 3.2.17. The STFT calculated using different window sizes are plotted in Figure 3.2.18. It can be seen that the optimal size of the window is the one that yields the sharpest image in both domains. In the case of Figure 3.2.18, the optimal window is represented in the middle image, with $B = 2$ Hz.

When describing a window $h(t)$, we define its properties by using the average time, average or center frequency, time duration, and bandwidth. As we have seen, the four quantities are not independent. We know from Fourier that if the window is finite in time, and its frequency components are infinite in number. This is summarized by saying that the signal has an infinite support in the frequency domain. However, in reality, we know that a signal will have a finite time duration and even if extremely large, and a finite frequency support, even if an event can have a fast transient. So we

Frequency

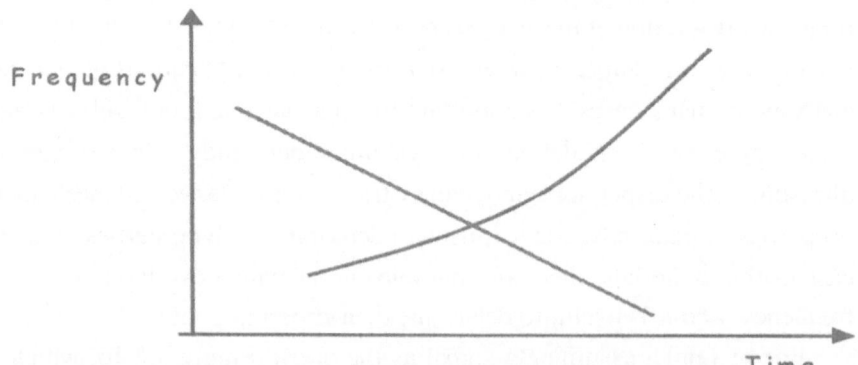

Time

Figure 3.2.17   Theoretical dispersion curve of the signal in Figure 3.2.16.

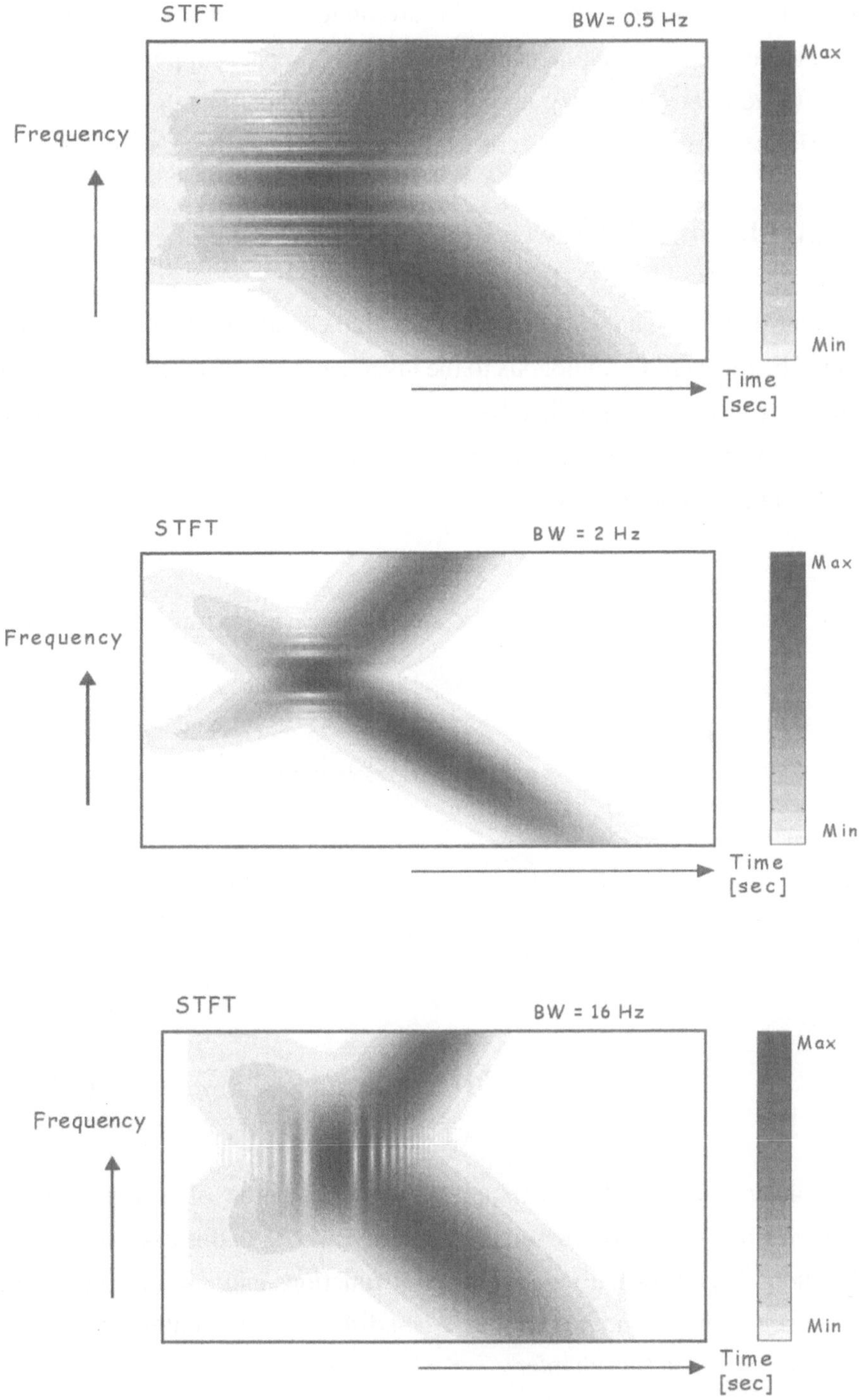

STFT   BW= 0.5 Hz

Frequency

                                                    Max

                                                    Min

                                        Time
                                        [sec]

STFT   B W = 2 Hz

Frequency

                                                    Max

                                                    Min

                                        Time
                                        [sec]

STFT   BW = 16 Hz

Frequency

                                                    Max

                                                    Min

                                        Time
                                        [sec]

Figure 3.2.18   STFT of the signal in Figure 3.2.16 calculated with windows
of different sizes.

now have a paradox, as both domains are finite. In order to eliminate it, we need to introduce the mathematical concept of compact support. A window $h(t)$ is said to be compactly supported if the signal is finite in time and if its Fourier transform decays to zero for large frequencies fast enough.

3.2.3   Inverse Windowed Fourier Transform

A natural question arises when dealing with the analysis of signals: Can the signal be recovered? Analogous to the inverse Fourier transform, is there an inverse transformation so that the signal $s(t)$ can be recovered from its short-time Fourier transform $S(\tau,\omega)$? The answer is yes, and the inversion formula is given here without proof:

$$s(t) = \frac{1}{2\pi} \int_{-\infty}^{+\infty} \int_{-\infty}^{+\infty} S(\tau,\omega) \, h(t-\tau) \, e^{j\omega t} \, d\omega \, dt \qquad (3.2.65)$$

The reconstruction formula can also be rewritten as

$$s(t) = \frac{1}{2\pi} \int_{-\infty}^{+\infty} \int_{-\infty}^{+\infty} <s,h_{\tau,\omega}> \, h_{\tau,\omega}(t) \, d\omega \, dt \qquad (3.2.66)$$

where

$$h_{\tau,\omega}(t) = h(t-\tau) \, e^{j\omega t} \qquad (3.2.67)$$

The inversion formula resembles the decomposition of a signal in an orthonormal basis as discussed in Chapters 4 and 6. In this case though, the functions $h_{\tau,\omega}(t)$ form a redundant set of functions and, as such, have very loose constraints. The same is true when discussing the continuous wavelet transform later in this chapter.

3.3   Continuous Wavelet Transform

When using the STFT, the time-bandwidth product is constant in the whole time-frequency plane, once the window h(*t*) is chosen. This means that the STFT at any point (t_0, ω_0) in the time-frequency plane provides information about the signal *s*(*t*) with an accuracy given by *T* and *B* in the time and in the frequency domain, respectively.   Perhaps "accuracy" could be better expressed as "uncertainty," given the nature of the *TB* product and the Uncertainty Principle.   This localization is uniform in the entire plane, resulting in a uniform tiling of this plane with a rectangular cell of fixed dimension $2B \times 2T$, as shown in Figure 3.3.1a.

If the signal *s(t)* has a transient component with a *support* (duration) smaller than *T*, it is difficult to locate the signal with a precision better than *T*.   The same is true for signals with a small support in the frequency domain.   As we choose smaller values of *T*, the time resolution increases while the frequency resolution *B* decreases.   Therefore, if the signal *s(t)* is composed of very short time transients and monochromatic (single frequency) sine waves, it is very hard to find an optimal window *h(t)*.   We can thus conclude that the STFT is suited for analyzing signals that have signal components with similar ranges for the temporal and frequency supports.   As we shall see in the following section, the wavelet transform was created specifically to address this limitation.   In the case of the CWT, the time-frequency resolution will vary according to the frequency of interest, as shown in Figure 3.3.1b, where at higher frequencies the time resolution is

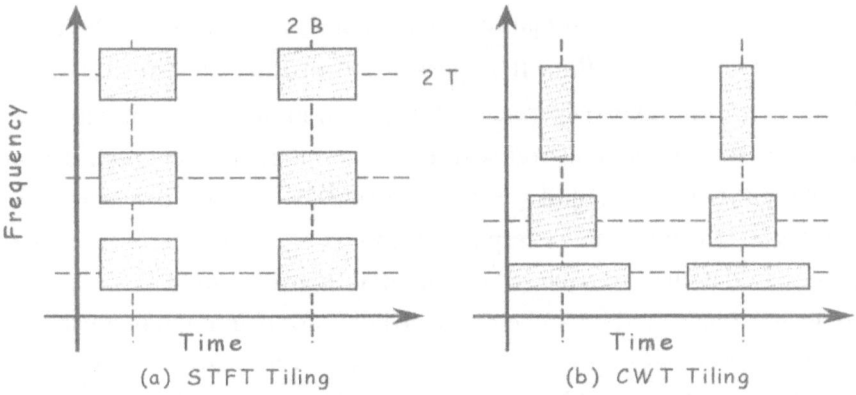

Figure 3.3.1 Time-Frequency resolution for the short-time Fourier transform (a) and the continuous wavelet transform (CWT). The area of each tile is constant and equivalent to the product 4*TB*.

Figure 3.3.2   Test signal with two harmonic signals and two impulse signals.

smaller at the cost of a larger frequency window. The Uncertainty Principle
ensures that the area of each tile is constant and equivalent to the TB as in
the STFT case [Dau91].

Let us begin this section with an example; consider a signal $s(t)$
composed of two narrow pulses in time (delta functions in time) and two
narrow pulses in frequency (two pure sine waves in frequency):

$$s(t) = \sin(2\pi f_1 t) + \sin(2\pi f_2 t) + \delta(t-t_1) + \delta(t-t_2) \quad . \quad (3.3.1)$$

We impose that the two pulses in time are separated by 32 msec, and the two
frequencies are 20 and 30 Hz, respectively. The signal would look like the
one in Figure 3.3.2. The signal $s(t)$ is analyzed with the STFT, and the result
is shown in Figure 3.3.3a. The window was chosen to separate the two
frequencies; that is, the frequency bandwidth B_h of the window was chosen
such that $\omega_2 - \omega_1 > B_{h_s}$. It is possible to distinguish between the two
frequencies, but not to separate the two pulses in time, since, in this case,
$T_h > t_2 - t_1$. As the temporal width of the window is decreased, the time
resolution is increased until $t_2 - t_1 > T_h$, and the two pulses can be resolved,
but it is no longer possible to separate the two frequencies. This is due to the
fact that the time-frequency resolution is constant in the whole time-
frequency plane, and it can be optimized only for one of the two variables.

In order to have a complete understanding of the signal $s(t)$, we need to
utilize at least two different window widths. Since higher-frequency
components need a smaller time window to be identified, it seems a good

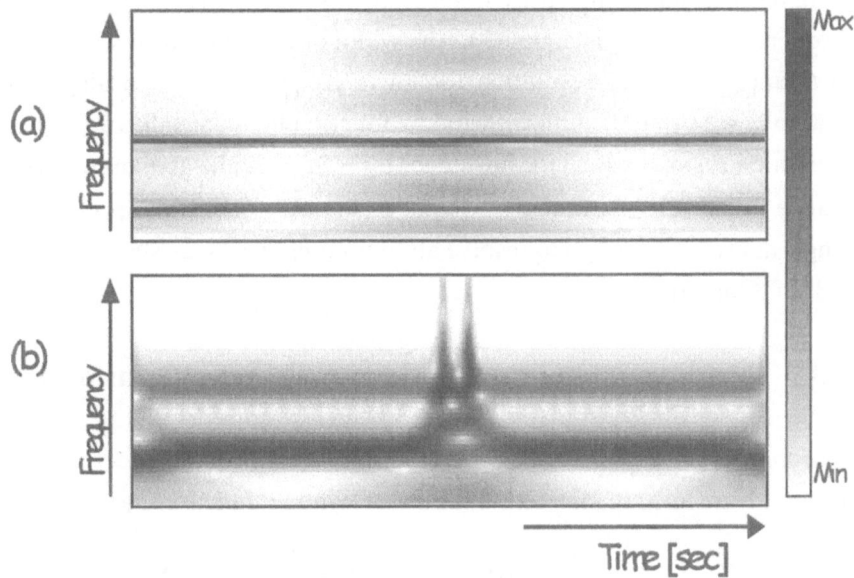

Figure 3.3.3 (a) Short-time Fourier transform of the signal in Figure 3.3.2. The two harmonics are easily identified, but it is not possible to discern the two separate pulses. (b) Continuous wavelet transform of the same signal, clearly showing the two frequency components and the two spikes in time.

approach to construct a window, say $\psi(t)$, such that its time support T_t is small when analyzing high-frequency components, and increases when considering lower frequencies. This different method of tiling the time-frequency plane is achieved using a window function $\psi(t)$ similar to $h(t)$, but instead of using a frequency shift $e^{j\omega t}$, we utilize a scaling in time $\psi(t/a)$. The scale a has the effect of dilating $(a > 1)$ or contracting $(a < 1)$ the window function $\psi(t)$ in time. This basic principle leads us to the definition of a continuous wavelet transform.

The continuous wavelet transform was introduced by Morlet in 1982 specifically to address this limitation of the STFT. Morlet was interested at looking at seismic signals constructed by superposition of long low-frequency components and very short transients. In seismology, it is important to separate in time the various pulses which characterize fine, closely spaced layers in the Earth's crust [Gou84,Gro86].

Since higher-frequency components need a smaller time interval to be identified, as their period is smaller, Morlet proposed constructing a window function $h(t)$ whose support is small when analyzing higher-frequency components, and it increases when analyzing lower frequencies. As the frequency resolution is coupled to the width of the window, it would result in a finer frequency resolution for lower frequencies. Returning to our example, we can thus utilize a window function h(t) which has a frequency resolution B smaller than the difference $\omega_2 - \omega_1 > B$, and at high frequencies we are still able to separate the two pulses in time. Using this approach, one single analysis is enough to resolve all four components, as shown in Figure 3.3.3b [Coh93].

3.3.1   Mathematics of the Continuous Wavelet Transform

The *continuous Wavelet transform (CWT)* of a function $s(t)$ is defined as the integral transform of $s(t)$ with a family of window functions $\psi_{a,b}(t)$:

$$W_s(a,b) = \int_{-\infty}^{+\infty} s(t)\, \psi_{a,b}^{*}(t)\, dt = \left\langle \psi_{a,b}^{*}, s(t) \right\rangle \qquad (3.3.2)$$

where the *Kernel functions* are constructed as follows:

$$\psi_{a,b}(t) = \frac{1}{\sqrt{|a|}}\, \psi\left(\frac{t-b}{a}\right) . \qquad (3.3.3)$$

The superscript $*$ in Equation 3.3.2 refers to the complex conjugate. The scale factor a represents the scaling of the function $\psi(t)$, and the shift factor b represents the temporal translation of the function. The normalization constant is chosen such that the norm of each function $\psi_{a,b}(t)$ is constant. Some other authors choose to use a different normalization (e.g., $1/a$), but in this case, the energy of each function (i.e., the square of the norm), is no longer a constant. We also note that

$$\left\| \psi_{a,b} \right\|^2 = \int_{-\infty}^{+\infty} \left| \psi_{a,b}(t) \right|^2 dt = \int_{-\infty}^{+\infty} \left| \psi(t) \right|^2 dt . \qquad (3.3.4)$$

This means that each dilated/compressed version of $\psi(t)$ has the same energy. The function $\psi(t)$ is commonly called the *mother wavelet* or simply the *wavelet*, whereas the family of functions $\psi_{a,b}(t)$ are called *daughter*

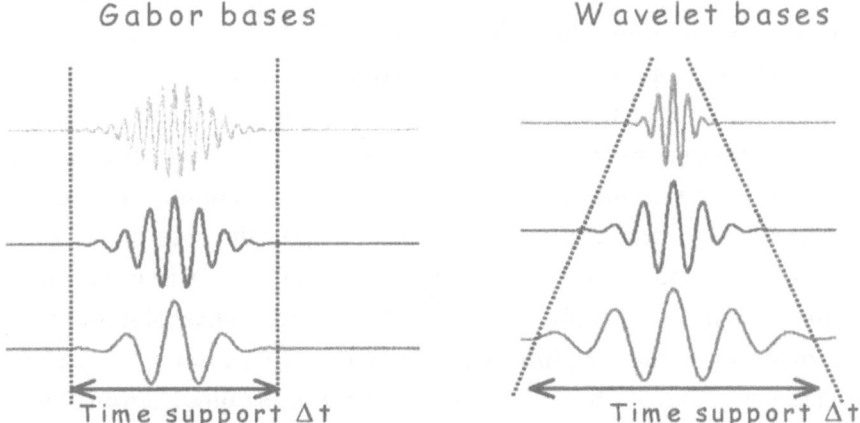

Figure 3.3.4 Plot of the Gabor bases and the wavelet bases, showing the scaling property of the wavelets.

wavelets because they are derived from scaling and shifting the mother wavelet [Chu95]. In contrast with the STFT, the temporal width (support) of the daughter wavelets is not constant; it varies with the scale parameter a. The frequency bandwidth varies in a similar fashion. The Fourier transforms of the mother wavelet and daughter wavelets are given by

$$\psi(t) \leftrightarrow \Psi(\omega)$$

$$\psi_{a,b}(t) = \frac{1}{\sqrt{a}}\, \psi\left(\frac{t-b}{a}\right) \leftrightarrow \sqrt{a}\, \Psi(a\omega)e^{-j\omega b} \quad . \tag{3.3.5}$$

The daughter wavelets thus form the basis used in decomposing the signal $s(t)$, and the CWT coefficients $W_s(a,b)$ represent the projections of the signal onto the basis. To properly represent the difference between the STFT basis and the CWT basis, we will use a pure harmonic signal modulated by a Gaussian window as the kernel function in both transformations. The basis function of the STFT can be written as

$$h_{\tau,\omega}(t) = h(t - \tau)\, e^{-j\omega t} \tag{3.3.6}$$

whereas in the case of the CWT, we have Equation 3.3.5. In Figure 3.3.4, the difference between the two bases is clearly displayed. The signals on the left represent the Gabor basis for different values of ω, whereas the right represents the same Gaussian basis with different values of *a*.

The dual representation in the frequency space is given in Figure 3.3.5. From comparison between the Figures 3.3.4 and 3.3.5, we can clearly see that in the case of the STFT, the time support is constant as well the frequency (*T*, *B* constants), whereas they vary in the case of the CWT [Rio91]. The signal in Figure 3.3.6 is composed by a Gaussian modulated harmonic signal, commonly called a tone burst, with a phase shift occurring approximately at the peak time of the Gaussian. Many times, it is possible to detect the anomaly by looking at the signal in the time domain (Figure 3.3.6a), but in many other cases, it is necessary to utilize a different approach. As the phase shift is a discontinuity in the signal, it must be detectable in the time-frequency plane. The short-time Fourier transform and the continuous wavelet transform of the signal are plotted in Figures 3.3.6b and 3.3.6c, respectively. It is possible to detect the discontinuity in both representations, but the CWT has the capability of zooming into its occurrence in time more accurately.

The main advantage of the CWT is to zoom in to discontinuities and sharp transients. As a rule of thumb, we can say that if the frequency range covered by the signal analysis is such that an hyperbolic or logarithmic scale is more appropriate, as, for example, in music where we utilize an octave scale, then the CWT will be most likely the best choice. Conversely, if the frequency range is limited to a linear scale, the STFT does offer some

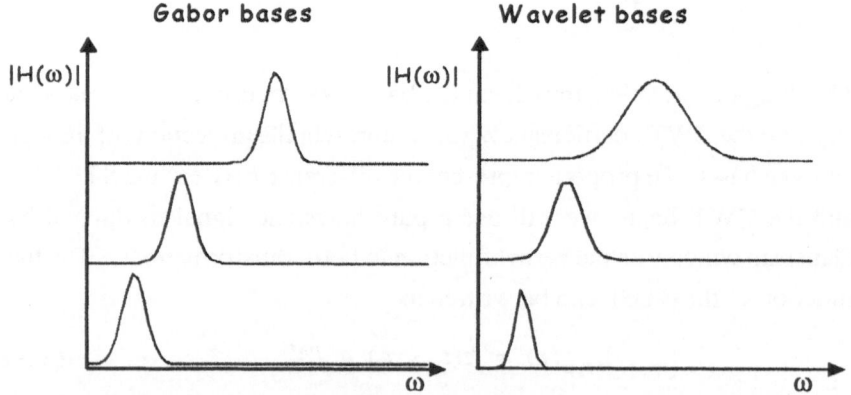

Figure 3.3.5 Magnitude of the Fourier transform of the Gabor and wavelet bases showing the change in bandwidth due to the scaling property of the wavelets.

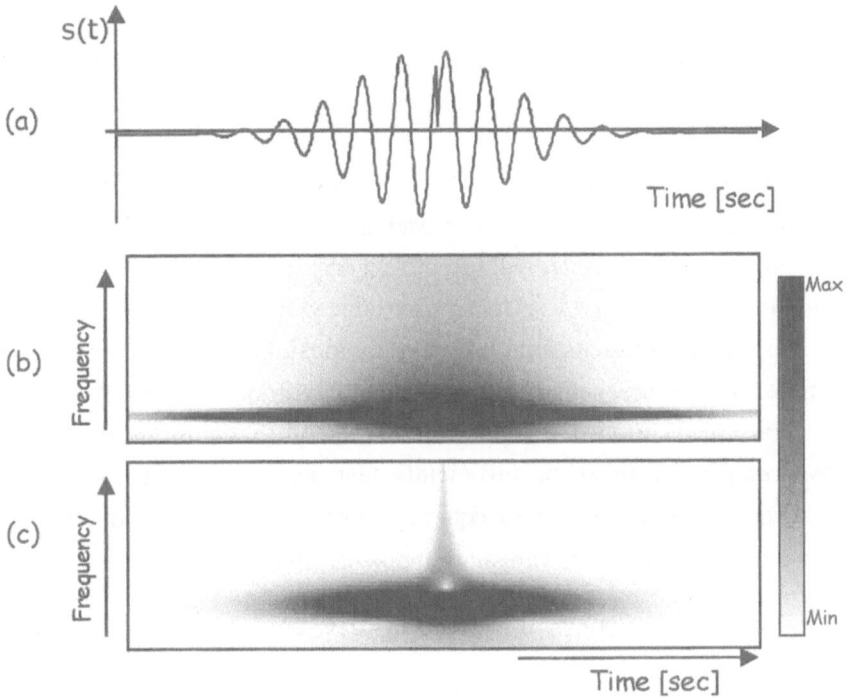

Figure 3.3.6 (a) Signal composed by a Gaussian modulated tone burst and a spike in the phase of the signal; (b) relative STFT; and (c) CWT.

advantages. We will see in later chapters that the discretized form of the wavelet transform has other advantages, such as multiresolution. The choice of the mother wavelet $\psi(t)$ is neither unique or arbitrary; $\psi(t)$ must be a function with finite energy. Another way to express this is to say that the mother wavelet must be square integrable (i.e., $\psi \in L^2(\mathbb{R})$):

$$\|\psi\|^2 = \int_{-\infty}^{+\infty} |\psi(t)|^2 \, dt < +\infty \qquad (3.3.7)$$

which satisfies the following *Admissibility Condition*:

$$C_\psi = \int_{-\infty}^{+\infty} \frac{|\Psi(\omega)|^2}{|\omega|} \, d\omega < +\infty \quad . \qquad (3.3.8)$$

For a band-limited mother wavelet $\psi(t)$, the admissibility condition is equivalent to imposing the condition that the function has zero mean

$$\int_{-\infty}^{+\infty} \psi(t)\, dt = 0 \qquad \text{or} \qquad \Psi(\omega)\big|_{\omega=0} = 0 \qquad (3.3.9)$$

The zero mean condition ensures that the total area under the function $\psi(t)$ is zero (i.e., that the function must oscillate between positive and negative values). Thus, the wavelet will resemble a ripple; from this feature, the term *Ondelette* (French for wavelet) was introduced in geophysics.

To ensure a more useful wavelet decomposition (from the point of view of time and frequency analysis), we commonly impose more restrictions to the choice of the mother wavelet. First, we require that the mother wavelet have *compact support*, or sufficiently fast decay in time to obtain good localization in both time and frequency. Second, it is useful to define $\psi(t)$ with a certain number of vanishing moments; that is,

$$\int_{-\infty}^{+\infty} t^n \psi(t)\, dt = 0 \qquad \text{for} \quad n = 0, 1, 2, ..., N-1 \quad . \qquad (3.3.10)$$

This property improves the efficiency of $\psi(t)$ at detecting singularities, since its average value for smoother curves is closer to zero. Many other conditions can be imposed to obtain different families of $\psi(t)$ with different properties; many of these conditions were originally defined by Daubechies. For example, in the time-frequency decomposition, we often choose an *analytic signal* as the mother wavelet:

$$\Psi(\omega) = 0 \qquad \text{for} \quad \omega < 0 \quad . \qquad (3.3.11)$$

The wavelet transform $W_s(a,b)$ is also an analytic signal; hence, its phase can be used to detect singularities more efficiently than its modulus. We will later discuss the Analytic Wavelet Transform in more detail.

3.3.2   Properties of the Continuous Wavelet Transform

Let us now review some of the important properties of the continuous wavelet transform. These properties derive from the properties of the mother wavelet $\psi(t)$. First, for proper time-frequency analysis, the functions $\psi(t)$

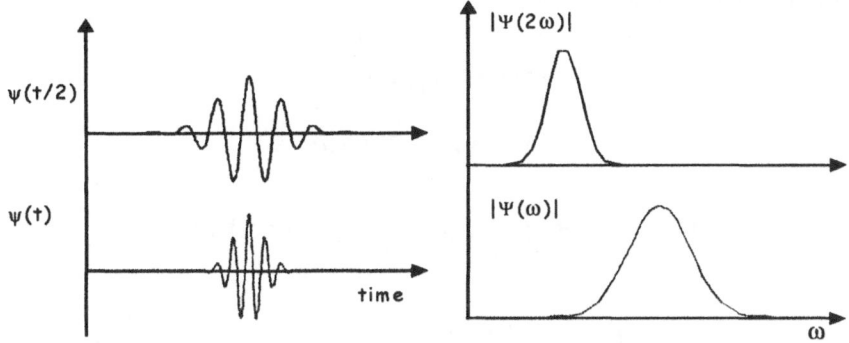

Figure 3.3.7 Mother (bottom) and daughter wavelet (top) with Fourier transforms.

and $\Psi(\omega)$ need to be well localized in both time and frequency. Once the mother wavelet is well defined, the properties of the daughter wavelets can be calculated as

$$\psi_{a,b}(t) \;=\; \frac{1}{\sqrt{a}} \, \psi\!\left(\frac{t-b}{a}\right)$$

$$\Psi_{a,b}(\omega) \;=\; \sqrt{a}\; \Psi(a\omega)e^{j\omega b} \quad .$$

(3.3.12)

Given an admissible mother wavelet $\psi(t)$, the continuous wavelet transform can be defined as an operator $\mathbf{T}_\psi$ which performs the following transformation: $\mathbf{T}_\psi : s(t) \rightarrow W_s(a,b)$, where $s(t)$ is a real function, [i.e., $s(t) \in \mathbb{R}$] and $W_s(a,b) \in \mathbb{R}_+^2$. The wavelet transform thus maps a real signal into the time-scale plane, where a is only positive ($a > 0$, $a \in \mathbb{R}_+$) and b can be any value ($b \in \mathbb{R}$). A plot of the wavelet $\psi(t)$ and its scaled version, $\psi(t/2)$, is given in Figure 3.3.7, together with their Fourier transforms.

Another important property of the CWT is linearity. Just like the STFT, the CWT is a linear transformation:

$$s(t) \leftrightarrow W_s(a,b)$$

$$c_1 s(t) + c_2 f(t) \leftrightarrow c_1 W_s(a,b) + c_2 W_f(a,b)$$

(3.3.13)

where c_1 and c_2 are constants. The time-shifting property of the CWT can be derived as follows:

$$s(t) \,\leftrightarrow\, W_s(a,b)$$

$$f(t) = s(t-\tau) \,\leftrightarrow\, W_s(a,b-\tau)$$

$$W_f(a,b) = \int \frac{1}{\sqrt{a}} \,\psi\!\left(\frac{t-b}{a}\right) s(t-\tau)\, dt \qquad (3.3.14)$$

$$W_f(a,b) = \frac{1}{\sqrt{a}} \int \psi\!\left(\frac{t'+\tau-b}{a}\right) s(t')\, dt' = W_s(a,b-\tau) \ .$$

In a similar manner, we can derive the scaling property of the CWT:

$$s(t) \,\leftrightarrow\, W_S(a,b)$$

$$f(t) = \frac{1}{\sqrt{a_0}} \, s\!\left(\frac{t}{a_0}\right) \,\leftrightarrow\, W_s(a/a_0, b/a_0)$$

$$W_f(a,b) = \frac{1}{\sqrt{a\,a_0}} \int \psi\!\left(\frac{t-b}{a}\right) \cdot s\!\left(\frac{t}{a_0}\right) dt \qquad (3.3.15)$$

$$W_f(a,b) = \sqrt{\frac{a_0}{a}} \int \psi\!\left(\frac{a_0 t' - b}{a}\right) s(t')\, dt' = W_s(a/a_0, b/a_0) \ .$$

Just as we derived Parseval's Theorem for conservation of energy in the Fourier transform, we can derive an analogous expression for energy conservation of the CWT:

$$\int_{-\infty}^{+\infty} |s(t)|^2 \, dt = \frac{1}{C_\psi} \int_{-\infty}^{+\infty} \int_{-\infty}^{+\infty} |W_s(a,b)|^2 \, \frac{da\,db}{a^2} \qquad (3.3.16)$$

Note here that the product $da\cdot db$ does not define a constant surface, whereas the area defined as $da\cdot db/a^2$ is invariant under time translation and time scaling (dilation/compression). An equivalent mathematical statement is that the mother wavelet $\psi(t)$ generates a *resolution of identity:*

$$C_\psi^{-1} \int_{-\infty}^{+\infty} \int_0^{+\infty} |\Psi_{a,b}\rangle \langle \Psi_{a,b}| \frac{da\,db}{a^2} = \mathbb{I} \ . \tag{3.3.17}$$

We can thus interpret the term $|W_s(a,b)|^2$ as the energy density of $s(t)$ in the time-scale plane (really in the half-plane $\mathbb{R}_+^2$, since the scale a is positive, $a>0$). A generalization of the above property is given by

$$\int_{-\infty}^{+\infty} f^*(t)\, g(t)dt = \frac{1}{C_\psi} \int_{-\infty}^{+\infty} \int_{-\infty}^{+\infty} W_f^*(a,b)\, W_g(a,b) \frac{da\,db}{a^2} \tag{3.3.18}$$

Continuing to follow the analogy of the STFT representation and its properties, we can define the following
- $C_\psi^{-1} \cdot |W_s(a,b)|^2 / a^2$ = Energy density of the signal $s(t)$ in the time-scale plane;
- $C_\psi^{-1} \cdot |W_s(a,b)|^2 \cdot (\Delta a \cdot \Delta b) / a^2$ = Energy on the scale interval Δa and time interval Δb centered around (a,b).

Assume for the mother wavelet the following values:

$$<t> = \frac{\int_{-\infty}^{+\infty} t\, |\psi(t)|^2\, dt}{\int_{-\infty}^{+\infty} |\psi(t)|^2\, dt} \tag{3.3.19}$$

$$T^2 = \frac{\int_{-\infty}^{+\infty} (t - <t>)^2\, |\psi(t)|^2\, dt}{\int_{-\infty}^{+\infty} |\psi(t)|^2\, dt} \tag{3.3.20}$$

$$B^2 = \frac{\int_{-\infty}^{+\infty} (\omega - \omega_o)^2\, |\Psi(\omega)|^2\, d\omega}{\int_{-\infty}^{+\infty} |\Psi(\omega)|^2\, d\omega} \tag{3.3.22}$$

To simplify, we will assume that $<t> = 0$. Then, the daughter wavelets will have the following properties:

- Mean or Average Time $< t_{a,b} >$:

$$<t_{a,b}> = \frac{\int_{-\infty}^{+\infty} t \left| \psi_{a,b}(t) \right|^2 dt}{\int_{-\infty}^{+\infty} \left| \psi_{a,b}(t) \right|^2 dt} \tag{3.3.23}$$

- Duration $T_{a,b}$:

$$T_{a,b}^2 = \frac{\int_{-\infty}^{+\infty} (t - <t>)^2 \left| \psi_{a,b}(t) \right|^2 dt}{\int_{-\infty}^{+\infty} \left| \psi_{a,b}(t) \right|^2 dt} \tag{3.3.24}$$

- Center frequency $\omega_{a,b}$:

$$<\omega_{a,b}> = \frac{\int_{-\infty}^{+\infty} \omega \cdot \left| \Psi_{a,b}(\omega) \right|^2 d\omega}{\int_{-\infty}^{+\infty} \left| \Psi_{a,b}(\omega) \right|^2 d\omega} = \omega_o \tag{3.3.25}$$

- Bandwidth $B_{a,b}$:

$$B_{a,b}^2 = \frac{\int_{-\infty}^{+\infty} (\omega - \omega_o)^2 \left| \Psi_{a,b}(\omega) \right|^2 d\omega}{\int_{-\infty}^{+\infty} \left| \Psi_{a,b}(\omega) \right|^2 d\omega} \tag{3.3.26}$$

Assuming that the wavelet function is a band-pass function with zero mean value, and using the values in Equations 3.3.19 to 3.3.22, the following results can be obtained:

$$\left\langle t_{a,b} \right\rangle = b \qquad\qquad T_{a,b} = a\,T$$

$$\left\langle \omega_{a,b} \right\rangle = \frac{\omega_0}{a} \qquad\qquad B_{a,b} = \frac{B}{a} \tag{3.3.27}$$

From the previous Eequation we can see that the time and frequency resolution of the CWT decomposition is not constant over the whole time-frequency plane, as in the STFT, but, instead, that it varies as a function of the scale parameter a. The time-frequency plane is thus *tiled* with resolution cells whose dimensions vary as a function of the scale a (and hence of the frequency). This localization is not uniform in the entire plane, but given a mother wavelet $\psi(t)$ whose time and frequency widths are T and B, we have:

If $a > 1 \; \rightarrow \; T_a = aT > T; \quad B_a = B/a < B$

If $a < 1 \; \rightarrow \; T_a = aT < T; \quad B_a = B/a > B$

The Uncertainty Principle always applies:

$$BT \leq \text{constant} \tag{3.3.28}$$

but now it is satisfied differently for each value of the scale a:

$$B_a T_a = \frac{B}{a} aT = BT \leq \text{constant} \tag{3.3.29}$$

We are now in a position to discuss the concept of *relative bandwidth* in relation to the CWT. In the short-time Fourier transform the *absolute bandwidth* B of the window is constant. In the wavelet transform, the daughter wavelets have constant relative bandwidths, so that they have a better time resolution at high frequencies; that is, at small scale a,

$$\frac{B_a}{\omega_a} = \frac{(B/a)}{(\omega_0/a)} = \frac{B}{\omega_o} \tag{3.3.30}$$

We can conclude that the CWT is suited for analyzing signals whose components do not have the same temporal and frequency support. Combining the localization properties of the CWT and the fact that the function $\psi_{a,b}(t)$ acts like a filter (due to its convolution with the signal), we see that the CWT performs a local filtering of the signal $s(t)$ both in time and frequency (scale). The wavelet transform can thus be seen as a bank of filters. The filter $\psi_{a,b}(t)$ constructed by the dilated version of the mother wavelet $\psi(t)$ processes the low-frequency information of the signal $s(t)$; hence, the output $W_s(a,b)$ represents the correlation between the signal $s(t)$ with $\psi_{a,b}(t)$ [Her93].   The value of $W_s(a,b)$ at scale a and shift b is highest when the wavelet $\psi_a(t)$ matches the component of the signal $s(t)$ which exists for time t, $b-T_a \le t \le b+T_a$; and at frequency ω, $\omega_a-B_a \le \omega \le \omega_a+B_a$. As we can see, a very useful approach interpreting the CWT is to consider it a time-scale transform [Rio91].

The CWT can also be written, using the change of variables, $u \rightarrow t/a$, as the following expression:

$$W_s(a,b) = \sqrt{a}\int_{-\infty}^{+\infty} s(au)\ \psi\left(u - \frac{b}{a}\right) du \quad . \tag{3.3.31}$$

In this equation, the wavelet (i.e., the window in time used to look at the signal) is constant, and the analysis is performed on scaled versions of the original signal $s(t)$.   This is equivalent to the process performed by a microscope; the glass window is the same, but we change the zoom by looking at different scales of the image (signal). This is not a new concept; in fact, it is very well understood and often utilized by cartographers when designing and reproducing maps [Fre93].

3.3.3   Inverse Wavelet Transform

At this point, a question naturally arises: Is it possible to reconstruct the signal from its continuous wavelet transform? The answer is yes; this process is known as *synthesis*. Webster's dictionary defines "synthesis" as "the putting of two or more things together so as to form a whole: opposed to analysis." The expression for recovering the signal is given by

$$s(t) = C_\psi^{-1}\int_{-\infty}^{+\infty}\int_{0}^{+\infty} \frac{1}{a^2} W_s(a,b)\psi_{a,b}(t)\ da\,db \tag{3.3.32}$$

where we define a new constant:

$$C_{\psi} = 2\pi \int_0^{+\infty} \frac{|\Psi(\omega)|^2}{\omega} d\omega < +\infty \quad . \tag{3.3.33}$$

Our proof of the inversion formula can be derived as follows:

$$W_s(a,b) = \int_{-\infty}^{+\infty} \psi_{a,b}^*(t)s(t)dt$$

$$\Psi_{a,b}(\omega) = \sqrt{a}\, e^{-jb\omega}\Psi(a\omega)$$

$$W_s(a,b) = \frac{1}{2\pi}\int_{-\infty}^{+\infty} \Psi_{a,b}(\omega)S(\omega)d\omega$$

so

$$W_s(a,b) = \frac{\sqrt{a}}{2\pi}\int_{-\infty}^{+\infty} \Psi^*(a\omega)S(\omega)e^{jb\omega}d\omega \quad .$$

Note that the last integral is proportional to the inverse Fourier transform of $\Psi^*(a\omega)S(\omega)$ as a function of b. Let us define the following

$$J(a) = \int_{-\infty}^{+\infty} W_s(a,b)\psi_{a,b}(t)db$$

$$J(a) = \frac{\sqrt{a}}{2\pi}\int_{-\infty}^{+\infty}\left(\int_{-\infty}^{+\infty}\Psi^*(a\omega)S(\omega)e^{jb\omega}d\omega\right)\psi_{a,b}(t)db$$

$$J(a) = \frac{\sqrt{a}}{2\pi}\int_{-\infty}^{+\infty}\Psi^*(a\omega)S(\omega)\int_{-\infty}^{+\infty}\psi_{a,b}(t)e^{j\omega b}db\,d\omega$$

Now

$$\int_{-\infty}^{+\infty}\psi_{a,b}(t)e^{jb\omega}db = \frac{1}{\sqrt{a}}\int\psi\left(\frac{t-b}{a}\right)e^{jb\omega}db$$

$$= \sqrt{a}\, e^{j\omega t}\int\psi(b')e^{-jb'\omega}db'$$

where $\quad b = \dfrac{t-b}{a} \quad .$

Thus,

$$\int_{-\infty}^{+\infty} \psi_{a,b}(t)e^{jb\omega}\,db = \sqrt{a}\,e^{j\omega t}\Psi(a\omega)$$

$$J(a) = \frac{\sqrt{a}}{2\pi}\int \Psi^*(a\omega)S(\omega)\,d\omega\,\sqrt{a}e^{j\omega t}\Psi(a\omega)$$

This gives

$$J(a) = \frac{|a|}{2\pi}\int_{-\infty}^{+\infty} |\Psi(a\omega)|^2 S(\omega)e^{j\omega t}\,d\omega$$

$$\int J(a)\frac{da}{a^2} = \frac{1}{2\pi}\int_{-\infty}^{+\infty}\int_{-\infty}^{+\infty} \frac{|\Psi(a\omega)|^2}{|a|}S(\omega)e^{j\omega t}\,d\omega\,da$$

$$\int_{-\infty}^{+\infty} \frac{|\Psi(a'\omega)|^2}{|a'|}\,da' = C_\psi \quad .$$

Finally:

$$\int_{-\infty}^{+\infty} J(a)\,\frac{da}{a^2} = \frac{1}{2\pi}\int_{-\infty}^{+\infty}\int_{-\infty}^{+\infty} S(\omega)e^{j\omega t}C_\psi\,d\omega = C_\psi\,s(t)$$

$$s(t) = C_\psi^{-1}\int_{-\infty}^{+\infty}\int_{0}^{+\infty} \frac{1}{a^2} W_s(a,b)\,\psi_{a,b}(t)\,da\,db \quad .$$

3.3.4   Examples of Mother Wavelets

In this section, we will give several examples of mother wavelets which have proven to be useful in a variety of signal processing applications [Dau92].

Haar Wavelet

This is the simplest of all wavelets, even though it is not very useful for time-scale decomposition:

$$\psi(t) = \begin{cases} 1, & 0 \leq t < 1/2 \\ -1, & 1/2 \leq t < 1 \\ 0, & \textit{otherwise} \end{cases}$$

(3.3.34)

$$\Psi(\omega) = je^{-j\frac{\omega}{2}} \frac{\sin^2(\omega/4)}{\omega/4}$$

A plot of the Haar wavelet and its Fourier transform is given in Figure 3.3.8.

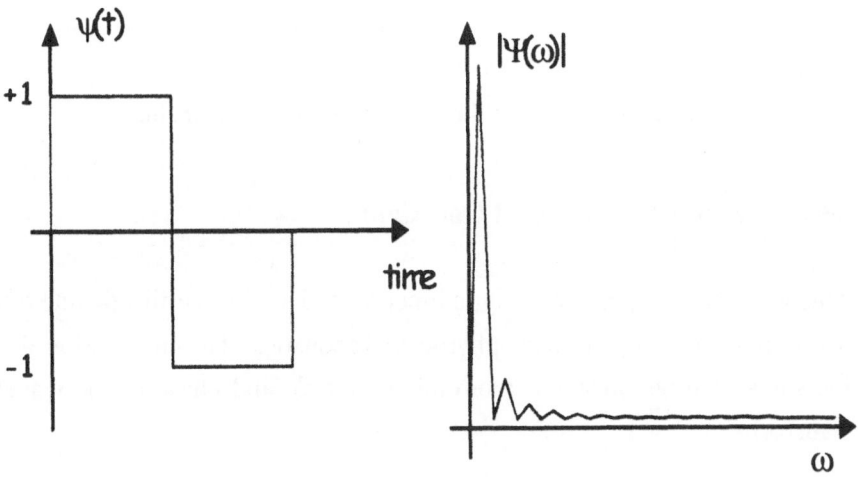

Figure 3.3.8 Haar wavelet in the time and frequency domain.

Shannon Wavelet

$$\psi(t) = \frac{\sin(\pi t/2)}{\pi t/2} \cos\left(\frac{3\pi t}{2}\right)$$

$$\Psi(\omega) = \begin{cases} 1, & \pi < |\omega| < 2\pi \\ 0, & \text{otherwise} \end{cases}$$

(3.3.35)

This wavelet is the dual of the Haar wavelet in the frequency domain. A plot of the Shannon wavelet and its Fourier transform is given in Figure 3.3.9.

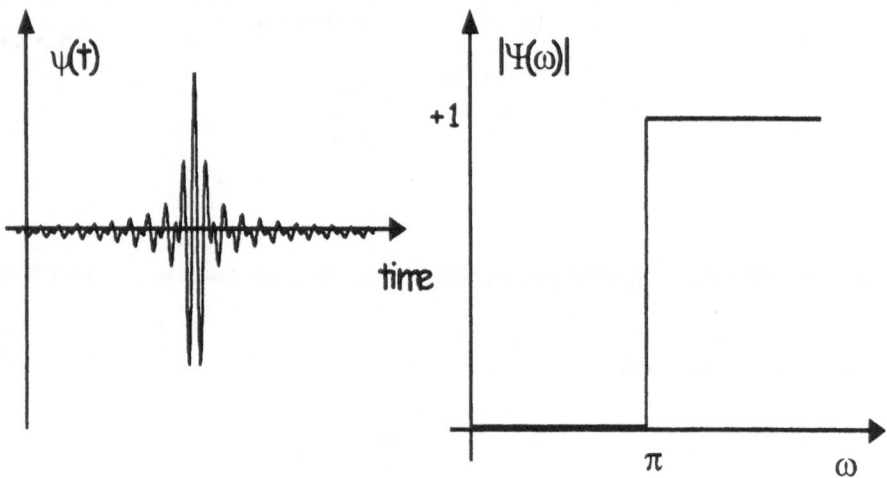

Figure 3.3.9 Shannon wavelet in the time and frequency domains.

Morlet Wavelet (Modulated Gaussian)

This wavelet is a complex analytic function and has been utilized quite often in the analysis of signals in geophysics and acoustics. The information about the signal can be extracted from the amplitude and phase of the wavelet transform:

$$\psi(t) = \pi^{-1/4} e^{-j\omega_0 t} e^{-t^2/2}$$

$$\Psi(\omega) = \pi^{-1/4} e^{-(\omega - \omega_0)^2/2}$$

(3.3.36)

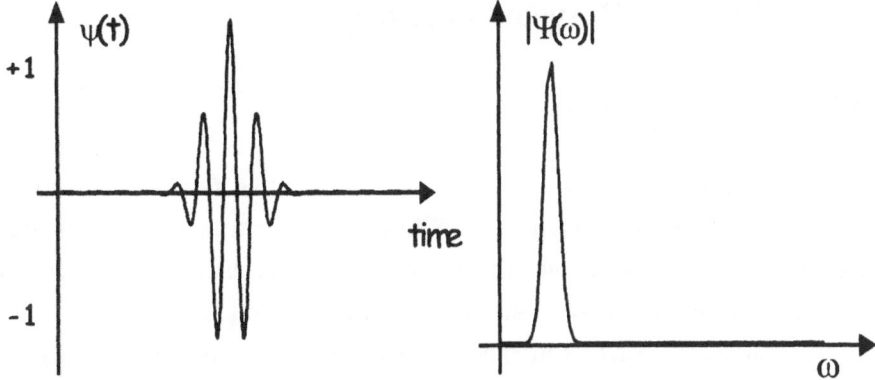

Figure 3.3.10 Morlet wavelet in the time and frequency domains.

The Morlet wavelet and its Fourier transform is given in Figure 3.3.10.

Mexican Hat (Second Derivative of Gaussian)

This function is admissible with two vanishing moments. This wavelet, being the second derivative of a commonly used smoothing function (Gaussian), has found application in edge detection:

$$\psi(t) = (1 - t^2)1e^{-t^2/2}$$

$$\Psi(\omega) = \omega^2 e^{-\omega^2} .$$

(3.3.37)

A plot of the Mexican hat wavelet and its Fourier transform is given in Figure 3.3.11.

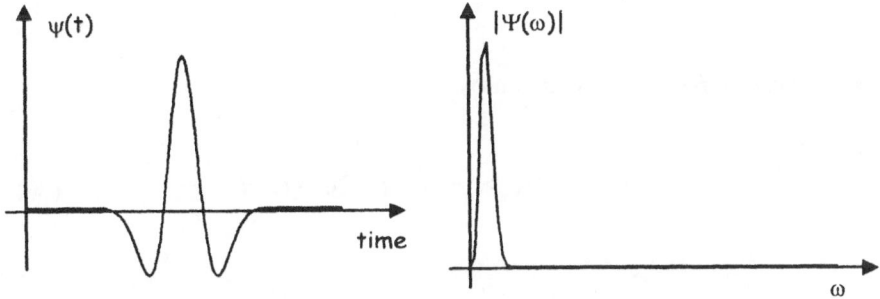

Figure 3.3.11 Mexican hat wavelet in the time and frequency domains.

3.4   Analytic Wavelet Transform

3.4.1   Analytic Signals

The continuous wavelet transform was introduced as an alternative to the short-time Fourier transform (or windowed Fourier transform) to analyze transient signals. The choice of the mother wavelet determines the way in which the information about the signal is represented [Kai95].

We have just seen that mother wavelets constructed using real functions are optimal for detecting sharp signal transitions. However, if we need to measure accurately the time evolution of frequency transients or tones, it is more appropriate to utilize as a mother wavelet a complex analytic function, in order to properly separate the phase and amplitude information of the signals [Mey93].

The utilization of analytic functions for analyzing transient signals was first proposed by Gabor. In the next section, we will introduce formally the concept of Analytic Wavelet Transform (AWT) and its applications are described in more detail in Part III of this book. The magnitude of the AWT is directly related to the instantaneous energy density of the signal in time and frequency; thus, it is an optimal representation for the determination of group delay or group velocity of transient signals [Gam81].

We have seen that given a signal $s(t) = A(t)e^{j\varphi(t)}$, it is useful to define the following quantities.

- *Instantaneous frequency* is given by

$$\omega_i(t) = -\varphi'(t) \tag{3.4.1}$$

- *Average frequency* is given by

$$<\omega> = <\omega_i> = \frac{1}{E} \int_{-\infty}^{+\infty} \omega_i(t) \, dt \tag{3.4.2}$$

- *Group delay* is given by

$$t_g(\omega) = -\Psi'(\omega) \tag{3.4.3}$$

- *Average delay* is given by

$$<t> = <t_g(\omega)> = \frac{1}{E} \int_{-\infty}^{+\infty} t_g(\omega) \, d\omega \tag{3.4.4}$$

Unfortunately, in the case of a real function or signal $s(t)$, the Fourier spectrum is symmetrical around the zero frequency [i.e., $S(-\omega) = S^*(\omega)$] and this results in a zero mean frequency:

$$<\omega> = \frac{1}{E} \int_{-\infty}^{+\infty} \omega \, |S(\omega)|^2 \, d\omega$$

$$<\omega> = \frac{1}{E} \int_{-\infty}^{0} \omega \, |S(\omega)|^2 \, d\omega \; + \; \frac{1}{E} \int_{0}^{+\infty} \omega \, |S(\omega)|^2 \, d\omega \tag{3.4.5}$$

$$<\omega> = -\frac{1}{E} \int_{0}^{+\infty} \omega \, |S(-\omega)|^2 \, d\omega \; + \; \frac{1}{E} \int_{0}^{+\infty} \omega \, |S(\omega)|^2 \, d\omega$$

Hence, $<\omega> = 0$. There are two ways to eliminate this problem. First, define the average frequency $<\omega>$ as

$$<\omega> = \frac{1}{E} \int_{0}^{+\infty} \omega \, |S(\omega)|^2 \, d\omega \tag{3.4.6}$$

or second, leave the terms of the integration the same, but define a new signal $z(t)$ which has the same spectrum of $s(t)$ for positive frequencies and zero spectrum for negative frequencies.

Let us define the Fourier transforms of $s(t)$ and $z(t)$, respectively, as $S(\omega)$ and $Z(\omega)$. We want to impose the condition that:

$$<\omega> = \frac{\int_{0}^{\infty} \omega \, |S(\omega)|^2 \, d\omega}{\int_{0}^{\infty} |S(\omega)|^2 \, d\omega} = \frac{\int_{-\infty}^{\infty} \omega \, |Z(\omega)|^2 \, d\omega}{\int_{-\infty}^{\infty} |Z(\omega)|^2} \tag{3.4.7}$$

We call $z(t)$ the *analytic signal* of $s(t)$ if

$$Z(\omega) = \begin{cases} 2S(\omega), & \omega \geq 0 \\ 0, & \omega < 0 \end{cases} \tag{3.4.8}$$

In this expression, the factor of 2 is introduced so that the real part of the analytic signal will be $s(t)$. We can calculate $z(t)$ from its Fourier transform

$$z(t) = 2 \frac{1}{\sqrt{2\pi}} \int_0^{+\infty} S(\omega) e^{j\omega t} \, d\omega$$

$$z(t) = 2 \frac{1}{2\pi} \int_0^{+\infty} \int_{-\infty}^{+\infty} s(t') e^{-j\omega t'} e^{-j\omega t} \, dt' \, d\omega$$

$$z(t) = \frac{1}{\pi} \int_0^{+\infty} \int_{-\infty}^{+\infty} s(t') e^{j\omega(t-t')} \, dt' \, d\omega \tag{3.4.9}$$

and using the relationship

$$\int_0^{\infty} e^{j\omega x} \, d\omega = \pi \delta(x) + \frac{j}{x} \tag{3.4.10}$$

we obtain

$$z(t) = \frac{1}{\pi} \int_{-\infty}^{+\infty} s(t') \left[\pi \delta(t - t') + \frac{j}{t - t'} \right] dt' \tag{3.4.11}$$

Equation 3.4.11 defines the *Analytic Operator $A\{\}$* as:

$$A\{s\} = z(t) = s(t) + \frac{j}{\pi} \int_{-\infty}^{+\infty} \frac{s(t')}{t - t'} \, dt' \tag{3.4.12}$$

Let us look at the above transformation using an example. Given a signal $s(t)$ defined as

$$s(t) = a \cos(\omega_0 t + \phi) \qquad (3.4.13)$$

Its Fourier transform is given by

$$S(\omega) = \pi a \left[e^{j\phi} \delta(\omega - \omega_0) + e^{-j\phi} \delta(\omega + \omega_0) \right] \qquad (3.4.14)$$

Its correspondent analytic signal $Z(\omega) = 2 S(\omega)$ for $\omega > 0$ is given by

$$Z(\omega) = 2S(\omega)\big|_{\omega \geq 0} = 2\pi a e^{j\phi} \delta(\omega - \omega_0) \qquad (3.4.15)$$

and by means of the inverse Fourier transform, we have

$$z(t) = \mathscr{F}\{Z(\omega)\} = a e^{j(\omega_0 t + \phi)} \qquad (3.4.16)$$

The signal $z(t)$ is thus a complex signal whose real and imaginary components are shifted in phase by 90 degrees:

$$\mathbf{Re}\{z(t)\} = a \cos(\omega_0 t + \phi) = s(t)$$

$$\mathbf{Im}\{z(t)\} = a \sin(\omega_0 t + \phi) = a \cos\left(\omega_0 t + \phi + \frac{\pi}{2}\right) \qquad (3.4.17)$$

The real and imaginary components of the signal $z(t)$ are plotted in Figure 3.4.1. The next question we need to ask is: What about its energy?

The energy of the analytic signal $z(t)$ is twice the energy of the original signal $s(t)$:

$$E_s = \int_{-\infty}^{+\infty} |S(\omega)|^2 \, d\omega = \int_0^{+\infty} \frac{|Z(\omega)|^2}{2} \, d\omega = \frac{E_z}{2} \qquad (3.4.18)$$

The convolution of a real signal $f(t)$ with an analytic signal $z(t)$ results in an analytic signal $y(t)$:

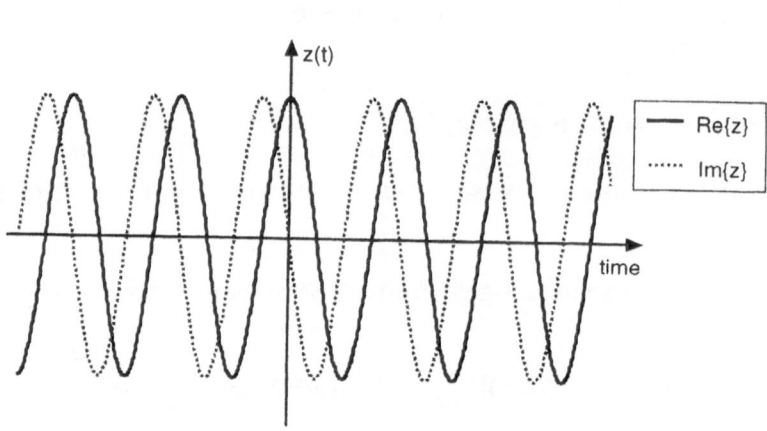

Figure 3.4.1 Real and imaginary components of the analytic signal $z(t)$.

$$y(t) = f(t) * z(t)$$

$$Y(\omega) = F(\omega) \, Z(\omega)$$

(3.4.19)

A practical way to create an analytic signal is to utilize the property of *imposed modulation*. If we have a signal that is band-limited between two frequencies, it is often necessary to shift the frequencies to higher values for the purposes of transmission. For example, a speech signal, limited to few thousands of hertz (Hz), is raised to the MHz range for better transmission. This is done by multiplying the original waveform by $\exp(j\omega_0 t)$ with $\omega_0 > 0$. The new signal is thus

$$s_{\text{new}}(t) = s(t) \, e^{j\omega_0 t} \,, \qquad \omega_0 > 0$$

$$S_{\text{new}}(\omega) = S(\omega - \omega_0)$$

(3.4.20)

Is this new signal an analytic signal? Assuming a real signal, $S_{\text{new}}(\omega)$ is analytic if $S(\omega)$ is bandlimited to $(-\omega_0, \omega_0)$. If $S(\omega) = 0$ for $\omega < -\omega_0$, then $S_{\text{new}}(\omega)$ will be zero for $\omega - \omega_0 < -\omega_0$ (i.e., $\omega < 0$). This is equivalent to the condition of the analytic signal expressed in Equation 3.4.8.

Let us consider the implications of creating an analytic signal. In particular, we are interested in understanding the relationship between the amplitude and phase of the original signal and the correspondent analytic signal. Our objective is to study the analytic signal $z(t)$ and still obtain information about the real signal $s(t)$. Since $z(t)$ is a complex signal, it can be written in its polar form:

$$z(t) = A(t)\, e^{j\varphi(t)} \ . \tag{3.4.21}$$

We define the Fourier transforms on the amplitude and phase of the signal as

$$S_A(\omega) = \frac{1}{\sqrt{2\pi}} \int_{-\infty}^{+\infty} A(t) e^{-j\omega t}\, dt$$

$$S_\varphi(\omega) = \frac{1}{\sqrt{2\pi}} \int_{-\infty}^{+\infty} e^{j\varphi(t)} e^{-j\omega t}\, dt \tag{3.4.22}$$

.

As the analytic signal $z(t)$ can be represented in the time domain as the product of two functions, the convolution theorem tells us that the Fourier transform of $z(t)$ can be represented as the convolution integral of the Fourier transforms of the two multiplicands:

$$Z(\omega) = \int_{-\infty}^{+\infty} S_A(\omega - \omega_1)\, S_\varphi(\omega_1)\, d\omega_1 \tag{3.4.23}$$

If $S_A(\omega)$ is bandlimited to ω_m [i.e., $S_A(\omega) = 0$ for $|\omega| \geq \omega_m$, and $S_\varphi(\omega) = 0$ for $\omega \leq \omega_m$] then we can create an analytic signal $z(t)$ using the imposed modulation of Equation 3.4.20.

The analytic procedure is thus equivalent to arranging the low frequency content of the signal $s(t)$ into its amplitude modulation $A(t)$ and the higher frequency content into the phase term $e^{j\omega(t)}$. This is represented by two signals for the low-frequency component, $S_A(\omega)$, and the high-frequency component, $S_\varphi(\omega)$:

$$\begin{cases} S_A(\omega) = 0 \ , & |\omega| \geq \omega_m \\ \\ S_\varphi(\omega) = 0 \ , & \omega \leq \omega_m \end{cases} \tag{3.4.24}$$

The low-frequency component can be easily extracted from the analytic signal as

$$|z(t)|^2 = A^2(t) \qquad (3.4.25)$$

and thus it can be used to represent the energy density distribution of the signal. This concept is extremely important, as the analytic wavelet transform is used to more easily analyze the energy distribution in the time-frequency plane.

Let us explain the above discussion with an example. If the original real signal is composed by a $\cos(\omega_0 t)$ signal whose amplitude is modulated by a Gaussian window

$$s(t) = \left(\frac{\alpha}{\pi}\right)^{1/4} e^{-\frac{\alpha}{2}(t-t_0)^2} \cos(\omega_0 t) \qquad (3.4.26)$$

then we can construct the analytic signal $z(t)$ by separating it into the product of two signals. If the Gaussian modulation represents the low-frequency component of the signal $s(t)$, and the cos function the high frequency, then we can easily represent the analytic signal as

$$z(t) = A(t) \, e^{j\omega_0 t} \qquad (3.4.27)$$

$$A(t) = \left(\frac{\alpha}{\pi}\right)^{1/4} e^{-\frac{\alpha}{2}(t-t_0)^2}$$

$$\text{Re}\left\{e^{j\omega_0 t}\right\} = \cos(\omega_0 t) \qquad (3.4.28)$$

$$\text{Re}\left\{z(t)\right\} = s(t)$$

3.4.2   Analytic Wavelet Transform on Real Signals

In many applications, the propagation channel is dispersive (i.e., the signals with different center frequency and bandwidth travel with different velocity).

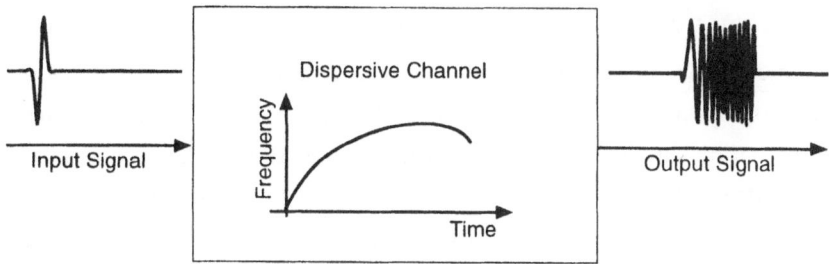

Figure 3.4.2   System model of a dispersive propagation channel showing input and output signals.

As a result, the various components of the signal are subjected to different time shifts, and the overall signal in time changes in shape. This is the case of the input signal $s(t)$ in Figure 3.4.2, after propagating through a dispersive channel as in the figure. The effect of the different phase delay for each frequency component is clearly observable by the spread in time of the waveform. The output signal $g(t)$ has the same spectra (magnitude of the Fourier transform) of $s(t)$, only with the frequency components not in phase as the input signal. The channel group delay as a function of frequency is plotted in Figure 3.4.3.

To separate the signal components, we will first utilize a real Gaussian wavelet with proper signal bandwidth as shown in Figure 3.4.4. The magnitude of the correspondent wavelet transform is shown in Figure 3.4.5.

To better understand the image in Figure 3.4.5, we have plotted, in Figure3.4.6, the correspondent $|W_s(a,b)|$ for three different values of a. For a fixed value of a, the correspondent wavelet transform is the output of the input signal filtered by a passband filter of center frequency $a\,\omega_0$ and

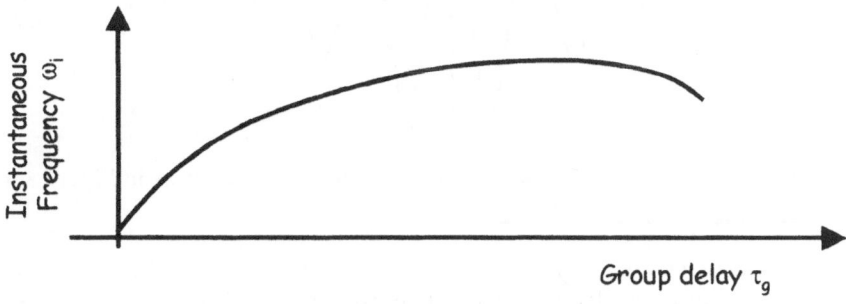

Figure 3.4.3 Plot of the instantaneous frequency ω_i as a function of the group delay τ_g.

bandwidth B / a, where ω_0 and B are the center frequency and bandwidth of the mother wavelet, respectively. For this reason, any of the signals plotted in Figure 3.4.6 represents a narrowband signal, and the CWT in Figure 3.4.5 shows ripples in the image.[***] The time of arrival of each narrowband signal in Figure 3.4.6 can be used to estimate the group delay of the channel [Gam81a]. In virtue of the convolution property of the analytic functions, if the wavelet is analytic, the correspondent CWT $W_s(a,b)$ is also analytic.

Furthermore, using the property of linearity of the wavelet transform, we have

$$W_s(a,b) = \sum_{k=1}^{N} \frac{1}{\sqrt{a}} \int_{-\infty}^{+\infty} A_k(t) \cos(\omega_k t + \phi_k) \, \psi^*\left(\frac{t-b}{a}\right) dt$$

$$\hspace{6cm} (3.4.29)$$

$$= \sum_{k=1}^{N} W_{s_k}(a,b)$$

where

$$W_{s_k}(a,b) = \frac{1}{\sqrt{a}} \int_{-\infty}^{+\infty} A_k(t) \cos(\omega_k t + \phi_k) \, \psi^*\left(\frac{t-b}{a}\right) dt \quad . \hspace{1cm} (3.4.30)$$

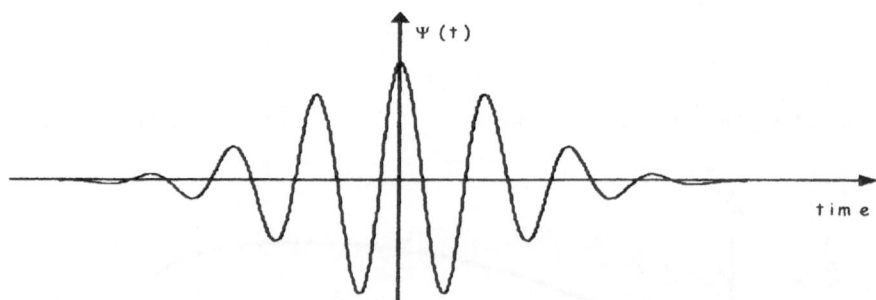

Figure 3.4.4   Real Gaussian wavelet used for the CWT analysis in Figure 3.4.5.

[***] In many of the images shown previously in the book, the CWT displayed was calculated using an analytic wavelet. This was done to better explain the key concepts. Only in this chapter are we actually showing in Figure 3.4.5 the CWT of a real wavelet, as they are used more in subband and in discrete wavelet transform applications.

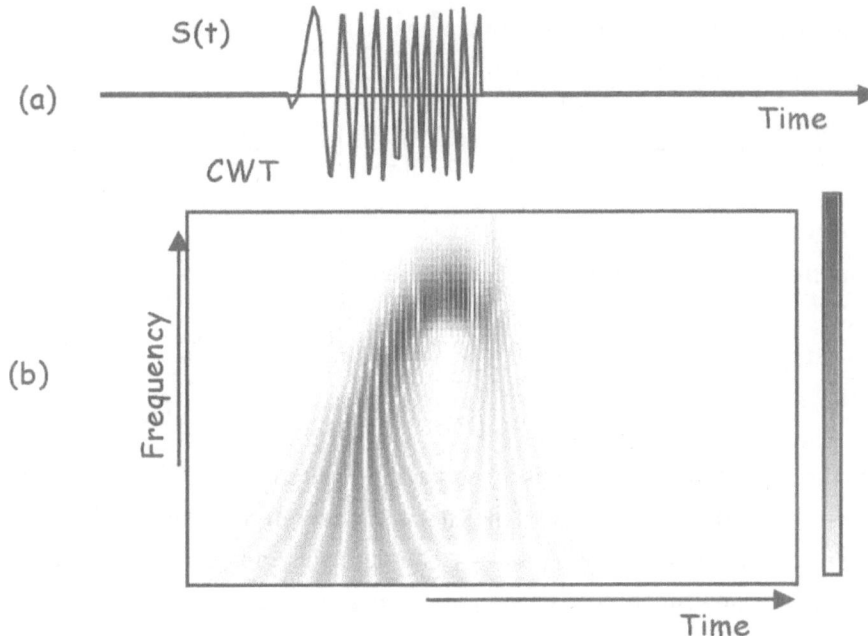

Figure 3.4.5 (a) Output signal $s(t)$ of the dispersive system in Figure 3.4.2 and (b) relative CWT calculated using the real Gaussian wavelet of Figure 3.4.4.

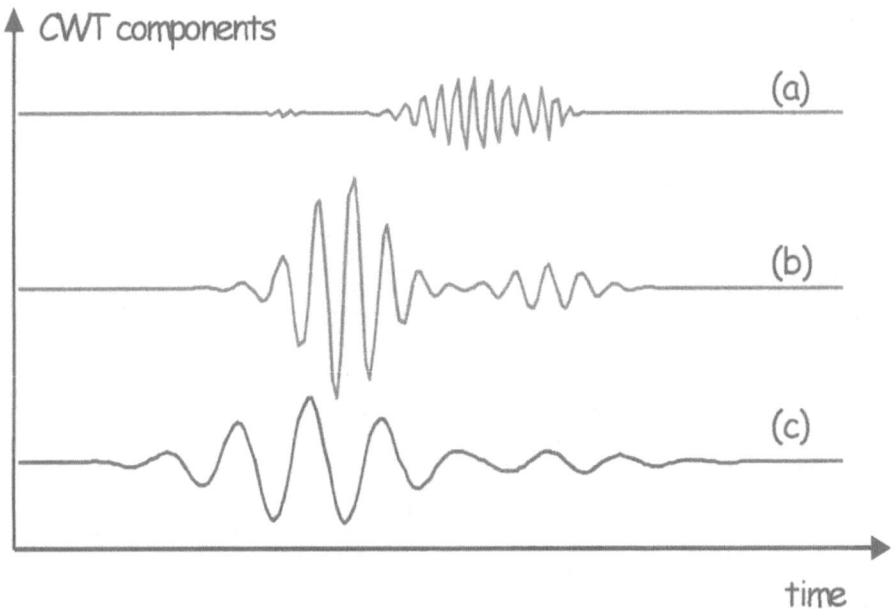

Figure 3.4.6 Signals extracted from the CWT of Figure 3.4.5 for three values of the scaling parameter a.

An analytic wavelet transform of a signal $s(t)$ depends only on the analytic part of the signal $s(t)$. For a real signal $s(t)$, it can be shown that

$$\int_{-\infty}^{+\infty} |s(t)|^2 dt = E_s = \frac{2}{C_\psi} \int_0^{+\infty} \int_{-\infty}^{+\infty} |W_s(a,b)|^2 \frac{db\,da}{a^2} \qquad (3.4.31)$$

and the real signal can be reconstructed as

$$s(t) = \frac{2}{C_\psi} \text{Re}\left\{ \int_0^{+\infty} \int_{-\infty}^{+\infty} W_s(a,b) \, \psi\left(\frac{t-b}{a}\right) \frac{db\,da}{a^2} \right\} \qquad (3.4.32)$$

In the case of the STFT, we have defined the spectrogram (Equation 3.2.12) to measure the energy of the signal $s(t)$ in the time-frequency tile. The same can be done with the AWT, and we can define a new local time-frequency energy density function $P_w(\tau, \omega)$ that measures the energy of the signal $s(t)$ in the time-frequency tile defined by the wavelet function $\psi_{a,b}(t)$:

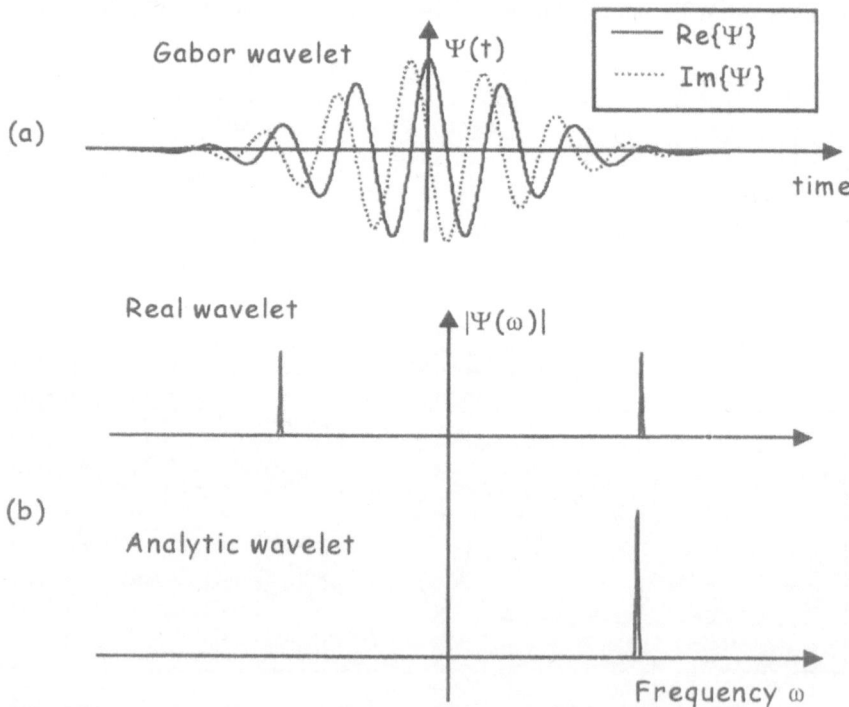

Figure 3.4.7   (a) Gabor analytic wavelet and correspondent (b) Fourier spectra. For comparison, the magnitude of the Fourier spectra of the real wavelet in Figure 3.4.4 is also shown.

$$P_w(\tau,\omega) = |W_s(a,b)|^2 = \left|W_s\left(a = \frac{\omega_0}{\omega},\ b = \tau\right)\right|^2 \qquad (3.4.33)$$

where ω_0 is the center frequency of the passband filter defined by the mother wavelet function $\psi(t)$. This new function $P_w(\tau, \omega)$ is called scalogram.

In order to create an analytic wavelet function $\psi(t)$, we can utilize the properties of modulation presented in the previous section. The wavelet function can thus be expressed by the product of a low-frequency amplitude-modulation window $h(t)$ and an higher-frequency harmonic function of center frequency ω_0:

$$\psi(t) = h(t)\ e^{j\omega_0 t}$$
$$\Psi(\omega) = H(\omega - \omega_0)\ . \qquad (3.4.34)$$

If $H(\omega) = 0$ for $|\omega| > \omega_0$, then $\Psi(\omega) = 0$ for $\omega < 0$ (i.e., the wavelet is an analytic function). Furthermore, if the function $h(t)$ is real and even, then $H(\omega)$ is also real and symmetric with its maximum value in magnitude occurring at $\omega = 0$. This means that the maximum value of $\upsilon(\omega)$ is located

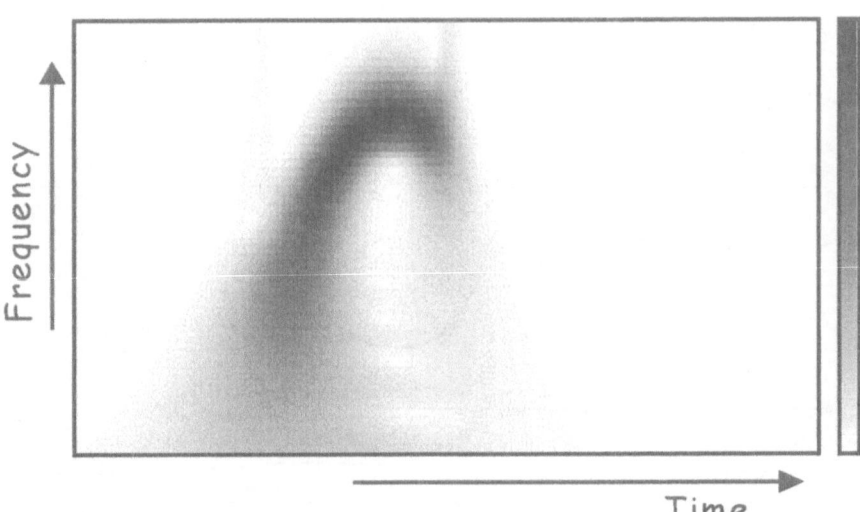

Figure 3.4.8   Analytic wavelet transform of the signal in Figure 3.4.5a, calculated using the analytic Gaussian wavelet of Figure 3.4.7.

at $\omega = \omega_0$. A Gabor wavelet $\psi(t)$ is obtained using a Gaussian window function $h(t)$:

$$h(t) = \frac{1}{(\sigma^2 \pi)^{1/4}} e^{-\frac{t^2}{2\sigma^2}} \tag{3.4.35}$$

and its Fourier transform is

$$H(\omega) = (4\pi\sigma^2)^{1/4} e^{-\frac{\sigma^2\omega^2}{2}} \tag{3.4.36}$$

Strictly speaking, the function $H(\omega)$ is never equal to zero, but if $(\sigma\omega_0)^2 \gg 1$, then the exponential term is almost zero, $H(\omega) \sim 0$, for $|\omega| > \omega_0$. In this case, the Gabor wavelet is considered analytic, as shown in Figure 3.4.7.

The Analytic Wavelet Transform of the signal in Figure 3.4.1, calculated using the Gabor wavelet, is plotted in Figure 3.4.8. Also the magnitude of the AWT calculated for three fixed values of the scaling parameter a, is plotted in Figure 3.4.9. In order to better understand Figure 3.4.9, let us look at some examples of AWT.

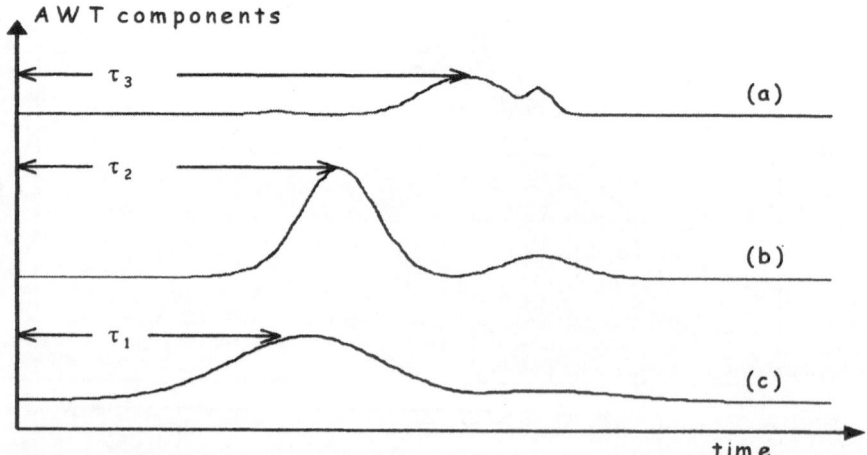

Figure 3.4.9   Signals extracted from the AWT of Figure 3.4.8 for three values of the scaling parameter a.

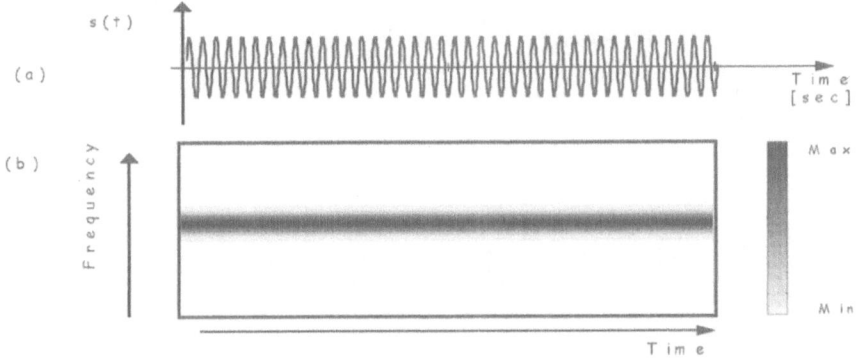

Figure 3.4.10 AWT of a single real harmonic signal calculated using Equation 3.4.43.

Analytic Wavelet Transform of an Harmonic Function

If the signal is an harmonic function

$$s(t) = A\ e^{j\omega_1 t} \tag{3.4.37}$$

then the AWT is expressed as

$$W_s(a,b) = \int_{-\infty}^{+\infty} A e^{-j\omega_1 t} \psi^*_{a,b}(t)\, dt \tag{3.4.38}$$

but Equation 3.4.38 resembles Equation 3.1.15, hence the wavelet transform of an harmonic signal is the Fourier transform of the wavelet functions:

Figure 3.4.11 Harmonic signal with low-frequency amplitude modulation.

$$W_s(a,b) = \sqrt{2\pi} \, A \, \Psi_{a,b}(\omega_1) \qquad (3.4.39)$$

$$\Psi_{a,b}(\omega_1) = \sqrt{a} \, \Psi(a\omega_1) \, e^{-j\omega_1 b} \qquad (3.4.40)$$

but
and by using Equation 3.4.34:

$$W_s(a,b) = A \sqrt{2\pi a} \, \Psi(a\omega_1) \, e^{-j\omega_1 b}$$
$$= A \sqrt{2\pi a} \, H(a\omega_1 - \omega_0) \, e^{-j\omega_1 b} \qquad (3.4.41)$$

and, finally, the scalogram is given by

$$P_w(\tau,\omega) = |W_s(a,b)|^2$$

$$P_w(\tau,\omega) = 2\pi A^2 \left| H\left[\frac{\omega_0}{\omega}(\omega_1 - \omega)\right] \right|^2 \left(\frac{\omega_0}{\omega}\right) \qquad (3.4.42)$$

Since the function $|H(\omega)|^2$ has a maximum at $\omega = 0$, the scalogram is maximum at $\omega = \omega_1$. Please note that since the signal $s(t)$ is an harmonic function, the time dependence of the scalogram (τ) is canceled in the calculations. The AWT of the harmonic signal is plotted in Figure 3.4.10.

The amplitude of the scalogram is scaled in frequency by ω; for this reason, it is common to find in literature the definition of scalogram as

$$\hat{P}_w(\tau,\omega) = \frac{|W_s(a,b)|^2}{a} = 2\pi A^2 \left| H\left[\frac{\omega_0}{\omega}(\omega_1 - \omega)\right] \right|^2 \qquad (3.4.43)$$

It is left to the reader to verify that the conclusions of the analysis presented here do not change.

The above discussion is generalized for the case that the amplitude A in Equation 3.4.37 is a function of time, $s(t) = A(t) \cdot \exp(j\omega_1 t)$. In this case, the wavelet transform can be seen as the Fourier transform of the product of two signals, $A(t)$ and $\psi_{a,b}(t)$, which results in a convolution of the two Fourier transforms. It can be shown [Mal98], that if the signal $s(t)$ is such that the amplitude and instantaneous frequency have small relative variations over the tile represented by the wavelet function $\psi_{a,b}(t)$ (i.e., they can be considered constants in the tile), the AWT can be expressed as

$$W_s(a,b) \approx \frac{\sqrt{a}}{2} A(b) \, e^{-j\omega_1 b} \, H(a\omega_1 - \omega_0) \qquad (3.4.44)$$

In this case, the spectrogram will have a maximum in frequency $\omega = \omega_1$, but this time its value is modulated in time by the function $A(b)$, with $\tau = b$. An example of such signal is given in Figures 3.4.11 and 3.4.12.

3.4.3 Physical Interpretation of an Analytic Signal

When dealing with most of the applications, an added requirement is imposed by the condition of causality. It is a common approach to describing the system under study as a black box, completely defined by input and output signals, an example being the dispersive channel system in Figure 3.4.2. If

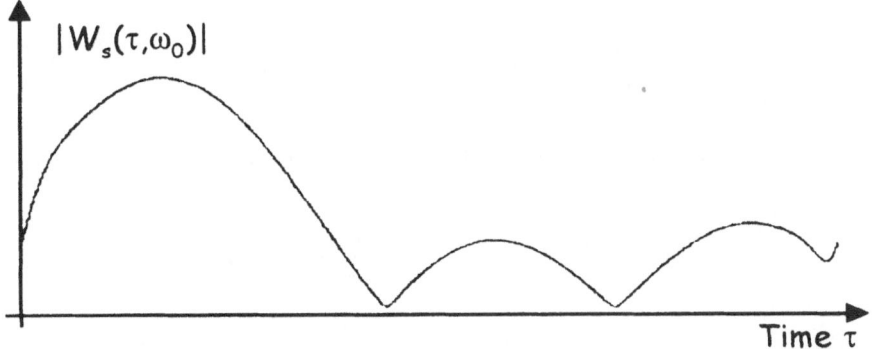

Figure 3.4.12 Magnitude of the AWT plotted as a function of the time shift at the frequency of the harmonic component. As expected, the magnitude of the AWT represents the low-frequency amplitude modulation $A(t)$.

the system is linear, it is completely characterized by the impulse response function $h(t)$. If the system is causal, it means that $g(t) = 0$ for $t \le 0$, if $s(t)=0$ for $t \le 0$. This simple condition states that no output can be present before the input is applied to the system. Using the step function $u(t)$

$$u(t) = \begin{cases} 1, & t \ge 0 \\ 0, & t < 0 \end{cases} \qquad (3.4.45)$$

and it can be shown that its Fourier transform is

$$U(\omega) = \pi \delta(\omega) + \frac{1}{j\pi} \qquad (3.4.46)$$

The condition of causality can be written as

$$h(t) = h(t)\, u(t)$$
$$H(\omega) = H(\omega) * U(\omega) \qquad (3.4.47)$$

If we define $R(\omega)$ and $Y(\omega)$ as the real and imaginary components of $H(\omega)$, $H(\omega) = R(\omega) + jY(\omega)$, then Equation 3.4.47 can be written as

$$[R(\omega) + jY(\omega)] * \left| \pi \delta(\omega) + \frac{1}{j\omega} \right| = [R(\omega) + jY(\omega)] \qquad (3.4.48)$$

The above equation can be split into two separate equations for the real and imaginary components:

$$\pi R(\omega) * \delta(\omega) + Y(\omega) * \frac{1}{\omega} = R(\omega)$$
$$\qquad (3.4.49)$$
$$\pi Y(\omega) * \delta(\omega) - R(\omega) * \frac{1}{\omega} = Y(\omega)$$

Equation 3.4.49 shows that the condition of causality of a system imposes a relationship between the real and imaginary components of its impulse response $h(t)$. This is a very important condition, since it allows one to define completely a system once only its real or imaginary response is known. Equation 3.4.49 can be manipulated and using the variable $f = \omega/2\pi$, it can be shown that

$$R(f) = \frac{1}{\pi} \int_{-\infty}^{+\infty} \frac{Y(\xi)}{f - \xi} \, d\xi$$

$$Y(f) = -\frac{1}{\pi} \int_{-\infty}^{+\infty} \frac{R(\xi)}{f - \xi} \, d\xi \qquad (3.4.50)$$

The above relationship between the real and imaginary components is known in physics as the Kramer-Kronig relationship. In signal processing, the integral in Equation 3.4.50 is known as the Hilbert transform of a signal:

$$s_H(t) = \frac{1}{\pi} \int_{-\infty}^{+\infty} \frac{s(\tau)}{t - \tau} \, d\tau \qquad (3.4.51)$$

If we recall the definition of the analytic operator given in Equation 3.4.12, we can see that the analytic equivalent of a real signal $s(t)$ is given by:

$$z(t) = s(t) + js_H(t) \qquad (3.4.52)$$

The analytic signal is a signal whose real and imaginary components are related through the Hilbert transform; hence, $z(t)$ represents the causal signal whose real component is given by the signal $s(t)$ that we commonly measure in our applications. The concept of analytic signals is extremely useful in describing mathematically real causal systems without having to carry around the causality condition as given in Equation 3.4.47. Furthermore, we have established the link between the energy of the analytic signal and its projection onto the real axis $\mathbb{R}$. All of this results in the analytic wavelet transform, which is better suited for analyzing the time-frequency behavior of dispersive systems. The application of the AWT will be discussed in Chapter 7.

3.5   Quadratic Time-Frequency Distributions

The wavelet and windowed Fourier transforms are computed by correlating the signal with families of time-frequency windows. The time-frequency resolution of these transforms is thus limited by the time-frequency resolution of the windows, and thus there is a loss in resolution due to the smearing effect of these windows [Rio92].

The *Wigner-Ville distribution* is a time-frequency energy density function computed by correlating the signal with a time and frequency translation of itself, resulting in a representation which is free of any window effect, avoiding any loss of time-frequency resolution. Unfortunately, this is achieved at a cost, namely the creation of interference terms, as we will show in the following. We will also present an approach to reduce such effects, by a local time-frequency averaging, which behaves similarly to a windowing effect [Coh89, Coh95].

The conventional definition of a *Power Spectral Density* (PSD) function of a signal $s(t)$ is given by

$$\mathbf{PSD}(\omega) = |S(\omega)|^2 = \left| \int_{-\infty}^{+\infty} s(t) e^{-j\omega t} dt \right|^2 \tag{3.5.1}$$

However, if we utilize the concept of an autocorrelation function $R(\tau)$

$$R(\tau) = \int_{-\infty}^{+\infty} s(t) \, s^*(t-\tau) dt = \int_{-\infty}^{+\infty} s\left(t+\frac{\tau}{2}\right) s^*\left(t-\frac{\tau}{2}\right) dt \tag{3.5.2}$$

we can express the PSD as

$$\mathbf{PSD}(\omega) = \int_{-\infty}^{+\infty} R(\tau) e^{-j\omega t} d\tau \tag{3.5.3}$$

The autocorrelation function $R(\tau)$ can thus be expressed as the time average of the instantaneous autocorrelation $s(t+\tau/2)s^*(t-\tau/2)$. By time-averaging we have thus lost the temporal information.

In this way, we can replace the time-average $R(\tau)$ by the instantaneous autocorrelation and obtain the Wigner-Ville Distribution:

$$\mathbf{WVD}(t,\omega) = \int_{-\infty}^{+\infty} s\left(t+\frac{\tau}{2}\right) s^*\left(t-\frac{\tau}{2}\right) e^{-j\omega t} d\tau \qquad (3.5.4)$$

A related concept is the notion of instantaneous frequency using the WVD. Let $s(t) = A(t)\, e^{j\varphi(t)}$. The first derivative of the phase $\varphi'(t)$ represents the instantaneous frequency, and can be calculated from the WVD using the following relationship:

$$\varphi'(t) = \frac{\dfrac{1}{2\pi} \displaystyle\int_{-\infty}^{+\infty} \omega\, \mathbf{WVD}(t,\omega)\, d\omega}{|s(t)|^2} \qquad (3.5.5)$$

The problem which arises when using the WVD is known as *cross-term interference*. For example, consider the case of a signal $s(t)$ constructed from two modulated Gaussian functions of non overlapping frequencies. Let us assume that the signal is constructed by the sum of a Gaussian centered at time T_1 and frequency Ω_1, and the second is at T_2 and Ω_2. From the time signal we can easily distinguish the two temporal and spectral components, but the Wigner-Ville analysis results in the creation of a cross-term interference positioned in time and frequency at $(T_1+T_2)/2$ and $(\Omega_1+\Omega_2)/2$. Although the average of the cross-term interference is nearly zero, and thus it does not contribute energy to the signal; in many cases, the magnitude of the cross-term can be twice as large as the signal terms. It is the cross-term interference that limits applicability of the Wigner-Ville distribution in real signal processing applications, even though it possesses many desirable properties for signal analysis.

Because the cross-term appears as high-frequency oscillations, a natural way of reducing its effect is to apply a 2-D low-pass filter to smooth the rapid oscillating terms:

$$\mathbf{SWVD}(t,\omega) = \int_{-\infty}^{+\infty} \Phi(t-\mu,\omega-\Omega) s\left(t+\frac{\tau}{2}\right) s^*\left(t-\frac{\tau}{2}\right) e^{-j\omega t} d\tau \qquad (3.5.6)$$

The *Smoothed Wigner-Ville distribution* (SWVD) is thus obtained by multiplying the instantaneous autocorrelation of the signal by a smoothing kernel $\Phi(t,\omega)$. This form was first introduced, in the area of quantum mechanics, by Leon Cohen, and so expressions of this type have been defined as *Cohen's class of distributions*. Note that if $\Phi(t,\omega) = 1$, we obtain the WVD [Coh95]. Another very common smoothing function is the *Choi-Williams kernel* function:

$$\Phi(t,\omega) = e^{-\frac{t^2\omega^2}{\sigma^2}}$$
(3.5.7)

The value of σ is used to optimize the trade-off between the resolution and the presence of cross terms. In the Choi-Williams distribution of two modulated Gaussian signals the cross terms virtually disappear.

3.6   References

[Abb95]   A. Abbate, M. Doxbeck, and P. Das, Applications of wavelet transform in signal processing, *Proc. of the International Conference on Signal Processing Applications and Technology*, pp. 652-655, 1995.

[Chu95]   C.K. Chui, *Wavelets: Theory, Algorithms, and Applications*, Academic Press, New York, 1995.

[Coh89]   L. Cohen, Time-frequency Distributions - A review, *Proc. IEEE*, vol. 77 no. 7, pp. 941-981, 1989.

[Coh93]   L. Cohen, The scale representation, *IEEE Trans. Signal Process.* vol. 41, pp. 3275-3292, 1993.

[Coh95]   L. Cohen, *Time-Frequency Analysis*, Prentice-Hall, Englewood Cliffs, NJ, 1995.

[Coh96]   A. Cohen and J. Kovačević, Wavelets: The mathematical background, *Proc.IEEE*, vol. 84, no. 4, pp. 514-522, 1996.

[Dau90]   I. Daubechies, The wavelet transform, time-frequency localization and signal analysis, *IEEE Trans. Inform. Theory*, vol. 36, pp. 961-1005, 1990.

[Dau91]   I. Daubechies, The wavelet transform: A method for time-frequency localization, in *Advances in Spectrum Analysis and Array Processing*, edited by S. Haykins, Prentice-Hall, Englewood Cliffs,

NJ, pp. 366-417, 1991.

[Dau92] I. Daubechies, *Ten Lectures on Wavelets*, SIAM, Philadelphia, 1992.

[Fre93] M. Freeman, Wavelets: signal representations with important advantages , *Opt. and Photon. News*, vol. 4, pp. 8-14, 1993.

[Gab46] D. Gabor, Theory of communication, *J. Inst. Elec. Eng.*, vol. 93, pp. 429-457, 1946.

[Gam81] P.M. Gammel, Improved ultrasonic detection using analytic signal magnitude, *Ultrasonics*, pp. 73-76, 1981.

[Gam81a] P.M. Gammel, Analogue implementation of analytic signal processing for pulse-echo systems, *Ultrasonics*, pp. 279-283, 1981.

[Gou84] P. Goupillaud, A. Grossmann, and J. Morlet, Cycle-octave and related transforms in seismic signal analysis, *Geoexploration*, vol. 23, pp. 85-102, 1984.

[Gro86] A. Grossmann, J. Morlet, and T. Paul, Transforms associated to square integrable group presentations II. Examples, *Am. Inst. Henry Poincare*, vol. 45, no. 3, pp. 293-309, 1986.

[Her93] C. Herley, J. Kovačević, K. Ramchandranard, and M. Vetterli, Tilings of the time-frequency plane: Construction of arbitrary orthogonal bases and fast tiling algorithms," *IEEE Trans. Signal Process.*, vol. 41, pp. 3341-3359, 1993.

[Kai95] G. Kaiser, *A Friendly Guide to Wavelets*, Birkhauser, Boston,1995

[Mal98] S. Mallat, *A Wavelet Tour of Signal Processing*, Academic Press, San Diego, 1998.

[Mey93] Y. Meyer, *Wavelets. Algorithms and Applications*, translated by R.D. Ryan, SIAM, Philadelphia, 1993.

[Pap84] A. Papoulis, *Probability, Random Variables, and Stochastic Processes*, McGraw-Hill Book Co., New York, 1984.

[Pap77] A. Papoulis, *Signal Analysis*, McGraw-Hill, New York,1977.

[Rab78] L.R. Rabiner and R.W. Schafer, *Digital Signal Processing of Speech Signals*, Prentice-Hall, Englewwod Cliffs, NJ, 1978.

[Rio91] O. Rioul and M. Vetterli, Wavelets and signal processing, *IEEE Signal Proc. Mag.*, pp. 14-38, 1991.

[Rio92] O. Rioul and P. Flandrin, Time-scale energy distributions: A general class extending wavelet transforms, *IEEE Trans. Signal Process.*, vol. 40, pp. 1746-1757, 1992.

Chapter 4

Discrete Wavelet Transform: From Frames to Fast Wavelet Transform

4.1   Introduction

Both the short-time Fourier transform and the continuous wavelet transform can be seen as operators that project the signal $s(t)$ from the one-dimensional time domain into the two-dimensional time-frequency plane. In the case of the continuous wavelet transform the scaling a and delay b are assumed to be continuous in value; that is, it is said that the CWT is defined in the $(\mathbb{R}^+)^2$ plane where the parameters a and b are continuous in value: ($a \in \mathbb{R}^+$ and $b \in \mathbb{R}$). Since no new information can be created by this transform, the same information contained in the signal $s(t)$ with $t \in \mathbb{R}$ is available with the CWT. The increase in complexity from $t \in \mathbb{R}$ to $(a, b) \in (\mathbb{R}^+)^2$ results only in a redundant representation of the signal. This redundancy can be reduced by discretizing the transform parameters (a, b). Care must be taken so that we can still achieve reconstruction without any loss of information. Thus, the first question we must answer is how do we sample the parameters (a,b)?

In the case of the short-time Fourier transform, the (ω, τ) plane is sampled with kernel or basis functions $h(t)$ of constant width in time and frequency. It is thus natural to consider a rectangular sampling grid of the form $(m\omega_0, n\tau_0)$, where m and n are integers [i.e., with $m, n \in Z$ or $(m,n) \in Z^2$]. Clearly, as we choose smaller values for ω_0 and τ_0, the sampling more closely resembles the continuous case. So, for the STFT, what would be the proper choice of ω_0 and τ_0? As always, we must choose a sampling that at least fully covers the entire time-frequency without leaving any hole or gap. This is achieved by imposing that $\omega_0 \tau_0 \le 2\pi$, as illustrated in Figure 4.1.1a.

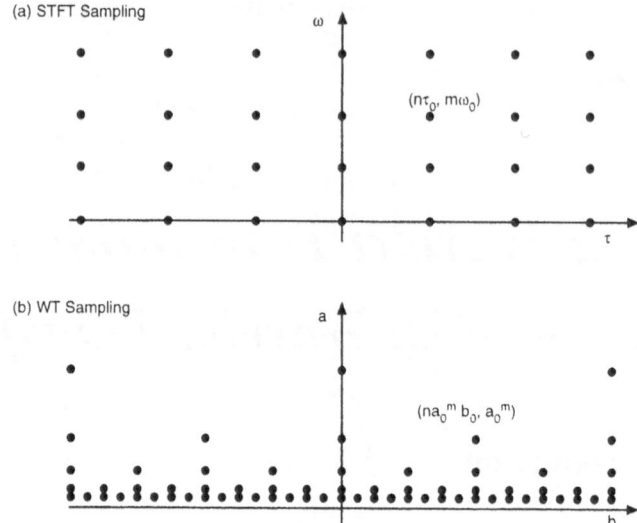

Figure 4.1.1 (a) Constant sampling grid of the time-frequency plane for the short-time Fourier transform. (b) Hyperbolic sampling grid for the wavelet transform.

In the wavelet case, an hyperbolic grid is used, and, in particular, one based on the power of 2 sampling (*dyadic sampling*). The (a, b) plane is thus sampled by first taking the scaling coefficient a sampled as

$$ a \rightarrow a_0^m , \qquad m \in Z , \qquad a_0 \neq 1 . \qquad (4.1.1) $$

Now, for $m = 0$, we sample the shift parameter b by taking integer multiples of a constant $b_0 > 0$. The step b_0 should be chosen in such a way that $\psi(t - nb_0)$ "covers" the whole time axis. Since the basis functions are rescaled for $m \neq 0$, also the step size b should be a function of the scaling. If we define the "width" of the wavelet $\psi(t)$ as ΔT

$$ \Delta T^2(\psi) = \frac{\int_{-\infty}^{+\infty} t^2 |\psi(t)|^2 \, dt}{\int_{-\infty}^{+\infty} |\psi(t)|^2 \, dt} \qquad (4.1.2) $$

and the "width" of $\psi_{a,m}(t)$, with $a = a_0^m$, times the "width" of $\psi(t)$:

$$\Delta T\left(\psi_{a_0^m,0}(t)\right) = a_0^m \Delta T(\psi(t)) \qquad (4.1.3)$$

then it is obvious that to cover the entire axis we need a step on the order of $b_0 a_0^m$. Hence, the sampling of the wavelet transform is commonly defined as

$$a = a_0^m, \qquad b = nb_0 a_0^m, \qquad m,n \in Z$$
$$\text{with} \qquad a_0 \neq 1, \qquad b_0 \neq 0. \qquad (4.1.4)$$

Large basis functions (a_0^m is large) are thus shifted in large steps ($b_0 a_0^m$), whereas small basis functions are shifted in small steps. This grid is shown in Figure 4.1.1b. In order for the sampling of the (a, b) plane to be sufficiently fine resolution, a_0 has to be chosen sufficiently close to 1 and b_0 close to 0. These discretized version of continuous transforms are examples of *frames* [Ben90].

The reconstruction of the original signal is possible using frames, but strongly depends on the sampling density and on the functions used for basis. Reconstruction is thus linked on the fact that the (ω,τ) plane is properly sampled without any aliasing or leakage in both domains [Uns00]. In general, the reconstruction Equation will require different synthesis functions than analysis functions, except in a special case, called *tight frames*. In this case, the frame behaves just as an orthonormal basis, except that the set of functions used to expand the signal is redundant and, thus, the functions are not independent.

Using the hyperbolic discretization of Equation 4.1.4, we obtain the following expression for the daughter wavelets:

$$\psi(a,b)\big|_{a=a_0^m, b=nb_0a_0^m} = \psi(a_0^m, nb_0a_0^m)$$

$$= \frac{1}{\sqrt{a_0^m}} \psi\left(\frac{t - nb_0a_0^m}{a_0^m}\right) \tag{4.1.5}$$

$$= a_0^{-m/2} \psi\left(a_0^{-m}t - nb_0\right) = \psi_{m,n}(t) \quad.$$

From this follows the expression for the *Discrete Wavelet Transform* (DWT):

$$W_s(m,n) = \int_{-\infty}^{+\infty} s(t)\,\psi_{m,n}(t)\,dt$$

$$\tag{4.1.6}$$

$$W_s(m,n) = a_0^{-m/2} \int_{-\infty}^{+\infty} s(t)\,\psi(a_0^{-m}t - nb_0)\,dt \quad.$$

Given a decomposition of the signal $s(t)$ using only a discretized set of wavelets

$$\left\{\psi_{m,n}(t)\right\}_{m,n \in Z} \subset \left\{\psi_{a,b}(t),\ a \in \mathbb{R}_+,\ b \in \mathbb{R}\right\} \quad. \tag{4.1.7}$$

A number of questions arise:

- Is this set of wavelets enough to characterize the function $s(t)$?
- Is it possible to reconstruct the signal $s(t)$ from its discrete wavelet transform (DWT) coefficients $W_{m,n}(t)$?
- Can any function $f(t)$ be written as a superposition of the "elementary building blocks" $\{\psi_{m,n}(t)\}$?

These are the same questions which form the basis of Shannon's Sampling Theorem. In [Sha49], Shannon defined a general approach in how to convert an analog signal into a sequence of numbers.

In the case of the continuous wavelet transform, these questions were answered using the resolution of identity, but in the discrete case, there is no analogous condition. Reconstruction is indeed possible; that is, for a wavelet $\psi(t)$ and an appropriate choice of a_0, b_0 there exist a function $\varphi_{m,n}(t)$ such that the signal $s(t)$ can be reconstructed as follows:

$$s(t) = \sum_m \sum_n \left\langle \psi_{m,n}, s \right\rangle \varphi_{m,n} \tag{4.1.8}$$

As $a_0 \to 1$ and $b_0 \to 0$, the double summation resembles a double integral. As with the continuous wavelet transform, we can construct a discrete wavelet transform which provides a very redundant description of the original signal $s(t)$. This redundancy can be exploited to ensure that there are positive answers to the three questions posed above. Also, as we will see later in this book, the redundancy may be useful also for practical purposes, such as reducing noise in signals.

An approach to eliminate the redundancy is to use an orthormal set of wavelets as the basis. In this case, orthonormality will guarantee the possibility of satisfying the three questions above, but at the cost of decreasing our freedom in the choice of a valid mother wavelet function. Of course, this approach also has many practical advantages, especially in communication systems in which we are extremely interested in signal compression and reduction of the number of data bits to be transferred in a given application.

Even though both approaches result in a discrete wavelet transform (DWT), we will further distinguish them as either the *Redundant Discrete Wavelet Transform* (R−DWT) or *Orthonormal Discrete Wavelet Transform* (ON−DWT). The DWT are also called Wavelets Frames.

4.2   Fundamentals of Frame Theory

By frames, we define a set of non independent vectors which can be used to write an expansion for every vector in the space (in the rest of the book we will use boldface to denote vector quantities). For example, given the vector

v in the 2–D space, **v** can be expressed as the linear combination of two orthonormal vectors $\mathbf{u}_1 = (1,0)$ and $\mathbf{u}_2 = (0,1)$, but also as the linear combination of three vectors $a_1 = (1,-1)$, $a_2 = (1,1)$ and $a_3 = (-1,1)$:

This representation of the vector **v** is redundant, but it may result in the

$$v = 2 \cdot u_1 + u_2 = 1.5 \cdot (a_1 + a_2) + a_3 \tag{4.2.1}$$

decomposition of the vector into components which have a physical meaning.

In order to continue with the concept of frames, we must introduce the concept of Hilbert space. The following discussion is not complete and it is suggested to consult the many references on the subject [Uns95].

A proper definition of an Hilbert space is that $\mathcal{H}$ is a vector space with an inner product, where the definition of inner product is $<f,g>$, which is complete with respect to the norm. What does this mean for us? The Hilbert space $\mathcal{H}$ is the domain of all functions or signals which are square integrable; that is, that their total energy is finite and satisfies the following relationship:

$$\| f \| = \left(\int_{-\infty}^{+\infty} | f(t) |^2 \, dt \right)^{1/2} = \sqrt{<f,f>} \; . \tag{4.2.2}$$

A sequence of functions $\{h_n\}_{n \in Z}$ in a Hilbert space $\mathcal{H}$ is called a frame if there exist two constants A and B, called *frame bounds*, that satisfy the condition $0 < A \le B < \infty$, so that all functions $f(t)$ in the Hilbert space satisfy the following relationship:

$$A\| f \|^2 \le \sum_n \left| <f,h_n> \right|^2 \le B\| f \|^2 \; . \tag{4.2.3}$$

In this expression, the constant $B < \infty$ guarantees that the transformation $f \rightarrow \{<f, h_n>\}$ is continuous. The constant $A > 0$ guarantees that this transformation is invertible, and that the inverse is a continuous function. To verify that a sequence of functions $\{h_n\}$ constitutes a frame, the bounds A and B must be calculated.

At this point, we may ask what conditions are necessary in order to have a stable reconstruction? Intuitively, the operator that maps a function $f(t)$ into coefficients $<\psi_{m,n}, f>$ has to be bounded; that is, if $f(t) \in L_2(\mathbb{R})$, then

$$\sum_{m,n} \left| \left\langle \psi_{m,n}, f \right\rangle \right|^2 \qquad (4.2.4)$$

has to be finite; furthermore, no function $f(t)$ which has $\| f \| > 0$ should be mapped to 0. These two conditions lead to frame bounds, which guarantee stable reconstruction.

Let us consider the first condition of Equation 4.2.2. For any wavelet with zero mean and with some decay in time and frequency, the above condition can be written as

$$\sum_{m,n} \left| \left\langle \psi_{m,n}, f \right\rangle \right|^2 \leq B \| f \|^2 \; . \qquad (4.2.5)$$

for any choice of $a_0 > 1$ and $b_0 > 0$. The above Equation states that the sequence $(<\psi_{m,n}, f>)_{m,n} \in L_2(\mathbb{R})$ (i.e., is square integrable).   The second condition in Equation 4.2.3 refers to the fact that if the double summation is small, then $\| f \|^2$ should be small as well:

$$A \| f \|^2 \leq \sum_{m,n} \left| \left\langle \psi_{m,n}, f \right\rangle \right|^2 \; . \qquad (4.2.6)$$

Both conditions impose

$$A \| f \|^2 \leq \sum_{m,n} \left| \left\langle \psi_{m,n}, f \right\rangle \right|^2 \leq B \| f \|^2 \; . \qquad (4.2.7)$$

If this condition is satisfied, then the family $(\psi_{m,n})_{m,n \in Z}$ constitutes a Frame [Duf52]. We can thus associate a *Frame Operator* F with the set of functions $\{\psi_n\}_{n \in Z}$ that maps any signal $s(t)$ into the square-summable sequence:

$$F : f(t) \rightarrow f_n = (<f,\psi_n>)_{n \in Z} \quad . \tag{4.2.8}$$

We would like to reconstruct the original function $f(t)$ as a linear combination of a set of functions $\{\varphi_n\}_{n \in Z}$, whose coefficients are the sequence $(f_n)_{n \in Z}$; that is,

$$f(t) = \sum_n f_n \varphi_n \tag{4.2.9}$$

$$F^* : f_n \rightarrow f(t) = \sum_n f_n \varphi_n \quad . \tag{4.2.10}$$

In general, the set of basis $\{\psi_n\}_{n \in Z}$ and $\{\varphi_n\}_{n \in Z}$ are not the same. In order to reconstruct the function $f(t)$ from the discrete sequence f_n, we need to define a new operator F^* that maps the sequence $\{f_n\}_{n \in Z}$ back into $f(t)$:

$$\varphi_n = (F^*F)^{-1} \psi_n \quad . \tag{4.2.11}$$

If the frame bound condition is satisfied (i.e., the two constant A and B exist for the set of functions $\{\psi_n\}_{n \in Z}$), then the set of functions $\{\varphi_n\}_{n \in Z}$ can be constructed from $\{\psi_n\}_{n \in Z}$.

The frame bound condition assures that the product F^*F is invertible and that $\{\varphi_n\}_{n \in Z}$ also constitutes a frame (it is called the *dual frame*) with bounds $A^{-1} \geq B^{-1} > 0$. The frame operator associated with this frame is F' and is obtained as

$$F' = F(F^*F)^{-1} \quad . \tag{4.2.12}$$

This means that the signal $s(t)$ can be expressed as a linear combination of either set of functions:

$$f(t) = \sum_{n \in Z} <f, \psi_n> \varphi_n = \sum_{n \in Z} <f, \varphi_n> \psi_n \ . \qquad (4.2.13)$$

Hence, once we have defined the function basis $\{\psi_n\}_{n \in Z}$, its dual frame $\{\varphi_n\}_{n \in Z}$ is constructed by inverting the product F^*F.

The frame bound condition can be written as

$$A <f, f> \ \leq \ < F^*F f, f > \ \leq \ B <f, f>$$

or

$$A \| f \|^2 \ \leq \ < F^*F f, f > \ \leq \ B \| f \|^2 \qquad (4.2.14)$$

or in operator form

$$A\mathrm{I} \ \leq \ F^*F \ \leq \ B\mathrm{I} \ . \qquad (4.2.15)$$

Hence, we can write

$$F^*F = \frac{A - B}{A + B} (\mathrm{I} - R) \qquad (4.2.16)$$

where $\mathbf{R}$ is the *residual operator* $R = \mathbb{I} - 2/A + B F^*F$, which satisfies the condition

$$\|R\| \ \leq \ \frac{A - B}{A + B} < 1 \qquad (4.2.17)$$

We can write the inverse operation as

$$(F^*F) = \frac{2}{A + B} (\mathrm{I} + R + R^2 + \dots) \qquad (4.2.18)$$

As the values of A and B grow closer together, $A - B \ll A + B$, then $R \to 0$, and the product $(F^*F)^{-1}$ converges to

$$(F^*F)^{-1} \to \frac{2}{A+B} \; \mathbb{I} \qquad\qquad (4.2.19)$$

and thus the dual frame is calculated as

$$\varphi_n = (F^*F)^{-1} \psi_n \to \frac{2}{A+B} \psi_n \qquad\qquad (4.2.20)$$

In general, a frame does not constitute an orthonormal basis, and thus provides a redundant representation of the function $f(t)$. This is the equivalent of representing a vector in the Euclidean plane using three or more basis vectors. However, the condition of frames ensures that any vector in the plane can be represented by the basis vector.

The ratio A/B is called the *redundancy ratio* or *redundancy factor*, and it is used to gauge how close the basis comes to an orthonormal representation. If $A = B$, the frame is called a *tight frame;* in this case $\varphi_n = A^{-1}\psi_n$ and the reconstruction formula becomes

$$f(t) = \frac{1}{A} \sum_n \langle f, \psi_n \rangle \, \psi_n \; . \qquad\qquad (4.2.21)$$

This formula is almost identical to the one obtained for an orthonormal basis. In this case, however, the set $\{\psi_n\}$ may not be linearly independent (i.e., the representation is still redundant). For a tight frame, if $A = B = 1$ and if $\|\psi_n\|^2 = 1$, then the set $\{\psi_n\}$ forms an *orthonormal basis*. In practice, it is difficult to have $A = B$, but it may be easier to have A close to B ($A \approx \underline{B}$), and we can define this closeness using a parameter ϵ defined as:

$$\epsilon = \frac{B}{A} - 1 \ll 1 \; . \qquad\qquad (4.2.22)$$

Daubechies calls these frames *snug frames* [Dau92]. The reconstruction formula is in this case given by

$$f(t) = \frac{2}{A + B} \sum_n <f, \psi_n> \psi_n + \gamma \qquad (4.2.23)$$

where the error γ is of the order of $\epsilon/(2+\epsilon) \|f\|^2$. In these results, we have used

$$\varphi_n = \frac{2}{A + B} \psi_n . \qquad (4.2.24)$$

In summary, then if $\{\psi_n\}$ constitute a frame, we have an algorithm to reconstruct the signal $s(t)$ from $<f, \psi_n>$ using the definition of the dual base:

$$\varphi_n = (F^* F)^{-1} \psi_n \rightarrow \frac{2}{A + B} \psi_n . \qquad (4.2.25)$$

There are really no strong constraints on $\psi(t)$, a_0, and b_0.

4.3   Sampling Theorem

In 1949, Shannon published his fundamental sampling theorem [Sha49]. It stated that if a signal $s(t)$ is band-limited (i.e., if it contains no frequencies above a maximum value f_{MAX}), then the signal is completely determined by taking a discrete set of values $s(kT)$ ($k \in Z$), uniformely spaced in time at a distance $T = f_{MAX}/2 = 1/f_N$, where the frequency f_N is called the Nyquist or sampling frequency.
The reconstruction formula is also defined as:

$$s(t) = \sum_{k \in Z} s(kT) \frac{\sin(t - kT)}{(t - kT)} \qquad (4.3.1)$$

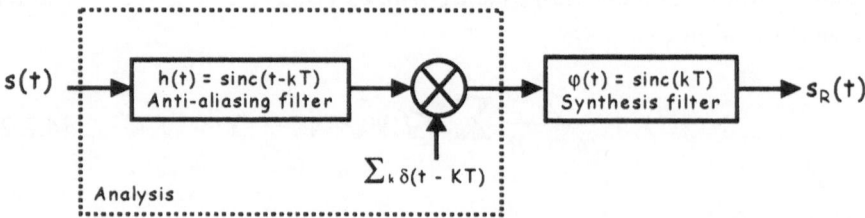

Figure 4.3.1 Schematic representation of the sampling algorithm originally described by Shannon. The anti aliasing and synthesis filters are now replaced by frames.

We can completely define the signal using its sampled values $s(kT)$ and a basis function $\delta(t)$:

$$s(kT) = \int_{-\infty}^{+\infty} s(t)\delta(t-kT)dt \quad . \tag{4.3.2}$$

The reconstruction basis $\sin(x)/x = \mathrm{sinc}(x)$ is also called the sinc function. Remember that in order for Equations 4.3.1 and 4.3.2 to be valid, the signal must be band-limited. Hence, we can defined the Shannon Theorem in block diagram as in Figure 4.3.1. Let's review Equation 4.3.1 using frame theory[Jer77],[Zib93]. From frame theory, we can write

$$s(t) = \sum_{k \in Z} c_k \, \varphi_k(t) \quad . \tag{4.3.3}$$

It follows that

$$\varphi_k(t) = \frac{\sin(t-kT)}{(t-kT)} = \mathrm{sinc}(t-kT)$$

$$c_k = s(kT) \quad . \tag{4.3.4}$$

In 1941, Hardy showed that the sinc functions are orthonormal, i.e.

$$< \varphi_k , \varphi_l > = \int_{-\infty}^{+\infty} \text{sinc}(t-kT) \, \text{sinc}(t-lT) dt$$

$$= \text{sinc}(kT-lT) = \delta(k-l) \quad . \tag{4.3.5}$$

From frame theory, $A=B=1$ for an orthonormal basis, and we can write

$$s(t) = \sum_{k \in Z} < s, \varphi_k > \varphi_k(t) \tag{4.3.6}$$

where

$$< s, \varphi_k > = \int_{-\infty}^{+\infty} s^*(t) \, \frac{\sin(t-kT)}{(t-kT)} \, dt \quad . \tag{4.3.7}$$

We can conclude that Shannon's Sampling Theorem is a particular set of orthogonal frames [Xia93, Uns00]. We will also show that wavelet frames can also be seen as an extension of Shannon Sampling Theorem [Wal92].

Before concluding we need to clarify the following: Shannon had used the delta Dirac function $\delta(t)$ as his basis, whereas we have shown that the sinc function is enough. The difference lies in the fact that Shannon had to add the requirement of band-limited signal, which is equivalent of saying that the signal is filtered by a low-pass filter. But this is equivalent to filtering the basis function $\delta(t)$ by a low-pass filter. The optimal choice of the ideal low-pass filter is the rectangular function:

$$U_T(\omega) = \begin{cases} 1 , & |\omega| < \dfrac{1}{T} \\ 0 , & \text{otherwise} \end{cases} \tag{4.3.8}$$

The impulse response of this filter is

$$u_T(t) = \frac{1}{T} \text{sinc}\left(\frac{t}{T}\right) \tag{4.3.9}$$

hence

$$\delta(t-kT) * u_T(t) = \text{sinc}\left(\frac{t-kT}{T}\right) \tag{4.3.10}$$

which is exactly what we have as the basis for the frame. Please note that the * symbol between two functions or signals represents the convolution operator. We want also to remind the reader that we have already seen the sinc function as the Shannon wavelet in Chapter 3.

4.4   Wavelet Frames

If we define as L the operator of the transformation

$$L: f(t) \rightarrow \left\{ \left\langle f, \psi_{m,n} \right\rangle \right\} \tag{4.4.1}$$

we can characterize the signal $s(t)$ through its wavelet transform coefficients $<s, \psi_{m,n}>$ provided that the basis constructed by the mother wavelet and its daughters satisfies the frame bound condition:

$$A\|s\|^2 \leq \sum_m \sum_n \left| \left\langle s, \psi_{m,n} \right\rangle \right|^2 \leq B\|s\|^2 . \tag{4.4.2}$$

If $\{\psi_{m,n}\}$ constitutes at least a snug frame, the signal $s(t)$ can be reconstructed using

$$s(t) = \frac{2}{A + B} \sum_m \sum_n \left\langle s, \psi_{m,n} \right\rangle \psi_{m,n}(t) + \gamma \quad . \tag{4.4.3}$$

if $\{\psi_{m,n}\}$ is a tight frame, $A = B$ and $\gamma = 0$.

Using the wavelet family $\{\psi_{m,n}\}$ with m and n integers ($m, n \in Z$), we have given an algorithm to reconstruct the signal $s(t)$ from the wavelet coefficients, given that $\{\psi_{m,n}\}$ constitutes at least a snug frame. A new question now arises: How can we calculated the frame bounds of the wavelet family? Daubechies [Dau92] has proven that if

$$\psi_{mn}(t) = a_0^{-m\,2} \psi(a_0^{-m}t - nb_0) \tag{4.4.4}$$

with $m, n \in Z$, constitute a frame in $L^2(\mathbb{R})$, then the frame bounds condition can be written as

$$\frac{b_0 \ln a_0}{2\pi} A \le \int_0^\infty \frac{|\Psi(\omega)|^2}{\omega} d\omega \le \frac{b_0 \ln a_0}{2\pi} B \quad . \tag{4.4.5}$$

It is very important to note that this condition is clearly based on the fact that the mother wavelet must be admissible!

If $\{\psi_{m,n}\}$ is a *tight frame* ($A=B$), then

$$A = \frac{2\pi}{b_0 \ln a_0} \int_0^\infty \frac{|\Psi(\omega)|^2}{\omega} d\omega = B \tag{4.4.6}$$

in particular, if $\{\psi_{m,n}\}$ forms an orthornormal basis ($A=1$), then

$$\frac{b_0 \ln a_0}{2\pi} = \int_0^\infty \frac{|\Psi(\omega)|^2}{\omega} d\omega \tag{4.4.7}$$

and if $a_0 = 2$ and $b_0 = 1$, the above condition becomes

$$\int_0^\infty \frac{|\Psi(\omega)|^2}{\omega} d\omega = \frac{\ln 2}{2\pi} \qquad (4.4.8)$$

Note that the symmetric Equation also applies:

$$\int_0^\infty \frac{|\Psi(\omega)|^2}{\omega} d\omega = \int_{-\infty}^0 \frac{|\Psi(\omega)|^2}{\omega} d\omega \qquad (4.4.9)$$

Not all the choices for $\psi(t)$, a_0, and b_0 lead to Frames, even if $\psi(t)$ is admissible. Daubechies has calculated a general criterion of existence for frame bounds under the assumption of a square integrable set of wavelets. If $\psi(t)$ and a_0 are such that:

$$\min_{1 \le |\omega| \le a_0} \sum_{m=-\infty}^{+\infty} \left| \Psi\left(a_0^m \omega\right) \right|^2 > 0$$

$$\max_{1 \le |\omega| \le a_0} \sum_{m=-\infty}^{+\infty} \left| \Psi\left(a_0^m \omega\right) \right|^2 < \infty \qquad (4.4.10)$$

where $\min(x)$ and $\max(x)$ are functions that return the minimum and maximum values of the variable x, respectively.
If the function $\beta(s)$ defined as:

$$\beta(s) = \sup_{1 \le |\omega| \le a_0} \sum_{m=-\infty}^{+\infty} \left| \Psi\left(a_0^m \omega\right) \right| \left| \Psi\left(a_0^m \omega + s\right) \right| \qquad (4.4.11)$$

decays at least as fast as $(1+|s|)^{-(1+\epsilon)}$, with $\epsilon > 0$, then there exists a threshold value of b_0 (defined as $(b_0)_{th} > 0$, for which the $\psi_{m,n}$ constitute a frame for all choices of $b_0 < (b_0)_{th}$. For $b_0 < (b_0)_{th}$ the frame bounds are given by

$$A = \frac{2\pi}{b_0} \left\{ \inf_{1 \leq |\omega| \leq a_0} \sum_{m=-\infty}^{+\infty} \left| \Psi \left(a_o^m \omega \right) \right|^2 \right.$$

$$\left. - \sum_{(k=-\infty) \text{and} (k \neq 0)}^{\infty} \left[\beta \left(\frac{2\pi}{b_0} k \right) \beta \left(-\frac{2\pi}{b_0} k \right) \right]^{1/2} \right\}$$

$$B = \frac{2\pi}{b_0} \left\{ \sup_{1 \leq |\omega| \leq a_0} \sum_{m=-\infty}^{+\infty} \left| \Psi \left(a_o^m \omega \right) \right|^2 \right.$$

$$\left. - \sum_{(k=-\infty) \text{and} (k \neq 0)}^{\infty} \left[\beta \left(\frac{2\pi}{b_0} k \right) \beta \left(-\frac{2\pi}{b_0} k \right) \right]^{1/2} \right\} .$$

The frame bounds are thus a function of the scaling constant a_0 and the translation constant b_0. An extensive description of the effect of a_0 and b_0 on these bounds is given in [Dau92].

To summarize, if $\psi(t)$ is a function with reasonable decay in time and frequency (*decent* as defined by Daubechies in her book) and if $\psi(t)$ is admissible, then there exist a whole range of values for a_0 and b_0 such that the corresponding $\{\psi_{m.n}\}$ constitute a frame.

For $a_0 \to 1$ and $b_0 \to 0$, the discretization resembles the continuous wavelet transform. Thanks to the resolution of identity of $\psi(t)$, we already know that reconstruction is possible; hence, the family $\{\psi_{m.n}\}$ for $a_0 \to 1$ and $b_0 \to 0$ is clearly a frame.

In practice, it is very nice to have $a_0 = 2$, since it divides the frequency range into *octaves*. Unfortunately, this imposes stronger limitations on the choice of $\psi(t)$. In order to remedy this situation, without having to give up too much of the freedom in the choice of a mother wavelet $\psi(t)$, Grossmann, Kronland–Martinet, and Morlet proposed to decompose each octave into several *voices* (a name borrowed from music) by choosing $a_0 = 2^{1/M}$, where M indicates the number of voices per octave [Gro89]. With such a choice, we obtain

$$\psi_{m,n}^{M}(t) = 2^{-m/2M} \, \psi \, (2^{-m/M}t - nb_0) \quad .$$

By properly choosing the number of voices M, we can obtain a good decay and achieve $B/A - 1 \ll 1$ (snug frames). Such a decomposition, usually referred to as a *multivoice frame*, enables us to cover the range of scales in smaller steps in order to obtain a more "continuous" picture of the signal. As an example, for $M = 4$, the set of scales used is:

$$\{ \, a = ..., \, 1, \; 21/4, \, 21/2, \, 23/4, \, 2, \, 25/4, \, 23/2, \, 27/4, \, 4, \, ...\}$$

as opposed to the standard dyadic $(M = 1)$:

$$\{ \, a = ..., \, 1, \, 2, \, 4, \, ...\}$$

The multivoice frame can be described as using several different wavelets, $\psi^1, ..., \psi^M$ and to consider the frame:

$$\{\psi_{m,n}^{v}; \; m,n \in Z, \, v = 1,...,M\}$$

which leads to estimates for the frame bounds of multivoice frames. By choosing the different wavelets to have slightly staggered frequency localization centers, coupled with a good decay at ∞, one can achieve $B/A - 1 \ll 1$.

4.5   Examples of Wavelets Frames

As an example of wavelet frames, consider the so-called *Mexican hat function* illustrated in Figure 4.5.1. The expression for the Mexican hat function, normalized to have norm 1 ($\|\psi\|^2 = 1$) is

$$\psi(t) = \frac{2}{\sqrt{3}} \, \pi^{-1.4} \, (1 - t^2) \, e^{-t^2 2} \tag{4.5.1}$$

Table 4.5.1 Frame bounds for the Mexican hat wavelet				
b_0	M	A	B	A/B
1	1	3.22	3.59	0.89
1	2	6.78	6.87	0.98
1	3	10.17	10.27	0.99
1	4	13.58	13.69	0.99
0.25	4	54.55	54.55	1
0.50	4	27.27	27.27	1
0.75	4	18.18	18.18	1
1	4	13.58	13.69	0.99
1.25	4	10.2	11.61	0.88
1.5	4	6.59	11.59	0.57

In Table 4.5.1, the values of the frame bounds A and B are given as a function of the sampling constant b_0 and of the number of voices M. Please note that we have utilized a dyadic sampling for the scale ($a_0 = 2$). From the table, we can see that for $b_0 = 1$, it is very hard to achieve the condition of a tight frame ($A/B = 1$), given the fact that the distance between the maximum of $\psi(t)$ and its zero is equal to 1 hence, for $b_0 = 1$, we are heavily undersampling the function. Even with $M = 4$, we almost have a redundancy ratio of 1! On the other hand, with an $M = 4$ and a sampling constant b_0 less than 1, we obtain a redundancy factor A/B equal to 1, but the constants A and B are very large. Remember that the closer A and B are to 1, the more the frame resembles a orthonormal representation [Mey93]. Note that for $b_0 < 1$, the frame is a tight frame, but not orthonormal since $A = B \neq 1$. For fixed M, and b_0 small enough, so that the frame is almost tight, Table 4.5.1 also shows that $A \simeq B$ is inversely proportional to b_0 which fits the intuition understanding that for tight frames of normalized vectors, $A = B$ measures the

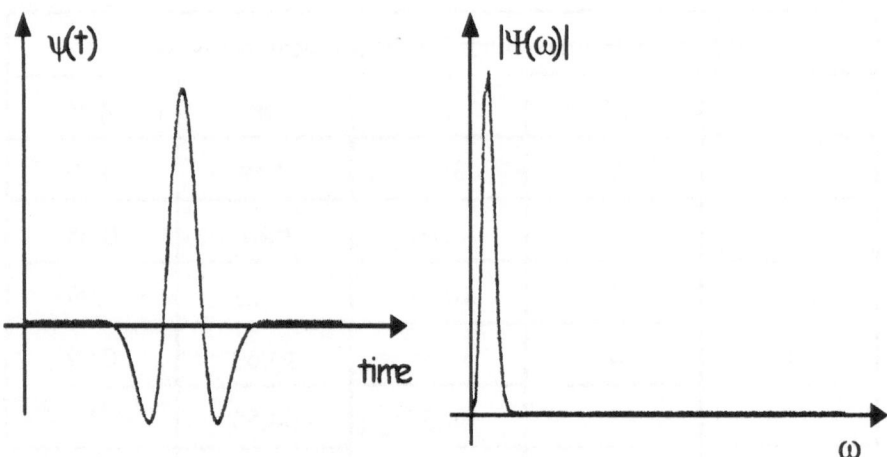

Figure 4.5.1 Mexican hat wavelet in time and frequency domain.

redudancy of the frame, which should indeed double if b_0 is reduced by half.

For another example of wavelet frames, consider the so-called *Morlet function* illustrated in Figure 4.5.2. The expression for the Morlet wavelet is given by

$$\psi(t) = \pi^{-1/4} \left(e^{-j\omega_0 t} - e^{-\omega_0^2/2} \right) e^{-t^2/2} . \tag{4.5.2}$$

Often, ω_0 is chosen so that the ration between the highest and second highest maximum of $\psi(t)$ is approximately 0.5; that is,

$$\omega_0 = \pi (2/\ln 2)^{1/2} \approx 5.3364... \tag{4.5.3}$$

In practice one chooses $\omega_0 = 0.5$, and the term $\exp(-\omega_0^2/2)$ can be neglected. The Fourier transform of the wavelet is

$$\Psi(\omega) \approx \pi^{-1/4} \left| e^{-(\omega-\omega_0)^2/2} \right| . \tag{4.5.4}$$

For $a_0 = 2$, the frame bounds as a function of the sampling b_0 and the voice number M are given in Table 4.5.2. For $b_0 = 1$, we need at least $M=4$ voices to obtain a frame bound condition which assures reconstruction. In practice, b_0 is much smaller than 1 or M is larger than 4. We will present more families of wavelet frames in the rest of the chapter.

b_0	M	A	B	A/B
		Table 4.5.1 Frame bounds for the Morlet wavelet		
0.5	2	6.01	7.82	0.77
1		3.00	3.92	0.76
1.5		1.94	2.67	0.72
2		1.17	2.28	0.51
2.5		0.48	2.28	0.21
0.5	4	13.8	13.8	1
1		6.92	6.92	1
1.5		4.54	4.68	0.97
2		3.01	3.91	0.76
2.5		1.70	3.82	0.44
3		0.59	4.01	0.14

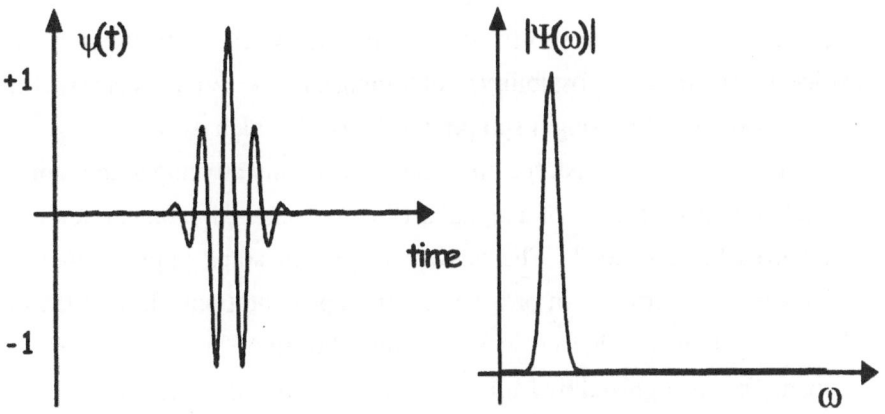

Figure 4.5.2 Morlet wavelet in time and frequency domain.

4.6   Time–Frequency Localization

Let us summarize the above discussion as follows [Hes96]. The discrete wavelet transform is obtained by discretizing the continuous wavelet transform using $a = a_0^m$ and $b = n\, a_0^m\, b_0$. If the frame bound condition is satisfied (i.e., we can determine the two constants A and B relative to the frame $\{\psi_{m,n}\}$). The choices for the constants a_0 and b_0 used to construct the frame $\{\psi_{m,n}\}$ determine the properties of the frame and, in particular, the existence of the *Inverse Discrete Wavelet Transform* (IDWT). A sampling grid with too coarse resolution obviously will not allow a "good" (or perfect) reconstruction. It is possible to choose a_0 and b_0 such that $A \sim B$; in this case, reconstruction of the signal is given by

$$s(t) \sim \frac{2}{A+B} \sum_m \sum_n \left\langle s, \psi_{m,n} \right\rangle \psi_{m,n}(t) \; . \qquad (4.6.1)$$

For tight frames $A = B$ and

$$s(t) = \frac{1}{A} \sum_m \sum_n \left\langle s, \psi_{m,n} \right\rangle \psi_{m,n}(t) \; . \qquad (4.6.2)$$

The frame forms an orthonormal basis if $A = B = 1$ and $\|\psi\|^2 = 1$. In many practical cases, a_0 is chosen to be equal to 2 ($a_0 = 2$). Unfortunately, this choice may restrict the possibility of obtaining a frame. A multivoice wavelet frame is obtained by using a sampling scheme in which $a_0 = 2^{1/M}$.

Now, we may well ask the question, Is there any advantage in having a redundant decomposition of a signal $s(t)$ using frames, in comparison to an orthonormal expansion? The answer is yes; in some applications, this redundancy is necessary in order to obtain representations close to the one obtained with the continuous wavelet transform (time–frequency analysis). Also, as first recognized by Morlet, redundancy also leads to robustness, in the sense that one could store the wavelet coefficients with low precision and still reconstruct the signal $s(t)$ with comparatively higher precision than what is normally obtained with a orthonormal decomposition. In fact, it has been shown that the reconstruction mean square error due to noise can be reduced by a factor which is precisely the oversampling rate [Coh96].

The redundancy and robustness derive from our previous vector representations of signals. In the Euclidean plane, an orthonormal basis is given by the two vector $u_1=(1,0)$ and $u_2=(0,1)$. Let us consider a tight frame given by the four vectors $a_1=(1/\sqrt{2},0)$; $a_2=(0,1/\sqrt{2})$; $a_3=(1/2,1/2)$, and $a_4=(1/2,-1/2)$. Any vector v can be written as

$$v=<v,u_1> u_1 + <v,u_2> u_2 \tag{4.6.3}$$

or

$$v=<v,a_1> a_1 + <v,a_2> a_2 + <v,a_3> a_3 + <v,a_4> a_4 \tag{4.6.4}$$

So far, we have discussed the limits on a_0 and b_0 for obtaining a frame, but nothing has been said about the range of the coefficients m and n necessary to obtain a practical representation of the signal $s(t)$. Clearly, we do not want to extend m and n to the complete set of $Z!$ The properties of localization in time and frequency of the frame $\{\psi_{m,n}\}$ generated using $\psi(t)$ derive from the properties of the mother wavelet itself. Assuming a mother wavelet $\psi(t)$ with mean time zero ($<t>=0$), and mean frequency ω_0, then the discretized daughter wavelet $\psi_{m,n}(t)$ will similarly be well localized around $<t_{mn}> = a_0^m n b_0$ in time, and $<\omega_m> = a_0^{-m} \omega_0$ in frequency.

Intuitively speaking, $<s, \psi_{m,n}>$ represents the *information content* in $s(t)$ near time $a_0^m n b_0$ and in the range of frequencies near $a_0^{-m} \omega_0$. If the value of $<s, \psi_{m,n}>$ is higher, then the features of the signal and the wavelet are better matched in that region of the time–frequency plane.

If the signal $s(t)$ is *essentially* localized in a time interval T and in a frequency range (Ω_1, Ω_2) then it is clear that we need to utilize values of m and n for which the generic sampling point $(a_0^m n b_0, a_0^{-m} \omega_0)$ belongs to the time–frequency domain of $s(t)$ (i.e., it lies within the region in the time-frequency plane defined by the interval $[-T, T] \times ([-\Omega_1, -\Omega_2] \cup [\Omega_1, \Omega_2])$. In practice, the sampling constant b_0 is defined by the sampling rate of the signal $s(t)$ (for $t = kb_0$). Also it is common to utilize $a_0 = 2$ in order to have a dyadic scale.

The choice of the mother wavelet will then affect the need for voices and thus the value of M as well as the range of values of the coefficient m. The range of the coefficient n is at this point defined from the other settings. Assuming that the wavelet has a support in time $(-T,T)$ and in frequency $(\omega_0 - B, \omega_0 + B)$, then the range of values of m and n are calculated as follows. The minimum scaling has to be such that the center frequency of the wavelet is close to the Nyquist frequency:

$$<\omega_m> = \omega_0 / 2^m \leq 2\pi / 2b_0 \qquad (4.6.5)$$

assuming that b_0 is the sampling period. It follows that

$$m_{min} \geq \log_2 (\omega_0 b_0 / \pi) . \qquad (4.6.6)$$

The maximum scaling is calculated considering the temporal support of the wavelet. If we define as T_w the maximum time window possible [for example, the length of the signal $s(t)$], the maximum scale will be limited by the width of the wavelet. Knowing that most of the wavelet energy is contained within a range $3T_m$ around $b = n\, a_0{}^m\, b_0$, then we must impose the conditions

$$T_w / 2 \geq 3\, T_m = 3\; 2^m\, T$$
$$m_{max} \leq \log_2 [(T_w / T) / 6] .$$

In this way, we can understand the time-frequency localization properties of wavelet frames.

4.7   Orthonormal Discrete Wavelet Transforms

In the case of orthonormal discrete wavelet transforms (DWT), the signal $s(t)$ is decomposed into a discrete set of wavelet coefficients using an orthonormal set of basis functions $\{\psi_{n.m}\}$. Let us consider the Discrete Wavelet Transform with $a_0 = 2$ and $b_0 = 1$; the wavelet functions can thus be written as:

$$\psi_{m,n}(t) = 2^{-m/2}\, \psi\,(2^{-m}t - n) = \frac{1}{\sqrt{2^m}}\, \psi\left(\frac{t - n2^m}{2^m}\right) \qquad (4.7.1)$$

and clearly $\psi_{0,0} = \psi(t)$. It is possible to construct a class of wavelets $\psi(t)$ such that the set of functions $\{\psi_{mn}(t)\}$ are orthonormal, i.e. they satisfy the following relationship:

$$\left\langle \psi_{m,n}\, ,\, \psi_{m',n'} \right\rangle = \int\limits_{-\infty}^{+\infty} \psi_{m,n}(t) \cdot \psi_{m',n'}(t)\, dt = \delta_{m,m'} \cdot \delta_{n,n'} \qquad (4.7.2)$$

where δ_{ij} is the *Kroneger delta function*:

$$\delta_{ij} = \left\{ \begin{array}{ll} 1, & \text{for } i = j \\ 0, & \text{for } i \neq j \end{array} \right. \qquad (4.7.3)$$

This condition implies that these wavelets are orthogonal to their dilated and translated versions (their dilates and translates). Furthermore, from frame theory, we know that the set of orthonormal functions $\{\psi_{mn}(t)\}$ form a complete orthonormal basis for all functions that are square−integrable [i.e., $f \in L^2(\mathbb{R})$ − function of finite energy].

As an example, consider the Haar basis, which is historically the first orthonormal wavelet basis (developed in 1910). The Haar wavelet $\psi(t)$ is defined as:

$$\psi(t) = \left\{ \begin{array}{cl} -1 & \text{for } 0 \leq t < 1/2 \\ +1 & \text{for } 1/2 \leq t < 0 \\ 0 & \text{elsewhere} \end{array} \right. \qquad (4.7.4)$$

Note that the Haar function does not have a good time−frequency localization: Its Fourier transform $\Psi(\omega)$ decays only as $|\omega|^{-1}$ for $\omega \to \infty$. Nevertheless, it is very useful for illustration purposes. If we impose $a_0 = 2$

and $b_0=1$, the set of Haar basis is defined as

$$\left\{ \psi_{m,n}(t) \right\} = \left\{ 2^{-m/2} \psi (2^{-m}t - n) \right\} . \tag{4.7.5}$$

Note that the following relationships are valid:

$$\psi_{00} = \psi(t), \quad \psi_{10} = \psi(t/2), \quad \psi_{-10} = \psi(2t), \quad \psi_{-11} = \psi(2t-1).$$

The wavelets represented in Equation 4.7.5 are plotted in Figure 4.7.1.

In order to prove that $\{\psi_{mn}\}$ constitutes an orthonormal basis, we need to establish two conditions:

a.   The functions $\psi_{mn}(t)$ are orthonormal.

b.   Any square integrable function can be approximated by a linear combination of the ψ_{mn}.

The first condition is easy established. The $\psi_{mn}(t)$ is not zero (has support) only in the time interval $[2^m n, 2^m(n+1)]$, hence two Haar wavelets of the scale must never overlap:

$$< \psi_{m,n} , \psi_{m,n'} > = \delta_{n\,n'} = \delta(n-n') . \tag{4.7.6}$$

For wavelets with different scales ($m \neq m'$ and $m > m'$) the overlap can exist, but it can be easily seen that the support of $\psi_{m'n'}$ lies always in a region for which ψ_{mn} is constant, hence the inner product of ψ_{mn} and $\psi_{m'n'}$ is then proportional to the integral of the function $\psi(t)$, which must be zero due to the admissibility condition.

Now, what happens if we consider a basis constructed with $a_0 = 2$ and $b_0 = \frac{1}{2}$? The function $\psi_{0\frac{1}{2}} = \psi(t - 1/2)$ is *not orthogonal* with ψ_{00}, but the set $\{\psi_{mn}\}_{\{a_0=2,\,b_0=1/2\}}$ does constitue a *tight frame*.

A different kind of orthonormal wavelet basis is the *Littlewood–Paley basis* (also known as the *Shannon wavelets*). This is an orthonormal wavelet basis with time–frequency properties complementary to the Haar basis; these wavelets are defined as

$$\psi(t) = \frac{\sin(2\pi t) - \sin(\pi t)}{\pi t}$$

$$\Psi(\omega) = \begin{cases} (2\pi)^{-1/2}, & \pi < |\omega| < 2\pi \\ 0, & \text{otherwise} \end{cases}$$

(4.7.7)

It is easy to check that the $\{\psi_{mn}\}$ constitutes indeed an orthonormal basis and that $\|\psi\|^2 = 1$ for all values of $m, n \in Z$. This basis has very good frequency localization, but it decays in time as $\psi(t) \sim |t|^{-1}$ for $t \to \infty$, which means poor time localization.

In the last 10 years, several other orthonormal wavelet bases have been constructed. These bases share the best features of both the Haar basis and the Shannon (or Littlewood–Paley) basis; that is, these new base have excellent localization properties both in time and frequency. We will list a few of them here, and the interested reader may look up details in the references. For example, in 1982, Stromberg contructed bases which have exponential decay in time and frequency. In 1985, Meyer constructed a base in which the $\Psi(\omega)$ is compactly supported in the frequency domain. In

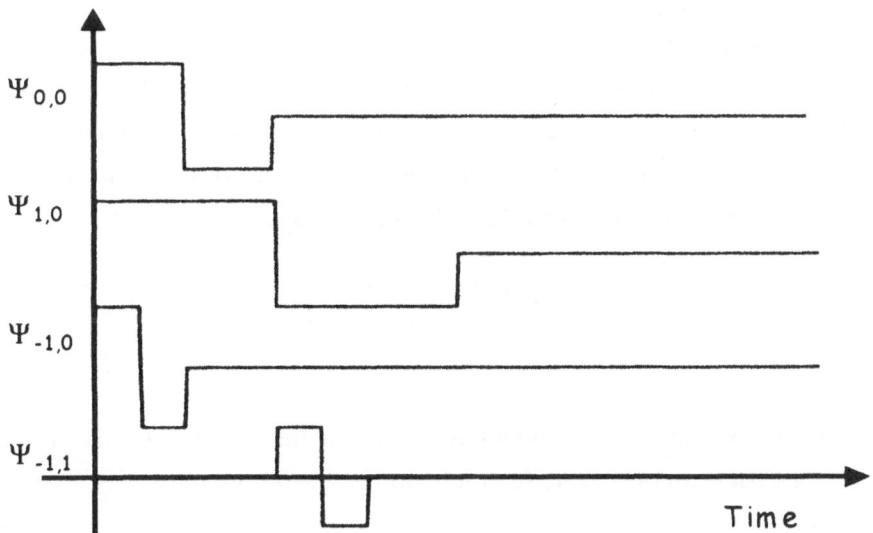

Figure 4.7.1 Haar wavelet basis plotted for different values of scale and time shift.

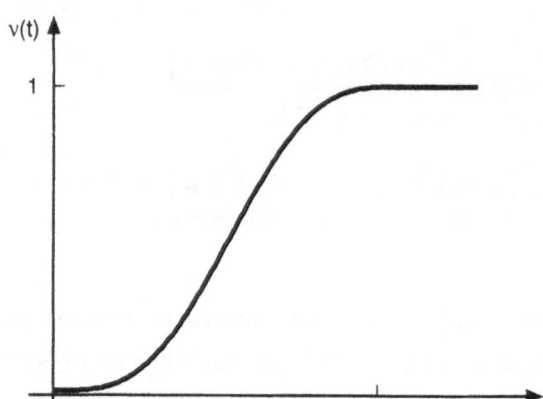

Figure 4.7.2 Plot of the function $n(t)$ described in Equations 4.7.8 and 4.7.5.

1987, Tchamitchian constructed the first example of biorthogonal wavelet bases. In 1987 and 1988, Battle and Lemarie independently constructed identical families of orthonormal wavelet bases with exponential decays. In 1986, Mallat and Meyer developed the *multiresolution analysis* framework which gives a satisfactory explanation of all previous constructions of bases and provides a tool for the construction of yet other bases. In 1989, Daubechies constructed a new family of compact supported wavelets and Coifman constructed a family of symmetric wavelets.

We will explain one of these constructions in more detail. The following construction was first developed by Meyer and was an extension of a tight frame developed in conjuction with Grossmann and Daubechies. Using a clever trick with the phase factors, Meyer was able to eliminate the "redundancy" of the frame and thus generate an orthonormal basis. Let $v(t)$ be a function which satisfies the following boundary conditions:

$$v(t) = \begin{cases} 0, & t \le 0 \\ 1, & t \ge 1 \end{cases} \tag{4.7.8}$$

Note that nothing is said about the value of the function in the interval $0<t<1$. An example of one such function is, for $0 < t < 1$:

$$v(t) = t^4 \left(35 - 84\, t + 70\, t^2 - 20\, t^3 \right) . \tag{4.7.9}$$

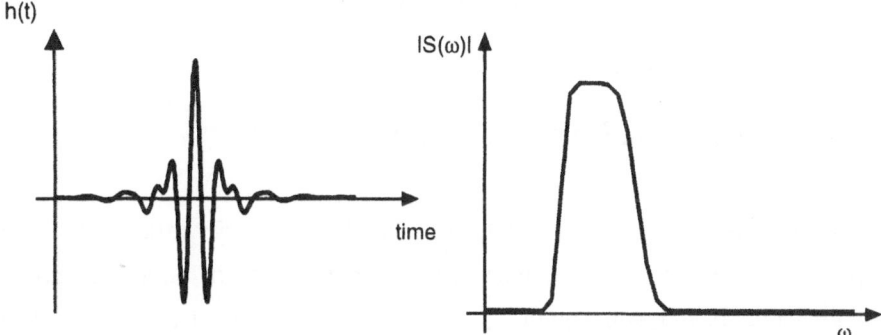

Figure 4.7.3 Meyer wavelet in time and frequency domain.

The function $v(t)$ is shown in Figure 4.7.2.

For an arrbitrary $a_0 > 1$ and $b_0 > 0$, we can define the function:

$$\Psi(\omega) = \begin{cases} (2\pi)^{-1/2} e^{j\omega/2} \sin\left[\frac{\pi}{2} v\left(\frac{3}{2\pi}|\omega|-1\right)\right], & \frac{2}{3}\pi \le |\omega| \le \frac{4}{3}\pi \\[3mm] (2\pi)^{-1/2} e^{j\omega/2} \cos\left[\frac{\pi}{2} v\left(\frac{3}{4\pi}|\omega|-1\right)\right], & \frac{4}{3}\pi \le |\omega| \le \frac{8}{3}\pi \\[3mm] 0, & \text{otherwise} \end{cases} \qquad (4.7.10)$$

and we add the property

$$v(t) + v(1-t) = 1 \qquad (4.7.11)$$

A plot of the Meyer wavelet is given in Figure 4.7.3.

In order to prove that the $\{\psi_{mn}\}$ constructed by Meyer constitute an orthonormal basis, we need to prove that the set verifies two properties: first, that $\psi(t)$ is admissible and $\|\psi\|^2 = 1$; second, that $\{\psi_{mn}\}$ constitutes a tight frame with frame bounds $A = B = 1$. The proof that Meyer's wavelets constitute an orthonormal basis relies on a number of "quasi–miraculous" cancellations, using the interplay between the phase of $\Psi(\omega)$ and the properties of the function $v(t)$. For the complete description of the proof, see

[Dau92]. It was clear that a different approach was needed to create wavelet basis. The breakthrough came with the work of Ingrid Daubechies, described in detail in the rest of this chapter [Dau88].

4.8   Multiresolution Analysis

Multiresolution analysis (MRA), formulated in 1986 by Mallat and Meyer, provided the natural framework for understanding the logic behind the wavelet basis and, subsequently, became the tool for constructing new ones. MRA can also be considered the link between the harmonic analysis (i.e., time–frequency analysis) and discrete signal processing, and, in particular, discrete multirate filter banks using conjugate or quadrature mirror filters [Kha93]. The concept of MRA is based on the analysis of a signal at different levels of resolution. Often, it is not necessary to observe a signal in great detail; a coarse approximation may be sufficient. For example, a low-pass version of the signal may be enough to estimate if the propagation channel is dispersive or not. Or, if we are looking for a particular section of a test sample under a microscope, it is more useful to look at all of the sample with a lower resolution (larger view) and only after we have found the right location to zoom in and look at the sample in greater detail. This concept is widely utilized in image processing and computer graphics, one example being downloading of images from the web. When a site containing hundreds of images is reached, it is common to download the thumbnail view of the images (a coarse version of the image which has a much smaller size and faster download time). Once we have decided which images are of interest, then we can download the full detailed file. It is interesting to note that the most recent versions of this thumbnail processing actually use wavelet-based image compression. Hence, it is not surprising that Mallat, while pursuing his Ph.D. in the field of image processing in 1986, formulated, along with Meyer, the mathematical framework for using MRA in the construction of wavelet bases. The wavelet decomposition obtained using MRA is a successive approximation method which adds more and more projections into "detail" spaces spanned by the wavelets and their shifts

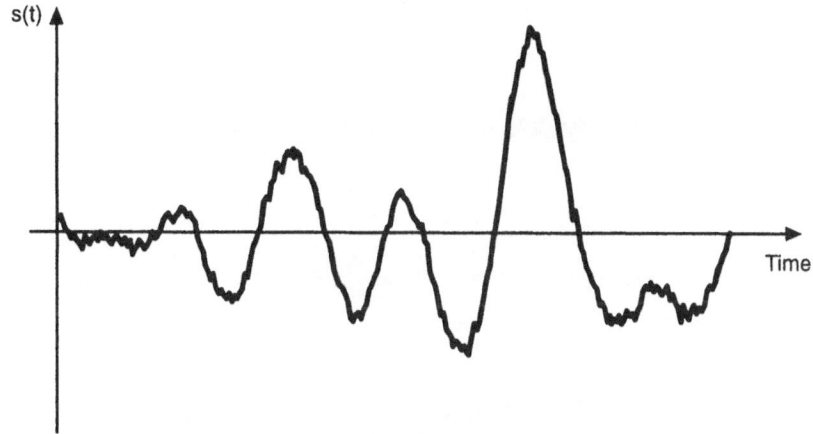

Figure 4.8.1 Test signal used to explain multiresolution analysis.

at different scales.

Given a signal $s(t)$ that belongs to this space and assuming that the set of wavelet functions $\psi_{mn}(t)$ constitutes an orthonormal basis for all square-integrable functions $[L^2(\mathbb{R})]$. The reconstruction formula is

$$s(t) = C \sum_{m=-\infty}^{+\infty} \sum_{n=-\infty}^{+\infty} \langle s, \psi_{n,m} \rangle \psi_{n,m}(t) \qquad (4.8.1)$$

where C is a constant. Without any loss of generality, we can assume that $b_0 = 1$. The signal $s(t)$ can thus be seen by the sum of a series of signals $d_m(t)$ calculated as

$$d_m(t) = C \sum_{n=-\infty}^{+\infty} \langle s, \psi_{n,m} \rangle \psi_{n,m}(t) \qquad (4.8.2)$$

and thus Equation 4.8.1 can be written as

$$s(t) = C \sum_{m=-\infty}^{+\infty} d_m(t) \ . \qquad (4.8.3)$$

Assuming the signal $s(t)$ shown in Figure 4.8.1, let us calculate the $d_m(t)$ terms in Equation 4.8.2 using the Haar wavelet. The respective $d_m(t)$ for

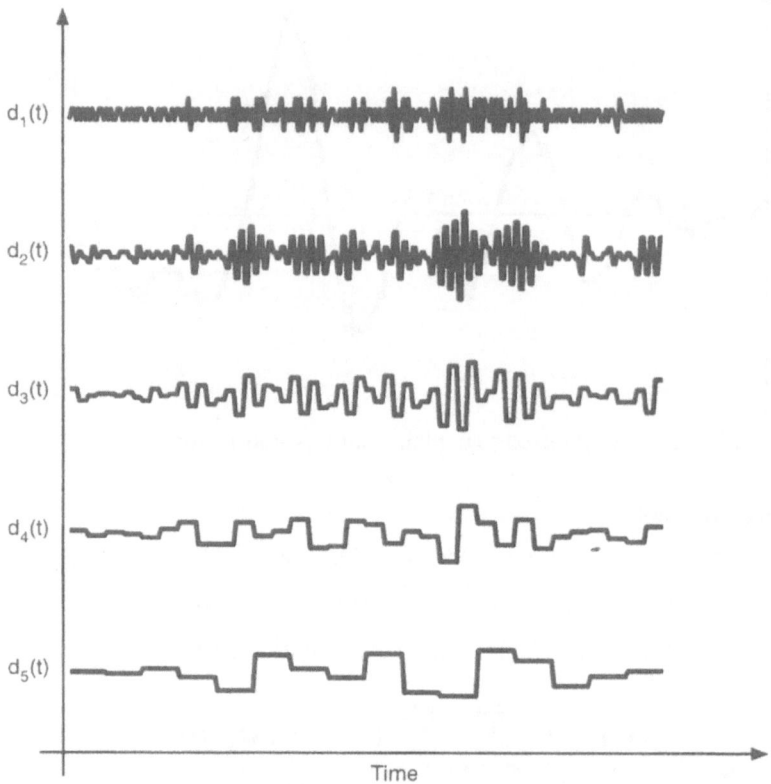

Figure 4.8.2 Successive $d_m(t)$ of the test signal on Figure 4.8.1. An Haar wavelet was used to calculate the detail signals. The high frequency information of the signal is contained in $d_1(t)$ and lower components are displayed in $d_5(t)$.

$m = 1, 2, 3, 4,$ and 5 is shown in Figure 4.8.2. As we have assumed that $b_0=1$, the Haar wavelet for $m = 0$ is only one step wide, two steps wide for $m = 1$, four for $m = 2$, and so forth.   We thus expect to see the highest-frequency components of the signal in Figure 4.8.1 for the lowest values of m, and the low-frequency components for the higher values of the dilation coefficient m. Equation (4.8.3) can thus be re-written as:

$$s(t) = C \sum_{m=1}^{+\infty} d_m(t) \ . \qquad (4.8.4)$$

What about the highest value of m? As the wavelet will always be zero for $\omega = 0$ as per the admissibility condition, in order to approximate the dc value of the signal, we will have very large values of m.

Mallat, when looking at the inverse wavelet transform, realized that the $d_m(t)$ of Equation 4.8.2 represents the projection of the signal $s(t)$ on the space defined by the $\psi_{mn}(t)$ daughter wavelet [Mal89]. If, as we have done with the Haar wavelet in our example, we use an orthonormal set of wavelets basis, reconstruction can be seen as the sum of successive projections of the signal $s(t)$. If we start adding projections, or details as defined in the signal and image processing community, with large values of m, we will obtain an approximate representation of $s(t)$, as shown in Figure 4.8.3, for $s_5(t)$, where

$$s_5(t) = \sum_{m=5}^{+\infty} d_m(t) \ . \tag{4.8.5}$$

A better approximation of the signal is given in Figure 4.8.3, with $s_4(t)$:

$$s_4(t) = \sum_{m=4}^{+\infty} d_m(t) = d_4(t) + s_5(t) \tag{4.8.6}$$

and even better for $s_3(t)$, $s_2(t)$ and, finally, $s_1(t)$. In each case we have

$$s_k(t) = d_k(t) + s_{k+1}(t) \tag{4.8.7}$$

and, finally, we have $s(t) = s_0(t)$, for $b_0 = 1$, in Figure 4.8.3.

Equation 4.8.7 is extremely important because it properly determines that it is not necessary to extend the summation in Equation 4.8.4 to $+\infty$, but instead only to a level L for which the approximation $s_k(t)$ is enough and no more decomposition is necessary.

We can thus summarize the above discussion using a new set of reconstruction Equations:

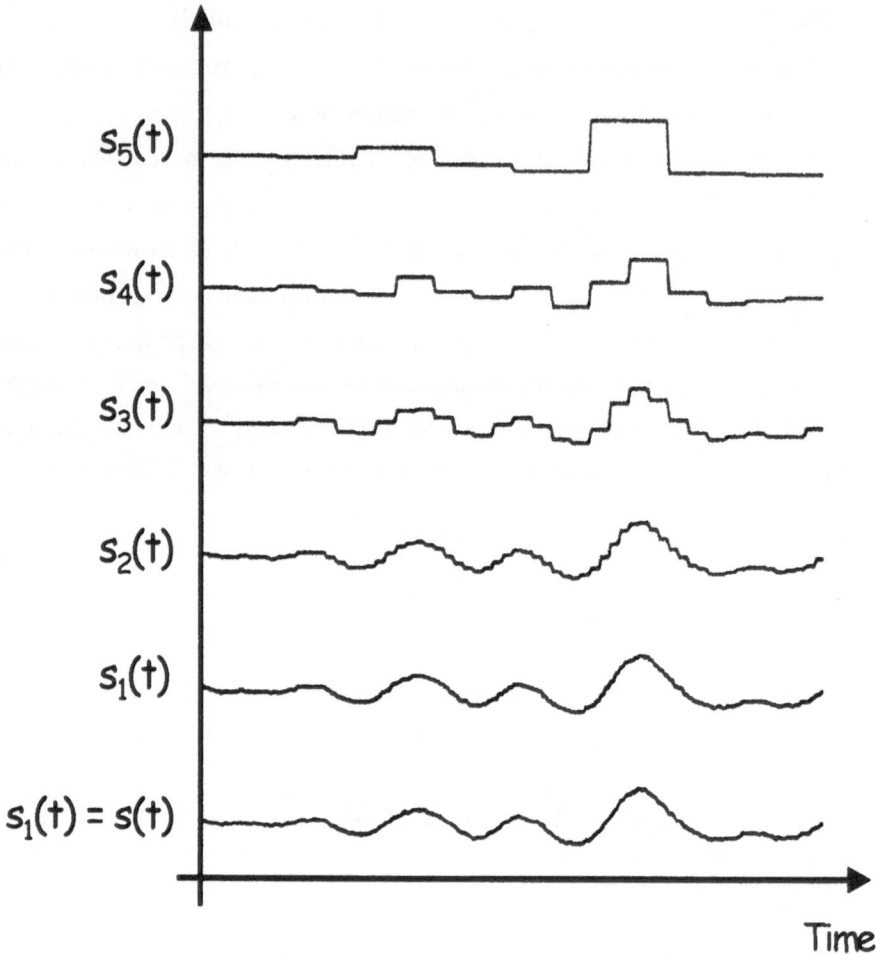

Figure 4.8.3 Successive approximations $s_k(t)$ of the test signal on Figure 4.8.1. As the level of detail information $d_k(t)$ is added to the reconstruction, the approximation resembles the original signal more.

$$s(t) = \sum_{m=0}^{L} d_m(t) + s_{L+1}(t) \qquad (4.8.8)$$

where $d_m(t)$ are given by Equation 4.8.2 and

$$s_{L+1}(t) = C \sum_{m=L+1}^{+\infty} \sum_{n=-\infty}^{+\infty} \langle s, \psi_{n,m} \rangle \, \psi_{n,m}(t) \; . \qquad (4.8.9)$$

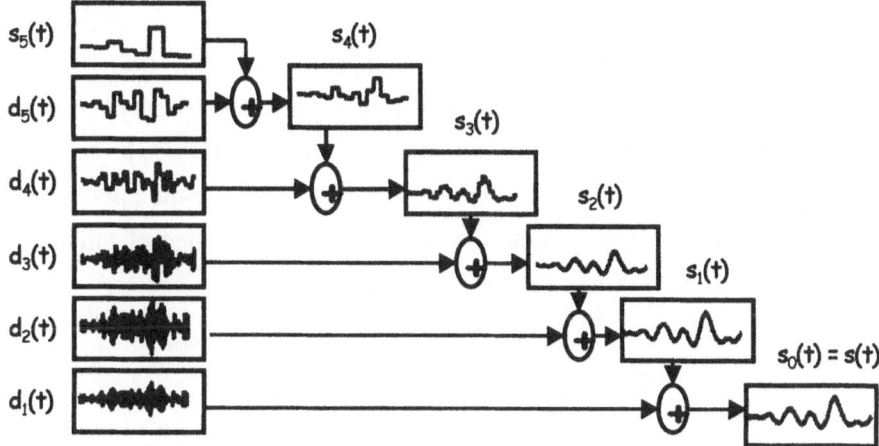

Figure 4.8.4 Block diagram of the signal synthesis described in the text.

To summarize, we have established that a signal can be represented by a coarser or approximate signal, provided that detail signals at higher resolution are added as more resolution is needed. This structure of the signal synthesis is shown in Figure 4.8.4 for the example previously discussed. Using Equation 4.8.7, we can also represent the synthesis filter bank as a cascade of filters that can be used to extract signal synthesized at different levels of resolution.

A note regarding the lower limit of the summation in m: By starting the summation for $m=0$, we are imposing that the wavelet $\psi(t)$ represents the filter with the highest frequency content of the signal. In order to have an orthonormal frame, the signal must be band-limited, and it is always possible to define a new wavelet function $h(t)$ such that $h(t) = \psi(t/2^{MAX})$, where MAX is the value of the scaling m that satisfies the condition of highest-frequency content. The multiresolution analysis is then performed using the basis functions $h_{mn}(t)$.

The above discussion represents the basis of MRA, in which a signal is computed as the sequence of various approximations of the signal itself at different resolutions. We will now introduce the proper definition of MRA and develop expressions for new orthonormal wavelet bases. In the field of computer vision, Burt and Adelson introduced an MRA pyramidal scheme that can be used to process a low-resolution image first and then selectively increase the resolution when necessary. The approximation of a signal $s(t)$ at a resolution m_0 is the equivalent to representing the signal by local averages over a range of size proportional to 2mo. An MRA approximation is thus composed of embedded grids of approximations [Djo97].

Mathematically, the basic idea of MRA is to divide the space of all possible square integrable functions $L^2(\mathbb{R})$ into a sequence of embedded subspaces V_j which satisfy the following properties:

$$\cdots V_2 \subset V_1 \subset V_0 \subset V_{-1} \subset V_{-2} \cdots \tag{4.8.10}$$

$$\bigcup_{j \in Z} V_j = L^2(\mathbb{R}) \tag{4.8.11}$$

$$\bigcap_{j \in Z} V_j = \{0\} \ . \tag{4.8.12}$$

The above Equations describe a set of subspaces V_j. For example, let us take the signal $s(t)$ in Figure 4.8.5 and its projection in the space V_0 (Figure 4.8.6a) defined as samples of s(t) every period to T_0:

$$\text{Proj}_{V_0} s(t) = s(kT) , \qquad k \in Z . \tag{4.8.13}$$

Now we define in Figure 4.8.6b the projection in the space V_1 as:

$$\text{Proj}_{V_1} s(t) = s(l2T) , \qquad l \in Z . \tag{4.8.14}$$

We can easily verify that the values $s(2lT)$ are also represented in the $s(kT)$ for $k = 2l$. This means that the signal sampled every $T(V_0)$ contains the signal sampled at a sampling rate of $2T$ (V_1). So we can say that the space V_1 is contained in V_0 (i.e., $V_1 \subset V_0$). The projection of a function on V_1 is thus contained in the projection of the same function on V_0. The second Equation requires that the choice of the subspaces V_j is such that all square-integrable functions can be represented, while the third condition implies orthogonality.

It is then possible to project a signal $s(t)$ onto the various subspaces. Formally, we can write that the approximation of a signal at a resolution m is defined as the orthogonal projection on a space $V_m \subset L^2(\mathbb{R})$.

The Multiresolution Theorem by Mallat and Meyer states that, given a sequence $\{V_m\}$ where m is an element of a closed subspace of $L^2(\mathbb{R})$, the sequence defines a multiresolution approximation of a function $f(t)$ if the following six properties are satisfied.

Property 1:

$$\forall\, m, n \in Z,\ f_m \in V_m \ \leftrightarrow\ f_m(t - 2^m k) \in V_m . \qquad (4.8.15)$$

This means that if $f_m(t)$ is an element of V_m, also its translated version $f_m(t - 2^m k)$ belongs to V_m. Note that not all possible translations are acceptable, only those multiples of 2^m.

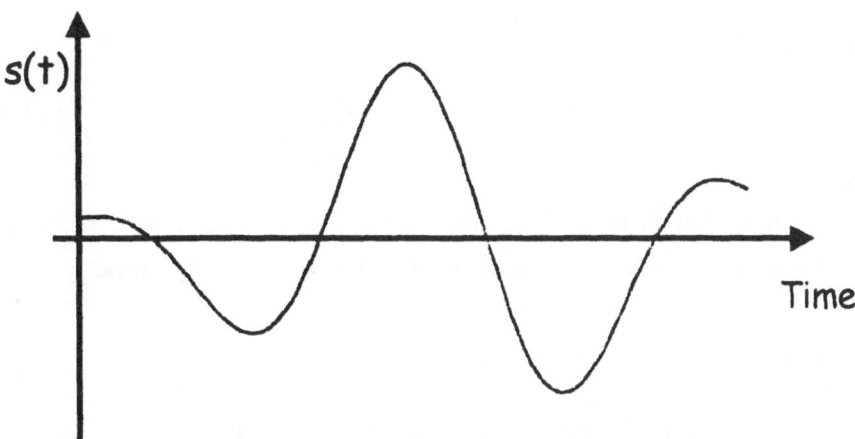

Figure 4.8.5 Test signal used to explain projections onto subpaces.

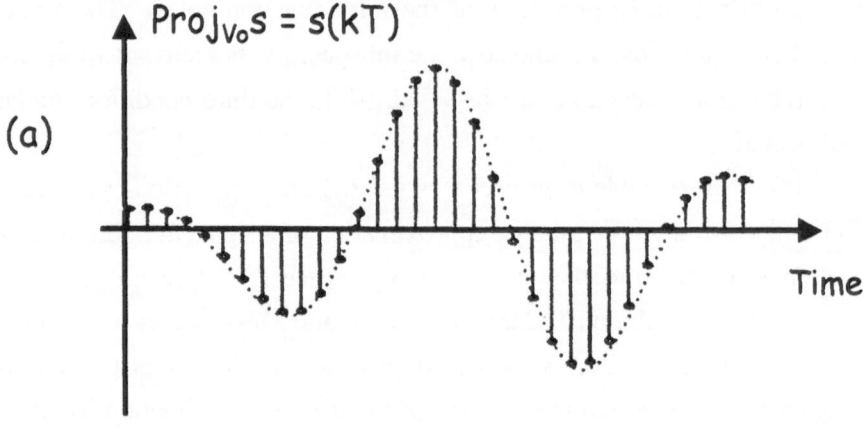

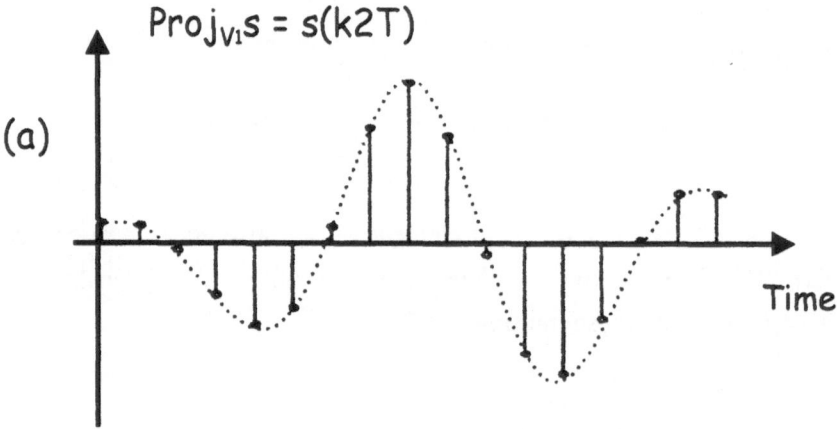

Figure 4.8.6 (a) Projection of the signal $s(t)$ onto the subspace V_0. (b) Projection of the signal $s(t)$ onto the subspace V_1.

Property 2:

$$\forall m \in Z, \quad V_{m+1} \subset V_m .$$
(4.8.16)

This means that the signal approximation at a scale m can also be considered a signal approximation at a scale $m+1$. This is a causality property.

Property 3:

$$\forall m \in Z, \quad f(t) \in V_m \quad \leftrightarrow \quad f(t/2) \in V_{m+1} .$$
(4.8.17)

Dilating the function in V_m by a factor of 2 enlarges the details and again can be considered an approximation at a scale $m+1$.

Property 4:

$$\lim_{m \to \infty} V_m = \bigcap_{m-\infty}^{+\infty} V_m = 0 \qquad (4.8.18)$$

If m tends to infinity, then the resolution 2^{-m} goes to zero and thus the projection of $f(t)$ on V_m as m tends to infinity is zero since we have lost all the details; in other words,

$$\lim_{m \to \infty} \| P_{V_m} f \| = 0 \qquad (4.8.19)$$

where P_{Vm} represents the projection of $f(t)$ onto the space V_m .

Property 5:

$$\lim_{m \to \infty} V_m = \text{Closure} \left(\bigcup_{m=-\infty}^{+\infty} V_m \right) = L^2(\mathbb{R}) \qquad (4.8.20)$$

When the resolution 2^{-m} goes to infinity, we can perfectly represent the signal; hence, the signal approximation is equivalent to $f(t)$.

Property 6:

Given a subspace V_0, there exist a function $\phi(t)$ so that $\{\phi(t-n)_{n\in Z},$ is an orthonormal basis for V_0. Furthermore, we will impose that the function $\phi(t)$ is such that

$$\phi_{m,n}(t) = 2^{-m/2} \phi(2^{-m}t - n) \qquad (4.8.21)$$

is a orthonormal basis of V_m. This is equivalent to saying that the family of functions $\{\phi_m(t) = \phi\,(t/2^m)\}_{m \in Z}$ represents a Riesz basis. A family of vectors $\{u_n\}$ is said to be a Riesz basis of the space $\mathcal{H}$ if it is linearly independent and there exist two constants $A > 0$ and $B > 0$ such that for any function $f \in \mathcal{H}$, we can represent the function as a linear combination of the bases:

$$f = \sum_{m=0}^{+\infty} c_n\, u_n \tag{4.8.22}$$

such that

$$\frac{1}{A}\,|f\,|^2 < \sum_{n=0}^{+\infty} |c_n|^2 < \frac{1}{B}\,|f\,|^2 \ . \tag{4.8.23}$$

Not only does the above condition ensure that a dual basis function $\{v_n\}$ exists and is such that

$$c_n = <f, v_n> \tag{4.8.24}$$

but also the signal can be represented in both bases as

$$f = \sum_{m=0}^{+\infty} <f, v_n> u_n = \sum_{m=0}^{+\infty} <f, u_n> v_n \tag{4.8.25}$$

and the dual basis v_n is such that:

$$A\,|f\,|^2 < \sum_{n=0}^{+\infty} | <f, u_n> |^2 < B\,|f\,|^2 \ . \tag{4.8.26}$$

Equations 4.8.23 and 4.8.26 are statements of energy equivalence, which guarantee that the signal expansions over the bases are numerically stable. It can be shown that if a basis function $\{\phi(t-n)\}$ is a Riesz basis for V_0, then also $\{\phi(2^{-m}t-n)\}$ is a Riesz basis for V_m with the same Riesz bounds A and

B at every scale 2^m.

The necessary and sufficient condition of Equation 4.8.23 can also be expressed in the Fourier domain. A family $\{\phi(t-n)\}$ is a Riesz basis of the space V_0 that it generates, if and only if there exist two bounds $A > 0$ and $B>0$ such that

$$\frac{1}{B} < \sum_{k=0}^{+\infty} |\Phi(\omega - 2\pi k)|^2 < \frac{1}{A} \qquad (4.8.27)$$

As an example, let us consider the Haar approximation. This is also called the piecewise continuous approximation, as it can be defined as a simple MRA approximation composed of piecewise constant functions. The space V_1 is the set of all functions $\{\phi\}$ within $L^2(\mathbb{R})$ such that $\phi(t)$ is constant for *t*, an element of the set $\{n2^j, (n+1)2^j\}$. The approximation at a resolution 2^{-j} is the closest piecewise constant function of intervals with size 2^j. The resolution cell can be chosen to be a box window. Clearly, $V_j \subset V_{j-1}$ since functions constant on intervals of size 2^j are also constant on intervals of size 2^{j-1}. Unfortunately, the piecewise constant functions result in approximations of the signals which are not smooth but have discontinuities; hence, the Haar approximation is used mainly for tutorial purposes.

Another more practical example is the Shannon approximation, which is piecewise constant in the frequency domain. This is possible thanks to the duality between time and frequency; we can construct band-limited functions in the space V_j, defined as the set of functions whose Fourier transforms have support within $\{-2^{-j\pi}, 2^{-j\pi}\}$. The function

$$\phi(t) = \frac{\sin(\pi t)}{\pi t} \qquad (4.8.28)$$

also provides an orthonormal basis of V_0.

A final example is the polynomial spline approximation, which constructs smooth approximations with fast asymptotic decay. The space V_j

of splines of degree $m > 0$ is the set of functions that are $(m-1)$ times continuously differentiable and equal to a polynomial of degree m on any interval $\{n^{2j}, (n+1)^{2j}\}$. From this definition, we can see that the piecewise constant approximation is equivalent to a spline approximation in the case of $m = 0$. If $m = 1$, the functions are thus piecewise linear and continuous. A Riesz basis of polynomial splines is the one constructed using box splines [i.e., using the Haar function convolved $(m+1)$ times]. In the frequency domain, the time convolution corresponds to a product; the Fourier transform of the function is

$$\Phi(\omega) = \left[\frac{\sin(\omega/2)}{\omega/2}\right]^{m-1} e^{-i\epsilon\omega/2} \tag{4.8.29}$$

note that i represents the imaginary number $i = (-1)^{\frac{1}{2}}$; if m is even then $\epsilon = 1$ and the function has support centered at $t = 1/2$. If m is odd, then $\epsilon = 0$ and the function is symmetric about $t = 0$.

4.9   Scaling Functions

Multiresolution Analysis thus ensures that if the properties given in the previous section are satisfied, there exists an orthonormal basis $\varphi_{m,n}(t)$ such that the signal detail on the subspace V_m is given by:

$$d_m(t) = \sum_{n=-\infty}^{+\infty} <s, \varphi_{m,n}> \cdot \varphi_{m,n} \tag{4.9.1}$$

and the projection of the signal onto the space V_m is linked to the projection on the space V_{m+1} by

$$P_m s = P_{m+1} s + d_{m+1} . \tag{4.9.2}$$

Equations 4.9.1 and 4.9.2 are similar to Equations 4.8.2 and 4.8.7, with the substitution of $\varphi = \psi$.

The approximation of a signal s at the resolution 2^{-m} is defined as the orthogonal projection on V_m, and it is represented as $P_{Vm} s = P_m s$. To compute this projection, we must find an orthonormal basis for V_m. It is interesting to find out that we can construct an orthogonal basis for each space V_m by dilating and scaling a single function $\phi(t)$ similar to the wavelet function. The function $\phi(t)$ is called the *scaling function*.

Assuming that the Riesz basis of the space V_0 is the family of functions $\{\theta(t-n)\}_{n \in Z}$, then $\{\theta(2^{-m}t - n)\}_{n \in Z}$ is a Riesz basis for V_m, and we have

$$\frac{1}{B} \leq \sum_{k=-\infty}^{+\infty} |\theta(\omega - 2\pi k)|^2 \leq \frac{1}{A} \qquad (4.9.3)$$

We can define the scaling function $\phi(t)$ using its Fourier transform $\Phi(\omega)$ and the following relationship:

$$\Phi(\omega) = \frac{\theta(\omega)}{\left(\displaystyle\sum_{k=-\infty}^{+\infty} |\theta(\omega - 2\pi k)|^2\right)^{1/2}} \qquad (4.9.4)$$

and the family of functions $\{\phi_{m,n}\}_{n \in Z}$ calculated as

$$\phi_{m,n}(t) = \frac{1}{\sqrt{2^m}} \cdot \phi\left(\frac{t-n}{2^m}\right) \qquad (4.9.5)$$

is then an orthonormal basis for V_m for all $m \in Z$.

The orthogonal projection of s over V_m is thus obtained with an expansion in the scaling orthogonal basis:

$$P_{V_m} s = \sum_{n=-\infty}^{+\infty} \langle s, \phi_{m,n} \rangle \phi_{m,n}(t) \qquad (4.9.6)$$

The inner products $a_{m,n} = <s, \phi_{m,n}>$ provide a discrete approximation at the scale 2^m. The inner product can also be written as a convolution integral:

$$a_{m,n} = \int_{-\infty}^{+\infty} s(t) \frac{1}{\sqrt{2^m}} \phi\left(\frac{t - 2^m n}{2^m}\right) dt = f * \phi_{m,n} \qquad (4.9.7)$$

where

$$\phi_m(t) = \sqrt{2^{-m}} \phi\left(2^{-m}t\right) \qquad (4.9.8)$$

Typically, the energy of the $\Phi(\omega)$ is limited to $|\omega| < \omega_1$ hence, the energy of the function $\Phi_m(\omega)$ is concentrated in $|\omega| < \omega_1 / 2^m$. The discrete approximation a_{mn} is therefore a low–pass filtering of the signal s sampled at intervals 2^m.

Let us look for example at a scaling function created using the polynomial spline functions in Equation 4.8.29, and substituting in Equation 4.9.3, we obtain

$$\Phi(\omega) = \frac{e^{-i\frac{\varepsilon\omega}{2}}}{\omega^{m+1}\sqrt{S_{2m+2}(\omega)}} \qquad (4.9.9)$$

where

$$S_n(\omega) = \sum_{k=-\infty}^{+\infty} \frac{1}{(\omega + 2k\pi)^n} \qquad (4.9.10)$$

and $\varepsilon = 1$ if m is even, and $\varepsilon = 0$ if m is odd. A closed-form expression of $S_{2m+2}(\omega)$ is obtained by computing the derivative of order $2m$ and we have for $m = 0$,

$$S_2(\omega) = \sum_{k=-\infty}^{+\infty} \frac{1}{(\omega + 2k\pi)^2} = \frac{1}{4 \sin^2\left(\dfrac{\omega}{2}\right)} \qquad (4.9.11)$$

We can calculate the scaling function:

$$\Phi(\omega) = \frac{e^{-i\frac{\varepsilon\omega}{2}}}{\omega^{m+1}\sqrt{S_2(\omega)}} = \frac{\sin\left(\dfrac{\omega}{2}\right)}{\left(\dfrac{\omega}{2}\right)} \qquad (4.9.12)$$

For $m = 0$, the spline is constant (i.e., the scaling function represented in Equation 4.9.12, is correspondent to the Haar wavelet function).

For $m = 1$, $n = 4$:

$$S_4(\omega) = \frac{1 + 2\cos^2\left(\dfrac{\omega}{2}\right)}{48 \sin^4\left(\dfrac{\omega}{2}\right)} \qquad (4.9.13)$$

and we can calculate the scaling function

$$\Phi(\omega) = \frac{4\sqrt{3} \sin^2\dfrac{\omega}{2}}{\omega^2\left(1 + 2\cos^2\dfrac{\omega}{2}\right)} \qquad (4.9.14)$$

Meyer's wavelet basis also fits this scheme [Mey89]. The idea behind Meyer's wavelet is to soften the ideal sinc wavelet. Again, we need to define a scaling function that satisfies the orthogonality and scaling requirements of the MRA. To do this, we define a smooth function in frequency that still satisfies the orthonormality condition in the Fourier domain, also known as Poisson's condition:

$$\sum_{k=-\infty}^{+\infty} |\Phi(\omega + 2\pi k)|^2 = 1 \ . \tag{4.9.15}$$

Let $v(t)$ be a function which satisfies the following boundary conditions:

$$v(t) = \begin{cases} 0 \ , & \text{for } t \le 0 \\ 1 \ , & \text{for } t \ge 1 \end{cases} \tag{4.9.16}$$

and such that

$$v(t) + v(1 - t) = 1 \ , \quad \text{for } 0 < t < 1 \ . \tag{4.9.17}$$

Many functions satisfy these conditions, for example

$$v(t) = \begin{cases} 0 \ , & t \le 0 \\ 3t^2 - 2t^3 \ , & 0 < t < 1 \\ 1 \ , & t \ge 1 \end{cases} \tag{4.9.18}$$

The multiresolution analysis of the signal $s(t)$ is thus completely defined by the scaling function $\phi(t)$ that generates the orthogonal basis for the subspaces V_m. If $\phi(t)$ is an orthonomal basis in V_0, then $1/\sqrt{2}\ \phi(t/2)$ is an orthonomal basis in V_1. But any function that belongs in V_1 can be represented by a linear combination of the basis of V_2, hence:

$$\frac{1}{\sqrt{2}} \phi\left(\frac{t}{2}\right) = \sum_{n=-\infty}^{+\infty} h_n\ \phi(t - n) \tag{4.9.19}$$

where

$$h_n = \left\langle \frac{1}{\sqrt{2}} \phi\left(\frac{t}{2}\right), \phi(t-n) \right\rangle \qquad (4.9.20)$$

The discrete sequence h_n represents the projection of the function $1/\sqrt{2}\phi(t/2)$ onto the basis $\phi(t-n)$. In signal processing, h_n can be seen as the taps or coefficients of a discrete filter. We will now show that these are, in many cases, the coefficients of a special kind of filter called a quadrature mirror filter. This is the link with subband theory.

In the frequency domain, Equation 4.9.20 becomes

$$\Phi(2\omega) = \frac{1}{\sqrt{2}} H(\omega)\Phi(\omega) \qquad (4.9.21)$$

where

$$\Phi(\omega) = \mathcal{F}\{\phi(t)\} \qquad (4.9.22)$$

and $H(\omega)$ is the discrete Fourier transform of h_n

$$H(\omega) = \sum_{n=-\infty}^{+\infty} h_n e^{-jn\omega} \qquad . \qquad (4.9.23)$$

Equation 4.9.21 can also be written as

$$\Phi(\omega) = \frac{1}{\sqrt{2}} H\left(\frac{\omega}{2}\right)\Phi\left(\frac{\omega}{2}\right) \qquad (4.9.24)$$

and using again Equation 4.9.21

$$\Phi(\omega) = \frac{H(\omega)}{\sqrt{2}} \frac{H(\omega)}{\sqrt{2}} \Phi\left(\frac{\omega}{4}\right) \qquad (4.9.25)$$

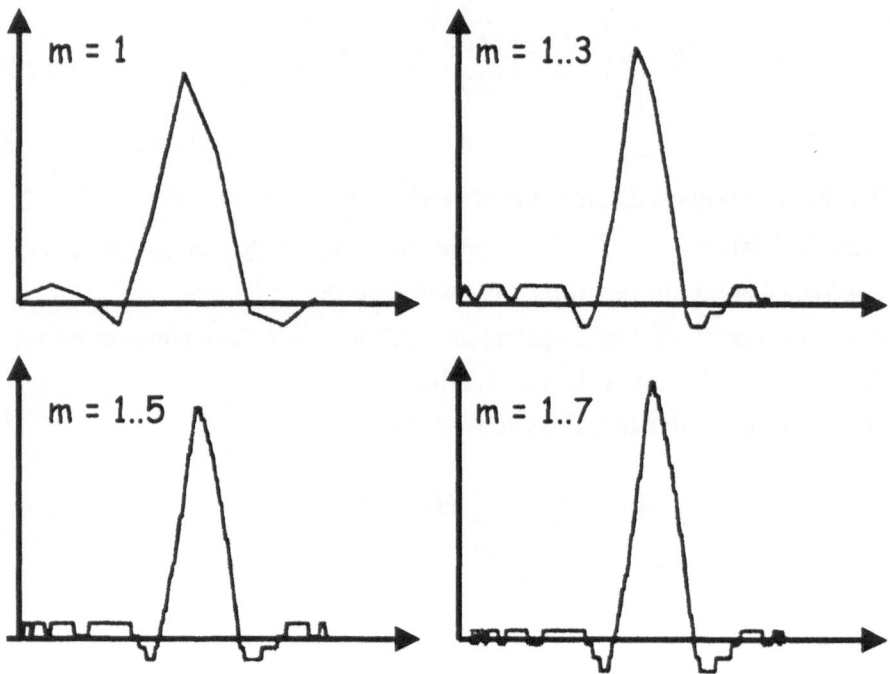

Figure 4.9.1 Plot of the scaling function for increasing values of m in Equation 4.9.30.

$$\Phi(\omega) = \left(\prod_{m=1}^{M} \frac{H(\omega/2^m)}{\sqrt{2}} \right) \Phi\left(\frac{\omega}{2^M} \right) \qquad (4.9.26)$$

If we choose a value of M large enough that

$$\lim_{M \to \infty} \Phi\left(\frac{\omega}{2^M} \right) = \Phi(0) \qquad (4.9.27)$$

then

$$\Phi(\omega) = \left(\prod_{m=1}^{M} \frac{H(\omega/2^m)}{\sqrt{2}} \right) \cdot \Phi(0) \qquad (4.9.28)$$

Mallat and Meyer [Mey89a] proved in a theorem that if the filter defined by the set of coefficients $\{h_n\}$ satisfies the condition

$$|H(\omega)|^2 + |H(\omega + \pi)|^2 = 2 \qquad (4.9.29)$$

and $H(0) = 2^{\frac{1}{2}}$, then

$$\Phi(\omega) = \prod_{m=1}^{+\infty} \frac{H(\omega/2^m)}{\sqrt{2}} \qquad (4.9.30)$$

For the proof of the theorem, we refer to [Mal98]. A plot of the scaling function $\phi(t)$ for four different values of upper limit in the product in Equation 4.9.30, is given in Figure 4.9.1. It is easily observed that as the number of factors in the product increases, higher frequency details are added. The discrete filters that satisfy condition 4.9.29 are called quadrature mirror filters (QMFs). These filters represent the link between wavelet transform and subbands[Mal90].

4.10 Construction of Wavelet Bases Using Multiresolution Analysis

Using the definition of multiresolution analysis, in the previous section we have constructed a set of nested spaces $\cdots V_3 \subset V_2 \subset V_1 \subset V_0$ and a scaling function $\phi(t)$ that can be used to construct an orthonormal basis for each subspace. Equation 4.9.19 tells us how to construct this basis using a recursive filter function. We have also seen that the projection of the signal $s(t)$ on the space V_m represents the approximation of this signal at the level m, whereas the wavelet transform represents the detail information that is added to create an approximation at level $(m-1)$. Hence, the wavelet transform of level m can also be considered as the projection of the signal $s(t)$ on a space W_m which is the orthogonal complement of V_m:

$$V_{m-1} = V_m \oplus W_m \tag{4.10.1}$$

and Equation 4.9.2 can be rewritten as

$$P_m^V(s) = P_{m+1}^V(s) + P_{m+1}^W(s) \tag{4.10.2}$$

where $P_m^V(s)$ and $P_m^W(s)$ are the projection of the signal s onto the V_m and W_m spaces, respectively:

$$P_m^W(s) = \sum_{n=-\infty}^{+\infty} <s, \psi_{m,n}> \psi_{m,n}(t) \quad . \tag{4.10.3}$$

Mallat and Meyer have proven that if $\phi(t)$ is the scaling function constructed from the filter h_n, then the function $\psi(t)$ has a Fourier transform $\Psi(\omega)$ given by:

$$\Phi(\omega) = \frac{1}{2} G(\omega) \Phi(\omega) \tag{4.10.4}$$

with

$$G(\omega) = e^{-j\omega} H^*(\omega + \pi) \quad . \tag{4.10.5}$$

The function $\psi(t)$ can be used to construct an orthonormal basis for W_m given that

$$\psi_{m,n}(t) = \frac{1}{\sqrt{2^m}} \psi\left(\frac{t - 2^m n}{2^m}\right) \tag{4.10.6}$$

This theorem is extremely important since it links the two functions, wavelet and scaling, by means of two filters, h_n and g_n, which are mirror conjugate of each other as defined in Equation 4.10.5.

As a lemma, Mallat and Meller also defined the necessary and sufficient conditions on $G(\omega)$ for designing and orthogonal wavelets. The family $\{\psi_{m,n}\}_{n \in Z}$ is an orthonormal basis of W_n if and only if

$$|G(\omega)|^2 + |G(\omega + \pi)|^2 = 2 \qquad (4.10.7)$$

and

$$G(\omega)\, H^*(\omega) + G(\omega + \pi)\, H^*(\omega + \pi) = 0 \qquad (4.10.8)$$

Please note that condition 4.10.7 is identical to condition 4.9.29 set for $H(\omega)$. The proof of the theorem can be found in [Mal98]. It has to be noted that as a result, the filter coefficients g_n can be expressed as

$$g_n = \left\langle \frac{1}{\sqrt{2}}\, \psi\!\left(\frac{t}{2}\right),\ \phi\,(t-n) \right\rangle \qquad (4.10.9)$$

which means that $\psi(t/2)$ can be expressed as

$$\frac{1}{\sqrt{2}}\, \psi\!\left(\frac{t}{2}\right) = \sum_{n=-\infty}^{+\infty} g_n\, \phi\,(t-n) \ . \qquad (4.10.10)$$

Calculating the inverse Fourier transform of Equation 4.10.5, we have

$$g_n = (-1)^{1-n}\, h_{1-n} \ . \qquad (4.10.11)$$

Equation 4.11.11 clearly defines the impulse response of the mirror filter of h_n. To construct the basis of the MRA of a signal $s(t)$ we will use either Equation 4.10.5 or Equation 4.11.1. Interesting enough, the wavelet basis constructed by Daubechies et al. utilized the QMF condition in the frequency domain, Equation 4.10.5, whereas subband filters are usually constructed starting from the condition on the tap filters, Equation 4.10.11.

Example: Spline Wavelets

Polynomial spline wavelets are widely used for their capability of constructing smooth approximation of the signals while mantaining good localization properties. The space V_m of spline functions of degree m is the set of functions that are $(l-1)$-times continously differentiable and for each interval in time $[n2^m, (n+1)2^m]$ of width 2^m, the function is approximable by an l-order polynomial. A Riezs basis $\theta(t)$ is usually constructed using multiple convolutions of the square rectangular function:

$$u_1(t) = u(t) = \begin{cases} 1, & 0 < t < 1 \\ 0, & \text{otherwise} \end{cases} \qquad (4.10.12)$$

The $\theta(t)$ for an m-spline basis is obtained by convolving the function $u(t)$ m-times with itself. Since the Fourier transform of $u(t)$ is a sinc function, we have:

$$\Theta(\omega) = \left[\frac{\sin(\omega/2)}{(\omega/2)} \right]^{m+1} e^{-j\frac{\omega}{2}} \qquad (4.10.13)$$

and the scaling function can be calculated using Equation 4.9.4

$$\Phi(\omega) = \frac{e^{-j\epsilon\omega/2}}{\omega^{m+1} \sqrt{S_{2m+2}(\omega)}} \qquad (4.10.14)$$

where

$$S_{2m+2}(\omega) = \sum_{k=-\infty}^{+\infty} \frac{1}{(\omega + 2k\pi)^{2m+2}} \qquad (4.10.15)$$

and the quadrature mirror filter h_n is

$$H(\omega) = 2^{-(m+1/2)} \cdot e^{-j\epsilon\omega/2} \cdot \sqrt{\frac{S_{2m+2}(\omega)}{S_{2m+2}(2\omega)}} \qquad (4.10.16)$$

For cubic splines, the filter h_n is numerically calculated to be approximately with 40 taps. There are the so-called Battle–Lemarie cubic spline wavelets. The wavelet function and the scaling function are given in Figure 4.10.1.

4.11   Wavelets Bases

Originally, the orthonormal families of functions used for wavelet bases have been constructed through shifting and scaling of a single prototype wavelet:
Thanks to Mallat and Meyer, we now have a procedure to construct the wavelet bases based on a direct approach from a continuous–time domain based on the axioms of multiresolution analysis, but we must define a method to choose proper functions for a given application. This is extremely important for data compression, denoising, and similar applications. In one

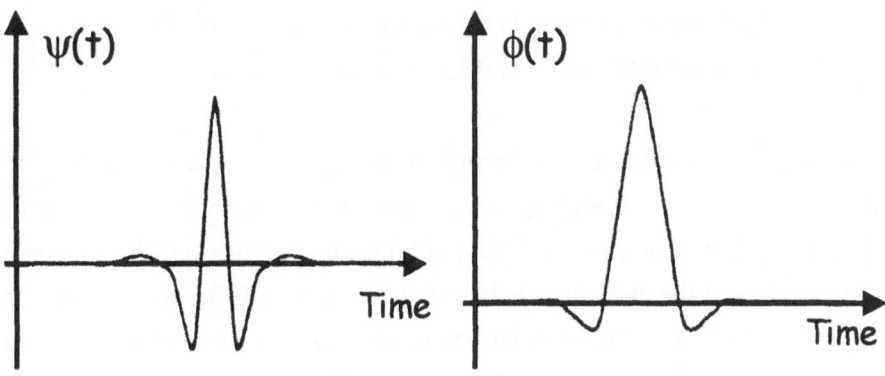

Figure 4.10.1 Wavelet and scaling spline functions.

application, for example, the design of $\psi(t)$ must be optimized to produce the maximum number of zero wavelet coefficients.

The properties of the wavelet that mostly affect the number of nonzero coefficients are the following:

1. Regularity of the signal $s(t)$

$$\psi_{m,n}(t) = 2^{-m/2} \psi(2^{-m}t - n), \qquad m,n \in Z \qquad (4.11.1)$$

2. Number of vanishing moments of the wavelet $\psi(t)$
3. Size and support of $\psi(t)$

Since we want to construct the wavelet function using a filter h_n, we must relate the above properties to the filter. The wavelet $\psi(t)$ has p vanishing moments if:

$$\int_{-\infty}^{+\infty} t^k \psi(t)\, dt = 0, \qquad \text{for } 0 \le k < p \qquad (4.11.2)$$

It can be shown that if the Fourier transform of the wavelet function $\psi(t)$ is p times continuously differentiable at $\omega = 0$, then the following holds:

1. The wavelet function has p vanishing moments.
2. $\Psi(\omega)$ and its first $(p-1)$ derivatives are zero at $\omega = 0$.
3. $\Psi(\omega)$ and its first $(p-1)$ derivatives are zero at $\omega = \pi$.

We have said that the size of the support is important. Let us assume that there is a large discontinuity at $t = t_0$. If ΔT is the support (or duration) of $\psi(t)$, then all the wavelet coefficient obtained using functions $\psi_{j,n}(t)$ which are not zero in the interval $t_0 \pm \Delta T$ will be large and definitely not zero. If the support of the wavelet functions is large, the number of the wavelet coefficients which contain the information of the discontinuity at $t = t_0$ will be large. By reducing the support, we thus can reduce the number of these coefficients. It can be proven that the scaling function ϕ has a compact support if and only if h has a compact support. It can also be shown that both ϕ and h have the same support. If this support is $[N_1, N_2]$ then the support of ψ is

$$\phi, h \leftrightarrow [N_1, N_2] \rightarrow \psi \leftrightarrow \left[\frac{N_1 - N_2 + 1}{2}, \frac{N_2 - N_1 + 1}{2} \right] \quad (4.11.3)$$

The support size of a function and the number of vanishing moments are, in theory, independent; however, we will soon see that in order to have a practical wavelet function, if ψ has p vanishing moments, then its support must be at least $2p-1$. Hence, the choice of a wavelet function often is based on the trade-off between the number of vanishing moments and the support of the basis. Daubechies wavelets are optimal in the sense that they have the minimum support size with the largest possible number of vanishing moments.

The last property we want to discuss is the regularity of the wavelet function ψ. This conditions is imposed by the fact that when reconstructing a signal or an image from its wavelet coefficients, any error ϵ added to the coefficients will result in an added function $\epsilon \, \psi_{j,n}$:

$$s = \sum_{j=0}^{+\infty} \sum_{n=-\infty}^{+\infty} \left(\langle s, \psi_{j,n} \rangle + \epsilon \right) \psi_{j,n}$$

$$(4.11.4)$$

$$s = \sum_{j=0}^{+\infty} \sum_{n=-\infty}^{+\infty} \langle s, \psi_{j,n} \rangle \psi_{j,n} + \epsilon \, \psi_{j,n}$$

If ψ is a smooth function, the added error $\epsilon \, \psi_{j,n}$ will also result in a smooth function. For image coding applications, this is highly desirable, since a smooth error is less detectable and thus better quality images can be obtained. For this reason, the Haar wavelet is not often used in signal and image processing.

The uniformity of the functions ϕ and ψ can be related to the number of zeros of $H(\omega)$ at $\omega = \pi$. However, it must be emphasized that even though the number of vanishing moments and the regularity of orthogonal wavelets are related properties, it is the number of vanishing moments and not the regularity which ultimately affects the number of wavelet coefficients which are not zero.

In the following, we will show how some of the wavelets we have been discussing in his book, can be constructed from their Fourier spectrum using the following Equation:

$$\Psi(\omega) = \frac{1}{\sqrt{2}} \, G\left(\frac{\omega}{2}\right) \Phi\left(\frac{\omega}{2}\right)$$

$$= \frac{1}{\sqrt{2}} \, e^{-i\frac{\omega}{2}} H^*\left(\frac{\omega}{2} + \pi\right) \Phi\left(\frac{\omega}{2}\right) \,.$$

(4.11.5)

4.11.1 Shannon Wavelet

The Shannon wavelet is constructed by the multiresolution approximation based in the frequency spectrum. The function is approximated using low-frequency intervals. It thus corresponds to

$$\Phi(\omega) = u_\pi(\omega) = \begin{cases} 1, & |\omega| < \pi \\ 0, & \text{otherwise} \end{cases}$$

(4.11.6)

and it follows that for $\omega \in [-\pi, \pi]$, we have

$$H(\omega) = \sqrt{2} \, u_{\pi/2}(\omega) = \begin{cases} \sqrt{2}, & |\omega| < \frac{\pi}{2} \\ 0, & \text{otherwise} \end{cases}$$

(4.11.7)

It follows from Equation 4.11.5 that the Fourier transform of the wavelet is

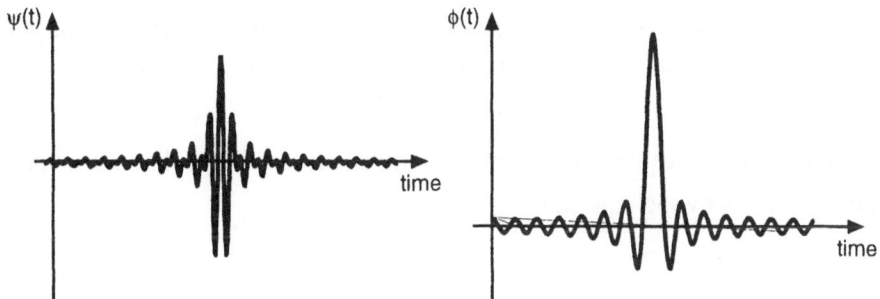

Figure 4.11.1 Wavelet and scaling Shannon function.

$$\Psi(\omega) = \begin{cases} e^{-i\frac{\omega}{2}}, & \text{if } \omega \in [-2\pi, -\pi] \cup [\pi, 2\pi] \\ 0, & \text{otherwise} \end{cases} \qquad (4.11.8)$$

Hence, the temporal representation of the Shannon wavelet is

$$\psi(t) = \frac{\sin 2\pi\left(t - \dfrac{1}{2}\right)}{2\pi\left(t - \dfrac{1}{2}\right)} - \frac{\sin \pi\left(t - \dfrac{1}{2}\right)}{\pi\left(t - \dfrac{1}{2}\right)} \qquad (4.11.9)$$

This wavelet is zero for $\omega = 0$, and all its derivatives are also zero for $\omega = 0$; hence, it has an infinite number of vanishing moments. Since the Fourier spectrum of ψ has compact support, the time signal must be of infinite extent [$\psi(t) \in C^\infty$]. However, it can be shown that $|\psi(t)|$ decays only like $|t|^{-1}$ at infinity because the Fourier transform $\Psi(\omega)$ is discontinuous at $\omega = \pm\pi$ and $\pm 2\pi$. The wavelet and scaling functions are shown in Figure 4.11.1.

4.11.2 Meyer Wavelet

A Meyer wavelet is a frequency band-limited function whose Fourier transform is smooth, unlike the Fourier transform of the Shannon wavelet. This smoothness provides a much faster asymptotic decay in time. These wavelets are constructed with conjugate mirror filters $H(\omega)$ that are C^n and satisfy the following equation:

$$
H(\omega) = \begin{cases} \sqrt{2}, & \text{if } \omega \in \left[-\dfrac{\pi}{3}, \dfrac{\pi}{3} \right] \\[3mm] 0, & \text{if } \omega \in \left[-\pi, -2\dfrac{\pi}{3} \right] \cup \left[2\dfrac{\pi}{3}, \pi \right] \end{cases}
\qquad (4.11.10)
$$

The only degree of freedom in the above equation lies in the behavior of the filter $H(\omega)$ in the transition bands $[-2\pi/3, -\pi/3] \cup [\pi/3, 2\pi/3]$. Of course, it still must satisfy the quadrature condition $|H(\omega)|^2 + |H(\omega + \pi)|^2 = 2$. The scaling function can be calculated using

$$
\Phi(\omega) = \begin{cases} \dfrac{1}{\sqrt{2}} H\left(\dfrac{\omega}{2} \right), & \text{if } |\omega| \le \dfrac{4\pi}{3} \\[3mm] 0, & \text{if } |\omega| > \dfrac{4\pi}{3} \end{cases}
\qquad (4.11.11)
$$

The resulting wavelet is thus

$$\Psi(\omega) = \begin{cases} 0, & \text{if } |\omega| \leq \dfrac{2\pi}{3} \\[2ex] \dfrac{1}{\sqrt{2}} G\left(\dfrac{\omega}{2}\right), & \text{if } \dfrac{2\pi}{3} \leq |\omega| \leq \dfrac{4\pi}{3} \\[2ex] \dfrac{1}{\sqrt{2}} e^{-i\frac{\omega}{2}} H\left(\dfrac{\omega}{4}\right), & \text{if } \dfrac{4\pi}{3} \leq |\omega| \leq \dfrac{8\pi}{3} \\[2ex] 0, & \text{if } |\omega| > \dfrac{4\pi}{3} \end{cases} \qquad (4.11.12)$$

The wavelet and scaling functions are shown in Figure 4.11.2. The functions ϕ and ψ are $C^\sim$ since their Fourier transform have a compact support. Since $\Psi(\omega) = 0$ in the vicinity of $\omega = 0$, all of its derivatives are zero at $\omega = 0$, which proves that $\psi(t)$ has an infinite number of vanishing moments. If $H(\omega)$ is C^n, then $\Psi(\omega)$ and $\Phi(\omega)$ are also C^n. The discontinuities of the derivative of $H(\omega)$ of order $(n+1)$ are generally at the junction of the transition band $|\omega| = \pi/3, 2/3\,\pi$, in which case one can show that there exist a constant A such that:

$$|\phi(t)| \leq A(1 + |t|)^{-n-1} \qquad (4.11.13)$$

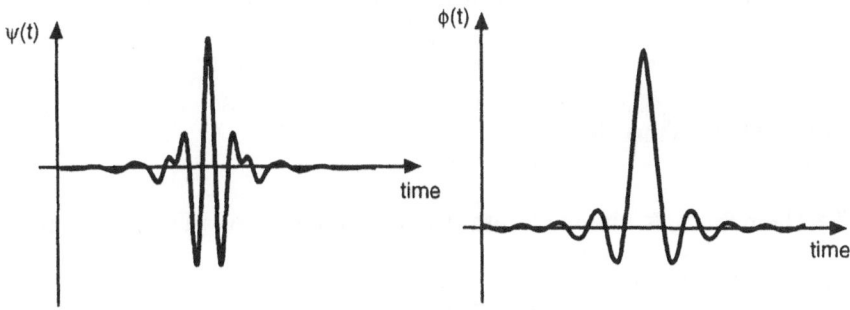

Figure 4.11.2 Wavelet and scaling Mayer's functions.

and

$$|\psi(t)| \leq A(1+|t|)^{-n-1} \quad . \tag{4.11.14}$$

Although the asymptotic decay of ψ is fast when n is large, its effective numerical decay may be relatively slow, which is reflected by the fact that A is quite large. As a consequence, a Meyer wavelet transform is generally implemented in the Fourier domain.

As an example, we can choose the $H(\omega)$ in the transition band as follows:

$$H(\omega) = \sqrt{2} \cos\left[\frac{\pi}{2}\beta\left(\frac{3|\omega|}{\pi}-1\right)\right], \text{ for } |\omega| \in \left[\frac{\pi}{3}, \frac{2\pi}{3}\right] \tag{4.11.15}$$

where $\beta(x)$ is a function that goes from 0 to 1 in the interval $[0,1]$ and satisfies the following condition for $\forall x \in [0,1]$:

$$\beta(x) + \beta(x-1) = 1 \quad . \tag{4.11.16}$$

An example of such function was provided by Daubechies [Dau92]:

$$\beta(x) = x^4(35 - 84x + 70x^2 - 20x^3) \quad . \tag{4.11.17}$$

The resulting $H(\omega)$ has $n = 3$ vanishing derivatives at $|\omega| = \pi/3, 2/3 \, \pi$.

4.11.3 Haar Wavelet

The Haar basis is obtained with a multiresolution of piecewise constant functions. The scaling function is $\phi = 1_{[0,1]}.$* The filter $h[n]$ is defined as follows:

$$h[n] = \begin{cases} \dfrac{1}{\sqrt{2}} & \text{if } n = 0,1 \\ 0 & \text{otherwise} \end{cases} \qquad (4.11.18)$$

has two nonzero coefficients at $n = 0$ and $n = 1$. Hence, we can easily write

$$\frac{1}{\sqrt{2}} \, \psi\left(\frac{t}{2}\right) = \sum_{n=-\infty}^{+\infty} (-1)^{1-n} \, h[1-n] \, \phi(t-n)$$

$$= \frac{1}{\sqrt{2}} \left(\phi(t-1) - \phi(t)\right) \qquad (4.11.19)$$

and it follows that the mother wavelet is

$$\psi(t) = \begin{cases} -1, & \text{if } 0 \le t < \dfrac{1}{2} \\ +1, & \text{if } \dfrac{1}{2} \le t < 1 \\ 0, & \text{otherwise} \end{cases} \qquad (4.11.20)$$

The Haar wavelet has the shortest support among all orthonormal wavelets. It is not well adopted to approximating smooth functions because it has only one vanishing moment. The wavelet and scaling functions are shown in Figure 4.11.3.

---

* The function $1_{[0,1]}$ corresponds to a rectangular function $u_1(x - 1/2)$, that has value equal to one only in the range [0, 1].

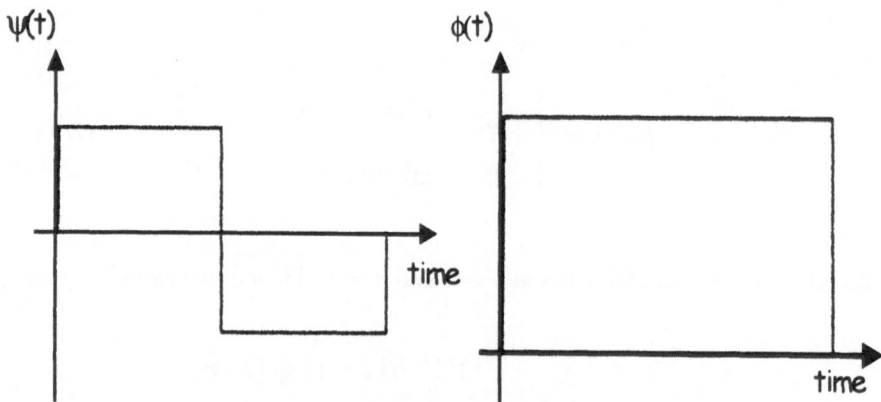

Figure 4.11.3 Wavelet and scaling Haar's functions.

4.11.4 Battle–Lemarié (Spline) Wavelets

Polynomial spline wavelets introduced by Battle and Lemarié are computed from spline multiresolution approximations. We have already calculated the expressions for $\Phi(\omega)$ and $H(\omega)$:

$$\Phi(\omega) = \frac{e^{-i\frac{\varepsilon\omega}{2}}}{\omega^{m+1}\sqrt{S_{2m+2}(\omega)}} \tag{4.11.21}$$

$$H(\omega) = e^{-i\frac{\varepsilon\omega}{2}}\sqrt{\frac{S_{2m+2}(\omega)}{2^{2m+1}\cdot S_{2m+2}(2\omega)}} \tag{4.11.22}$$

We can see that for splines of degree m, $H(\omega)$ and its first m derivatives are zero at $\omega = \pi$ and that $\psi(t)$ has $m+1$ vanishing moments. It follows that

$$\Psi(\omega) = \frac{e^{-i\frac{\varepsilon\omega}{2}}}{\omega^{m+1}} \sqrt{\frac{S_{2m+2}\left(\frac{\omega}{2} + \pi\right)}{S_{2m+2}(\omega)\, S_{2m+2}\left(\frac{\omega}{2}\right)}} \qquad (4.11.23)$$

This wavelet has an exponential decay and it is $m-1$ times continuously differentiable. Polynomial spline wavelets have a faster decay than Meyer's wavelet but also are less regular. It can be shown that for m odd, $\psi(t)$ is symmetric about 1/2, and for m even, $\psi(t)$ is anti symmetric about ½.

4.12   Daubechies Compactly Supported Wavelets

Daubechies wavelets are extremely important because it can be shown that they have the minimum support size for any given number p of vanishing moments [Dau88]. Daubechies constructed her wavelets starting from the finite response of the conjugate mirror filter $h_n = h[n]$. She considered a causal filter $h[n]$, which implies that its Fourier series is represented by the trigometric polynomial

$$H(\omega) = \sum_{n=0}^{N-1} h[n]\, e^{-in\omega} \qquad (4.12.1)$$

To ensure that the function $\psi(t)$ has p vanishing moments, we know that $H(\omega)$ must have a zero of order p at $\omega = \pi$. We can thus write the function $H(\omega)$ as a product of an exponential factor with p zeros at $\omega = \pi$ and a residual function $R(\omega)$:

$$H(\omega) = \sqrt{2}\left(\frac{1 + e^{-j\omega}}{2}\right) R\left(e^{-j\omega}\right) . \qquad (4.12.2)$$

We now must define a polynomial $R(\omega)$ of minimum degree m such that the correspondent $H(\omega)$ satisfies the condition

$$|H(\omega)|^2 + |H(\omega + \pi)|^2 = 2 \qquad (4.12.3)$$

As a result, the filter function $h[n]$ has $N = m + p + 1$ nonzero coefficients, where p is the number of zeros at $\omega = \pi$, and m is the degree of the polynomial R. Daubechies has proven that m is the minumum degree of R. Furthermore, she also demonstrated that if a conjugate mirror filter $H(\omega)$ has p zeros at $\omega = \pi$, then $h[\mathbf{n}]$ has at least $2p$ nonzero coefficients. Thus, Daubechies filters have the minimum number of nonzero coefficients, which is $2p$. The proof of this theorem is used to construct such filters. Since $h[n]$ is real, $|H(\omega)|^2$ is an even function and can be written as a polynomial of an even function such as $\cos(\omega)$. It follows that also $|R(e^{-j\omega})|^2$ is also a polynomial in $\cos(\omega)$. But $\cos(\omega)$ can be written as function of $\sin^2(\omega/2)$; hence, we can define a new polynomial $P(\sin^2(\omega/2))$ and thus

$$|H(\omega)|^2 = 2 \left(\cos \frac{\omega}{2} \right)^{2p} P \left(\sin^2 \frac{\omega}{2} \right) \qquad (4.12.4)$$

and the quadrature condition becomes

$$(1 - y)^p P(y) + y^p P(1 - y) = 1 \qquad (4.12.5)$$

for $\forall y = \sin^2(\omega/2) \in [0, 1]$. To minimize the number of nonzero terms of the finite Fourier series $H(\omega)$, we must find the solution $P(y) \geq 0$ of minimum degree.

Daubechies has proven that such a polynomial is given by

$$P(y) = \sum_{k=0}^{p-1} \binom{p - 1 + k}{k} y^k \qquad (4.12.6)$$

Clearly, $P(y) \geq 0$ for $y \in [0, 1]$; hence, $P(y)$ is the polynomial of minimum degree which satisfies the above quadrature condition. Now that we have calculated the polynomial $P(y)$, we need to calculate the polynomial $R(\omega)$ such that $P(\sin^2\omega/2) = |R(e^{-j\omega})|^2$:

$$R(e^{-j\omega}) = \sum_{k=0}^{m} r_k e^{-jk\omega} = r_0 \prod_{k=0}^{m} (1 - a_k e^{-j\omega}) \quad . \tag{4.12.7}$$

Since the coefficients of the filter are real, this means that $R^*(e^{j\omega}) = R(e^{j\omega})$, hence,

$$|R(e^{-j\omega})|^2 = P\left(\frac{2 - e^{j\omega} - e^{-j\omega}}{4}\right) = Q(e^{-j\omega}) \tag{4.12.8}$$

or by going into the z plane,

$$R(z)R(z^{-1}) = r_0^2 \cdot \prod_{k=0}^{m} (1 - a_k z)(1 - a_k z^{-1})$$

$$= Q(z) = P\left(\frac{2 - z - z^{-1}}{4}\right) \quad . \tag{4.12.9}$$

$R(z)$ is often called the *spectral factor* of $P'(z)$ and the technique to extract it is called *Spectral Factorization*:

$$P(1 - y) = |R(e^{j\omega})|^2 = R(e^{j\omega}) R(e^{-j\omega})$$
$$P'(z) = P(1-y) = R(z) R(z^{-1}) \quad . \tag{4.12.10}$$

These spectral factors are not unique and are obtained by assigning one zero of each pair to $R(z)$. The choice of which zero to assign to $R(z)$ leads to different spectral factors. In general, we factor the polynomial $P'(z)$ into its zeros as follows:

$$P'(z) = \alpha \prod_{i=1}^{N} (1 - z_i z^{-1}) \prod_{i=1}^{N} (1 - z_i^* z) \qquad (4.12.11)$$

where the z_i are the zero outside/inside the unit circle, $|z_i| < 1$. To get a minimum phase solution, we consistently choose the zeros inside the unit circle; hence,

$$R(z) = \sqrt{\alpha} \prod_{i=1}^{N} (1 - z_i z^{-1}) . \qquad (4.12.12)$$

Example: Case $N=2$ and $Q(y) = 0$

$$P(y) = \left. \sum_{j=0}^{N-1} \binom{N-1-j}{j} y^j + y^N Q(Y) \right|_{N=2}$$

$$P(y) = \sum_{j=0}^{1} \binom{1+j}{j} y^j = 1 + 2y$$

$$P(1-y) = 3 - 2y , \qquad y = \cos^2\left(\frac{\omega}{2}\right)$$

$$P(1-y) = 3 - 2\cos^2\left(\frac{\omega}{2}\right) = 2 - \cos\omega = R(e^{j\omega})R(e^{-j-\omega})$$

$$|R(e^{j\omega})|^2 = 2 - \frac{e^{j\omega}}{2} - \frac{e^{-j\omega}}{2} . \qquad (4.12.13)$$

The roots of $R(e^{j\omega})$ are:

$$r_1 = 2 + \sqrt{3}$$
$$r_2 = 2 - \sqrt{3} = \frac{1}{r_1} . \qquad (4.12.14)$$

A possible expression for $R(e^{j\omega})$ is:

$$R(e^{j\omega}) = \frac{1}{\sqrt{3}-1}\left[e^{j\omega} - (2-\sqrt{3})\right] = \frac{1}{2}\left[(1+\sqrt{3})\,e^{j\omega} + 1 - \sqrt{3}\right]$$

$$M_0(e^{j\omega}) = \left[\frac{1}{2}(1+e^{j\omega})\right]^N R(e^{j\omega}) \qquad (4.12.15)$$

$$M_0(e^{j\omega}) = \left[\frac{1}{2}(1+e^{j\omega})\right]^2\left\{\frac{1}{2}\left[(1+\sqrt{3})\,e^{j\omega} + 1 - \sqrt{3}\right]\right\}$$

and, thus,

$$M_0(e^{j\omega}) = \frac{\left[(1+\sqrt{3})e^{j3\omega}+(3+\sqrt{3})e^{j2\omega} + (3-\sqrt{3})e^{j\omega} + (1+\sqrt{3})\right]}{8}$$

$$(4.12.16)$$

However,

$$\Phi(\omega) = \prod_{k=1}^{\infty} M_0\left(\frac{\omega}{2^k}\right) = M_0\left(\frac{\omega}{2}\right)\Phi\left(\frac{\omega}{2}\right)$$

$$(4.12.17)$$

$$M_0\left(\frac{\omega}{2}\right) = \frac{1}{\sqrt{2}}\,G_0(e^{j\omega/2})$$

hence,

$$G_0(z) = \frac{1}{4\sqrt{2}}\left[(1+\sqrt{3})\,e^{j3\omega} + (3+\sqrt{3})\,e^{j2\omega}\right.$$

$$(4.12.18)$$

$$\left.+ (3-\sqrt{3})\,e^{j\omega} + (1+\sqrt{3})\right]$$

and thus the filter coefficients are

$$h[0] = (1 + \sqrt{3})/4\sqrt{2}, \quad h[1] = (3 + \sqrt{3})/4\sqrt{2}$$

$$h[2] = (3 - \sqrt{3})/4\sqrt{2}, \quad h[3] = (1 - \sqrt{3})/4\sqrt{2}.$$

(4.12.19)

For the case $p = 1$, the Daubechies wavelet becomes the Haar wavelet. The coefficients of other Daubechies filters are given in Table 4.12.1. In the table, N is the number of zeros and equals $L/2$ where L is the length of the filter. The low-pass filter coefficients $h[n]$ are here given. These are obtained from a minimum phase factorization of $P(x)$. A plot of the scaling and wavelet functions defined by the filter coefficients in Table 4.12.1 is given in Figure 4.12.1.

Daubechies wavelets are very asymmetric because they are constructed by selecting the minimum phase square root of $Q(e^{-j\omega})$. It can be shown that filters corresponding to a minimum phase square root have their energy optimally concentrated at the beginning of their duration (support), resulting in signals that are highly asymmetric. To obtain a symmetric wavelet, the filter h must be symmetric as well; hence, $H(\omega)$ must have a linear complex phase. Daubechies has proven that the only real compactly supported conjugate mirror filter that has a linear phase is the Haar wavelet. The *symmlet* wavelets are constructed by optimizing the choice of $R(\omega)$ such that its square [i.e., $Q(\omega)$] has an almost linear phase. Complex conjugate mirror filters with compact support and a linear phase can be constructed, but they produce complex wavelet coefficients whose real and imaginary parts are redundant when the signal is real [Dau89]. For an application in numerical analysis, Coifman and Daubechies have constructed a family of wavelets ψ that have p vanishing moments and a minimum size support, but whose scaling function also satisfy [Coi92]

$$\int_{-\infty}^{+\infty} \phi(t)\, dt = 1$$

(4.12.20)

$$\int_{-\infty}^{+\infty} t^k \phi(t)\, dt = 0, \quad \text{for} \quad 1 \le k < p.$$

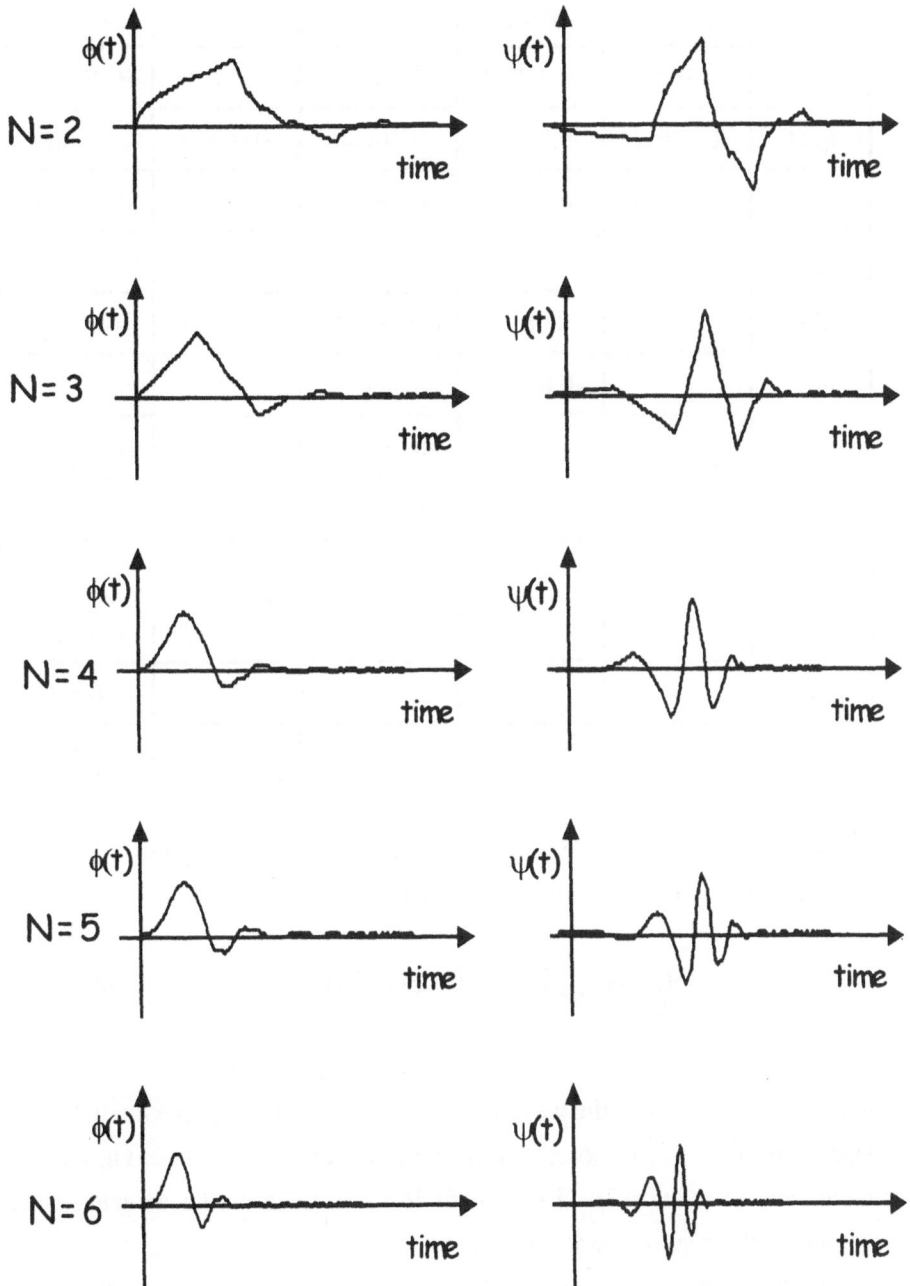

Figure 4.12.1 Plot of Daubechies scaling and wavelet functions for different values of the length N of the filters.

h [n]	N = 2	N = 3	N = 4	N = 5	N = 6
\multicolumn{6}{c}{Table 4.12.1 First few maximally flat Daubeches filters}
$h[0]$	+0.4829	+0.33267	+0.23037	+0.16010	+0.11154
$h[1]$	+0.83651	+0.80689	+0.714845	+0.60382	+0.49462
$h[2]$	+0.22414	+0.45987	+0.63088	+0.72430	+0.751133
$h[3]$	−0.1294	−0.13501	−0.027983	+0.13842	0.31525
$h[4]$		−0.08544	−0.18703	−0.24229	−0.22626
$h[5]$		+0.03522	+0.03084	−0.03224	−0.12976
$h[6]$			+0.03288	+0.07757	+0.097501
$h[7]$			−0.01059	−0.00624	+0.02752
$h[8]$				−0.01258	−0.03158
$h[9]$				+0.00333	+0.00055
$h[10]$					+0.00477
$h[11]$					−0.00107

Such scaling functions are useful in the case of smooth signals $s(t)$. If we choose a fine scale J such that for very large J

$$\left\langle s,\ \phi_{-J,n}\right\rangle \approx 2^{J/2} \cdot s(2^{-J}n) \tag{4.12.21}$$

at fine scales (J large), the wavelet coefficients are thus approximated by the signal samples, and the order of approximation increases with p. The Coiflets have a support of size $3p-1$ instead of $2p-1$, as the previous wavelets, since we have added some more constraints.

Even if not originally defined as wavelet filters, we should mention the first conjugate mirror filters developed by Smith and Barnwell in 1986. These filters satisfy the quadrature condition:

$$|H(\omega)|^2 + |H(\omega + \pi)|^2 = 2$$

which is a necessary and sufficient condition for a filter bank reconstruction. The main difference is given by the fact that the filters designed by Smith and Barnwell, and later by Vaidyanathan and Hoang, have a minimal transition band, where $|H(\omega)|$ decays from nearly $\sqrt{2}$ to nearly 0 near $\pm\pi/2$, but do not yield a wavelet basis of $L^2(\mathbb{R})$ since $H(0) \neq \sqrt{2}$. Because of this, these filters are very useful in compressing and coding audio signals, but cannot be used in a large cascade structure. We will discuss these filters in more detail in next chapter on subbands.

4.13 Fast Wavelet Transform

Up to this point, we have considered the synthesis of the signal $s(t)$ starting from its wavelet coefficients, but we have said little about the analysis.

An attractive feature of the wavelet series expansion is that the underlying multiresolution structure leads to an efficient discrete-time algorithm based on a filter bank implementation. This connection was first pointed out by Mallat in [Mal89].

In multiresolution analysis, we have defined a set of nested subspaces:

$$\cdots V_3 \subset V_2 \subset V_1 \subset V_0$$

$$V_{m-1} = V_m \oplus W_m \qquad (4.13.1)$$

$$V_m \cap W_m = \{0\}$$

and the basis functions $\phi(t)$ and $\psi(t)$ to determine the projection of the signal on these spaces [Sel99]. In particular, using the scaling function $\phi(t)$ and the wavelet function $\psi(t)$, we can define the $a_{n,m}$ and $d_{n,m}$ coefficients as

$$a_{n,m} = \left\langle s(t), \phi_{n,m}(t) \right\rangle$$
$$d_{n,m} = \left\langle s(t), \psi_{n,m}(t) \right\rangle$$

(4.13.2)

$$a_{n,m} = a_{n,m+1} + d_{n,m+1} \quad .$$

(4.13.3)

Since we will consider the shifting parameter n as variable, we can rewrite Equation 4.13.3 without any loss of generality as follows:

$$a_m[n] = a_{m+1}[n] + d_{m+1}[n]$$

(4.13.4)

If we define the space V_0 as the space of signals that completely contains $s(nT)$, with T to be better defined later, for this is equivalent to say that $s \in V_0$ then

$$a_0[n] = \left\langle s(t), \phi(t-n) \right\rangle , \quad n \in Z$$

(4.13.5)

and since the projection onto W_0 is null, then we have

$$s(t) = \sum_{n=-\infty}^{+\infty} a_0[n] \, \phi(t-n) \quad .$$

(4.13.6)

The coefficients $a_0[n]$ are a weighted average of $s(t)$. The projection of $s(t)$ onto V_1 can be calculated as

$$a_1[n] = \left\langle s(t), \frac{1}{\sqrt{2}} \phi\left(\frac{t}{2} - n\right) \right\rangle$$

(4.13.7)

Using Equation 4.9.19, the tap coefficients h_k can be used to calculate $a_1[n]$:

$$a_1[n] = \sum_{k=-\infty}^{+\infty} h_k \langle s(t), \phi(t-2n-k) \rangle \quad . \tag{4.13.8}$$

Using the definition of $a_0[n]$ in Equation 4.13.5, we can calculate the signal projection $a_1[n]$ using the previous projection $a_0[n]$ as follows:

$$a_1[n] = \sum_{k=-\infty}^{+\infty} h_k \, a_0[2n + k] \tag{4.13.9}$$

or using a change of variable $k' = 2n + k$, and $h_n = h[n]$, we have

$$a_1[n] = \sum_{k=-\infty}^{+\infty} h[k-2n] \, a_0[k] \quad . \tag{4.13.10}$$

The above equation allow us to calculate the approximation of the signal at level $m + 1 = 1$, starting from its approximation at level $m = 0$. Generalizing, we have

$$a_{m+1}[n] = \sum_{k=-\infty}^{+\infty} h[k-2n] \, a_m[k] \quad . \tag{4.13.11}$$

The same can be achieved from the projection onto W_0 using

$$\psi(t) = \sqrt{2} \sum_{k=-\infty}^{+\infty} g[k] \, \phi(2t - k) \tag{4.13.12}$$

and, thus, the projection is calculated as

$$d_{m+1}[n] = \sum_{k=-\infty}^{+\infty} g[k-2n] \, a_m[k] \tag{4.13.13}$$

Starting from $a_m[k]$, we can calculate the lower-level approximation and details $a_{m+1}[k]$ and $d_{m+1}[k]$ using Equations 4.13.11 and 4.13.13.

Reconstruction of the approximation at the higher level m is obtained from the lower-level approximation and details using

$$a_m[n] = \sum_{k=-\infty}^{+\infty} h[n-2k] \cdot a_{m+1}[k]$$

$$+ \sum_{k=-\infty}^{+\infty} g[n-2k] \cdot d_{m+1}[k] .$$

(4.13.14)

Please note that the filters h and g are inverted in time in respect to the analysis Equations 4.13.11 and 4.13.13.

All the above equations are in the discrete time domain. The question is thus, how do we start from a continuos time domain signal?

We have started the discussion already having the signal $s(t)$ projected onto the V_0, imposing the condition $s \in V_0$. This condition ensures that $a_0[n]$ $= \langle s(t), \phi(t-n) \rangle$. This means that $a_0[n]$ is a local average of $s(t)$ around $t = nT$, but it is strictly not equal to $s(nT)$. So, we can say that the initial step in the fast wavelet transform is to obtained a sampled and local average of the signal $s(t)$, that we will define as $a_0[n]$. The basic assumption is that the sampling process used to determine $a_0[n]$ is such that no property of $s(t)$ is altered. In most of real life applications, the discrete input signal $a_0[n]$ is obtained by finite analog to digital conversion using an instrument such as a recorder. For example, a CCD camera averages the intensity over

Analysis - Fast Wavelet Transform

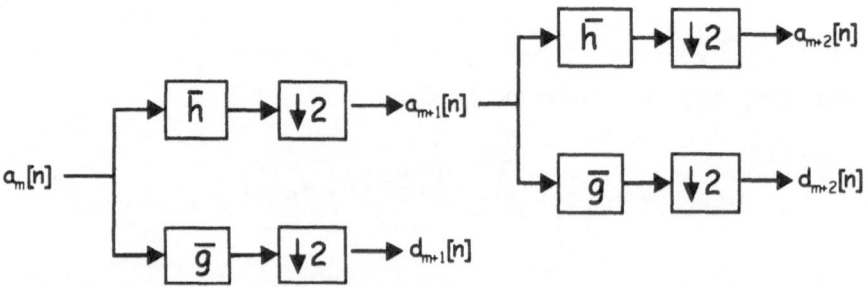

Figure 4.13.1 Block diagram of the analysis filter bank or fast wavelet transform.

exposure time, and the resulting signal or image is discretized using the readout sampling of the device.

We can thus summarize the above discussion using the concept of analysis and synthesis filters [Vet90a]:

- Analysis Filters

$$a_{m+1}[n] = a_m * \bar{h}$$

$$d_{m+1}[n] = a_m * \bar{g}$$

(4.13.15)

Please note that the analysis filters are reversed in time with respect to the synthesis filters:

$$\bar{h}[n] = h[-n]$$

$$\bar{g}[n] = g[-n]$$

(4.13.16)

- Synthesis Filters

$$a_m[n] = \dot{a}_{m+1} * h + \dot{d}_{m+1} * g$$

(4.13.17)

where:

$$\dot{a}[n] = \begin{cases} a[k], & \text{if } n = 2k \\ 0, & \text{if } n = 2k+1 \end{cases}$$

(4.13.18)

Synthesis - Inverse Fast Wavelet Transform

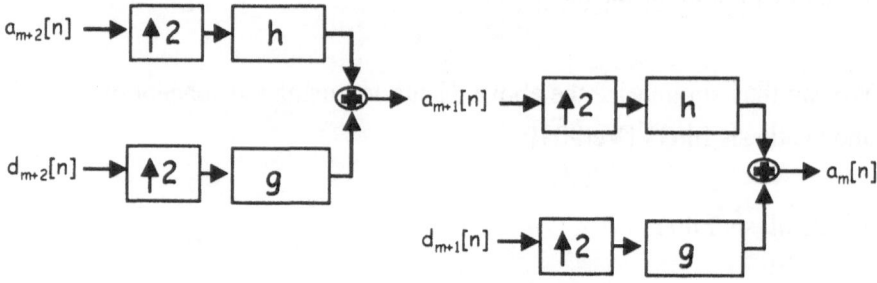

Figure 4.13.2 Block diagram of the synthesis filter bank or inverse wavelet transform.

Equation 4.3.18 represents an interpolator in which every point of the input signal is interpolated by a zero. Equations 4.3.15 and 4.3.17 can be represented by block diagrams, as in Figures 4.13.1 and 4.13.2, respectively [Vet92]. The symbols with the up and down arrows with 2 represent a decimator and interpolator, respectively. These concepts were introduced in Chapter 2 and a more detail discussion will follow in Chapter 5 [Ve86]. A single-stage synthesis filter is plotted in Figure 4.13.3.

The synthesis of the signal in Figure 4.8.1 is now performed as plotted in Figure 4.13.4. By comparison with Figure 4.8.4, we can see that we have successfully eliminated the redundancy present in the Figure 4.8.4 by decimating and later interpolating by the factor 2. This is the great benefit of the fast wavelet transform. By using the concepts of multiresolution analysis, we are capable of removing the redundancy of the original discrete

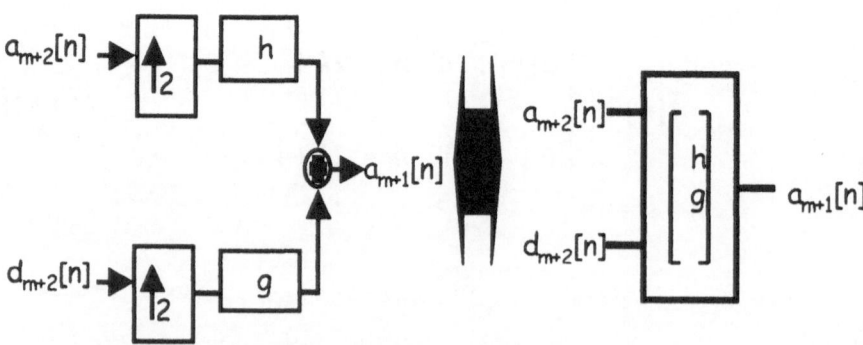

Figure 4.13.3 Block diagram of the single-stage synthesis filter bank.

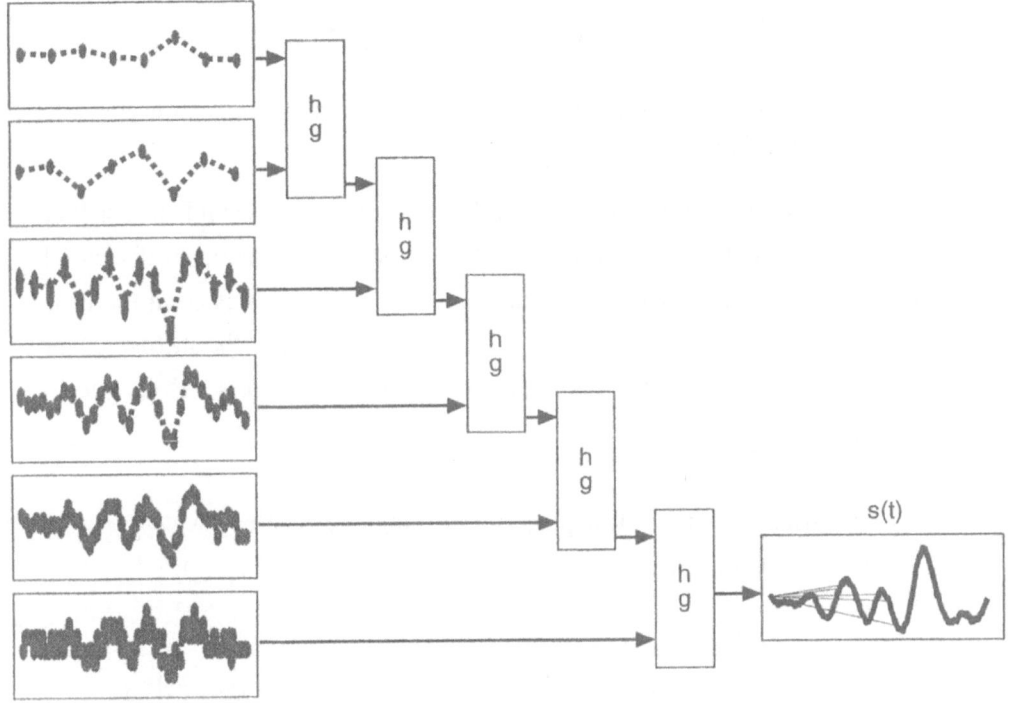

Figure 4.13.4 Block diagram of the signal synthesis using the Inverse Fast Wavelet transform.

wavelet transform without any loss of information [Vet95]. As we will show in more detail in next chapter, the fast wavelet transform is a special case of the perfect reconstruction conjugate mirror filters used in subband signal processing [Her93a].

4.14 Biorthogonal Wavelet Bases

Until now, we have calculated the conditions that link the filter h and g to the

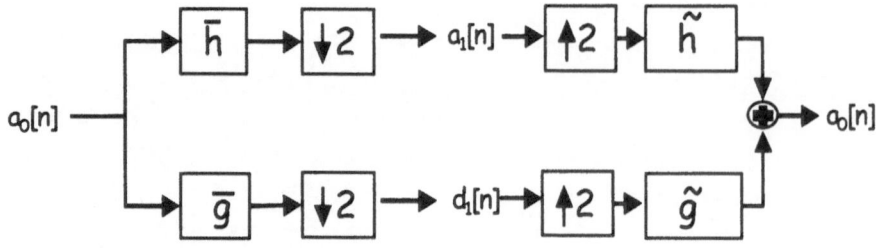

Figure 4.14.1 Block diagram of a single stage analysis and synthesis fast wavelet transform with biorthogonal filters.

wavelet and scaling functions. Furthermore, we have established a link between the analysis and synthesis filters. For the wavelet bases discussed until now, the only major difference between the filters has been the time reversal. An extension of the fast wavelet transform is the utilization of different set of filters for the analysis and synthesis filters. Referring to Figure 4.14.1, we have two sets of filters:

- Analysis filters (low-pass filter h_1, high-pass filter g_1)
- Synthesis filters (low-pass filter h_2, high-pass filter g_2)

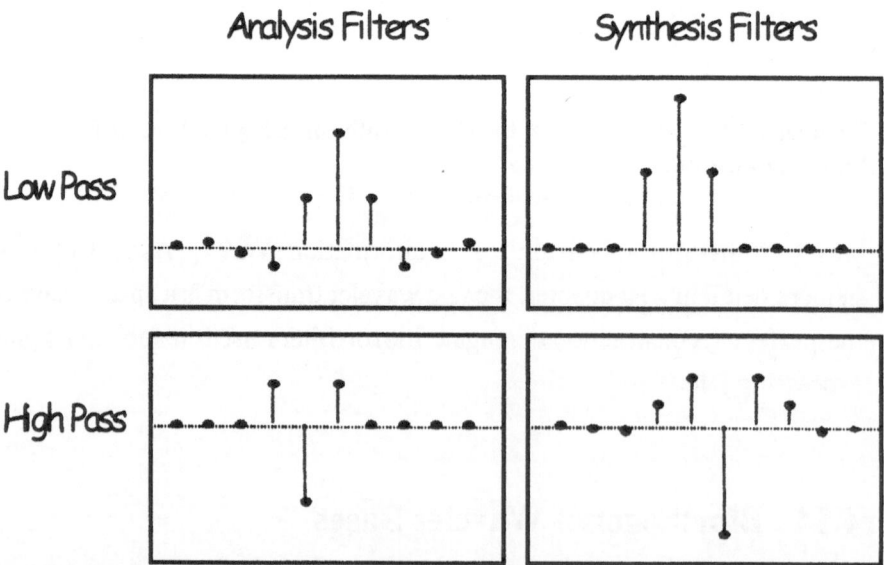

Figure 4.14.2 Analysis and synthesis filter coefficients for a biorthogonal wavelet with order (p=2, p~=4).

The conditions for perfect reconstruction in this case are

$$H_1^*(\omega + \pi)\, H_2(\omega) + G_1^*(\omega + \pi)\, G_2(\omega) = 0$$

$$H_1^*(\omega)\, H_2(\omega) + G_1^*(\omega)\, G_2(\omega) = 2 \ .$$

(4.14.1)

These conditions were originally developed in the context of subband decomposition; for this reason, we refer to Chapter 5 for a more detailed discussion. As the filter pairs (h_1, g_1) and (h_2, g_2) create basis for the representation of any signal $s[n] \in L^2(Z)$, then the resulting wavelet and scaling functions represent biorthogonal Riesz bases in $L^2(Z)$.

An infinite cascade of filters (h_1, g_1) and (h_2, g_2) will yield now two different scaling and wavelet functions whose Fourier transform will satisfy

$$\Phi_1(2\omega) = \frac{1}{\sqrt{2}}\, H_1(\omega)\, \Phi_1(\omega)$$

$$\Psi_1(2\omega) = \frac{1}{\sqrt{2}}\, G_1(\omega)\, \Phi_1(\omega)$$

(4.14.2)

$$\Phi_2(2\omega) = \frac{1}{\sqrt{2}}\, H_2(\omega)\, \Phi_2(\omega)$$

$$\Psi_2(2\omega) = \frac{1}{\sqrt{2}}\, G_2(\omega)\, \Phi_1(\omega)$$

with the following conditions

$$H_1^*(\omega)\, H_2(\omega) + H_1^*(\omega + \pi)\, H_2^*(\omega + \pi) = 2$$

$$G_1(\omega) = e^{-j\omega}\, H_2^*(\omega + \pi)$$

(4.14.3)

$$G_2(\omega) = e^{-j\omega}\, H_1^*(\omega + \pi) \ .$$

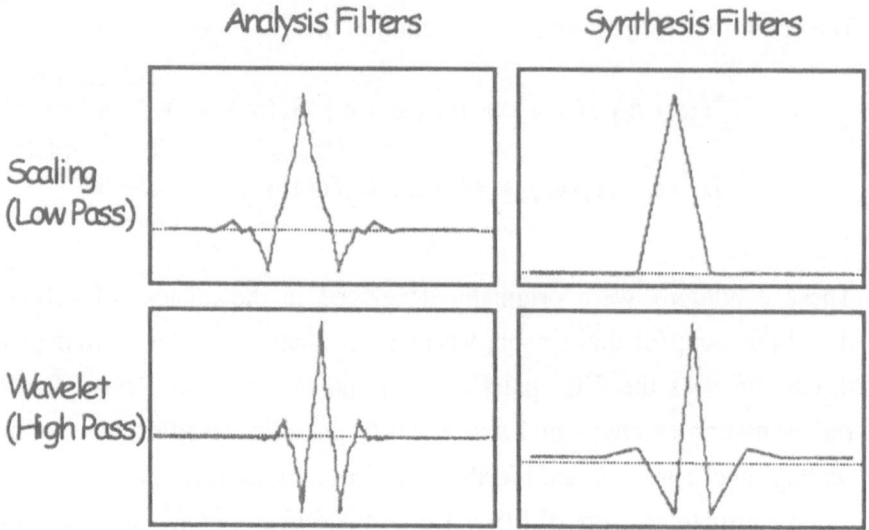

Figure 4.14.3 Scaling and wavelet function calculated using the filters of Figure 4.14.2.

Cohen, Daubechies, and Feauveau have established a design procedure for biorthogonal wavelets using the same approach of spectral factorization Daubechies used for her wavelets [Coh92]. By choosing

$$H_1(\omega) = \sqrt{2} \, e^{-j\frac{\epsilon\omega}{2}} \left(\cos\frac{\omega}{2} \right)^p \qquad (4.14.4)$$

if p is an even number, then $\epsilon = 0$, and $\epsilon = 1$ for p odd.

The scaling function is then a box-spline function of degree $(p-1)$:

$$\Phi_1(\omega) = e^{-j\frac{\epsilon\omega}{2}} \left(\frac{\sin\frac{\omega}{2}}{\frac{\omega}{2}} \right)^p \qquad (4.14.5)$$

The wavelet function is also a polynomial combination of the spline functions with a number of vanishing moments r.

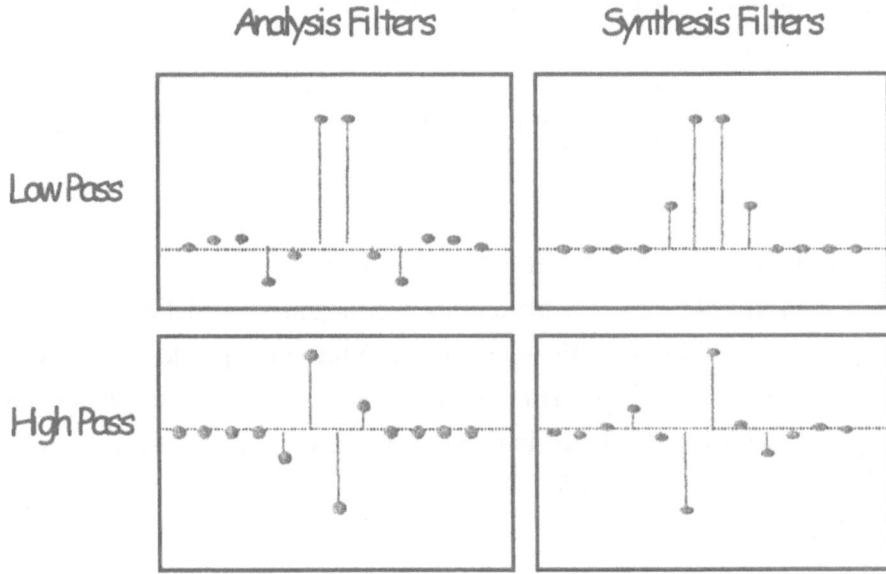

Figure 4.14.4 Analysis and synthesis filter coefficients for a biorthogonal wavelet with order (3,5).

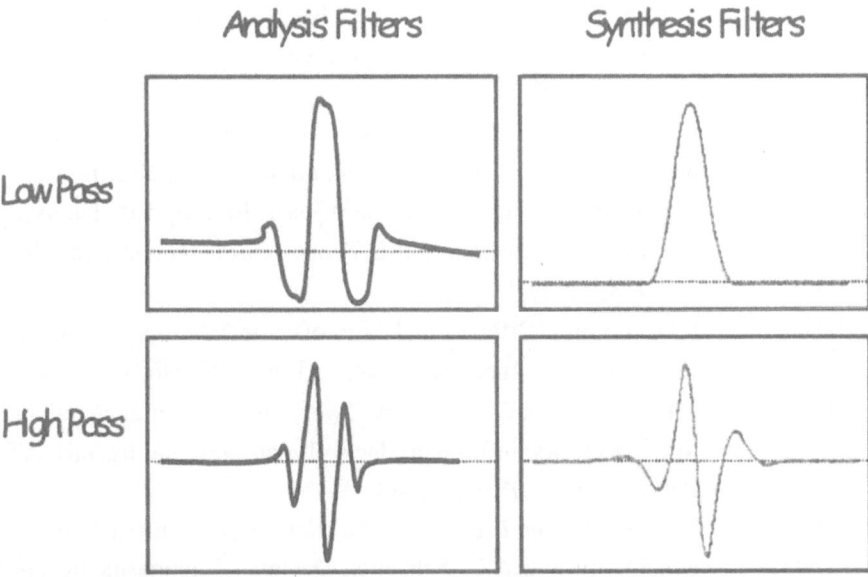

Figure 4.14.5 Scaling and wavelet functions for the biorthogonal filter coefficients of Figure 4.14.4.

If we define $q = p + r$, we can define the synthesis low-pass filter as:

$$H_2(\omega) = \sqrt{2}\, e^{-j\frac{\epsilon\omega}{2}} \left(\cos\frac{\omega}{2} \right)^p \sum_{k=0}^{q-1} \binom{q-1+k}{k} \left(\sin\frac{\omega}{2} \right)^{2k} \qquad (4.14.6)$$

The biorthogonal wavelet bases can thus be calculated.

The filters used to calculate spline biorthogonal wavelet bases for $p = 2$ and $r = 4$ are plotted in Figure 4.14.2, and the correspondent wavelet and scaling functions are given in Figure 4.14.3. For comparison, the filters for $p = 3$ and $r = 5$ are plotted in Figure 4.14.4, with the wavelet and scaling functions in Figure 4.14.5.

4.15 References

[Ben90]   J.J. Benedetto and W. Heler, Irregular sampling and the theory of frames, *Math. Note*, vol. 10, pp. 103-125, 1990.

[Coh92]   A. Cohen, I. Daubechies, and J.C. Feauveau, Bi-orthogonal bases of compactly supported wavelets, *Commun. Pure Appl. Math.* vol. 45, pp. 485-560, 1992.

[Coh96]   A. Cohen and J. Kovačević, Wavelets: The mathematical background, *Proc. IEEE*, vol. 84, no. 4, pp. 514-522, 1996.

[Coi92]   R.R. Coifman and M. V. Wickerhauser, Entropy-based algorithms for best basis selection, *IEEE Trans. Inform. Theory*, vol. 38, pp. 713-718, 1992.

[Dau88]   I. Daubechies, Orthonormal bases of compactly supported wavelets, *Commun. Pure Appl. Math*, vol. XLI, pp. 909-996, 1988.

[Dau89]   I. Daubeches and J. Lagarias, Two scale differential equations, 11 local regularity, infinite products of matricies, and fractals, AT&T Bell Labs Tech. Report, 1989.

[Dau92]   Daubechies, *Ten Lectures on Wavelets*, SIAM, Philadelphia, 1992.

[Duf52]   R.J. Duffin and A.C. Schaeffer, A class of nonharmonic Fourier series, *Trans. Am. Math. Soc.*, Vol. 72, pp. 314-366, 1952.

[Djo97]   I. Djokovic and P.P. Vaidyanathan, Generalized sampling theorems in mutiresolution subspaces, *IEEE Trans. Signal Proc.*, vol. 45, pp. 583-599, 1997.

[Gou84]   P. Goupillaud, A. Grossmann, and J. Morlet, Cycle-octave and related transforms in seismic signal analysis, *Geoexploration*, vol. 23, pp. 85-102, 1984.

[Gro86]   A. Grossmann, J. Morlet, and T. Paul, Transforms associated to square integrable group presentations II. Examples, *Am. Inst. Henry Poincare*, vol. 45, no. 3, pp. 293-309, 1986.

[Gro89]   A. Grossmann, R. Kronland-Martinet, and J. Morlet, Reading and understanding continuous wavelet transforms, pp. 2-20 in *Wavelets*, edited by J.M. Combes, A. Grossmann, and Ph. Tchamitchian, Springler-Verlag, Berlin, 1989.

[Her93a]   C. Herley and M. Vetterli, Wavelets and recursive filter banks, *IEEE Trans. Signal Proc.*, vol. 41, pp. 2536-2556, 1993.

[Hes96]   N. Hess-Nielsen and M.V. Wickerhauser, Wavelets and Time-Frequency Analysis, *Proc. IEEE*, vol. 84, no. 4, pp. 523-540, 1996.

[Jer77]   A.J. Jerri, The Shannon sampling theorem - Its various extensions and applications: A tutorial review, *Proc. IEEE*, vol. 65, pp. 1565-1596, 1977.

[Kha93]   M.R.K. Khansari and A. Leon-Garcia, Subband decomposition of signals with generalized sampling, *IEEE Trans. Sig. Process.*, vol. 41, no. 12, pp. 3365-3376, 1993.

[Mal89]   S. G. Mallat, A theory for multiresolution signal decomposition: the wavelet representation, *IEEE Trans. Pattern Anal Machine Intell.*, vol. 11, no. 7, pp. 674-693, 1989.

[Mal90]   S. Mallat, Multifrequency channel decompositions of images and wavelet models, *IEEE Trans. Acoust. Speech Signal Process.*, vol. 37, pp. 2091-2110, 1990.

[Mal98]   S. Mallat, *A Wavelet Tour of Signal Processing*, Academic Press, San Diego, 1998.

[Mey89]   Y. Meyer, Orthonormal wavelets, in *Wavelets: Time-Frequency, Methods and Phase Space*, edited. by J.M. Combes, A. Grossman and, Ph.. Tchamitchain, Springer-Verlag, New York, 1989.

[Mey89a]   Y. Meyer, Wavelets and operators, *Proc. Special Year in Modern Analysis, Urbana 1986-87*, Cambridge University Press, Cambridge, 1989.

[Mey93]   Y. Meyer, *Wavelets. Algorithms and Applications*, translated by R.D. Ryan, SIAM, Philadelphia, 1993.

[Sel99]   I.W. Selesnick, Interpolating multiwavelets bases and the sampling theorem, *IEEE Trans. Signal Proc.* vol. 47, pp. 1615-1621, 1999.

[Sha49]   C.E. Shannon, Communication in the presence of noise, *Proc. IRE*, vol. 37, pp. 10-21, 1949.

[Uns95]   M. Unser, A general Hilbert space framework for the discretization of continuous signal processing operators, in *Proc. SPIE Conf. on Wavelet Applications in Signal and Image Processing III*, pp. 51-61, 1995.

[Uns00]   M. Unser, Sampling - 50 years after Shannon, *Proc. IEEE*, vol. 88, pp. 569-587, 2000.

[Vet86]   M. Vetterli, Filter banks allowing perfect reconstruction, *Signal Process.*, vol. 10, no. 3, pp. 219-244, 1986.

[Vet90a]   M. Vetterli and C. Herley, Wavelets and filter banks: relationships and new results, *Proc. ICASSP*, vol. 3, pp. 1723-1726, 1990.

[Vet92]   M. Vetterli and C. Herley, Wavelets and filter banks: theory and design, *IEEE Trans. Signal Process.*, vol. 40, no. 9, 1992.

[Vet95]   M. Vetterli and J. Kovačević, *Wavelets and Subband Coding*, Prentice Hall, Englewood Cliffs, NJ, 1995.

[Wal92]   G.G. Walter, A sampling theorem for wavelet subspaces, *IEEE Trans. Inform. Theory*, vol. 38, pp. 881-884, 1992.

[Xia93]   X.G. Xia and Z. Zhang, On sampling theorem, wavelets and wavelet transforms, *IEEE Trans. Signal Process.* vol. 41, pp. 3524-3535, 1993.

[Zib93]   M. Zibulski and Y.Y. Zeevi, Oversampling in the Gabor scheme, *IEEE Trans. Signal Process.*, vol. 41, pp. 2679-2687, 1993.

Chapter 5

Theory of Subband Decomposition

5.1   Introduction

One of the most practical and successful applications of multirate filters is in video or audio compression using subband coding. Consider an example of audio subband coding shown in Figure 5.1.1. Let us say that one needs to sample the signal at a 10-kHz rate which corresponds to a bandwidth of 5 kHz.* If we need an accuracy of 16 bits, then we need to transmit or store a total of 160 Kb/s. As shown in Figure 5.1.1, in practice, the energy near the cutoff frequency is very small. Hence, we can take the time signal and, using multirate filters, split the signal into two bands, one containing the signal components below 5 kHz and the other band containing the components in the 5 kHz to 10 kHz change. Note that due to downsampling by 2, the total number of samples still remains the same.   We can now use 16-bit accuracy for the lower subband with more energy and 8-bit accuracy for the higher band without losing any fidelity, as the high pass band has very little energy. Thus, for this case, we need only a total of 80 Kb/s + 40 Kb/s = 120 Kb/s. Thus, we have achieved, for this example, a compression ratio of 4/3. Of course, one need not be limited to only two subbands. Actually, for speech signals, which need 8 kHz sampling, 16 Kb/s is enough or, on the average 2 bit/sample.

---

* It is common in signal processing to normalize all discussion to a sampling period $T=1$, and thus $f_s=1$. In this case, 5 kHz of the above discussion corresponds to $\omega = 2 \pi f_s / 2 = \pi$.

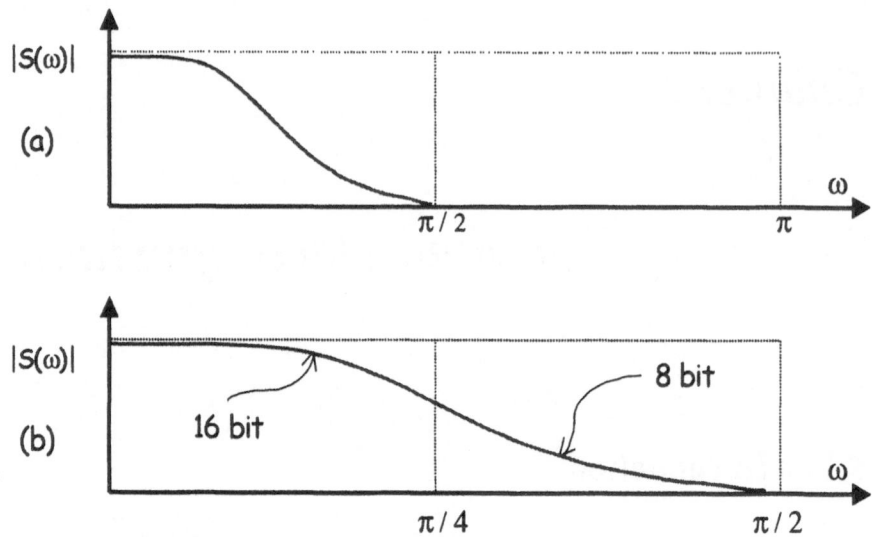

Figure 5.1.1 Example of audio subband coding: (a) a typical audio signal spectra; (b) expanded view showing two subbands.

For the high-fidelity case, 32 Kb/s or 4-bit/sample produces excellent results. These concepts can be extended to image and video compression; typically, in place of 8 bit per pixel, one obtains excellent results using only 0.5 bit per pixel or a compression ratio of 16. It should be mentioned that without compression techniques, digital television or high-definition television (HDTV) will not be a reality. This approach is also very important for multimedia Internet applications, video conferencing, medical images, etc.

Another important application of multirate filters is the design of the transmultiplexer used in digital telephony and also in specialized digital communication such as spread spectrum, wireless, etc. The advantage of multirate systems is based on the capability to efficiently process signals of different bandwidth without the need of oversampling. For example, many applications of multirate processing are based on the principle that if an analog signal is sampled at a rate much higher than that specified by the sampling theorem, then a simpler anti aliasing filter can be used before digitalization. By oversampling, the quantization noise is spread over a wider frequency range, resulting in a lower equivalent in-band noise. This

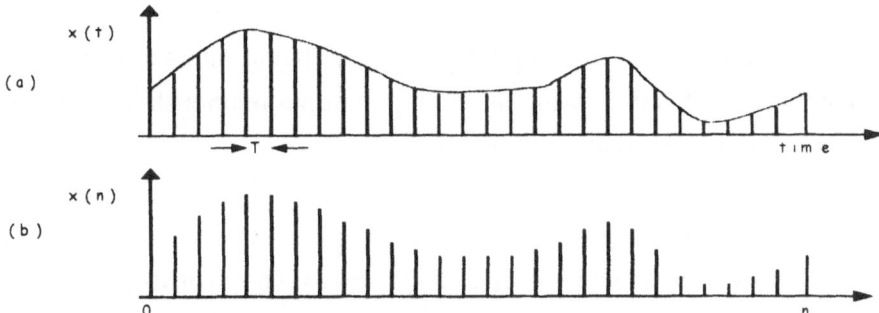

Figure 5.2.1 Discrete−time sampling $x(n)$ obtained by sampling the analog signal $x(t)$: (a) the analog signal with discrete values sampled with sampling period T; (b) normalized discrete sequence $x(n)$.

in turn results in a higher signal-to-noise ratio. Many of the high resolution ADCs (Analog to Digital Converters) (18, 20, 24 bits) in use today utilize internal multirate algorithms. Examples of such devices include the D8P56ADCX by Motorola and the CS532X from Crystal Semiconductors. Another area of application of the multirate is in the efficient implementation of narrow-band digital finite impulse response (FIR) filters. Using the traditional approach, an FIR filter will require a very large number of coefficients to meet their frequency response specifications. Utilization of multirate techniques allow the utilization of shorter filter lengths by performing the filtering of signal at a much lower rate. These and other topics will be discussed after we understand the fundamentals of multirate filters and the connection to wavelets, subbands, and multiresolution. Examples are given in the following sections.

5.2   Fundamentals of Digital Signal Processing

It is instructive to review the usual one-rate digital signal processing building blocks before we discuss multirate filters. Consider the example shown in Figure 5.2.1; the analog signal $x(t)$, which is bandlimited, is sampled at f_s, the sampling rate. The discrete signals are denoted by $x(nT)$ where T is the sampling time period. For convenience, it is customary to normalize $T = 1$, giving rise to normalized signals. The normalized angular bandwidth of the signal is thus π.

The fundamental building blocks for the digital filters are shown in Figure 5.2.2, with their symbols and input/output relationships. They are

- Amplifier, $y(n) = A \cdot x(n)$
- Adder, $y(n) = x_1(n) + x_2(n)$
- Multiplier, $y(n) = x_1(n) \cdot x_2(n)$
- Delay element by one or m-shifts, $y(n) = x(n-1)$ or $y(n) = x(n-m)$.

The transform of the single delay element is given by z^{-1}. For the readers not familiar with z-transform, it is recommended that they refer to the Appendix and the references therein for the details.

A typical FIR (finite impulse response) filter with tap weights $h(0)$, $h(1)$, .. , $h(N-1)$ etc. are shown in Figure 5.2.3. The output of this filter $y(n)$ is given by

$$Y(n) = \sum_{m=0}^{N-1} h(m)\, x(n-m) = \sum_{m=0}^{N-1} x(m)\, h(n-m) \qquad (5.2.1)$$

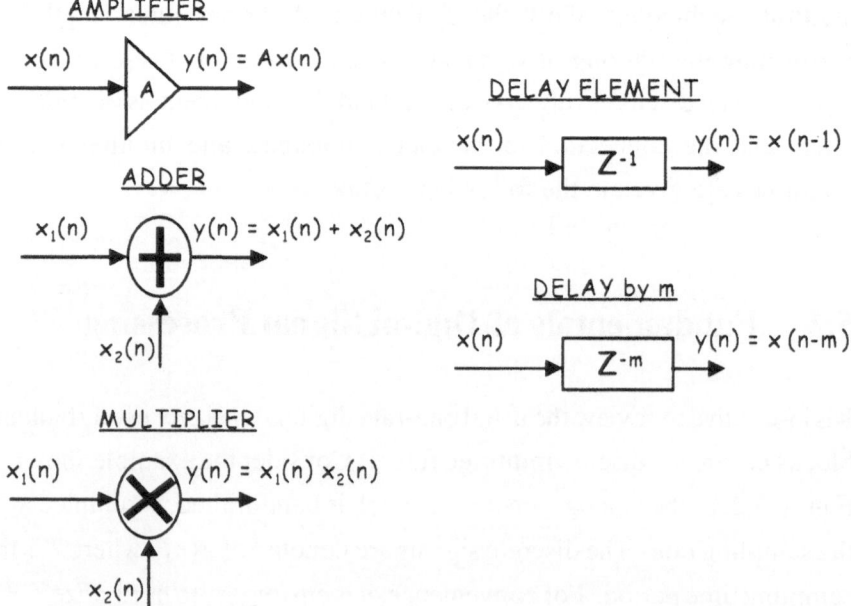

Figure 5.2.2. Elements of single-rate digital signal processing.

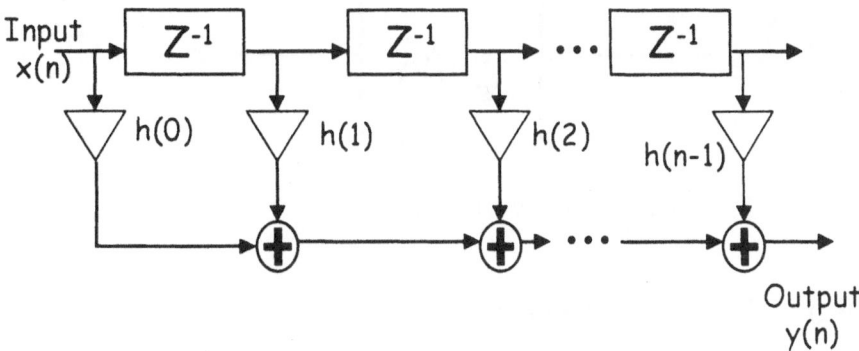

Figure 5.2.3. FIR filter with tap coefficients given by h(0), h(1),..., h(n).

where N is the number of taps. The above equation is the discrete version of the convolution equation and we have assumed a time-invariant linear system, which means that the tap coefficients are constant. In the z-transform domain Equation 5.2.1 becomes $Y(z) = H(z)X(z)$, where $H(z)$ is the z-transform of the filter impulse response $h(n)$. $H(z)$ is given by:

$$H(z) = \sum_{n=0}^{N-1} h(n)\, z^{-n} \quad . \tag{5.2.2}$$

The impulse response of the FIR filter is shown in Figure 5.2.4.

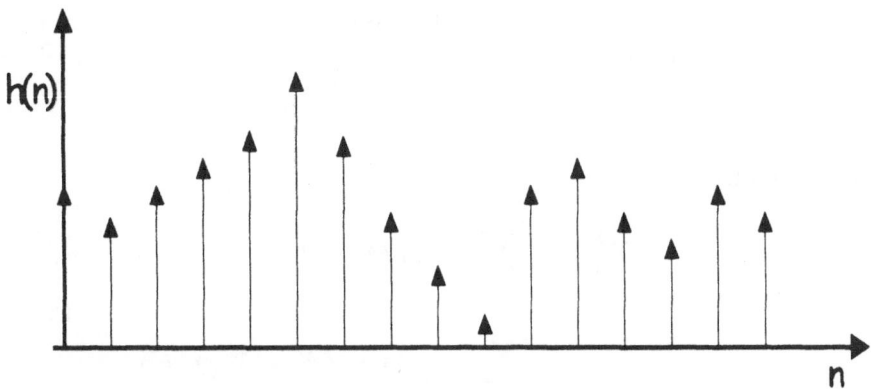

Figure 5.2.4. Delta response of FIR filter shown in Figure 5.2.3.

Figure 5.3.1. Decimation by a factor M.

5.3   Multirate Systems

In order to properly define the multirate filters, we need to introduce two more processing blocks:

- The decimator or down-sampler
- The interpolator or up-sampler or expander

Figure 5.3.1 shows the symbol for a M-fold decimator, with M a positive integer. The output $y(n)$ of the down-sampler is a compressed replica of the input signal $x(n)$, in which only every other M value of the input sequence is retained; that is,

$$y(n) = x(nM) , \qquad n \in Z . \tag{5.3.1}$$

Please note that both sequences extend for $n = -\infty$ to $n = +\infty$. An example of decimation by $M=2$ is given in Figure 5.3.2. If we define a new sequence $x_l(n)$ that is equal to $x(n)$ for values of n multiple of M and zero, otherwise

$$x_1(k) = \begin{cases} x(k) , & k = nM , \ n \in Z \\ 0 , & \text{otherwise} \end{cases} \tag{5.3.2}$$

then we have

$$Y(z) = \sum_{n=-\infty}^{+\infty} x_1(nM) \, z^{-n}$$

$$= \sum_{k=-\infty}^{+\infty} x_1(k) \, z^{-kM} = X_1(z^{1M}) . \tag{5.3.3}$$

If we define the function W_N as

$$W_N = e^{-j2\pi N}$$

(5.3.4)

it is easy to demonstrate that it satisfies the following properties:

$$(W_N)^N = \left(e^{-j2\pi N}\right)^N = 1$$

$$W_N^{kN+i} = W_N^i \;\;,\;\; \text{with}\;\; k, i \in Z$$

$$\sum_{k=0}^{N-1} W_N^{-kn} = \begin{cases} N\,, & n = lN\,,\;\; l \in Z \\ 0\,, & \text{otherwise} \end{cases}$$

(5.3.5)

We can thus write

$$x_1(n) = \left(\frac{1}{M} \sum_{k=0}^{M-1} W_M^{-kn}\right) x(n)$$

(5.3.6)

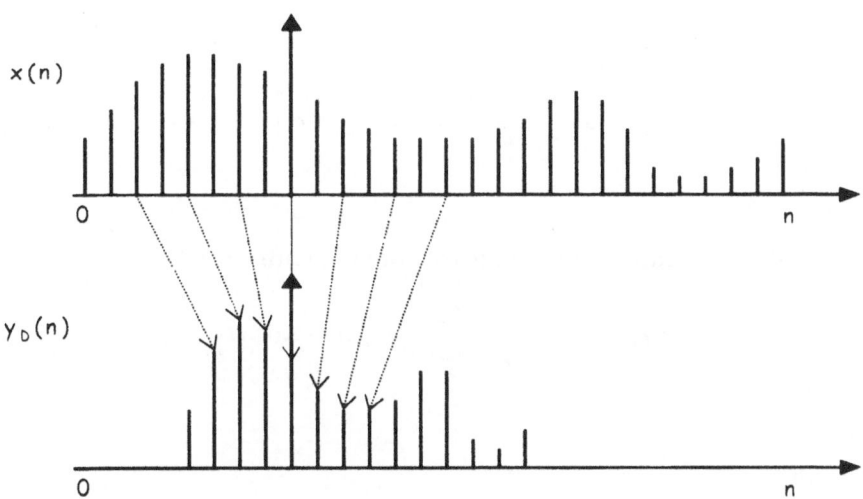

Figure 5.3.2. Decimation by $M=2$. Note that the samples $x(0)$, $x(2)$, $x(-2)$, etc. are retained.

and the z-transform is

$$X_1(z) = \sum_{n=-\infty}^{+\infty} x_1(n)z^{-n} = \sum_{n=-\infty}^{+\infty} \frac{1}{M} \sum_{k=0}^{M-1} W_M^{-kn} x(n) z^{-n}$$

$$X_1(z) = \frac{1}{M} \sum_{k=0}^{M-1} \sum_{n=-\infty}^{+\infty} x(n) \left(W_M^k z \right)^{-n} .$$

(5.3.7)

Equation 5.3.7 represents the z-transform with the substitution of $W_M^k z$ instead of z,

$$X_1(z) = \frac{1}{M} \sum_{k=0}^{M-1} X(W_M^k z) .$$

(5.3.8)

From the definition of $x_1(n)$, we can also see that $y(n)$ as defined in Equation 5.3.1 is also equivalent to $y(n) = x(nM) = x_1(nM)$, $n \in Z$; hence,

$$Y(z) = \sum_{n=-\infty}^{+\infty} x_1(nM) z^{-n} = \sum_{k=-\infty}^{+\infty} x_1(k) z^{-k/M} .$$

(5.3.9)

It follows that $Y(z) = X_1(z^{1/M})$ and, finally,

$$Y(z) = \frac{1}{M} \sum_{k=0}^{M-1} X (W_M^k z^{-k/M}) .$$

(5.3.10)

The above equation in the Fourier domain, with $z = e^{j\omega}$, becomes

$$Y(e^{-j\omega}) = \frac{1}{M} \sum_{k=0}^{M-1} X(e^{j(\omega - 2\pi k)/M})$$

(5.3.11)

with

$$W_M^k z^{1/M} = e^{-j2\pi k/M} e^{j\omega/M} = e^{-j(\omega - 2\pi k)/M} .$$

(5.3.12)

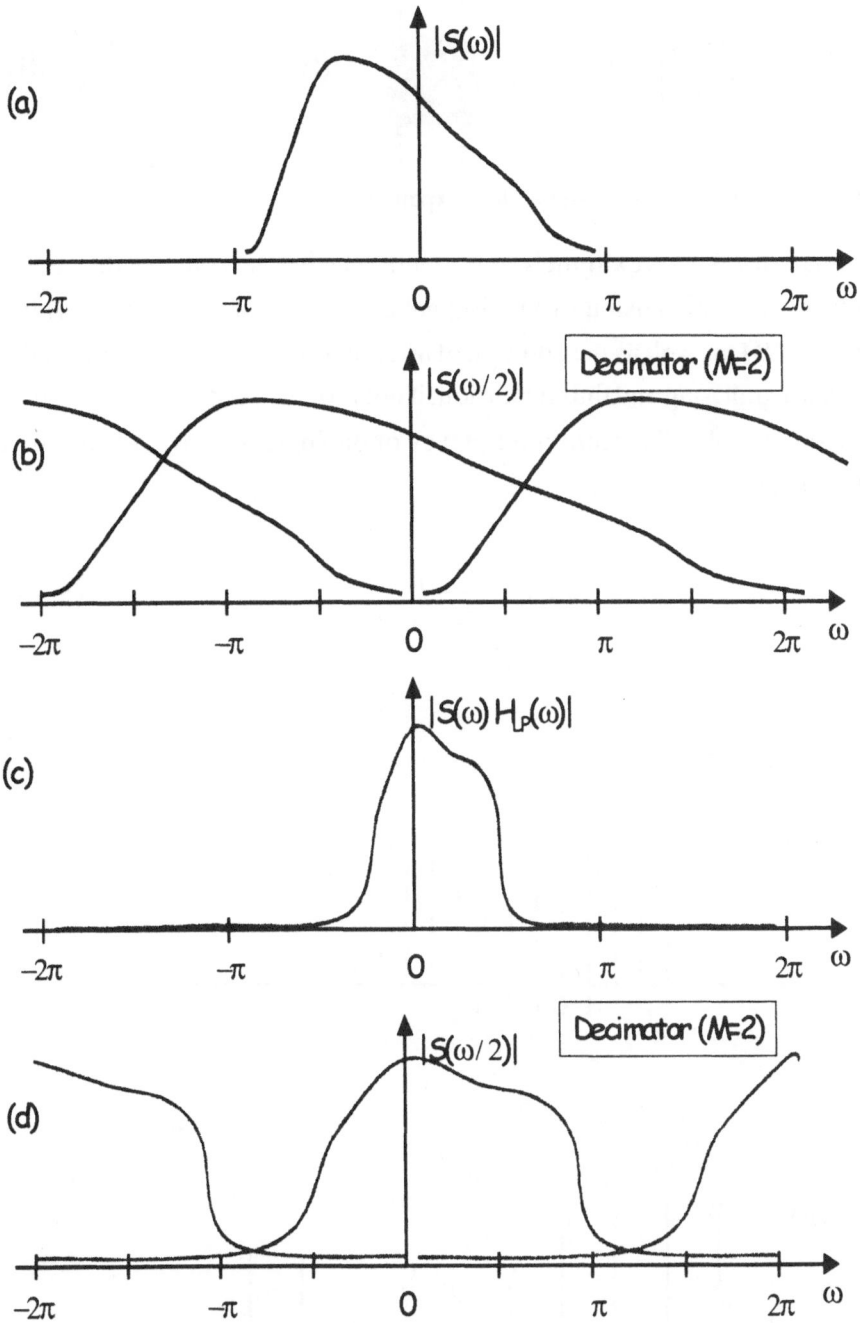

Figure 5.3.3 Decimation process in the frequency domain. If the original signal (a) has a bandwidth of the same order of the Nyquist band, decimation by 2 will result in aliasing as shown in (b). If a low-pass antialiasing filter is used prior to decimation (c), the correspondent output from the decimation is not aliased (d).

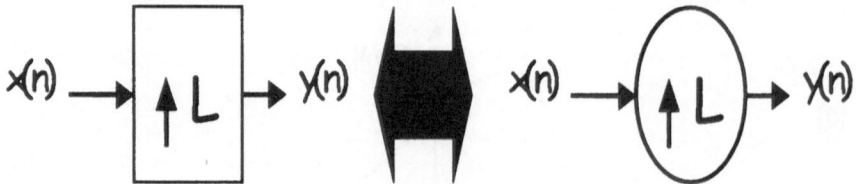

Figure 5.3.4 Symbols for the L-fold expander.

Equation 5.3.12 is extremely important, since it points out that decimation results in an expansion of the original bandwidth of the signal $x(n)$ by a factor $1/M$, as well as the addition of M-shifted versions. This will result in extensive aliasing and thus distortion, if not properly addressed, as shown in Figure 5.3.3b. The shift in frequency of the images of $X(\omega/M)$ occurs at frequencies

$$\omega_k = \frac{2\pi}{M} k \ .$$

(5.3.13)

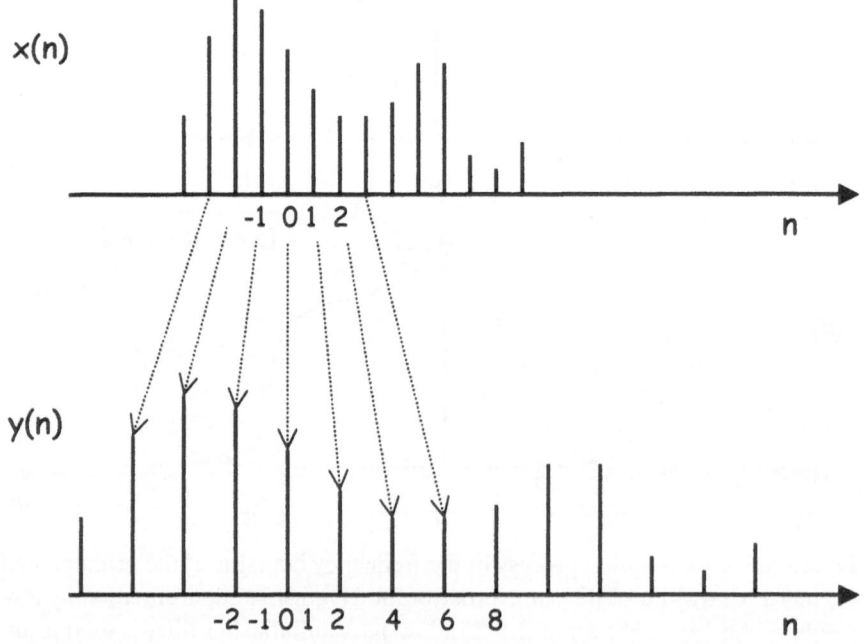

Figure 5.3.5 Graphical demonstration of the L-fold expander with $L=2$.

For the case in Figure 5.3.3, $M = 2$, the shift in frequency occurs at $\omega_1 = \pi$. If we add a low pass filter before the decimation with a stopband $\omega_{LP} < \pi/M$, then the images can be easily separated from the base signal.

The symbol for interpolation is shown in Figure 5.3.4, where L is a positive integer; it is also called an L-fold sampling rate expander. The output $Y(n)$ is given by $Y(n) = x(n/L)$ if n is an integer multiple of L, and $Y(n) = 0$ otherwise. For this case, the z-transform is given by

$$Y(z) = \sum_{n=-\infty}^{+\infty} y(n)\, z^{-n} = \sum_{n=\text{ multiple of } L} y(n)\, z^{-n} \tag{5.3.14}$$

$$Y(z) = \sum_{k=-\infty}^{+\infty} y(kL)\, z^{-kL} = \sum_{k=-\infty}^{+\infty} x(k)\, z^{-kL} \;. \tag{5.3.15}$$

Thus,

$$Y(z) = X(z^L) \tag{5.2.16}$$

or

$$Y(e^{j\omega}) = X(e^{j\omega L}) \;. \tag{5.3.17}$$

Examples of a decimator for $M = 2$ and an expander for $L = 2$ are shown in Figures 5.3.4 and 5.3.5, respectively. Figure 5.3.6 shows the expander in the frequency domain. Due to upsampling, the spectrum of $X(\omega)$ is contracted by a factor L. In this case, as shown in Figure 5.3.6, in addition to the original spectrum compressed by a factor L, we also have images created by the interleaving of zeros. Thus, we need a filter to suppress the unwanted extra images; this is also a low-pass filter with cutoff frequency given by $|\omega| < \pi/L$, similar to that used in the case of the decimator. The main difference is that the low-pass filter is used after interpolation. It is also to be mentioned that decimators and interpolators are linear systems but they are not time invariant. For the analysis of the multirate systems we will

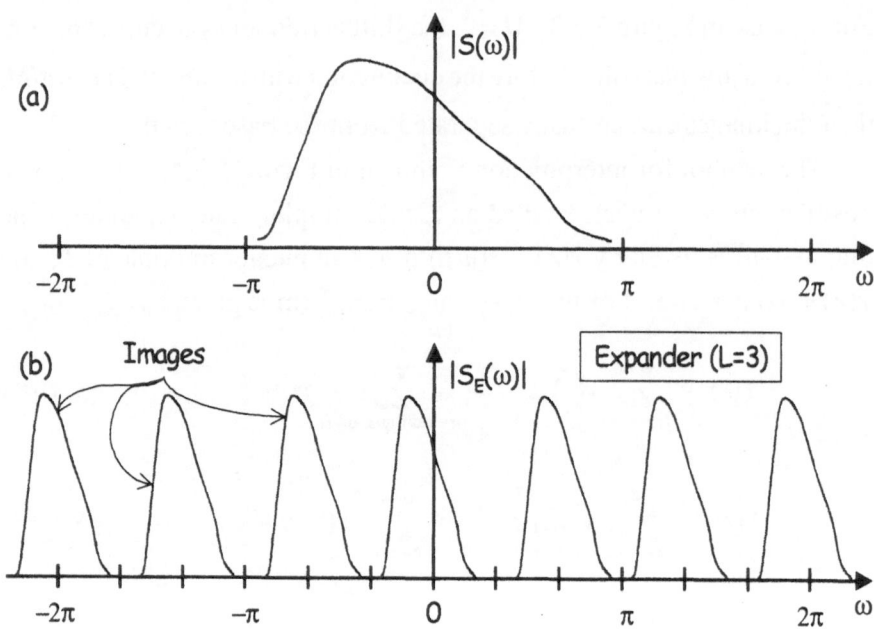

Figure 5.3.6 L-Fold expansion in the frequency domain with $L=3$.

consider later, it is convenient to derive the equivalent structures for the interpolators and decimators. They are also called *noble identities* and will be used often. Consider the decimator followed by a filter $H(z)$ as shown in Figure 5.3.7a. The output is $y(n)$. $Y(z)$ is given by

$$Y(z) = H(z)\frac{1}{M}\sum_{k=0}^{M-1} X\left(z^{1/M} W_M^k\right) . \tag{5.3.18}$$

In the equivalent structure in Figure 5.3.7b, the filter is in front, followed by the decimator. The filter z-response is, however, given by $H(z^M)$. To prove that the two structures are identical we note that $V(z)$, the output of the filter, is given by

$$V(z) = H(z^M) X(z) \tag{5.3.19}$$

and

(a)

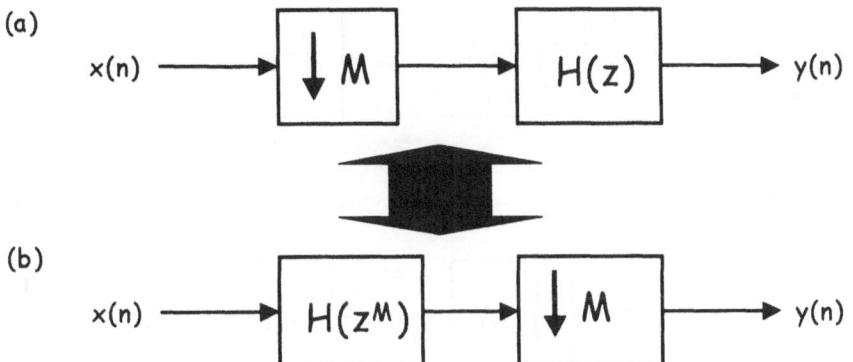

(b)

Figure 5.3.7 Noble indentities for M-fold decimator.

$$Y(z) = \frac{1}{M} \sum_{k=0}^{M-1} V\left(z^{1/M} W_m^k\right) .$$
(5.3.20)

Substituting into the above equations, we obtain

$$Y(z) = (1/M) \sum_{k=0}^{M-1} H(z^{1/M} W_m^k)^M X(z^{1/M} W_m^k)$$
(5.3.21)

$$Y(z) = (1/M) \sum_{k=0}^{M-1} H(z W_M^{kM}) X(z^{1/M} W_M^k)$$

or

$$Y(z) = (1/M) \sum_{k=0}^{M-1} H(z) X(z^{1/M} W_M^k)$$
(5.3.22)

as

$$W_M^{KM} = e^{jKM2\pi/M} = e^{jK2\pi} = 1 .$$
(5.3.23)

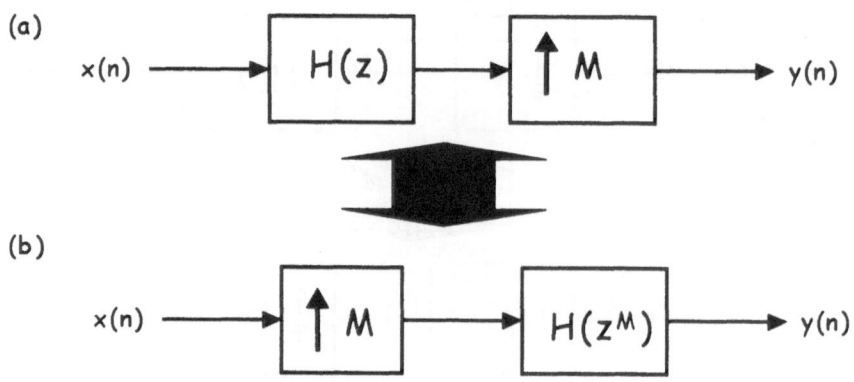

Figure 5.3.8 Noble identities for M-fold expander.

Equivalent structures can readily be proven to exist for the interpolator as well. It is to be mentioned that noble identities are only valid if $H(z)$ is rational. It can be the ratio of polynomials in z or z^{-1}. However, if it is irrational, i.e., $H(z) = z^{-1/2}$, the identities in Figures 5.3.7 and 5.3.8 are not valid.

5.4   Polyphase Decomposition

The importance of polyphase decomposition is best described by quoting Bellanger et al. (1976) as follows: "One of the reasons why multirate processing became practically attractive is the invention of the polyphase decomposition. This enables the designer to perform all computations at the lowest rate permissible within the given context, and reduces the speed requirements on the processors. Polyphase decomposition is useful in virtually every application of multirate signal processing, and often results in dramatic computational efficiency. It is valuable in theoretical study, practical design, and actual implementation of filter banks."

Let us explain the importance of polyphase decomposition by considering a system where a decimation by 2 is used; see Figure 5.4.1. We notice that at the output of such a system, we have actually computed has f_s samples per second, whereas $y(n)$ has only $f_s/2$ samples per second. Thus,

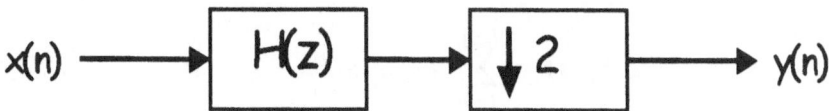

Figure 5.4.1 System with output decimated by a factor 2.

we are throwing away $f_s/2$ samples per second; this is computationally inefficient. As we shall see, what polyphase decomposition does is not to process the thrown away samples at all.

There are two types of polyphase decomposition, denoted as type I and type II. The idea behind this architecture can be explained by considering $H(z)$, a FIR filter response given by

$$H(z) = h_0 + h_1 z^{-1} + \cdots + h_{N-1} z^{N-1} . \qquad (5.4.1)$$

We can rewrite the above equation by putting together every Mth tap where we assume $L = N/M$ is an integer:

$$
\begin{aligned}
H(z) = & (h_0 + h_M z^{-M} + \cdots + h_{2M} z^{-2M} + \ldots) \\
& + (h_1 z^{-1} + h_{M+1} z^{-(M+1)} + h_{M+2} z^{-(M+2)} + \cdots) + \cdots \\
& + (h_{M-1} z^{-(M-1)} + h_{2M-1} z^{-(2M-1)} + \cdots) .
\end{aligned}
\qquad (5.4.2)
$$

We can thus write

$$
\begin{aligned}
H(z) = & (h_0 + h_M z^{-M} + h_{2M} z^{-2M} + \cdots + h_{(L-1)M} z^{-(L-1)M}) \\
& + z^{-1} \cdot (h_1 + h_{M+1} z^{-M} + \cdots + h_{(L-1)M+1} z^{-(L-1)M}) + \cdots \\
& + z^{-(M-1)} \cdot (h_{M-1} + h_{2M-1} z^{-M} + \cdots + h_{LM-1} z^{-(L-1)M})
\end{aligned}
\qquad (5.4.3)
$$

Let us define a polynomial function $E_k(z)$ such that

$$E_k(z) = \sum_{l=0}^{L-1} h_{k+lM}\, z^{-lM} \qquad (5.4.4)$$

hence,

$$E_0(z) = h_0 + h_M\, z^{-1} + \cdots + h_{(L-1)M}\, z^{-(L-1)} \qquad (5.4.5)$$

and

$$E_0(z^M) = h_0 + h_M\, z^{-M} + \cdots + h_{(L-1)M}\, z^{-(L-1)M} \ . \qquad (5.4.6)$$

Equation 5.4.3 thus becomes

$$H(Z) = \sum_{K=0}^{m-1} z^{-1} E_k(z^m)$$

$$= E_0(z^M) + z^{-1}E_1(z^M) + \cdots + z^{-(M-1)}E_{M-1}(z^M) \ . \qquad (5.4.7)$$

Equation 5.4.7 represents the type I polyphase decomposition, which is shown in Figure 5.4.2. Using the noble identity, we obtain the final type I polyphase decomposition, where the decimators are used before the filters $E_k(z)$, which reduces the computation burden enormously, as shown in Figure 5.4.3. It is instructive to consider an example where $M = 4$ and $N = 8$. For this case, we have $L = N/M = 2$:

$$H(z) = h_0 + h_1\, z^{-1} + \cdots + h_7\, z^{-7} \qquad (5.4.8)$$

$$H(z) = (h_0 + h_4\, z^{-4}) + (h_1\, z^{-1} + h_5\, z^{-5}) +$$
$$+ (h_2\, z^{-2} + h_6\, z^{-6}) + (h_3\, z^{-3} + h_7\, z^{-7}) \ . \qquad (5.4.9)$$

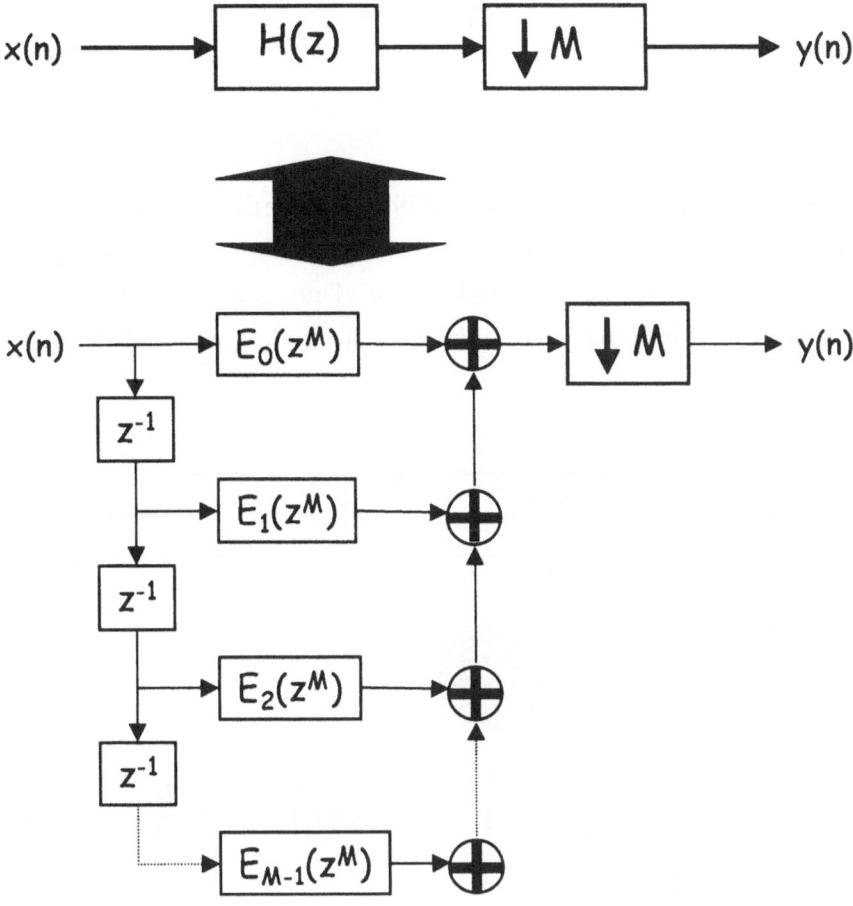

Figure 5.4.2 Type I polyphase decomposition of the filter $H(z)$.

The type I polyphase filter is thus given by

$$H(z) = E_0(z^4) + z^{-1}E_1(z^4) + z^{-2}E_2(z^4) + z^{-3}E_3(z^4) \quad (5.4.10)$$

where

$$E_0(z) = h_0 + h_4z^{-1} \quad\quad\quad (5.4.11)$$

$$E_1(z) = h_1 + h_5 \, z^{-1} \qquad (5.4.12)$$

and so on.

We can compare the computational savings for this case with and without polyphase decomposition. With polyphase decomposition, we need only eigth multiplications and four additions per sample. Without polyphase decomposition, we need 32 multiplications. Thus, polyphase decomposition provides a 4–fold savings in multiplications by requiring only four extra additions.

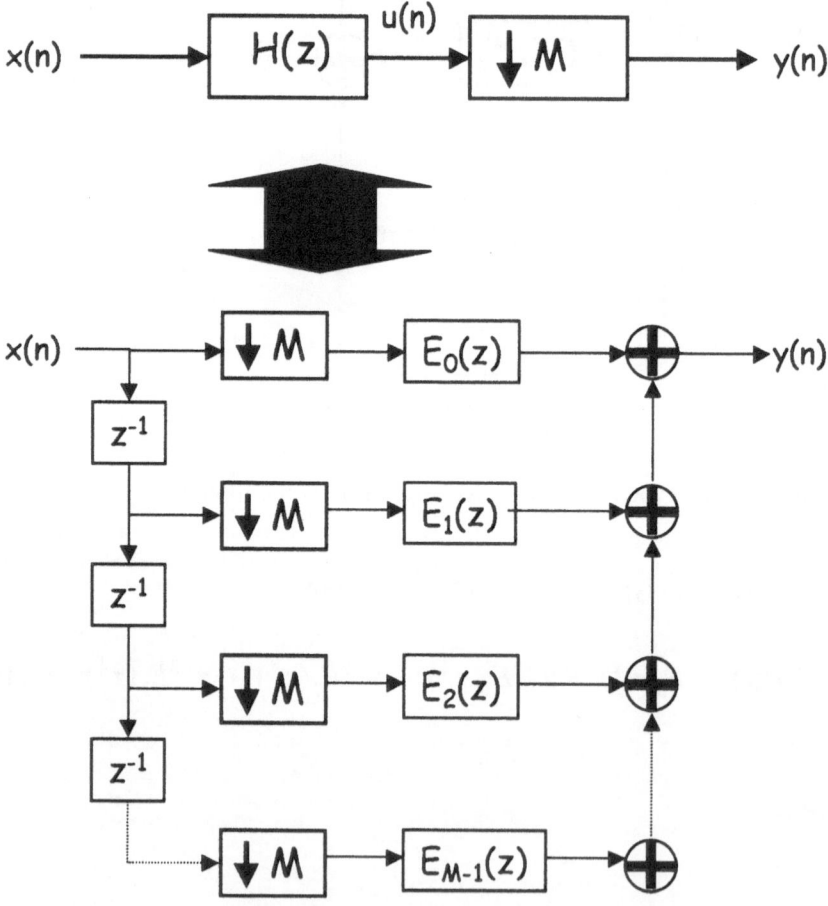

Figure 5.4.3 Type I polyphase decomposition of the filter $H(z)$ after using the noble identity.

We note that $H(z)$ can also be grouped starting from the end of the delay line as follows:

$$H(z) = \sum_{k=0}^{M-1} z^{-(M-1-k)} R_k(z^m)$$

$$= z^{-(M-1)} R_0(z^M) + z^{-(M-2)} R_1(z^M) + \cdots$$

(5.4.13)

where

$$R_K(z) = E_{M-1-K}(z)$$

(5.4.14)

Again using the noble identity for the decimator, we obtain the final result for the type II polyphase decomposition, as shown in Figure 5.4.4. For the example we considered earlier, we have

$$R_0(z) = E_{4-1-0}(z) = E_3(z)$$

$$R_1(z) = E_2(z)$$

$$R_2(z) = E_1(z)$$

$$R_3(z) = E_0(z) \ .$$

(5.4.15)

The type II polyphase filter is

$$H(z) = z^{-(4-1-3)} R_3(z^4) + z^{-(4-1-2)} R_2(z^4)$$

$$+ z^{-(4-1-1)} R_1(z^4) + z^{-(4-1-0)} R_0(z^4)$$

(5.4.16)

A type I polyphase representation of an analysis filter bank is shown in

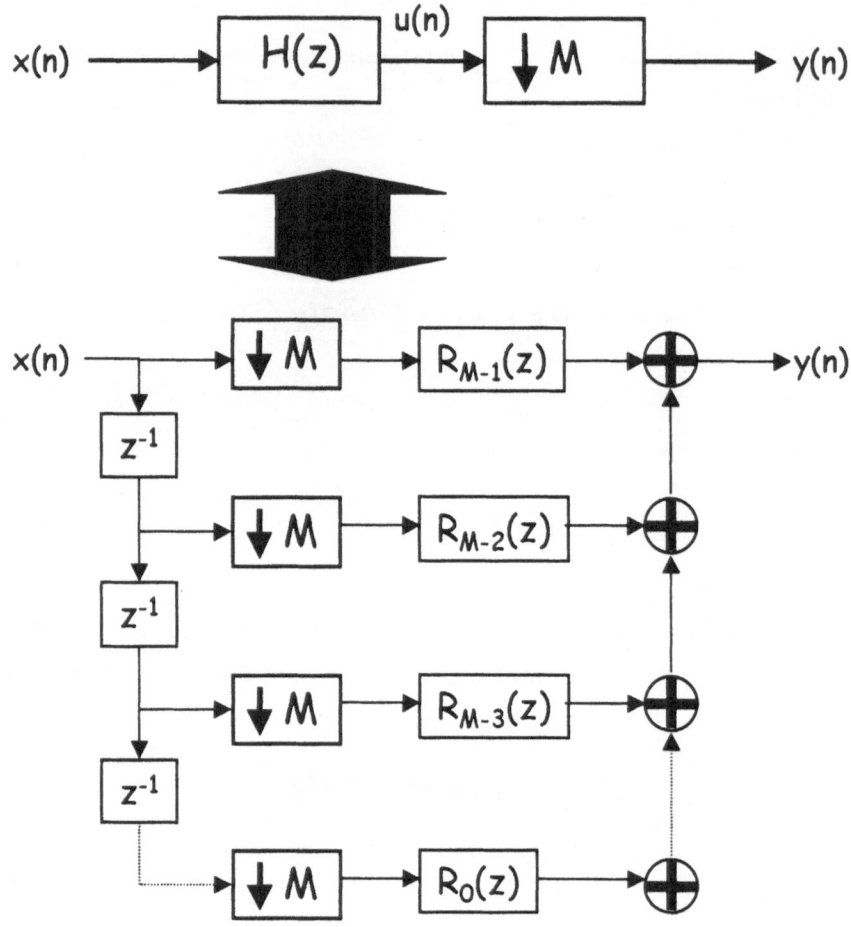

Figure 5.4.4 Type II polyphase decomposition of the filter $H(z)$ after using the noble identities.

Figure 5.4.6. Examples of decimation filters are given in Figures 5.4.7 and 5.4.8.

It is instructive to consider some practical examples related to improving the efficiency of fractional sampling mentioned earlier. The so-called M-band digital radio problem relates to splitting up the total bandwidth into M-bands as shown in Figure 5.4.9. Each band has identical narrow bandwidths. This is sometimes referred to as a channelized receiver. As each band is quite narrow, we need a very long FIR filter; typically, it can be $N = 1024$ taps or higher. Thus, we have for the mth band, the bandpass filter response $H_m(z)$ centered around a frequency of $2m\pi/M$ given by

$$H_m(z) \;=\; H_0(ze^{-j2\pi m/M})\qquad\qquad (5.4.17)$$

where

$$H_0(z) \;=\; \sum_{k=0}^{N-1} h_k\, z^{-k}$$

$$h_0(n) \;=\; \sum_{k=0}^{N-1} h_k\, \delta(n-k)\ \ .\qquad\qquad (5.4.18)$$

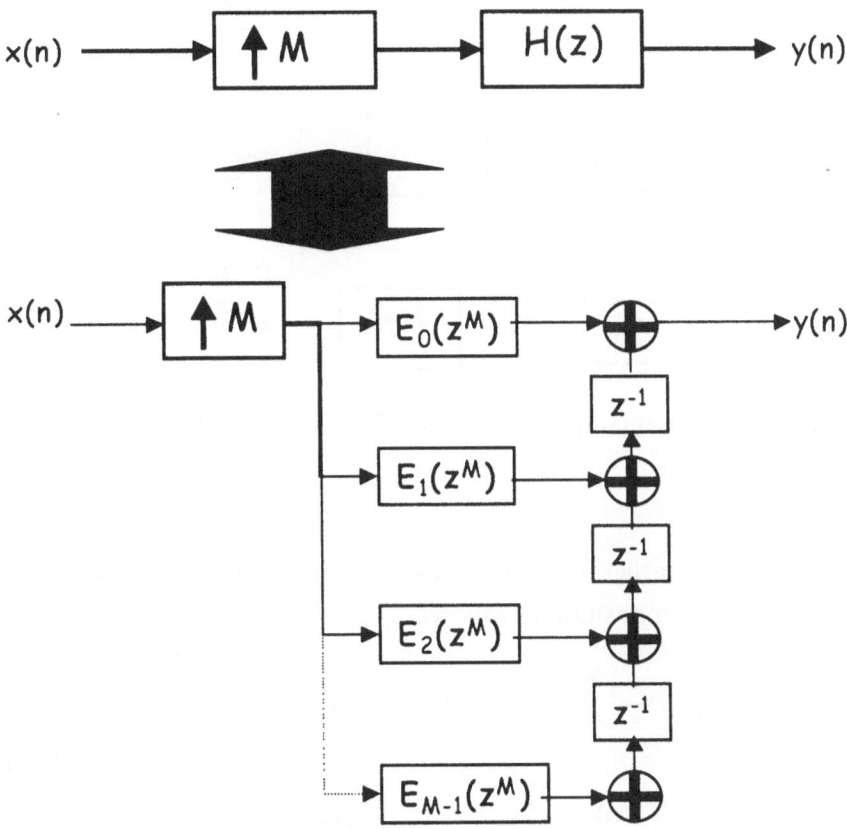

Figure 5.4.5 Type I polyphase decomposition of the filter $H(z)$ for an M-fold interpolator.

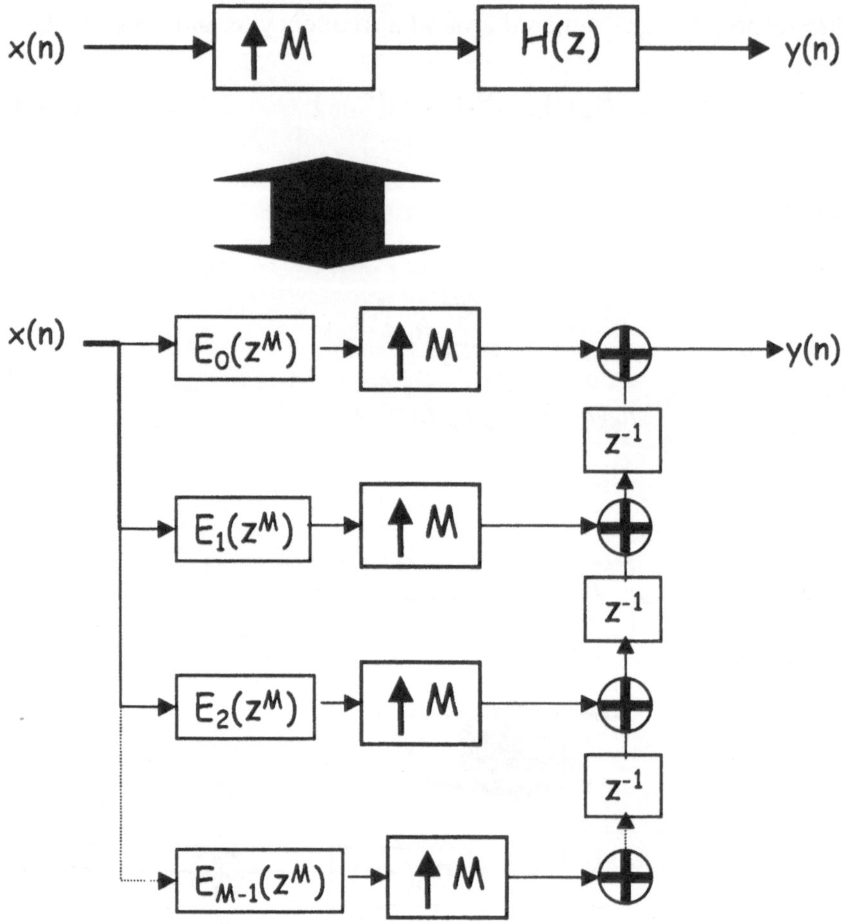

Figure 5.4.6 Type II polyphase decomposition of the filter $H(z)$ after using the noble identities.

As mentioned earlier, in general this N is very large. We can use type I polyphase decomposition as follows:

$$H_m(z) = \sum_{k=0}^{M-1} (ze^{-j2\pi m/M}) E_k(ze^{-j2\pi m/M})^M$$

(5.4.19)

$$H_m(z) = \sum_{k=0}^{M-1} z^{-k} W_M^{-mk} E_k(z^M) \ .$$

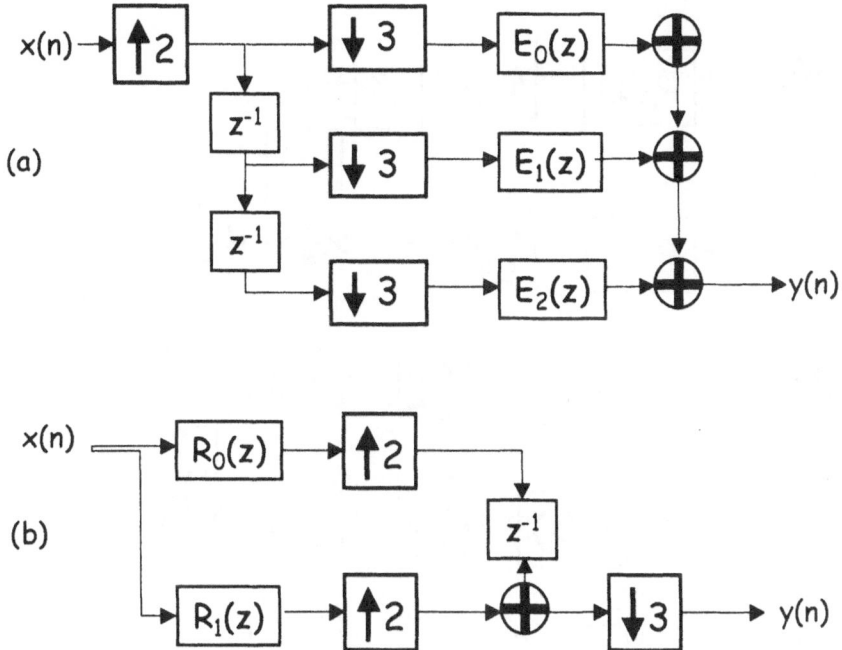

Figure 5.4.7 Two ways to improve the efficiency of the fractional decimation filter.

Equation 5.4.19 can be written in a matrix form, which contains the discrete Fourier transform (DFT) matrix for any N or the fast Fourier transform (FFT) matrix if $N = 2^p$, where p is an integer. Thus, the polyphase implementation involves the DFT matrix as shown in Figure 5.4.10. Actually, for practical implementation it is convenient to use a commutator or polyphase switch, which is shown in Figure 5.4.11. The commutator connects the i, $(i+M)$, $(i+2M)$, ..., samples to the first tap filter. It connects the $(i+1)$, $(i+M+1)$,..., to the second filter and so on. Also, note that E_k filters has N/M number of taps. Thus, the computational savings can be enormous. If $N = 1024$ taps and $M = 16$, the E's have $1024/16 = 32$ taps each. To implement it, for the ordinary case we need 1024×1024 multiplications whereas for the polyphase case, we need 1024 multiplications followed by a 16-point FFT. However, the crucial component for this implementation is the fast commutator.

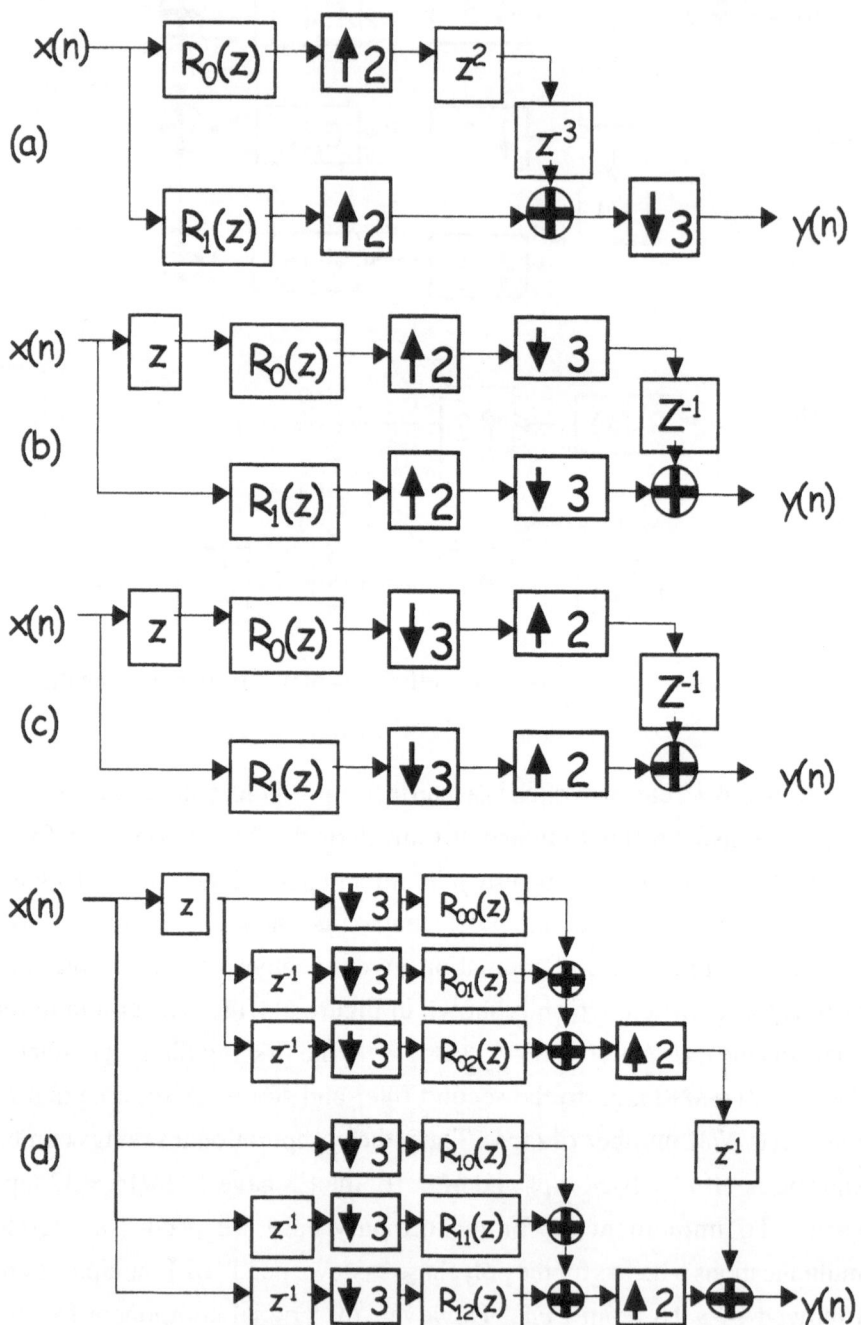

Figure 5.4.8 Possible implementations of the fractional decimation circuit, designed to maximize computational efficiency.

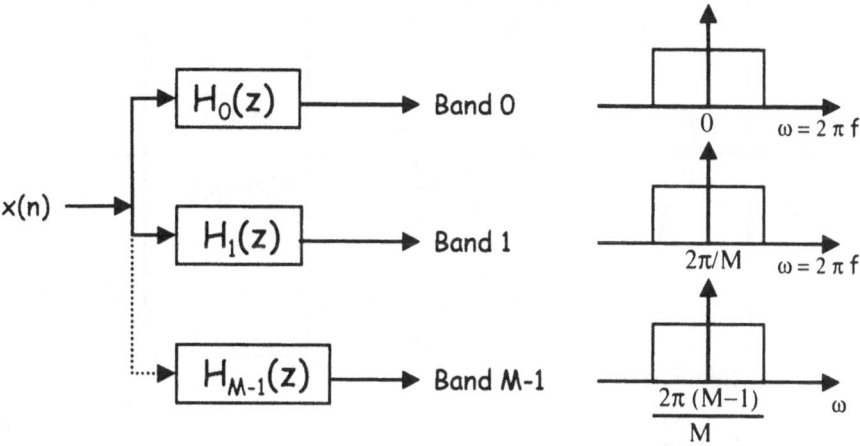

Figure 5.4.9 Block diagram and frequency response of the *M*-band digital radio. All bands have same bandwidth, with center frequency 0, $2\pi/M$, $4\pi/M$,..., $(M-1)2\pi/M$.

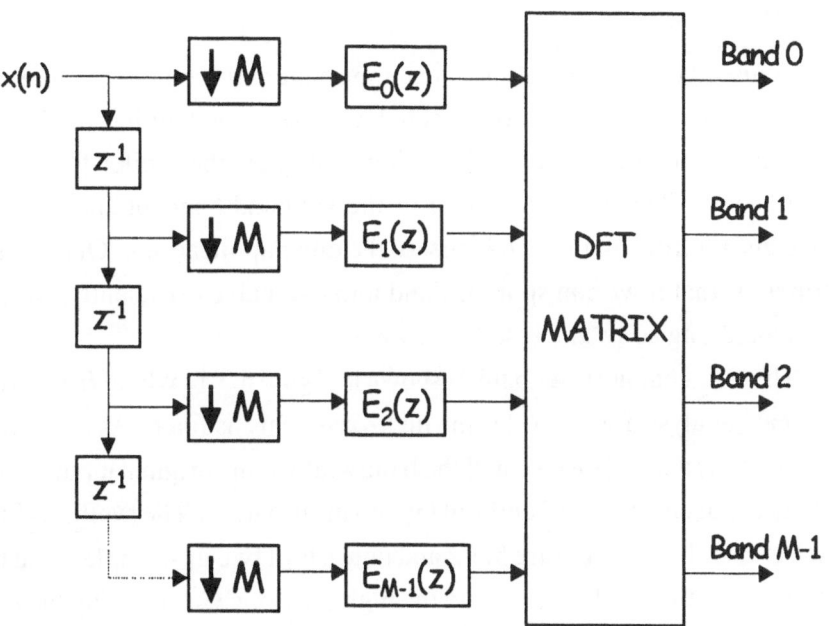

Figure 5.4.10 Block diagram of the *M*-band digital radio.

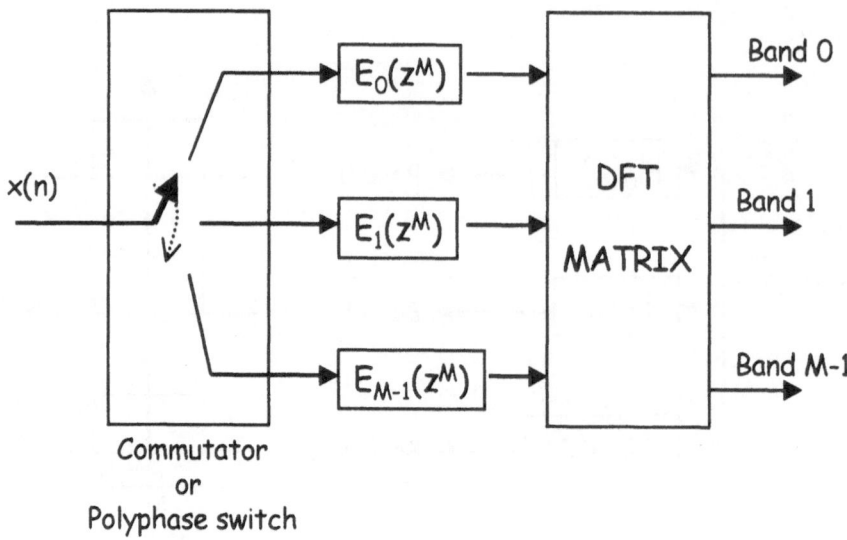

Figure 5.4.11 *M*-band digital radio implemented using polyphase decomposition and DFT matrix operation.

5.5   Two-Channel Filter Bank/PR Filter

As discussed earlier, one of the most important multirate filters is the one which splits a signal into two equal halves where one half has the lower-frequency components and the other half has the upper-frequency components. This is fundamental to the subband concept and leads to wavelets, multiresolution, and most of the other applications. Of course it is obvious that if we can split the band into two halves efficiently, we can continue doing so to any resolution we desire.

The two-channel filter bank is shown in Figure 5.5.1, where $H_0(z)$ is the low-pass analysis filter $H_1(z)$ is the high-pass analysis filter. As the output signals $\theta_0(n)$ and $\theta_1(n)$ have half the bandwidth of the original input signal, we can use decimation by 2 and still lose no information. The synthesis filter bank is also shown in Figure 5.5.2 and consists of two up-samplers and the low-pass synthesis filter $G_0(z)$ and the high-pass one $G_1(z)$. Using the type I polyphase decomposition of Figure 5.4.2, we obtain the following relations for the quantities defined in Figure 5.5.1:

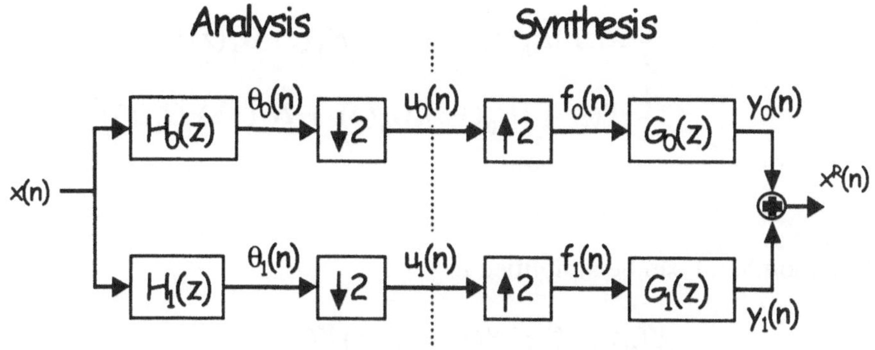

Figure 5.5.1 Two channel filter–bank structure showing analysis and synthesis stages.

$$\Theta_0(z) = H_0(z) \cdot x(z)$$

$$Y_0(z) = G_0(z) \cdot F_0(z)$$

$$V_0(z) = \frac{1}{2}[\Theta_0(z^{1/2}) + \Theta_0(-z^{1/2})]$$

$$F_0(z) = V_0(z^2) \ .$$

(5.5.1)

Substituting for $F_0(z)$ and $V_0(z)$, we obtain

$$Y_0(z) = \frac{1}{2} G_0(z) [H_0(z) X(z) + H_0(-z) X(-z)] \qquad (5.5.2)$$

and

$$Y_1(z) = \frac{1}{2} G_1(z) [H_1(z) X(z) + H_1(-z) X(-z)] \ . \qquad (5.5.3)$$

Finally, we obtain for the output $\hat{X}(z)$,

$$\hat{X}(z) = \frac{1}{2} X(z) [H_0(z) G_0(z) + H_1(z) G_1(z)]$$
$$+ \frac{1}{2} X(-z) [H_0(-z) G_0(z) + H_1(-z) G_1(z)] \tag{5.5.4}$$

Equation 5.5.4 can be simplified to

$$\hat{X}(z) = T(z)X(z) + S(z)X(-z) \tag{5.5.5}$$

where we have defined

$$T(z) = \frac{1}{2} [H_0(z)G_0(z) + H_1(z)G_1(z)]$$
$$S(z) = \frac{1}{2} [H_0(-z)G_0(z) + H_1(-z)G_1(z)] \quad . \tag{5.5.6}$$

In order to have perfect reconstruction (PR) at the synthesis, we must impose that

$$\hat{X}(z) = c \, X(z) \, z^{-n_0} \tag{5.5.7}$$

where c is a constant and n_0 is a fixed delay.
Thus, the conditions for PR are

$$S(z) = 0 = H_0(-z)G_0(z) + H_1(-z)G_1(z) \tag{5.5.8}$$

and

$$T(z) = H_0(z)G_0(z) + H_1(z)G_1(z) = cz^{-n_0} \quad . \tag{5.5.9}$$

From the above equations, we obtain

$$\frac{G_0(z)}{G_1(z)} = - \frac{H_1(-z)}{H_0(-z)} \tag{5.5.10}$$

Before we find the PR solution, it is of interest to note that $T(z)$ represents distortion and $S(z)$ represents aliasing. Equation 5.5.7 can also be written in matrix form as follows:

$$2\,\hat{X}(z) = [X(z) \quad X(-z)] \begin{bmatrix} H_0(z) & H_1(z) \\ H_0(-z) & H_1(-z) \end{bmatrix} \begin{bmatrix} G_0(z) \\ G_1(z) \end{bmatrix} \tag{5.5.11}$$

The 2×2 matrix on the right-hand side of the equation is called *alias component matrix*, or *A-C matrix*, $H_{AC}(z)$ given by

$$H_{AC}(z) = \begin{bmatrix} H_0(z) & H_1(z) \\ H_0(-z) & H_1(-z) \end{bmatrix} \tag{5.5.12}$$

As we will see later, this matrix formulation is very convenient for the M-band case and the above equations are a special case for $M=2$. Historically, Esteban et al. derived the results of the two-channel filters discussed above and chose

$$H_1(z) = H_0(-z) \tag{5.5.13}$$

which can be rewritten for real coefficients as

$$|H_1(e^{j\omega})| = |H_0[e^{j(\pi-\omega)}]| \tag{5.5.14}$$

which is the *quadrature mirror filter* (QMF) solution.

Note that for the case in which $H_0(z) = \sum h_n z^{-n}$, the highpass filter $H_1(z)$ is also a FIR filter that has the same tap weight but with alternating sign i.e., $H_1(z) = \sum (-1)^n h_n z^{-n}$. Furthermore, to cancel the aliasing term, they chose

$$G_0(z) = H_1(-z) = H_0(z)$$

$$G_1(z) = -H_1(z)$$

(5.5.15)

which allow us to define the *distortion transfer function, T(z)*, as

$$T(z) = \frac{1}{2}\left[H_0^2(z) - H_1^2(z)\right].$$

(5.5.16)

Thus, the Esteban solution is not PR, but it has no aliasing error. Actually if $T(z)$ is nonzero, we can write

$$T(z) = |T(e^{j\omega})| \; e^{j\phi(\omega)}$$

(5.5.17)

where $|T(e^{j\omega})|$ represents the amplitude distortion function and $\phi(\omega)$ represents the phase distortion function. For the linear phase case, $\phi(\omega) = a + b\omega$.

Before the PR solution was obtained by Smith et al. (which we discuss shortly), the design of filters involved reducing the amplitude and phase distortion function. Actually, for many practical purposes, these filters, although not PR, are still very useful and include the well-known Johnston filter.

One of the first PR solutions was introduced by Smith et al.; in this case, they chose

$$G_0(z) = -H_1(-z)$$

$$G_1(z) = H_0(-z)$$

(5.5.18)

To cancel the aliasing term, we substitute these into the previous expressions for $T(z)$ and obtain

$$T(z) = \frac{1}{2} [H_0(z)G_0(z) + H_1(z)G(z)]$$

$$T(z) = \frac{1}{2} [-H_0(z)H_1(-z) + H_1(z)H_0(-z)] \qquad (5.5.19)$$

$$T(z) = \frac{1}{2} [H_0(-z)H_1(z) - H_0(z)H_1(-z)] \ .$$

They also chose the *FIR para-unitary solution* for N tap, with N even as

$$H_1(z) = z^{-(N-1)} H_0(-z^{-1}) \ . \qquad (5.5.20)$$

Note that this is also a QMF solution. Note that for this case, the tap coefficients of $H_1(z)$ have alternating flips; that is, first flip the coefficients from either end and then flip the sign for every other one. Substituting into the above equations, we can obtain

$$T(z) = \frac{1}{2}[H_0(-z)H_0(-z^{-(N-1)})z^{-(N-1)} - H_0(z)H_0(-z^{-1})z^{-(N-1)}]$$

$$T(z) = \frac{1}{2}z^{-(N-1)}[H_0(z)H_0(z^{-1}) - H_0(-z)H_0(-z^{-1})]$$

$$T(z) = \frac{1}{2}z^{-(N-1)}Q(z) \qquad (5.5.21)$$

where

$$Q(z) = [H_0(z) H_0(z^{-1}) - H_0(-z) H_0(-z^{-1})]$$

$$Q(z) = R(z) + R(-z) \ . \qquad (5.5.21a)$$

The PR condition is to make $Q(z)$ a constant. This condition should give equations to determine the tap weights where the spectral density function, $R(z)$, is given by

$$R(z) = H_0(z) \, H_0(z^{-1}) \ . \tag{5.5.22}$$

We can write, in general,

$$R(z) = \gamma_{N-1} z^{N-1} + \gamma_{N-2} z^{N-2} + \cdots + \gamma_0 z^0$$
$$+ \ \gamma_1 z^{-1} + \cdots + \gamma_N z^{-(N-1)} \ . \tag{5.5.23}$$

To clarify the meaning of the above equation, consider a four-tap case given by

$$H_0(z) = a_0 + a_1 z^{-1} + a_2 z^{-2} + a_3 z^{-3}$$
$$H_0(z^{-1}) = a_0 + a_1 z + a_2 z^2 + a_3 z^3 \ . \tag{5.5.24}$$

$R(z)$ for this case is given by

$$R(z) = a_0 a_3 z^{-3} + (a_0 a_2 + a_1 a_3) z^{-2} + (a_0 a_1 + a_1 a_2 + a_2 a_3) z^{-1}$$
$$+ \ (a_0^2 + a_1^2 + a_2^2 + a_3^2)$$
$$+ \ (a_0 a_1 + a_1 a_2 + a_2 a_3) z + (a_0 a_2 + a_1 a_3) z^2 + a_0 a_3 z^3 \tag{5.5.25}$$

From these expressions, we get

$$R(-z) = -\gamma_{(N-1)} z^{N-1} + \gamma_{N-2} z^{N-2} + \cdots + \gamma_0 z^0$$
$$- \ \gamma_1 z^{-1} + \cdots - \gamma_{N-1} z^{-(N-1)} \ . \tag{5.5.26}$$

as we have chosen N to be even. Thus, $Q(z)$ is given by

$$Q(z) = R(z) + R(-z) = 2 \left[\gamma_{N-2} z^{N-2} + \cdots \right] \tag{5.5.27}$$

and only contains even powers of z. Thus for PR condition, we make $r_0 \neq 0$ but all other r's equal to 0. This makes

$$Q(z) = \gamma_0 = \text{constant}. \tag{5.5.28}$$

To understand the meaning of the equations given by the PR condition, that is,

$$\gamma_{N-2} = \gamma_{N-4} = \cdots = 0 \tag{5.5.29}$$

let us consider the case for $N = 4$ mentioned earlier. For this case, we have

$$a_0^2 + a_1^2 + a_2^2 + a_3^2 = 1$$
$$a_0 a_2 + a_1 a_3 = 0 \quad . \tag{5.5.30}$$

Note that we have 4 unknowns (a_0, a_1, a_2, a_3). However, we have only two equations. Thus, many PR solutions for this are possible. However, we can choose two other conditions for this case to optimize some other performance or criteria. One particular case to be discussed shortly deals with regularity or smoothness.

In general, for N-tap solution discussed above, we get $N/2$ equations and only $N/2$ conditions can be imposed. The equations discussed above can also be derived from the time domain analysis and are highly instructive. For the four-tap case, we have

$$h(n) = a_0 \delta(n) + a_1 \delta(n-1) + a_2 \delta(n-2) + a_3 \delta(n-3) \tag{5.5.31}$$

Using the definition of the autocorrelation function

$$\rho(u) = h(u) \otimes h(n) \tag{5.5.32}$$

we obtain

$$\rho(u) = \sum h(k)\, h(k + u)$$

$$= (a_0^2 + a_1^2 + a_2^2 + a_3^2)\, \delta(n)$$

$$+ (a_0 a_1 + a_1 a_2 + a_2 a_3)\, \delta(n-1)$$

$$+ (a_0 a_2 + a_1 a_3)\delta(n-2) + a_0 a_3 \delta(n-3) + \cdots$$

$$\text{(5.5.33)}$$

The PR condition is equivalent to impose that all values of $\rho(2n)$ for $n>0$, are zero:

$$\rho(2n) = \sum_{K=0}^{N-1} h(k)\, h(k+2n) = 0 \quad . \tag{5.5.34}$$

We mentioned earlier that the regularity condition or smoothness condition is often used to provide the other $N/2$ equations. Smoothness or regularity relates to the derivative of the filter response at $\omega = \pi$. If we expand $H(\omega)$ as a Taylor series, then we can demand that the first nonlinear term will occur at $N/2 = p$. We have

$$\frac{d^P H(\omega)}{d\omega^P} = (-j)^P \sum_{h=0}^{N-1} h_n\, n^P\, e^{-jn\omega} \tag{5.5.35}$$

and evaluating the pth derivative at π, this condition corresponds to

$$\frac{d^P H(\omega)}{d\omega^P}\bigg|_{\omega=\pi} = (-j)^P \sum_{h=0}^{N-1} h_n\, n^P\, (-1)^n \quad . \tag{5.5.36}$$

In order to have a smooth function of the order $N/2$, we must impose that all the derivatives calculated in Equation 5.5.36 be zero at $\omega = \pi$ for $p = 0, ...,$ $N/2 - 1$:

$$\sum_{h=0}^{N-1} h_n\, n^P\, (-1)^n = 0 \quad , \qquad \text{for} \quad p = 0,1, ... , \frac{N}{2} - 1 \tag{5.5.37}$$

For the four-tap case, we thus have the following set of four equations in four unknowns:

$$a_0 - a_1 + a_2 - a_3 = 0$$

$$-a_1 + 2a_2 - 3a_3 = 0$$

$$a_0^2 + a_1^2 + a_2^2 + a_3^2 = 1$$

$$a_0 a_2 + a_1 a_3 = 0$$

(5.5.38)

5.6   Biorthogonal Filters

Let us summarize what we have done so far to obtain the PR solution of the two-channel filter. To avoid distortion, we must satisfy Equation 5.5.9, repeated here:

$$T(z) = cz^{-n_0}$$

$$H_0(z)G_0(z) + H_1(z)G_1(z) = cz^{-n_0} .$$

(5.6.1)

To avoid aliasing errors, we must also impose that:

$$S(z) = 0$$

$$H_0(-z)G_0(z) + H_1(-z)G_1(z) = 0 .$$

(5.6.2)

In the last section we have discussed the orthogonal solution to the above conditions. However, a more general choice can be obtained, which is rather important from the practical point of view. The orthogonal case can be shown to give solutions which can never be linear phase. However, that is not true for the biorthogonal case. For this case, let us choose

$$G_0(z) = H_1(-z)$$
$$G_1(z) = -H_0(-z)$$

(5.6.3)

which are the same conditions we have imposed in the orthogonal case (Equation 5.5.18) except for sign changes. For this case, $T(z)$ becomes

$$T(z) = G_0(z)H_0(z) - G_0(-z)H_0(-z) = P_0(z) - P_0(-z) \quad (5.6.4)$$

having imposed

$$P_0(z) = H_0(z)G_0(z) \quad . \tag{5.6.5}$$

The PR conditions is given by

$$P_0(z) - P_0(-z) = z^{-l} \quad . \tag{5.6.6}$$

We note that substituting $-z$ for z, we have

$$P_0(-z) - P_0(z) = -z^{-l} \quad . \tag{5.6.7}$$

Thus, l must be odd for this case. The design procedure for the biorthogonal filters is as follows:

- Design the low-pass filters $H_0(z)$ and $G_0(z)$ using Equation 5.5.10.
- Factor $P_0(z)$ into $H_0(z)$ and $G_0(z)$.

If we define $P(z) = z^l P_0(z)$, then the PR condition becomes

$$P(z) + P(-z) = 2 \tag{5.6.8}$$

Thus all the even powers of z in $P(z)$ must be zero except the z^0 term. It is to be noted that this is the same condition for the half-band filter. We can choose

$$P_0(z) = (1+z^{-l})^N Q_{N-2}(z) \ . \tag{5.6.9}$$

Then, we find

$$P(z) = z^{-l} P_0(z) \tag{5.6.10}$$

for some value l. Finally, we can choose $Q_{N-2}(z)$ to cancel the even powers. The following example will clarify the procedure. Let us define $P_0(z)$ as

$$P_0(z) = (1+z^{-l})^4(-1+4z^{-l} - z^{-2}) \frac{1}{16}$$

$$= \frac{1}{16} (-1 + 9z^{-2} + 16z^{-3} + 9z^{-4} - z^{-6}) \ . \tag{5.6.11}$$

$P_0(z)$ has six roots, four of which are zeros, one is $c = 2 - \sqrt{3}$, and the other is $1/c = 2 + \sqrt{3}$. Thus, $P_0(z)$ can be written as

$$P_0(z) = (1+ z^{-1})^4 (z^{-1}-c) (z^{-1}-1/c) = H_0(z) G_0(z) \tag{5.6.12}$$

The factorization of $P_0(z)$ into $H_0(z)$ and $G_0(z)$ can be performed in many ways. The following is a list of these possible choices, using the labeling scheme of defining the order of H_0 and G_0:

A. Impose $G_0(z) = 1$ (order $N = 0$), then $H_0(z) = P_0(z)$.
B. Impose $G_0(z) = (1 + z^{-1})$ (order $N = 1$), then $H_0(z) = P_0(z)/G_0(z)$.
C. Impose $G_0(z) = (1 + z^{-1})^2$ or $G_0(z) = (1 + z^{-1}) \cdot (z^{-1} - c)$ (order $N = 2$), then $H_0(z) = P_0(z)/G_0(z)$.
D. Impose $G_0(z) = (1 + z^{-1})^3$ or $G_0(z) = (1 + z^{-1})^2 \cdot (z^{-1} - c)$ (order $N = 3$), then $H_0(z) = P_0(z)/G_0(z)$.

Let us consider case C. In this case, we can impose

$$H_0(z) = (1+z^{-1})^2 \, (z^{-1}-c) \, (z^{-1}-1/c) \qquad (5.6.13)$$

which has five taps and

$$G_0(z) = (1+z^{-1})^2 \qquad (5.6.14)$$

which has three taps. This is called a 5/3 filter.
On the other hand, we can choose

$$H_0(z) = (1+z^{-1})^2$$
$$G_0(z) = (1+z^{-1})^2 \, (z^{-1}-c) \, (z^{-1}-1/c) \quad . \qquad (5.6.15)$$

This is called a 3/5 filter.
Case D represents the 4/4 filter

$$H_0(z) = (1+z^{-1})^3$$
$$G_0(z) = (1+z^{-1}) \, (z^{-1}-c) \, (z^{-1}-1/c) \quad . \qquad (5.6.16)$$

This filter is symmetric and it has a linear phase. It can be shown that for a biorthogonal linear phase filter band with two channels, the filter lengths are all odd or all even.
The analysis filters can be
    (i) both symmetric of odd length
    (ii) one symmetric and the other antisymmetric of even length.

5.7   Lifting Scheme

In the last section, we solved the following PR condition to obtain different biorthogonal filters:

$$H_0(z)G_0(z) = -G_0(-z)H_0(-z) = 2z^{-l}, \quad \text{for } l = \text{odd}. \quad (5.7.1)$$

Actually, we only used the particular solution for the above equation. By keeping $G_0(z)$ fixed, once we have a solution $H_0(z)$, and we can find other solutions by adding the homogeneous term. Consider this $H_0^{new}(z)$ to be given by

$$H_0^{new}(z) = H_0(z) + G_0(-z) S(z^2) \quad (5.7.2)$$

for any $S(z)$. Then, the PR condition becomes

$$
\begin{aligned}
&H_0^{new}(z)G_0(z) - G_0(-z)H^{new}(-z) \\
&= H_0(z)G_0(z) + G_0(z)G_0(-z)S(z^2) \\
&\quad - G_0(-z)H_0(-z) - G_0(-z)G_0(z)S(z^2) \\
&= H_0(z)G_0(z) - G(-z)H(-z) = 2z^{-l}.
\end{aligned}
\quad (5.7.3)
$$

Thus, if $H_0(z)$ is a solution, then $H_0^{new}(z)$ must also be a solution. The particular solution corresponds to the $2z^{-l}$ term, whereas $G_0(z) S(z^2)$ term corresponds to the 0 term or the homogeneous solution. The *lifting scheme* is to generate new solutions starting from the most elementary solution, which is the lazy filter, one tap with tap weight 1. The other filters can be obtained by choosing different $S(z^2)$.

We have not discussed the connection between these filters and the scaling function and wavelet equation. For the lifting scheme, the scaling function remains the same. However, the wavelet equation is modified

$$\psi^{new}(t) = \psi(t) - \sum S(k) \phi(t-n). \quad (5.7.4)$$

We can also perform what is known as *dual lifting*. In place of keeping $G_0(z)$ fixed, we can fix $H_0(z)$ and then we can have a new solution $G_0^{new}(z)$ to be given by

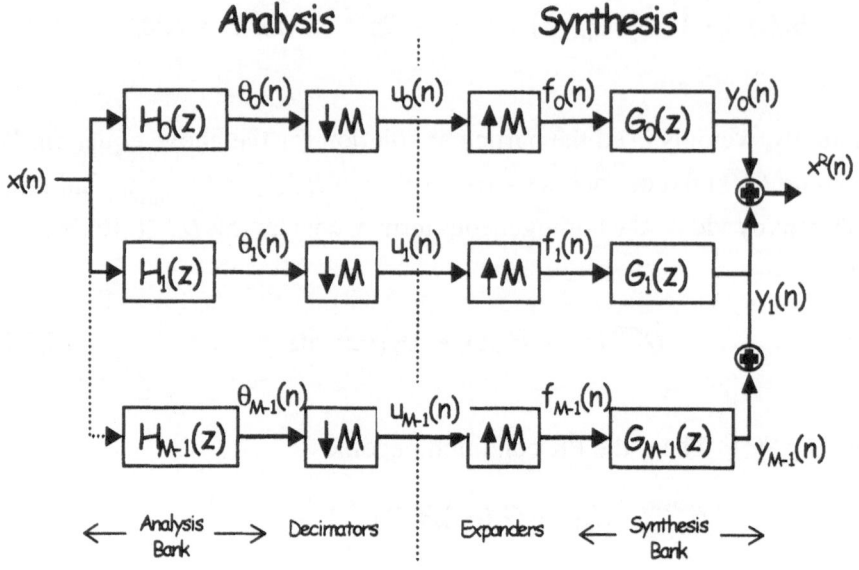

Figure 5.8.1 General M-band analysis and synthesis filterbank structure.

$$G_0^{\text{new}}(z) = G_0(H) + H_0(-z)\, T(z^2) \qquad (5.7.5)$$

for any $T(z)$. These liftings and dual lifting schemes will be used to generate filters with binary tap coefficients, which are easier to implement and are quite useful in the image compression.

5.8   M-Band Case

Most of our discussion has concentrated on two-channel filter banks. In this section, we consider the general M-channel filter bank. Of course, for $M = 2$, we will get the results already discussed. The general M-band case is mathematically quite complex. However, by considering the general case, many interesting insights of the subband filters can be gained.

Figure 5.8.1 shows general M-bank analysis and synthesis filter bank. The output $Y(z)$ is given by

$$Y(z) = \frac{1}{M} \sum_{l=0}^{M-1} X(z\omega^l) \sum_{k=0}^{M-1} H_k(z\omega^l) \, G_k(z)$$

$$= \sum_{l=0}^{M-1} A_c(z) \, X(z\omega^l) \tag{5.8.1}$$

where $A_l(z)$ is given by

$$A_c(z) = \frac{1}{M} \sum_{k=0}^{M-1} H_k(z\omega^l) \, G_k(z) \quad . \tag{5.8.2}$$

We note that $X(z\omega^l)$ is given by

$$X(e^{j\omega} W^l) = X(e^{j(W - 2\pi L/M)}) \tag{5.8.3}$$

for $l \neq 0$; thus, $X(e^{j\omega} W^l)$ represents a shifted version of the spectrum $X(e^{j\omega})$. Thus, $Y(\omega)$ is a linear combination of $X(e^{j\omega})$ and its $M-1$ uniformly shifted version. Thus, $X(z\omega^l)$ for $l = 0$ represents the lth aliasing term and $A_{cl}(z)$ is the gain of the lth aliasing term.

Since the aliasing error must be zero, we must have $A_l(z) = 0$ for all l except $l = 0$. For this case, keeping the $l = 0$ term, we obtain

$$Y(z) = T(z) \, X(z) \tag{5.8.4}$$

where

$$T(z) = \frac{1}{M} \sum_{k=0}^{M-1} H_k(z) \, G_k(z) \quad . \tag{5.8.5}$$

$T(z)$ represents the amplitude and phase distortions. As in the two-channel case, the PR condition is given by

$$T(z) = c\, z^{-n_0} \tag{5.8.6}$$

where c is a constant and n_0 is an integer. It is convenient to write $A_i(z)$ in matrix form as follows:

$$
\begin{bmatrix} A_0(z) \\ A_1(z) \\ A_{N-1}(z) \end{bmatrix}
=
\begin{bmatrix}
H_0(z) & H_1(z) & H_{M-1}(z) \\
H_0(zw) & H_1(zw) & H_{M-1}(zw) \\
H_0(zw^{M+1}) & H_1(zw^{M-1}) & H_{M-1}(zW^{M-1})
\end{bmatrix}
\cdot
\begin{bmatrix} G_0(z) \\ G_1(z) \\ G_{M-1}(z) \end{bmatrix}
\tag{5.8.7}
$$

or

$$A = H_{AC} \cdot G \tag{5.8.8}$$

where

$$
\begin{aligned}
A(z) &= \begin{bmatrix} A_0(z) & A_1(z) & \cdots & A_{N-1}(z) \end{bmatrix}^T \\
G(z) &= \begin{bmatrix} G_0(z) & G_1(z) & \cdots & G_{M-1}(z) \end{bmatrix}^T
\end{aligned}
\tag{5.8.9}
$$

and H_{AC}, which is called aliasing component matrix, is given by

$$
H_{AC}(z) =
\begin{bmatrix}
H_0(z) & H_1(z) & \cdots & H_{M-1}(z) \\
H_0(zw) & H_1(zw) & \cdots & H_{M-1}(zw) \\
\cdots & \cdots & \cdots & \cdots \\
H_0(zw^{M+1}) & H_1(zw^{M-1}) & \cdots & H_{M-1}(zW^{M-1})
\end{bmatrix}
\tag{5.8.10}
$$

The equation for the aliasing cancellation can be written as

$$A = H_{AC} \cdot G = t(z) \tag{5.8.11}$$

where

$$t(z) = [MA_0(z) \ 0 \ 0 \ ...]^T = [MT(z) \ 0 \ 0 \ 0 \ ...]^T \qquad (5.8.12)$$

For the PR condition, $T(z)$ is given by Equation 5.8.6. In principle, one can solve the matrix equation and obtain

$$G(z) = H_{AC}^{-1}(z) \ t(z) \qquad (5.8.13)$$

provided that $[\det H(z)] \neq 0$. However, the usual problems with matrix inversion makes the above solution highly impractical. For an example, if we design $H(z)$ to be a FIR, $G(z)$ becomes Infinite Impulse Response (IIR) filter.

To get out of this dilemma, we use polyphase decomposition as previously discussed in Section 5.4. We repeat the fundamental equations in the following for convenience. For the type I decomposition, we have

$$H(z) = E(z^M) \ Z^M \qquad (5.8.14)$$

where we have defined the ($1 \times M$) matrix as $H(z)$, ($M \times M$) matrix as $E(z^M)$, and ($1 \times M$) delay matrix as Z^M. Similarly, we expand $G(z)$ in terms of polyphase type II decomposition:

$$G(z) = R(z^M) \ Z^M \qquad (5.8.15)$$

where $G_0(z)$ is a ($1 \times M$) matrix, Z^M is a ($1 \times M$) matrix and $R(Z^M)$ is an $M \times M$ matrix.

We obtain Figure 5.8.2a, which shows both the analysis and synthesis filters combined for the M-channel case. Using noble identities, we can convert Figure 5.8.2a to Figure 5.8.2b where the downsampling is performed at the input and upsampling is performed at the output. This reduces computation by a factor of $2M$ for the combined filter. Finally, we can combine the $E(z)$ and $R(z)$ matrices by defining the $P(z)$ matrix given by

$$P(z) = E(z) \ R(z) \ . \qquad (5.8.16)$$

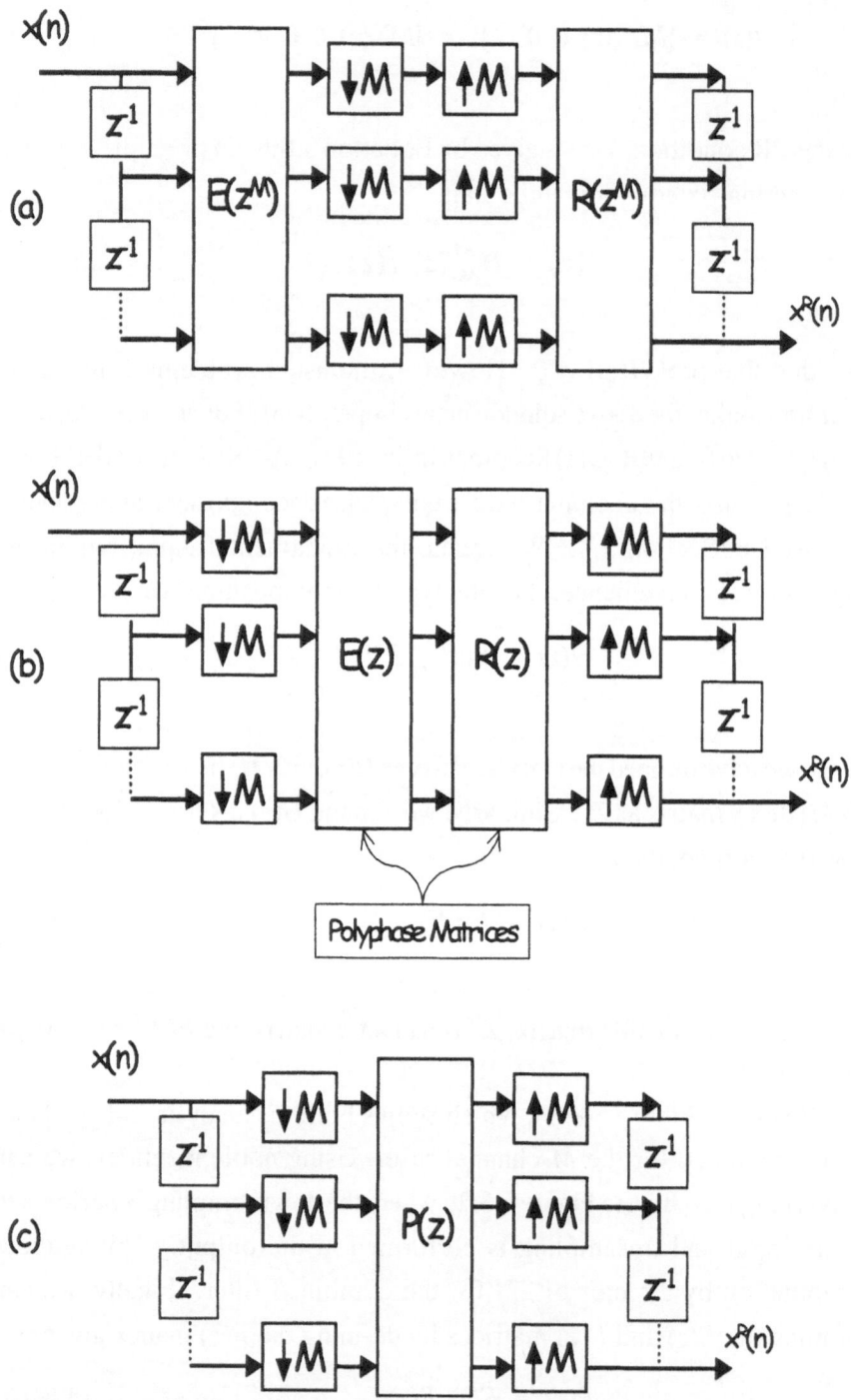

Figure 5.8.2 *M*-band analysis and synthesis filter bank after the applications of the noble identities.

This is shown in Figure 5.8.2c.

Before we proceed further, it is illuminating to consider a special case where $H_k(z)$ and $G_k(z)$ are simple delays given by

$$H_k(z) = z^{-k}$$

$$G_k(z) = z^{-(M-1-k)}$$

(5.8.17)

Thus, for this simple case, we have directly, without using polyphase decomposition

$$Y(z) = z^{-(M-1)}X(z) = H_k(z) \, G_k(z) \, X(z)$$

(5.8.18)

and in the discrete-time domain

$$y(n) = x(n-M+1)$$

(5.8.19)

Using polyphase decomposition, we can write

$$Y(z) = \frac{1}{M}\sum_{l=0}^{M-1} X(zw^l) \sum_{k=0}^{M-1} H_k(zw^l) \, G_k(z)$$

$$Y(z) = \frac{1}{M}\sum_{l=0}^{M-1} X(zw^l) \sum_{k=0}^{M-1} z^{-k} \, W^{lk} \, z^{-(M-l-k)}$$

(5.8.20)

where we have substituted for $H_k(z)$ and $G_k(z)$:

$$Y(z) = \frac{1}{M} \sum_{l=0}^{M-1} X(zw^l) z^{-(M-1)} \sum_{k=0}^{M-1} W^{lk}$$

(5.8.21)

and we have used the following identity:

$$\sum_{k=0}^{M-1} W^{lk} = \sum_{k=0}^{M-1} e^{j2\pi l/M^k} = \sum_{\nu=-\infty}^{+\infty} \delta(l-\nu M)$$

$$= \begin{cases} M, & \text{for } l = 0, \pm M, \pm 2M, \dots \\ 0, & \text{otherwise} \end{cases} \tag{5.8.22}$$

It is of interest to interpret these equations in the time domain. The nth channel passes the subset of input samples $x(nM-k)$. The analysis filter merely splits the input $x(n)$ into M subsequences given by

$$x(nM-k), \quad 0<k<M-1 \quad . \tag{5.8.23}$$

Consider now the design of the M-band PR filter. If we can design $R(z)$, $E(z)$, and $P(z)$ such that

$$R(z)E(z) = P(z) = Icz^{-m_0} \tag{5.8.24}$$

then, by inserting this identity matrix I in the simple problem just considered, we see that we have achieved PR output. Here, c is a constant, m_0 is an integer, and z^{-m_0} makes the filters causal. We note that

$$[\det E(z)] \cdot [\det R(z)] = cz^{-m_0} \quad . \tag{5.8.25}$$

Thus, $c \neq 0$ is a PR condition.

Note that for a FIR filter, we have $\det E(z) = \alpha z^{-k}$, and $a \neq 0$, and k is an integer. In placing a diagonal identity matrix, one can choose antidiagonal matrix as follows:

$$R(z)E(z) = P(z) = \begin{bmatrix} 0 & I_{M-\nu} \\ z^{-1}I_{\nu} & 0 \end{bmatrix} cz^{-m_0}, \quad \text{for } 0<\nu<M-1 \tag{5.8.26}$$

where $I_{M-\nu}$ and I_{ν} are identity matrices. For this case, we will have

$$y(n) = cx(n-n_0) \qquad (5.8.27)$$

where $n_0 = M m_0 + \nu + M - 1$.

Consider the two-channel case i.e., $M = 2$. For this case, we have

$$P(z) = cz^{-m_0} \begin{bmatrix} 1 & 0 \\ 0 & 1 \end{bmatrix} = cz^{-m_0} \begin{bmatrix} 0 & 1 \\ z^{-1} & 0 \end{bmatrix}. \qquad (5.8.28)$$

As mentioned earlier, the solution which determines $E(z)$ and $R(z)$ is the design of an m-channel PR filter. Thus it appears that one needs to use matrix inversion. However, using para-unitary matrices, this is achieved readily and elegantly.

A para-unitary matrix is a unitary matrix only for $|z| = |e^{j\omega}| = 1$. For this case, one has

$$E(z) \, E^T(z^{-1}) = 1 \,. \qquad (5.8.29)$$

Para-unitary matrices occur in lossless network theory. If $E(z)$ is para-unitary, then by choosing

$$R(z) = cz^{-k} E^T(z^{-1}) \qquad (5.8.30)$$

with $c \neq 0$ and k is chosen such that $R(z)$ is causal, we can satisfy Equation 5.8.29. The above equation implies that

$$G_k(z) = z^{-r} H_k(z^{-1}) \qquad (5.8.31)$$

where $r = Mn_0$ approaches $M-1$, or

$$g_k(n) = h_k(r-n) \qquad (5.8.32)$$

Actually, H_{AC}, the aliasing matrix, and the polyphase matrices must be connected, as they both provide the same input–output relationship. This relationship is useful for the design of PR filters, as will be evident shortly. We can then write

$$Y(z) = \sum_{l=1}^{M-1} A_c(z)\, X(z) \tag{5.8.33}$$

where

$$A = H_{AC}(z)\, G$$

$$Y = X^T A = \frac{1}{M} X^T H_{AC}\, G \ . \tag{5.8.34}$$

For zero aliasing, we must have

$$Y = X^T \frac{1}{M} [MT(z)\ 0\ 0\ ...]^T$$

$$H_{AC}(z)\, G = [MT(z)\ 0\ 0\ ...]^T \tag{5.8.35}$$

We can substitute successively for $z, zw, zw^2, ... zw^{M-1}$ in the above equation, and combining the above equations and rearranging them, we obtain:

$$H_{AC}(z)\, G_{AC}^T(z) = M\,\mathbf{diag}[T(z)\ T(zw)\ \cdots\ T(zw^{M-1})] \ . \tag{5.8.36}$$

where

$$G_{AC}(z) = \begin{bmatrix} G_0(z) & ... & G_{M-1}(z) \\ G_0(zw) & ... & G_{M-1}(zw) \end{bmatrix} . \tag{5.8.37}$$

Thus, the PR condition becomes

$$H_{AC}(z)\, G_{AC}^T(z) = M z^{-n_0}\,\mathbf{diag}[1\ w^{-n_0}\ \cdots\ w^{-(M-1)n_0}] \tag{5.8.40}$$

We obtain the following relationship between the H_{AC} matrix and the polyphase matrix:

$$H_{AC}(z) = W^* D(z) E^T(z^M) \qquad (5.8.41)$$

where

$$D(z) = \text{diag} \; [1 \; z^{-1} \; \cdots \; z^{-(M-1)}] \qquad (5.8.42)$$

If $E(z)$ is para-unitary, then one can show that $H_{AC}(z)$ is also para-unitary and thus lossless, or

$$H_{AC}(z) \, H_{AC}^T(z^{-1}) = I \;\;. \qquad (5.8.43)$$

To design M-channel PR filters one can thus design the AC matrix to be lossless and para-unitary. Expanding Equation 5.8.43, we obtain

$$
\begin{bmatrix}
H_0(z^{-1}) & H_0(z^{-1}W^{-(M-1)}) \\
H_{M-1}(z^{-1} & H_{M-1}(z^{-1})W^{-(M-1)})
\end{bmatrix}
$$

$$
\cdot
\begin{bmatrix}
H_0(z) & H_{M-1}(z) \\
H_0(zW^{M-1} & H_{M-1}(z)W^{(M-1)})
\end{bmatrix}
= K_{rs}(z) = I
\qquad (5.8.44)
$$

where

$$K_{rs}(z) = \frac{1}{M}\sum_{j=0}^{M-1} H_r z^{-1} W^{-j} H_s z W^j = \delta_{rs}$$

$$= \frac{1}{M}\sum_{j=0}^{M-1} \varphi_{rs}(zW^j) = \delta_{rs} \qquad (5.8.45)$$

and

$$\varphi_{rs} = H_r(z^{-1}) \, H_s(z^{-1}) \quad .$$ (5.8.46)

So the para-unitary condition becomes

$$\phi_{rs}(z) = \delta_{rs} \quad .$$ (5.8.47)

These applications can also lead to different designs and implementations of the filters.

5.9   Applications of Multirate Filtering

One of the most important practical applications of a multirate filter is the serial to parallel converter or demultiplexer and the correspondent parallel to serial converter or the multiplexer. Its development in telecommunication and data communication is the backbone of the networking which has lead to the present-day so-called Internet revolution. It has many other applications (e.g., the interleaved A/D converter).

To understand its importance, consider the simple case of demultiplexing shown in Figure 5.9.1. The input signal $x(n)$ is a signal which might be the output of a very high-bandwidth channel [e.g., a fiber-optic cable with a bandwidth of 1 Terabit/second (Tb/s)], that carries many channels of TV,

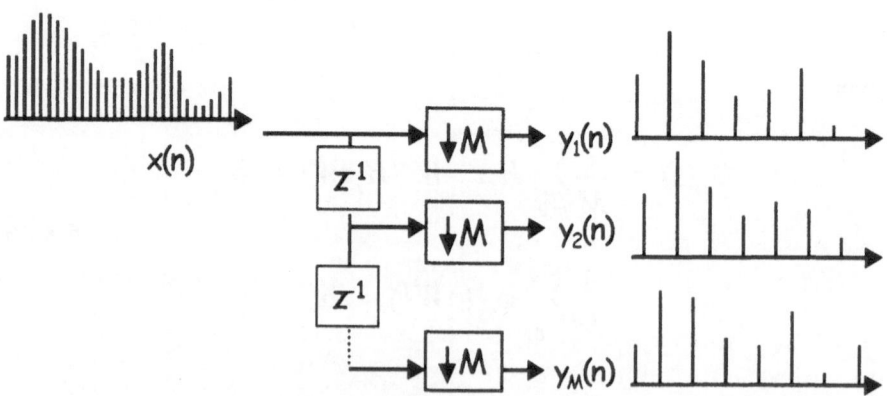

Figure 5.9.1 Block diagram of an M-fold demultiplexer.

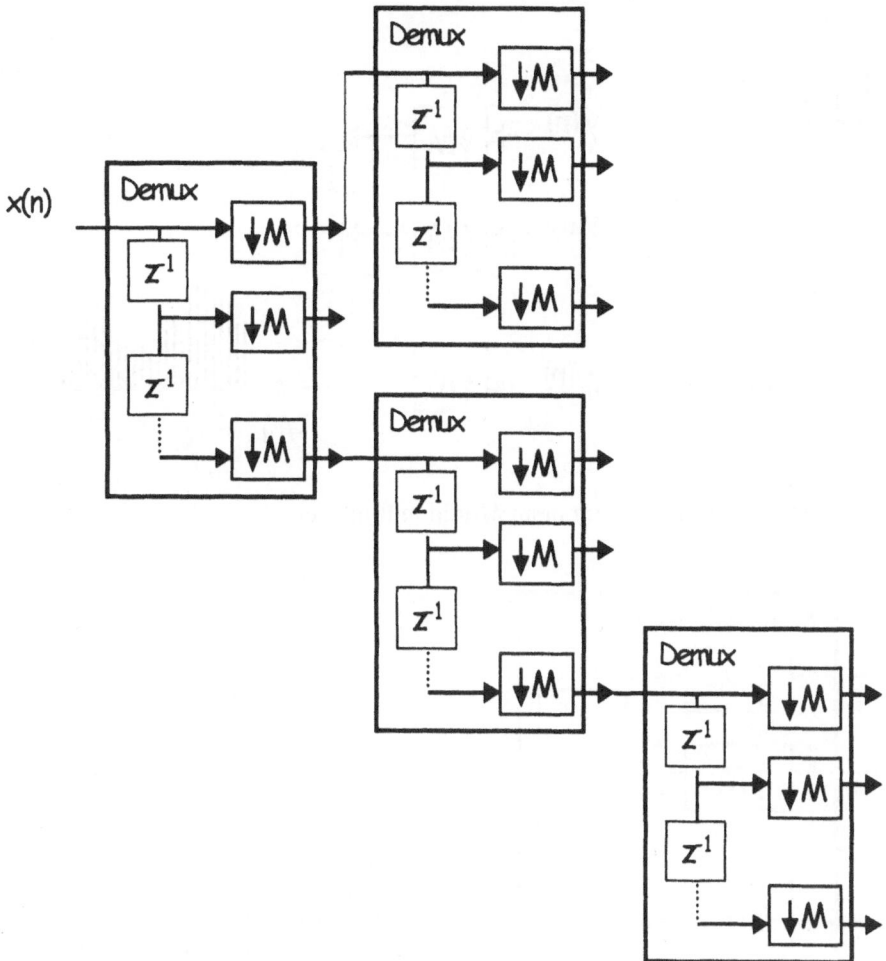

Figure 5.9.2 Block diagram of a cascade of M-fold demultiplexers.

voice, data, etc. [Vel98].

The signal is first split into M channels, where the M outputs $y_k(n)$, with $k = 0, 1, ..., M-1$ are given by

$$y_k(n) = x(nM+k) \qquad (5.9.1)$$

For the example chosen with $x(n)$ at a rate of 1 Tb/s and $M = 128$, $y(n)$ is at a rate of $1/128 = 7.8$ Gb/s. The process can be repeated again in a second and third stage and so on to reach, for example, an audio range of 30 kb/s. This cascade of demultiplexers is shown in Figure 5.9.2, whereas the correspondent multiplexer are given in Figures 5.9.3 and 5.9.4.

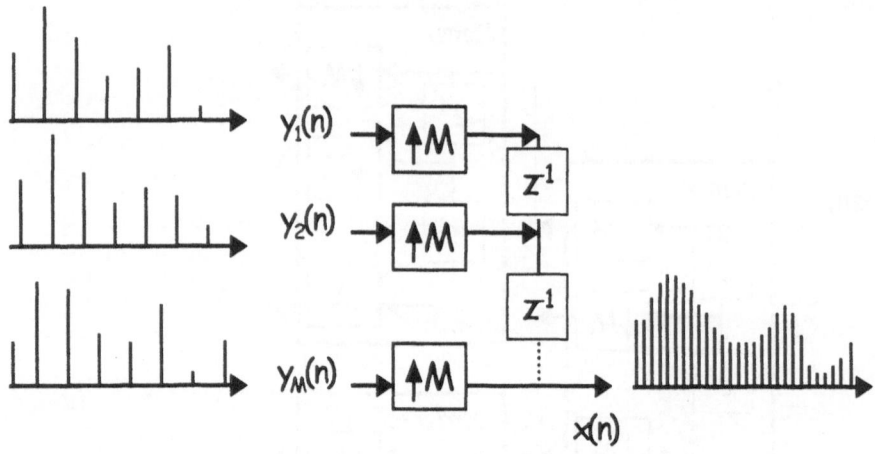

Figure 5.9.3 Block diagram of an M-fold multiplexer.

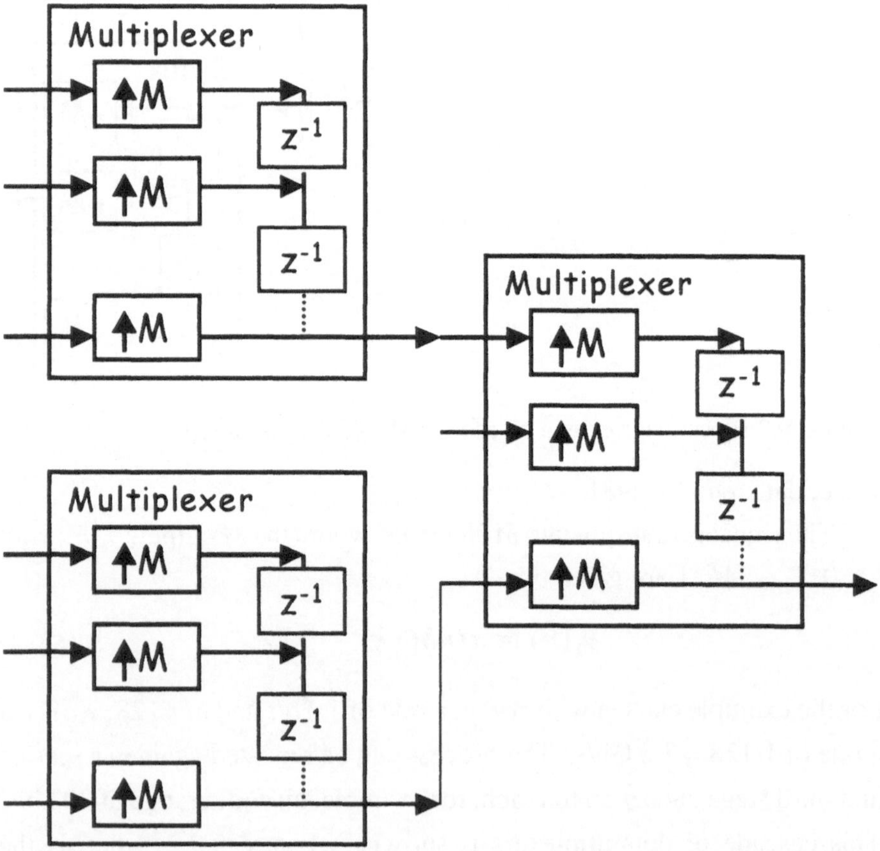

Figure 5.9.4 Block diagram of a cascade of M-fold multiplexers.

The multiplexer and demultiplexer briefly described here are the networking interface between many low-bandwidth channels and one high-bandwidth channel. In the most general case, one also includes two filters: the analysis filter $H_k(z)$ and the synthesis filter $F_k(z)$ as shown in Figure 5.9.5. The processing can include, in general, many functions such as narrow-band filtering in communication application or A/D conversion. We will discuss some of these applications in detail [Pet92, Fra97].

It is difficult to implement and design a narrow-band filter with less than 1% relative bandwidth in relation to the center frequency using single-rate filters, especially when working with very high-frequency signals. However, using multirate filters, it is much easier to obtain such narrow-band filters. An example of such filter is given in Figure 5.9.6, which involves both analog and digital processing. The input signal has a center frequency of the order of hundreds of kilohertz with a bandwidth of a few hundred hertz.

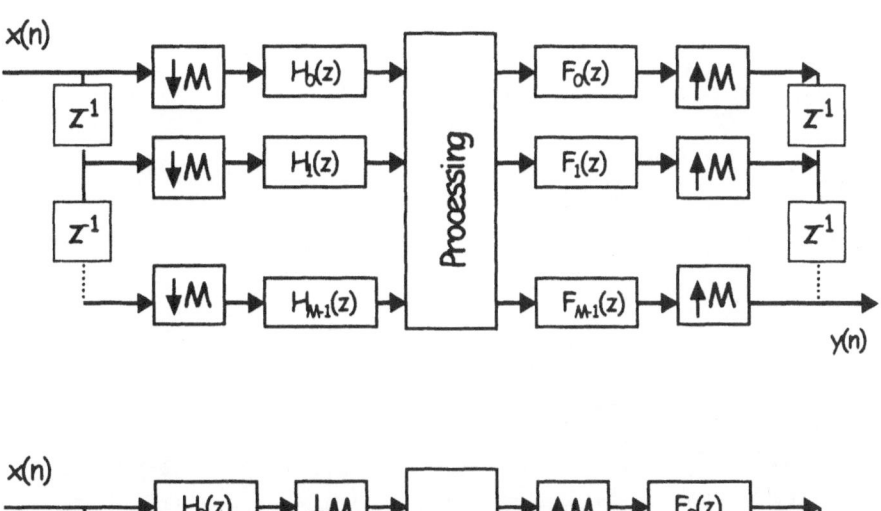

Figure 5.9.5 General case of M-band signal processing.

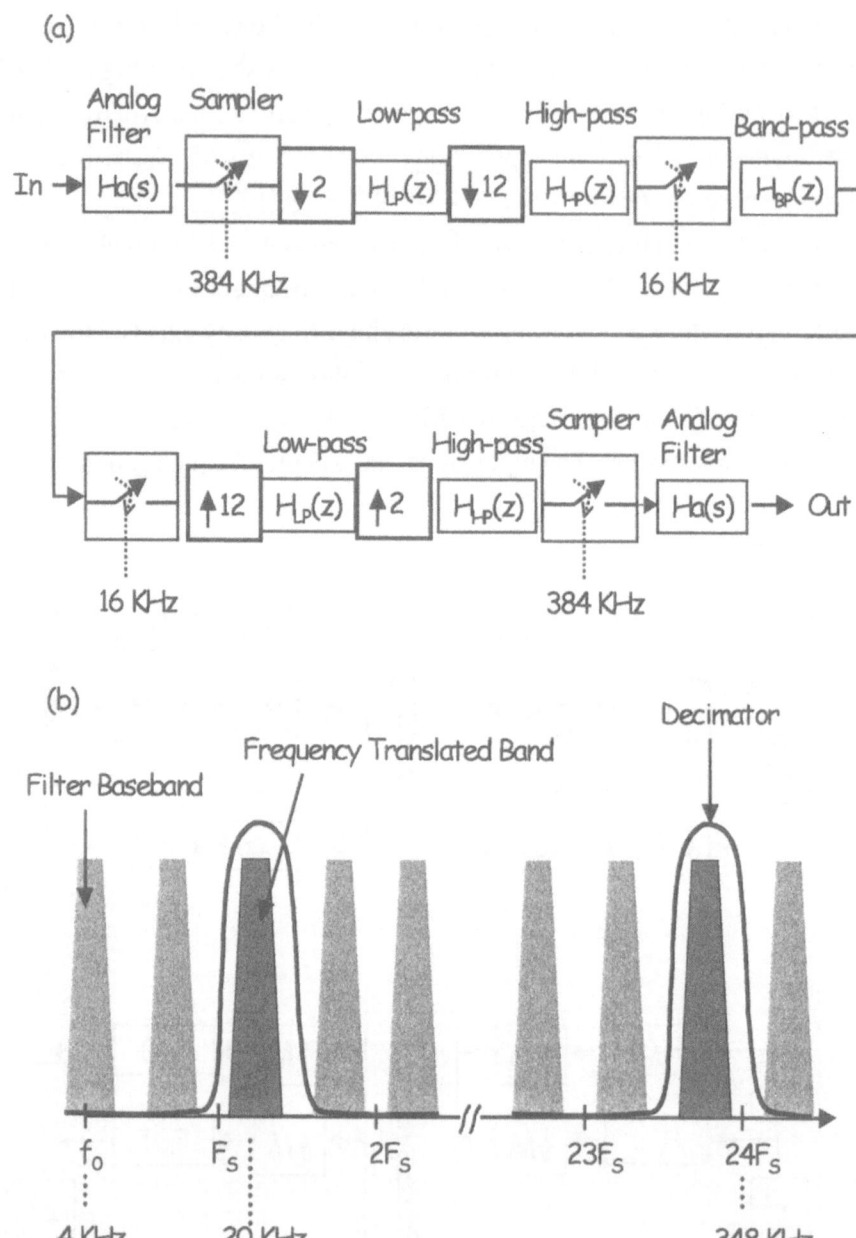

Figure 5.9.6 (a) Example of a narrow-band filtering performed using multirate filters. (B) Frequency domain description of the filter bank in (a).

The signal is sampled at 384 kHz after going through an antialiasing filter $H_a(\omega)$.

The switched-capacitor decimator consists of downsampling by a factor 2, followed by a low-pass filter $H_{LP}(z)$. The output of $H_{LP}(z)$ is downsampled again by a factor 12, bringing the signal down to $384 / 2 / 12 = 16$ kHz. A switched-capacitor bandpass filter is used to perform the bandpass filtering with a few hundred Hertz bandwidth. Using a switched-capacitor bandpass interpolator, the process is repeated in the output section to obtain the filtered continuous signal output. Note that the whole operation of this system depends on the frequency translation of the input from high frequency to a lower processing frequency and back to an output signal with the original center frequency. The conditions for avoiding aliasing and imaging associated with the frequency translation operations of the discrete signals can be met by pass-band discrete filters. Figure 5.9.6b shows the choice of the bandpass filters for the case of Figure 5.9.6a. Narrow-band filtering with more than 70 dB of dynamic range has been reported using this multirate solution.

Multirate filtering is also heavily utilized in A/D conversions. The first application we will discuss is the utilization of interleaving to increase M-fold the effective bandwidth of the A/D conversion process using M lower-speed A/D converter [Pet92]. The second application to be discussed is the increase in the resolution (i.e., the number of bits), using oversampling. These concepts have been fundamentals in the development of sigma–delta[**] A/D converters, which utilize noise-shaping or error-diffusion algorithms.

Figure 5.9.7 shows the time-interleaved A/D conversion using N lower-rate A/D converters. Although theoretically using this multirate architecture one can make N very large resulting in a theoretical very high bandwidth, in reality the harmonic distorsion due to mismatches between the different A/D converters poses a severe problem for implementation. Any mismatch causes aliasing which is not canceled out at the output and this resulting in a upper limitation to N. The situation can be improved significantly by inserting QMF analysis and synthesis filters as hown in Figure 5.9.7, where $H_k(z)$ and $F_k(z)$

** Please note that in many references, the sigma–delta conversion is also referred to as delta–sigma with no difference ($\Sigma-\Delta$ or $\Delta-\Sigma$).

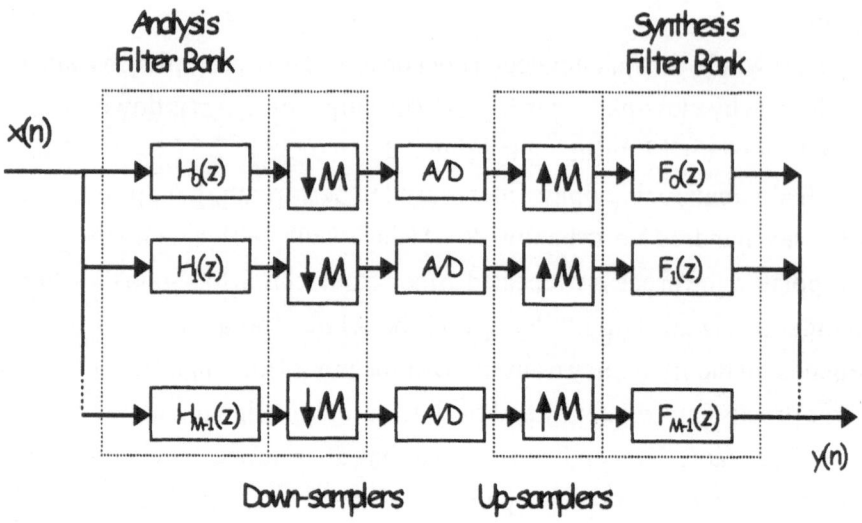

Figure 5.9.7 Time-interleaved A/D conversion using M lower-rate A/D converters.

are not simple delays [Kho97].

The input and output relationship for this case is given by

$$Y(z) = \frac{1}{M} \cdot \sum_{r=0}^{M-1} X(zW^r) \cdot \sum_{k=0}^{M-1} H_k(zW^r) \cdot F_k(z) \qquad (5.9.2)$$

The filters $H_k(z)$ and $F_k(z)$ are chosen such that they satisfies perfect reconstruction (PR) except for a delay. The PR condition is satisfied when

$$\frac{1}{M} \sum_{k=0}^{M-1} H_k(zW^r) F_k(z) = \begin{cases} z^{-n}, & \text{if } r = 0 \\ 0, & \text{otherwise} \end{cases} \qquad (5.9.3)$$

and n is an integer ($n \in Z$). The filters $H_k(z)$ and $F_k(z)$ are chosen such that they have sufficient large stop-band attenuation and small transition bandwidth. It has been shown that by incorporating the QMF filters, the mismatch effects can be drastically reduced.

For a particular four-channel case the reduction is of the order of 28 dB. This design uses switched capacitor implementation and it can achieve 12 bits of accuracy for video applications. In place of using switched-capacitor filters, one can also use a hybrid filter bank when the analysis filter is analog but the synthesis filters are digital. It has been shown that this hybrid approach can extend the effective parameters for A/D conversion to 14 bits and 300 megasamples/second.

Although we have discussed A/D converters earlier, it is of interest to point out the importance of A/D conversion before we discuss the application of multirate filters in the oversampling A/D conversion. In the past decade, the world has become digital and it is projected that this trend will continue. The relentless advances in computer speed fueled by near doubling of size and speed of VLSI circuits every 2 years is expected to make the information technology revolution in the near future. This is expected to have an impact on society, surpassing a similar effect produced by the industrial revolution in the last century. However, many aspects of the world are still analog and, in many cases, a very fast and accurate A/D converter is the last element holding up this digital revolution for many systems like all digital radio and TV, radar, etc. The electronic A/D converters have progressed significantly. For an example, using very high-frequency transistors like GaAs, HBT, InPDHBT, and others, one has achieved a few gigahertz sampling rate with 4-bit resolution. To achieve 100 GHz bandwidth with 4- or 8-bit resolution, it is expected that the electronic solution is not a near term one although one cannot rule out an electronic solution completely. So, it natural to look for some alternative approach: photonic, quantum optic, or other mechanism which might lead to a solution in the near future. It is well known that optics or photonics provide extremely high-speed (THz sampling rate is practical using femtosecond pulses) and enormous parallelism [Das00]. However, optical analog processing cannot provide the high dynamic range and linearity needed for many applications. It appears that oversampled A/D converter architecture can provide an effective large dynamic range by trading high sampling rate with a greater number of bits. One can augment the performance of the oversampled A/D converter further by using parallelism [Sho98].

Faster device speed increases the sampling rate of a A/D converter. However, device speed itself does not determine the actual bandwidth, as interconnect parasitics and architecture (number of parallel loads) come into play for actual devices. Device speed is related to the unity–gain cut–off frequency, f_T, which is also dependent on its breakdown voltage. Indium phosphide double heterojunction bipolar transistors (InPDHBT) appear to have the best performance compared to InPHBT, GaAs HBT, SiGe HBT, and Si BJT. Each technology has f_T = 100 GHz, but for the InPDHBT, the cut-off voltage at f_T = 100 GHz is the highest at approximately 17 V. Using this technoloy, it is expected that 10-bit 3-Gsps (Gigasamples per second) monolithic A/D converters will be developed. Using GaAs HBT technoloy, 8-bit 2.8-Gsps A/D converters have been demonstrated. Even 10-Gsps converters are possible to implement if one does not demand more than a few bits of accuracy. Compared to the processing speed of computers, A/D converters not only have the sampling rate but also resolution of the number of bits achievable at a given sampling rate.

Unfortunately, over the last 8 years, the performance of A/D converters has improved by only 1.5 bits for high-performance applications. As a result, there has recently been a renewed interest in new and innovative approaches to A/D conversion with significant emphasis on photonic techniques. The potential advances of using photonics technology come in the form of high-speed clocking, broad-band sampling, reduced mutual interference of signals, and compatibility with existing photonic-based systems. Another advantage of processing signals in the optical domain is the parallelism obtained by performing signal processing in both the space and time domains simultaneously. Photonic approaches to A/D conversion have been considered in the past with varying degrees of success. Some of the approaches

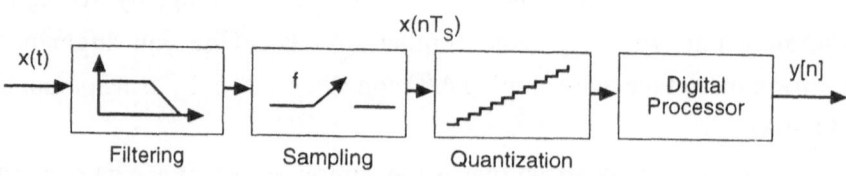

Figure 5.9.8 Block diagram of a generic analog-to-digital converter.

employed Mach-Zehnder interferometers, acousto-optic modulators, and, recently, multiple quantum well modulators have been incorporated into non traditional architectures. We will now propose a new technique which uses optical space-time data using wide-bandwidth modulators or diffractive elements. The spatial signal is then processed in a fashion similar to image processing using digital halftoning and spatial error diffusion filtering using a neural network architecture.

Analog-to-digital conversion is the process by which a continuous-time, continuous-amplitude signal is converted to a discrete-time, discrete-amplitude or digital signal. This process typically employs the four distinct functions shown in Figure 5.9.8. We can describe the overall operation of the generic A/D converter by following a signal as it progresses through each element in Figure 5.9.8. The analog input signal $x(t)$ is first band-limited to the range of frequencies $0 \leq f_X \leq f_B$ (Hz) by an analog filter to ensure protection against aliasing that could occur during the subsequent sampling operation. The sampling operation in a conventional Nyquist-rate A/D converter is chosen to satisfy the minimum Nyquist criterion: $f_S = f_N = 2f_B$, where f_N is the Nyquist frequency and f_B is the constrained signal bandwidth.

Oversampling is another alternative in which $f_S \gg f_N$ and subsequent signal processing techniques are used to provide an advantage. The output from the sampler is $x_n \equiv x(nT_S)$, where T_S is the uniform sampling period $T_S = 1/f_S$. The scalar quantization process maps each continuous-amplitude input x_n to one value in a discrete amplitude ensemble $q_n \equiv q(nT_S)$. Based on the result of this mapping, the digital processor subsequently generates the digital code of the level that most closely approximates the input analog signal value. The output $y_n \equiv y(nT_S)$ is then the multibit, digital word representing the input analog input value.

Oversampling converters such as the delta-sigma ($\Delta - \Sigma$) and sigma-delta ($\Sigma - \Delta$), or error-diffusion modulators, instead sample the input analog signal at rates which are typically much higher that that required by Nyquist criterion. In this case, liner filtering and signal processing techniques are employed to improve the overall performance of the A/D converter. Oversampling A/D converters fundamentally trade sampling bandwidth for improved amplitude resolution. Here, a low-resolution quantizer and a liner

(a)

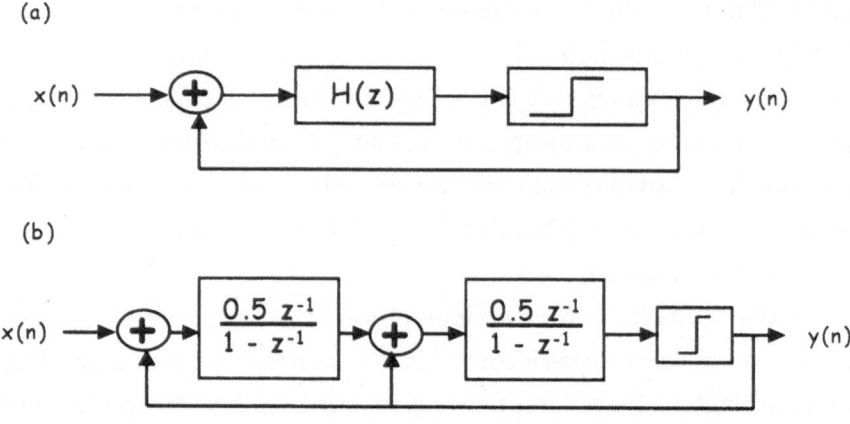

(b)

Figure 5.9.9 (a) Generic block diagram of a sigma-delta ($\Sigma-\Delta$) converter. (b) Details of a second-order ($\Sigma-\Delta$) converter.

filter are embedded in a feedback architecture in order to spectrally shape the quantization error resulting from a low-resolution quantizer. This spectral noise shaping forces the quantization noise to frequencies above the baseband information bandwidth. In a subsequent operation, a digital post processor removes the spectrally shaped noise and decimates the resulting digital output, thereby improving the overall signal-to-noise (SNR) ratio and converter performance and providing the high-resolution digital representation at the Nyquist conversion rate of the input signal.

Figure 5.9.9a shows the block diagram of the oversampled quantizer and Figure 5.9.9b shows the details of the filter $H(z)$ for a second-order interpolator. For higher-order filters, more sections are added. Note that the output $y(n)$ in Figure 5.9.9 is 1 bit. The digital post processing which includes the low-pass filtering and decimation uses the multirate filters. A typical decimation filter converts the 1-bit high-bit- rate (f_s/second) sequence to a lower-bit-rate sequence (f_N/second) but with much higher resolution b_{eff} $\gg 1$. For an example, it can be shown that for a Nth-order quantizer, the number of effective bits, b_{eff}, is given by

$$b_{eff} = \log_2 \left[\frac{\sqrt{2N+1}}{\pi^N} M^{\left(N+\frac{1}{2}\right)} \right] \qquad (5.9.4)$$

where M is the oversampling ratio and only quantization noise is considered.

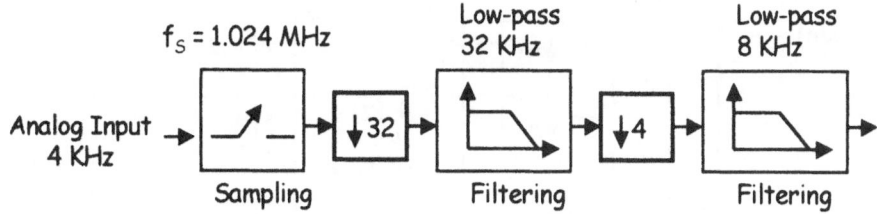

Figure 5.9.10 Decimation filter block diagram showing two stages of decimation. The signal with 4 kHz bandwidth is sampled at an oversampling ratio of 128 and then decimated by 32 and 4 to obtain $f_N = 8$ kHz.

The situation becomes quite complex if other practical noise sources such as thermal, sampling jitter, quantization hysteresis, etc. are also included. A typical decimation filter is shown in Figure 5.9.10, it generally consists of two stages of decimation due to practical limitations. For an example, to translate from $f_S = 1.024$ MHz to $f_N = 8$ kHz, one first performs a decimation to 32 kHz using sinc filters whose response is given by

$$H(z) = \frac{1}{N} \frac{1 - z^{-N}}{1 - z^{-1}}$$

(5.9.5)

for the case the Nth-order filter. These filters can be easily implemented using accumulate and store circuitry. A typical value of N is 2 or 3. The low-pass filter in the final stage is, in general, a more complex digital filter.

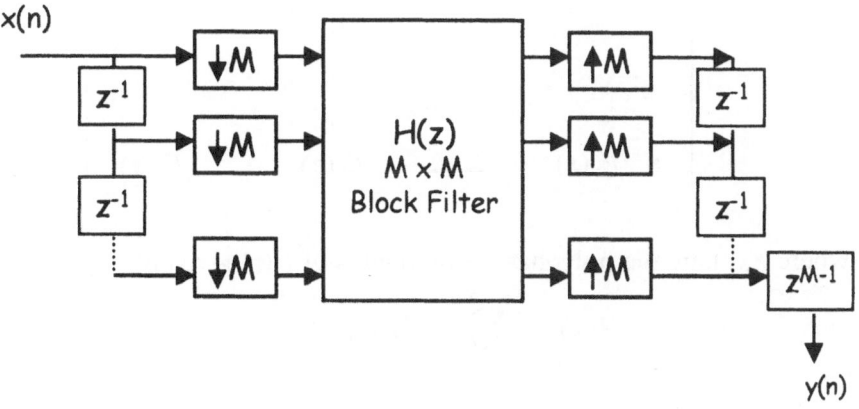

Figure 5.9.11 Equivalent block filtering for the single-input single-output transfer function $H(z)$.

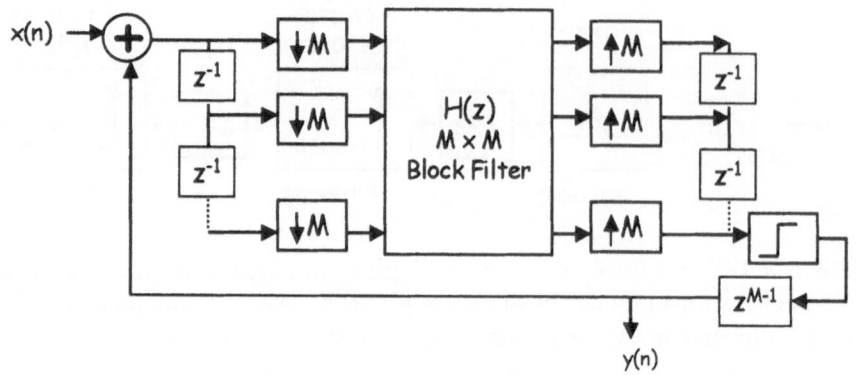

Figure 5.9.12 Oversampling converter with a digital block filter.

The time-interleaving architecture can be elegantly interpreted using block digital filtering concepts. Block digital filtering is a multirate filter as shown in Figure 5.9.11 for a single input-single output case. Note that Figure 5.9.11 is similar to Figure 5.9.7, where the analysis and synthesis blocks have been lumped together into a block filter matrix $\hat{H}(z)$. For the case of the oversampling converter

$$
\hat{H}(z) =
\begin{bmatrix}
E_0(z) & E_1(z) & E_2(z) & \cdots & E_{M-1}(z) \\
z^{-1}E_{M-1}(z) & E_0(z) & E_1(z) & \cdots & E_{M-2}(z) \\
z^{-1}E_{M-2}(z) & E_{M-1}(z) & E_0(z) & \cdots & E_{M-3}(z) \\
\cdot & \cdot & \cdot & \cdot & \cdot \\
\cdot & \cdot & \cdot & \cdot & \cdot \\
\cdot & \cdot & \cdot & \cdot & \cdot \\
z^{-1}E_1(z) & E_2(z) & E_3(z) & \cdots & E_0(z)
\end{bmatrix}
\tag{5.9.6}
$$

where $E_l(z)$ are the polyphase components of $H(z)$ given by:

$$
H(z) = \sum_{l-0}^{M-1} z^{-1}E_l(z^M) \ . \tag{5.9.7}
$$

Using the noble identities, one can go from Figure 5.9.12 to the final implementation shown in Figure 5.9.14, through the intermediate step shown

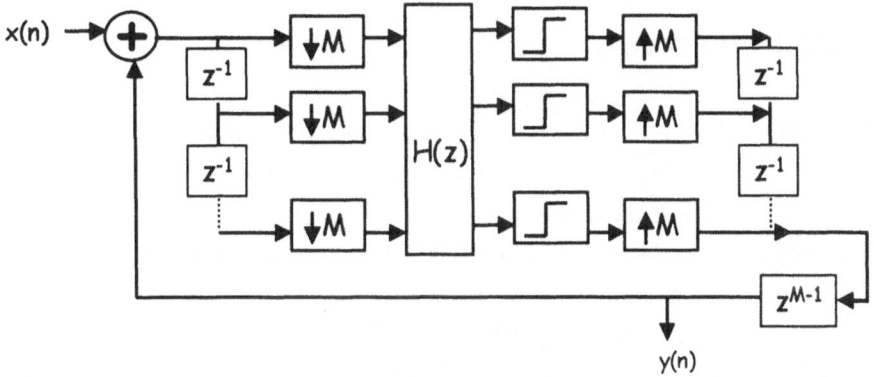

Figure 5.9.13 Oversampling converter with a digital block filter and quantizers at the lower-rate side.

pictorially in Figures 5.9.13.

There are many more applications of multirate filtering in digital signal processing. Some of these are subband coding of speech and image signals, digital audio systems, analog voice privacy system, and so forth. Some of these are discussed in other parts of this book; for a detailed review, the readers should review many of the references given at the end of the chapter.

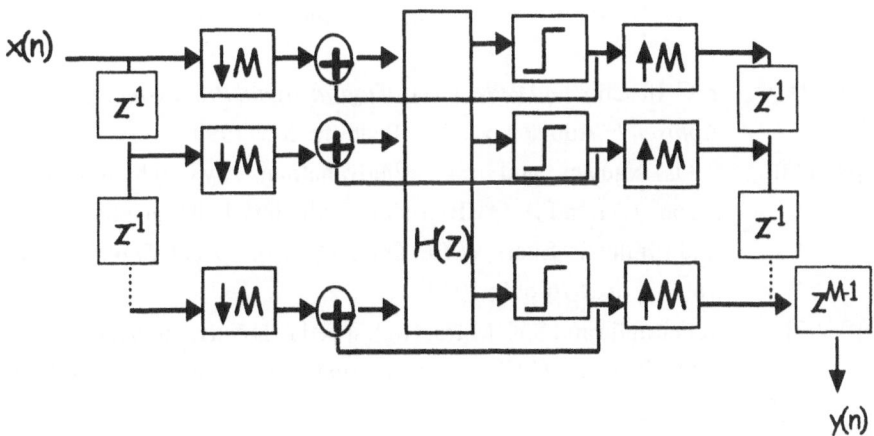

Figure 5.9.14 Time-interleaved oversampling converter.

5.10 References

[Aka94]   A. N. Akansu and R.A. Haddad, *Multiresolution Signal Decomposition Transforms, Subbands, Wavelets*, Academic Press, New York, 1994.

[Aka98]   A.N. Akansu, P. Duhamel, X. Lin, and M. de Courville, Orthogonal transmultiplexers in communication: A review, *IEEE Trans. Signal Proc.*, vol. 46, pp. 979-995, 1998.

[Aka99]   A.N. Akansu and M.J. Medley eds., *Wavelet, Subband and Block Transforms in Communications and Multimedia*, Kluwer, Dordrecht, 1999.

[Cro83]   R.E. Crochiere and L.R. Rabiner, *Multirate Digital Signal Processing*, Prentice Hall, Englewood Cliffs, NJ, 1983.

[Das00]   P. Das, B.L. Shoop, and D.M. Litynski, High speed A/D conversion using photonic implementation of an error diffusion neural network and oversampling techniques, to be published.

[Fra97]   J. Franca, A. Petraglia, and S. K. Mitra, Multirate Analog Digital systems for Signal Processing and Conversion, *Proc. IEEE*, vol. 85, pp. 242-262, 1997,

[Her93a]   C. Herley and M. Vetterli, Wavelets and recursive filter banks, *IEEE Trans. Signal Process.*, vol. 41, pp. 2536-2556, 1993.

[Kho97]   R. Khoni-Poorfard, L.B. Lim, and D.A. Johus, Time-interleaved oversampling A/D converters: Theory and Practice, *IEEE Trans. Circuits Syst. II: Analog Digital Signal Proc.*, vol. 44, pp. 634-645, 1997.

[Ife93]   E.C. Ifeachor and B.W. Jervis, *Digital Signal Processing. A Practical Approach*, Addison-Wesley, Reading, Ma, 1993.

[Ngu89]   T.Q. Nguyen and P.P. Vaidyanathan, Two channel perfect reconstruction FIR QMF structures which yield linear-phase analysis and synthesis filters, *IEEE Trans. Acoust. Speech Signal Process.*, vol. 37, pp. 676-690, 1989.

[Pet92]   A Petraglia and S.K. Mitra, High speed A/D conversion incorporating a QMF bank, *IEEE Trans. Instrum. Meas.*, vol. 41, pp. 427-431, 1992.

[Sho98]   B.L. Shoop, P. Das, and D.M. Litynski, Improved resolution of photonic A/D conversion using oversampling techniques, *Proc. SPIE* vol. 3463, pp. 192-199, 1998.

[Smi84]   M.J. Smith and T.P. Barnwell III, A procedure for designing exact reconstruction filter banks for tree structured subband coders, *Proc. IEEE Int. Conf. ASSP*, pp. 27.1.1-27.1.4, 1984.

[Smi86]   M.J. Smith and T.P. Barnwell III, Exact reconstruction for tree-structured subband coders, *IEEE Trans. Acoust. Speech Signal Process.* vol. 34, pp. 434-441, 1986.

[Sut97]   B.W. Suter, *Multirate and Wavelet Signal Processing*, Academic Press, New York,1997.

[Vai87]   PP. Vaidyanathan, Quadrature mirror filters banks, M-band extensions and perfect-reconstruction technique, *IEEE ASSP Mag.*, vol. 4, pp. 4-20, 1987.

[Vai88]   P.P. Vaidyanathan and P.Q. Hoang, Lattice structures for optimal design and robust implementation of two-band perfect reconstruction QMF banks, *IEEE Trans. Acoust. Speech Signal Process.*, vol. 36, pp.81-94, 1988.

[Vai90]   P.P. Vaidyanathan, Multirate digital filters, filterbanks, polyphase networks and applications: A tutorial, *Proc. IEEE*, vol. 78, pp. 56-93, 1990.

[Vai92]   J. Vaisey and A. Gersho, Image compression with variable block size segmentation, *IEEE Trans. Signal Process.*, vol. 2040-2W, Aug. 1992.

[Vai93]   P.P. Vaidyanathan, *Multirate Systems and Filter Banks*, Prentice Hall, Englewood Cliffs, N.J., 1993.

[Van64]   A. VanderLugt, Signal detection by complex spatial filtering, *IEEE Trans. Inform. Theory*, vol. IT-10, p. 2, 1964.

[Vel98]   S.R. Velzquez, T.Q. Nguyen, and S.R. Broadstone, Design of hybrid filterbanks for analog/digital conversion, *IEEE Trans. Signal Process.*, vol. 46, pp. 956-967, 1998.

[Vet86]   M. Vetterli, Filter banks allowing perfect reconstruction, *Signal Process.*, vol. 10, n. 3, pp. 219-244, 1986.

[Vet87]   M. Vetterli, A theory of multirate filter banks, *IEEE Trans. Acoust. Speech Signal Process.*, vol. 35, pp. 356-372, 1987.

[Vet89]   M. Vetterli and D. LeGalli, Perfect reconstruction FIR filter banks: some properties and factorizations, *IEEE Trans. Acoust. Speech Signal Process.*, vol. 37, pp. 1057-1071, 1989.

[Vet90]   M. Vetterli, J. Kovačević and D. J. LeGall, Perfect reconstruction filter banks for HDTV representation and coding, *Image Commun.*, vol. 2, pp. 349-364, Oct. 1990.

[Vet90a]   M. Vetterli and C. Herley, Wavelets and filter banks: relationships and new results, *Proc. ICASSP*, vol. 3, pp. 1723-1726, 1990.

[Vet92]   M. Vetterli and C. Herley, Wavelets and filter banks: theory and design, *IEEE Trans. Signal Process.*, vol. 40, no. 9, 1992.

[Vet95]   M. Vetterli and J. Kovačević, *Wavelets and Subband Coding*, Prentice-Hall, Englewood Cliffs, NJ, 1995.

Chapter 6

Two-Dimensional Wavelet Transforms and Applications

6.1   Introduction

In this chapter, we describe various methods of applying the wavelet transform to a two-dimensional signal or image. In particular, we will treat subband image decomposition and encoding, image enhancement, and wavelet techniques for image compression and video encoding. Optical image processing techniques using wavelets will be the subject of a later chapter.

It is well known from linear systems theory that any linear image transform can be used to express an image as a weighted sum of basis functions. The choice of transform (i.e., the choice of basis functions) can be made in many ways, with the objective of optimizing some property such as signal-to-noise ratio of the image or edge enhancement; the transform can also be used to encode or compress an image. A great deal of recent work in image coding has involved basis functions that are localized in both space and spatial frequency. Such transforms are very useful for various kinds of image analysis, because they provide information about intensity changes over different scales and yet retain information about where the events are occurring [Eff91]. There is also evidence that the human visual system performs a similar image decomposition in the early stages of its processing [Equ91].

Many approaches have been suggested for achieving the proper basis functions. One method of achieving localization in both space and spatial frequency is to divide the image into small blocks and to compute a discrete Fourier transform or a discrete cosine transform on each block (this class of functions are generally known as *block transforms* or *short space transforms*). Unfortunately, this approach introduces artificial block boundaries within the image; these can be somewhat arbitrary and will interrupt the continuity of the original image. Also, due to the sharp edges of the blocks, the frequency localization of the basis functions tend to be poor compared with the image as a whole. Working in the time domain, Gabor [Gab46] showed that functions produced as the products of complex sinusoids and Gaussian windows had optimal joint time-frequency localization. He developed a transform using such functions as a basis set, the so-called *Gabor functions* [Dau85, Vai92]. At each position in the image, a set of basis functions analyzes a Gaussian-weighted patch into frequency components; the process is then repeated for each patch. The Gabor functions do not form an orthogonal basis set but they can form a complete representation. Because the Gabor transforms also deal with dividing an image into subsections, but do not require a sharp boundary between these sections, they are sometimes known as *"soft block" transforms*, as opposed to the *"hard block" transforms* discussed earlier.

A second desirable property of the basis functions is scaling, or "self-similarity," which means that the basis functions should all have essentially the same shape but should be scaled versions of each other. The fundamental argument is that images contain information at various scale values, and so it makes sense to capture this information in a uniform way. Scale invariance is violated by typical blocked or so-called "short space" transforms. Block transforms inherently impose an arbitrary scale on the image analysis because the block is chosen to be some particular size. The problem of arbitrary scale can be avoided in self-similar transforms, in which the basis functions come in many sizes. Examples of self-similar transforms include the *Gaussian pyramids* and *Laplacian pyramids*, which have found applications in image coding [Bur83]. These basis functions are low-pass kernels repeated at a series of positions and appear at scales varying by factors of 2. The sampling functions are bandpass kernels, which are placed at corresponding positions and scales. The pyramid transform gives exact reconstruction, but the basis set is not orthogonal, and the number of transform coefficients exceeds the number of original pixels by a factor of 4/3 (the basis set is said to be *overcomplete*). A similar pyramid approach has also been suggested using an oriented pyramid whose basis functions resemble Gabor functions and attain a spatial frequency and orientation tuning similar to that inferred for channels in the human visual system [Ade87]. Again, the nonorthogonal basis set is overcomplete by 4/3, but the reconstruction is nearly exact. More recent efforts have focused on developing pyramids that use orthogonal basis functions (and therefore have the same number of transform coefficients as pixels), that incorporate orientation tuning, and that have good localization in space and spatial frequency. Such transforms should be useful for a variety of image processing tasks [Ans88, Tod95, Vet90].

6.2   Orthogonal Pyramid Transforms

We will illustrate this concept using the simplest self-similar orthonormal image transform, the so-called *Haar transform*. It is a good example, since it is a special case of a *quadrature mirror filter (QMF) pyramid*. The Haar transform is easy to understand and simple to compute, but its poor frequency tuning limits its application to image processing. The Haar basis in the one-dimensional domain was extensively discussed in Chapter 4.

A two-dimensional Haar transform is readily computed by combining the one-dimensional basis transforms for both dimensions. The simple application of separability leads to basis functions of widely varying shapes in the image plane; some basis will be rectangular along either the horizontal or vertical axes and others will be square. This is contrary to the original premise of self-similarity. A more suitable two-dimensional basis set may be produced by combining the various one-dimensional primitives to form a set of four two-dimensional primitives. These can then be applied recursively, subdividing only the low-frequency basis functions from one stage to the next. Since the low-pass basis sets are always subdivided, the final pyramid basis set is composed of only three basic kernel shapes, which one may describe as taking vertical differences, horizontal differences, and diagonal differences. Although the Haar basis functions have poor frequency selectivity, it is possible to develop an idealized frequency domain interpretation of the construction of the basis set. Each step of the process breaks the original band into four subbands: one tuned for low frequency, one tuned for vertical high frequency, one tuned for horizontal high frequency, and one tuned for both orientations of diagonal high frequency.

We will next derive some transforms which are better than the Haar transform at manipulating images. For simplicity, we consider the example of a one-dimensional discrete image $e[n]$, where n is an index referring to specific pixels in the image. An image transform expresses $e[n]$ as a sum of basis functions, f, weighted by a set of coefficients, p, as follows

$$e[n] = \sum_i p_i f_i[n]$$

(6.2.1)

The coefficients can be derived by computing the inner product of the image and a set of sampling functions, g, according to

$$p_i = \sum_n g_i[n] e[n]$$

(6.2.2)

Of course, this may also be expressed in matrix notation. If the signal is a column vector, E, and the basis matrix F is composed of the columns from the above expression for $f[n]$, we may rewrite the preceding equations as

$$E = F \cdot P$$

(6.2.3)

and

$$P = G^T \cdot E$$

(6.2.4)

where P is a column vector of coefficients, and G is the sampling matrix composed of the columns of $g[n]$. It follows that

$$E = F \cdot G^T \cdot E \tag{6.2.5}$$

so that

$$G = (F^{-1})^t \tag{6.2.6}$$

In the special case of an orthonormal transform, the sampling functions are identical to the basis functions ($G = F$). In particular, we will consider basis functions that can be partitioned into a few classes, where the functions within a given class are shifted versions of each other, so that the matrix takes the following form:

$$F = \begin{bmatrix} f_a & x & f_b & x & x \\ x & f_a & x & f_b & x \\ x & x & f_a & x & f_b \end{bmatrix} \tag{6.2.7}$$

An alternate description of this sort of transform is shown in the diagram in Figure 6.2.1. The image is given by a sequence $e[n]$; the kernels $ga, gb,...$ are convolved with the image; downsampling by factors of $ka,...$ leads to the subimages $pa[i]$, $pb[i]$,... These subimages form the transform representation. For reconstruction, the subimages are upsampled and convolved with the basis kernels $fa, fb, ...$ leading to the expanded subimages $qa[n]$, $qb[n]$,... which are summed to retrieve the original image $e[n]$.

One particular example of this class of transforms is the case of so-called *band-splitting transforms*. In this case, there are two classes of kernels: highpass and lowpass; then if there are n pixels in the starting image, there will be $n/2$ basis functions in each class of kernels. The Haar transform is built recursively from a band-splitting transform, as are many others which are far more practical for image processing. Such transforms can be applied recursively to build a pyramid, but generally band-splitting transforms are built from one primitive kernel. The lowpass basis set is

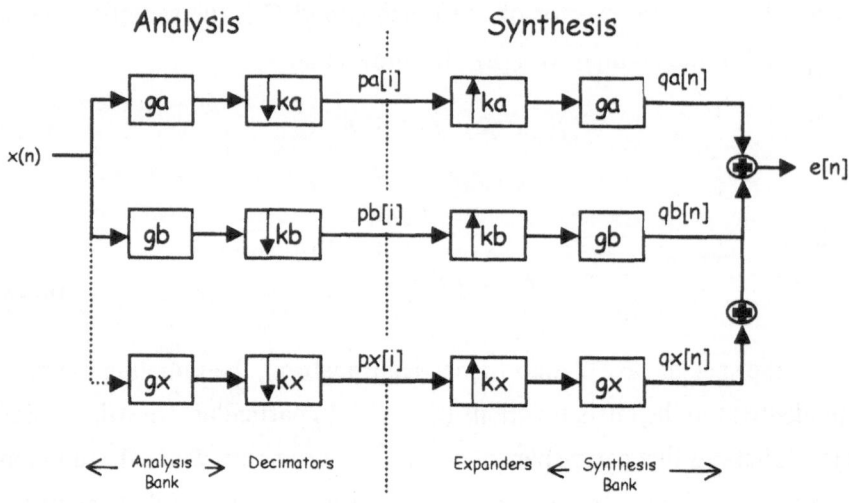

Figure 6.2.1 Block diagram corresponding to the application of repeated and translated kernels using multirate signal processing.

achieved by translating this kernel in steps of two pixels. The highpass kernel is derived from the low-pass kernel by multiplication with the modulating sequence

$$s[n] = -1^n \qquad (6.2.8)$$

The Haar basis functions are compact in space but do a poor job of separating the high- and low-frequency bands. In contrast, sinc functions [$\sin(x)/x$] offer the opposite extreme: They split the band sharply in frequency, but they are poorly localized in space. One can implement a self-similar sinc transform by successive subdivision of a discrete Fourier transform. Unfortunately, the resulting basis functions, which are sinc functions modulated by cosine functions, produce such poorly damped ringing that they are of little value for image analysis or image data compression.

Another approach, the so-called quadrature mirror filters or QMF [Est77], were introduced as a method of splitting a one-dimensional signal into subbands, efficiently decimating the subband signals, and achieving alias-free reconstruction of the original signal. It can be shown that the QMF filter requirements are equivalent to the requirement that a band-splitting transform have an orthogonal transformation matrix. The QMF basis functions can offer good localization in both space and spatial frequency and may be cascaded to form a self-similar orthonormal basis set. They can also be tailored for efficient computation and tuned for orientation as well as frequency. QMF techniques have been extended to two-dimensional signals [Che84], and pyramid versions of QMF decomposition have also been applied to image encoding [Gha87, Say92, Tra87]. Related formulations have been developed independently for applications outside of digital signal processing. Mallat [Mal90], working in the context of machine vision, has derived similar pyramids based on the wavelet theory of Meyer and other mathematicians [DeV92]. Note that the problem can be approached in many ways, including the viewpoints of matrix operations and orthonormal basis sets as well as the notation of digital signal processing, but the results can readily be mapped into different terminologies.

The vast majority of work on QMF representations has been based on low-pass kernels that have an even number of taps. This leads to high-pass kernels with odd symmetry. In general, these kernels must be rather large (12 taps or more) in order to obtain satisfactory performance. It has been demonstrated [Ade87] that odd-tap kernels (which lead to even symmetric high-pass kernels) can be far more compact and still give good performance. These kernels have largely been overlooked, perhaps because they require a nonobvious manipulation: The high- and low-pass kernels must be staggered with respect to each other by one pixel. In signal processing terms, the high-pass decimation must be preceded by a one-sample delay [Wit87].

Within an image of size N, odd-tap kernels cannot offer a perfectly orthogonal basis set unless they have at least $(N+4)/2$ taps. However, they can approach orthogonality even with a rather small number of taps (five to seven in some cases). However, the spatial shape and rather broad frequency tuning of the five- and seven-tap kernels limit their value for image coding applications. For image coding it is preferential to use nine-tap kernels, which are still quite compact; the mean square error in image reconstruction. The error can be reduced at the expense of poorer frequency tuning; but for image processing applications, this error is insignificant in any case. For the general case of an m-tap kernel, the values may be derived by replicating the kernels within an image of size $m+1$ (with wraparound mod $m+1$) and imposing a set of constraints such as orthonormality or zero dc gain for the high-pass kernel. When the QMF kernels are cascaded hierarchically to create a multiscale pyramid, the equivalent kernel at each pyramid level grows in scale by factors of 2.

The most straightforward way to generate two-dimensional QMF transforms is to apply two one-dimensional transforms for each axis. This leads to an image decomposition of the form described previously; at a given level, the image is decomposed into a low-pass subimage, a vertical subimage, a horizontal subimage, and a diagonal subimage. This is known as a *separable pyramid*. Various types of decimation patterns may be used to reduce the required sampling density of the image. In this manner, an original image is reduced to four subimages, each sampled at one-fourth the density of the original image. The fact that the number of transform samples equals the number of image transforms follows immediately from the use of an orthogonal transform. In the case of odd-tap kernels, there will be four such decimation grids, each offset to a different position; for even-tap kernels, the four decimation grids coincide. Note that the diagonal subimage contains high frequencies in the corners of the frequency spectrum, corresponding to diagonal tilt in both directions. Other methods of image decimation and segmentation also exist; for example, it is possible to divide the image spectrum into a pattern of nested diamonds and squares [Wal92]. Successive application of this rule leads to non separable, non oriented QMF pyramids known as *quincunx pyramids*, since the transform requires successive decimation on a quincunx grid. With this type of decimation, the linear scale of the kernal increases by the square root of 2 at each stage, and the number of samples is cut in half at each stage. Another interesting variation are those pyramids which can be built using hexagonally symmetric kernals, known as *hexagonal pyramids*. These have the unique property that all the high-pass kernels are oriented and all have the same shape (as opposed to the case of the separable pyramid, in which horizontal and vertical kernels are the same, but the diagonal one is different).

In transforms based on QMF kernels, the basis set is orthonormal and so the sampling functions are identical to the basis functions. But orthogonality is not required for image coding: it is only necessary that the transform be invertible. In general, then, the sampling functions can be quite different from the basis functions. The sampling matrix for a non-orthogonal basis set is readily derived as the transpose of the inverse of the basis matrix. Suppose that one wishes to use an extremely simple decoder at the expense of a more complex encoder. A very simple band-spitting transform can be built from the three-tap low-pass and high-pass kernels, as described in [Ade87]:

$$[0.25 \quad 0.5 \quad 0.25] \quad \text{and} \quad [-0.25 \quad 0.5 \quad -0.25]$$

These kernels are not a proper QMF pair and do not lead to an orthogonal basis set. Nonetheless, when placed on a staggered grid, they do form a linearly independent basis set; that is, the basis matrix F is non singular, and therefore the corresponding sampling matrix, G, is derived as the inverse of F. The true inverse sampling function has non zero taps over the entire image; however, other versions may be derived by truncating and scaling the taps to achieve orthogonality and a dc system gain of unity. The three-tap kernels are extremely easy to compute, using only shifts and adds, with no multiplications. They can be combined to produce two-dimensional kernels, also known as *band-splitting inverse pairs (BIP filters)*. In this case, the corresponding two-dimensional sampling functions are just the separable products of the one-dimensional sampling functions.

The QMF transforms perform very well for many image compression applications. It has been shown that a nine tap separable pyramid used with simple entropy coding gives performance equal to a blocked transform in terms of mean square error, and superior performance in terms of perceptual degradation. In some cases, it is desirable that a low-resolution version image become available quickly, and that higher resolution becomes available as time goes on; this is known as *progressive transmission*. Many image coding techniques can be modified to allow progressive transmission. However, in the case of pyramids, which are inherently multiscale representations, progressive transmission is achieved by simply sending information from successive pyramid levels in sequence, without adding any overhead to the information. Thus, QMF and BIP pyramids are especially well suited to progressive transmission [Tak94].

6.3   Progressive Transforms for Lossless and Lossy Image Coding

The subband decomposition techniques described so far have some limitations, as well; for example, they do not always result in a lossless (reversible) image compression, which is required by many important applications such as image subtraction, filtering, or contrast enhancement. When an imaging application requires only a quick visual inspection, lossy image compression methods are perfectly adequate and easier to implement in some cases. Lossless methods are also necessary for complex images in which it is unwise to discard any information that may later prove to be useful (e.g., medical images or photographs from a space telescope). In those cases, lossy compression methods may destroy some of the information required during processing or add artifacts to the image that lead to erroneous interpretations.

Traditionally, a user had to choose different coding methods depending whether the highest compression or fast inspection was desired [Rab91, Abb97a]. Historically, some of the most effective methods for lossless compression use linear predictive coding [Kud92], which has been adopted for lossless compression in the Joint Photographic Experts Group (JPEG) still-picture compression standard [Wal91]. This form of compression is usually defined for a single resolution level, and in a way that the image can only be recovered in its entirety, which impedes fast inspection. Several ad hoc methods have been proposed for lossless compression [Rab92]. These are based on first observing an image at its full size, then transmitting only the main features of the image and using some form of interpolation to cover the missing details. These details are progressively added, improving the image quality until near-perfect reconstruction is achieved. This approach is known as *progressive fidelity* transmission. However, the performance of progressive fidelity systems is much inferior to the lossy compression methods. Recently, a tree-structured vector quantizer was proposed [Eff92] for progressive fidelity transmission, which should provide good quality images at low bit rates but is not efficient for lossless compression, More efficient fast inspection can be obtained with the so-called *lossy-plus-residual* methods.   Another alternative is the so-called *progressive resolution* method, in which an image of reduced resolution is transmitted first, then the information required to obtain higher-resolution versions from the original image is transmitted. As an example, this is useful when several small images are displayed together for comparison and later magnified into a larger area or more pixels per unit area [Sai96b].

Excellent lossy compression results have been obtained using the wavelet transform [Woo91, Hee90]. In an image context, it produces a multiresolution representation, which has been shown to be naturally suited for progressive transmission. One multiresolution transform for lossless compression is known in the medical imaging community as the *sequential transform* (S-transform) [Sai96, Ran88].   Another method that enables progressive resolution transmission is called *hierarchical interpolation* (HINT) [Roo88].

These transformations are fairly efficient. but some studies show that they may not be as effective as predictive encoding. However, using recently developed methods based on progressive image processing, it is possible to have a compression scheme that simultaneously allows fast inspection and, only when necessary, exact recovery of the image. We will describe how this can be achieved using a new multiresolution transformation for both lossless and lossy compression called the *sequential plus prediction* or S+P-transform, following the development of [Sai96].

First, however, we will describe the S-transform, which is similar to the Haar multiresolution image representation in [Ade87]. There are different definitions of the S-transforrn in the literature, but most differ only in some implementation details.

A sequence of integers (which may represent a vector image) $e[n]$,

$n = 0,..,N-1$, with N even, can be represented by the two sequences

$$l[n] = \frac{e[2n] + e[2n+1]}{2} \qquad (6.3.1)$$

$$h[n] = e[2n] - e[2n-1] \qquad (6.3.2)$$

for all values of n from 0 to $N/2-1$, truncated downward.

The sequences $l[n]$ and $h[n]$ form the S-transform of $e[n]$. Since the sum and difference of two integers correspond to either two odd or two even integers, the truncation is used to remove the redundancy in the least significant bit. The division and downward truncation calculation can be performed with a single bit shift and the same computer memory used for $e[n]$ can be reused for $l[n]$ and $h[n]$, making for a numerically efficient algorithm implementation. The inverse transformation is given by

$$e[2n] = l[n] + \frac{h[n] + 1}{2} \qquad (6.3.3)$$

$$e[2n+1] = e[2n] - h[n] \quad . \qquad (6.3.4)$$

Advantages of the S-transform include its simplicity and the fact that it significantly reduces the first-order entropy [Sai96a].

The two-dimensional transformation is done by applying the one-dimensional transformations above in a sequential manner to the rows and columns of the image, as shown in Figure 6.3.1. Assuming a image of size $N \times N$, the first decomposition will result in four subimages of size $N/2 \times N/2$, respectively. Hence, the coefficients of each subimage corresponding to LL in Figure 6.3.1 are the mean of 2×2 pixel blocks, and they form another image with half the resolution. The same transformation is applied to these reduced resolution "mean images" to form a hierarchical pyramid. Note that the maximum number of bits required to represent each pixel in the LL

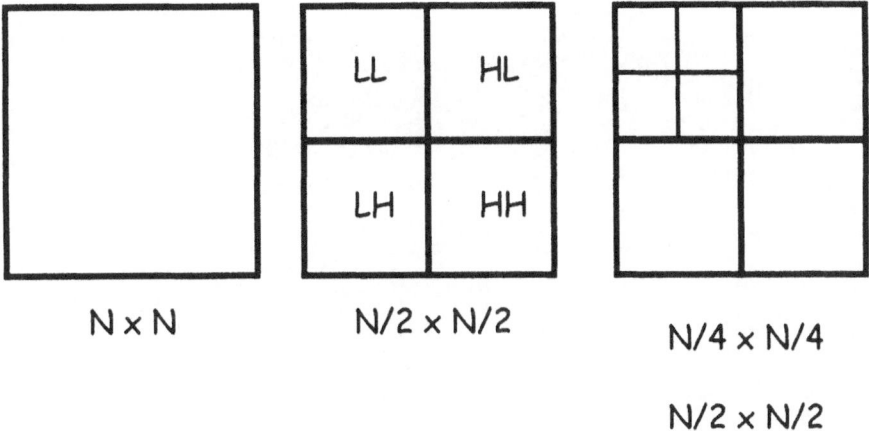

Figure 6.3.1 Block diagram of the construction of an image multiresolution pyramid.

images does not change with each transformation. For example, if the gray-level original image has six bits per pixel, the reduced LL image also has 6 bits per pixel. On the other hand, the other pixels require a signed representation with a larger number of bits. Except for the truncations performed on the discrete signals, this transformation corresponds to a subband decomposition. The low-resolution (LL) images are formed with mean values (a form of low-pass filtering), which reduces aliasing, and is superior to unfiltered subsampling used by linear interpolation methods.

An example of a two-dimensional wavelet transform is given in Figures 6.3.2 to 6.3.3. The original image is displayed in Figure 6.3.2 and the result of a single level decomposition is plotted in Figure 6.3.3a. The effect of a second decomposition of the LL image is then displayed in Figure 6.3.3b. It is clear from the images that the LL component at each decomposition level represents the "mean image."

Note that the S-transform leaves a residual correlation between the high-pass components, which is due to aliasing from the low-frequency components of the original image. Although one could expect an improvement if better filters were used, arithmetic operations with integer numbers create a statistical dependence in the least significant bits, which is

Figure 6.3.2   Original image to be decomposed using the approach depicted in Figure 6.3.1.

irrelevant for lossy compression, but that must be removed for efficient lossless compression. This means that for lossless compression, we must always pay attention to the truncation. By contrast, predictive coding methods do not have to be linear for perfect reconstruction, and the prediction value can be truncated to an integer. In this manner, we can improve upon the S-transform with predictive coding. Instead of using prediction in the final S-transformed pyramid, in the S+P-transform during each one- dimensional transformation, some values of $l[n]$ and $h[n]$ are employed to estimate the value of a given estimate $h[n]$. The differences given by

$$h_d[n] = h[n] - (h[n] + 1/2)$$

(6.3.5)

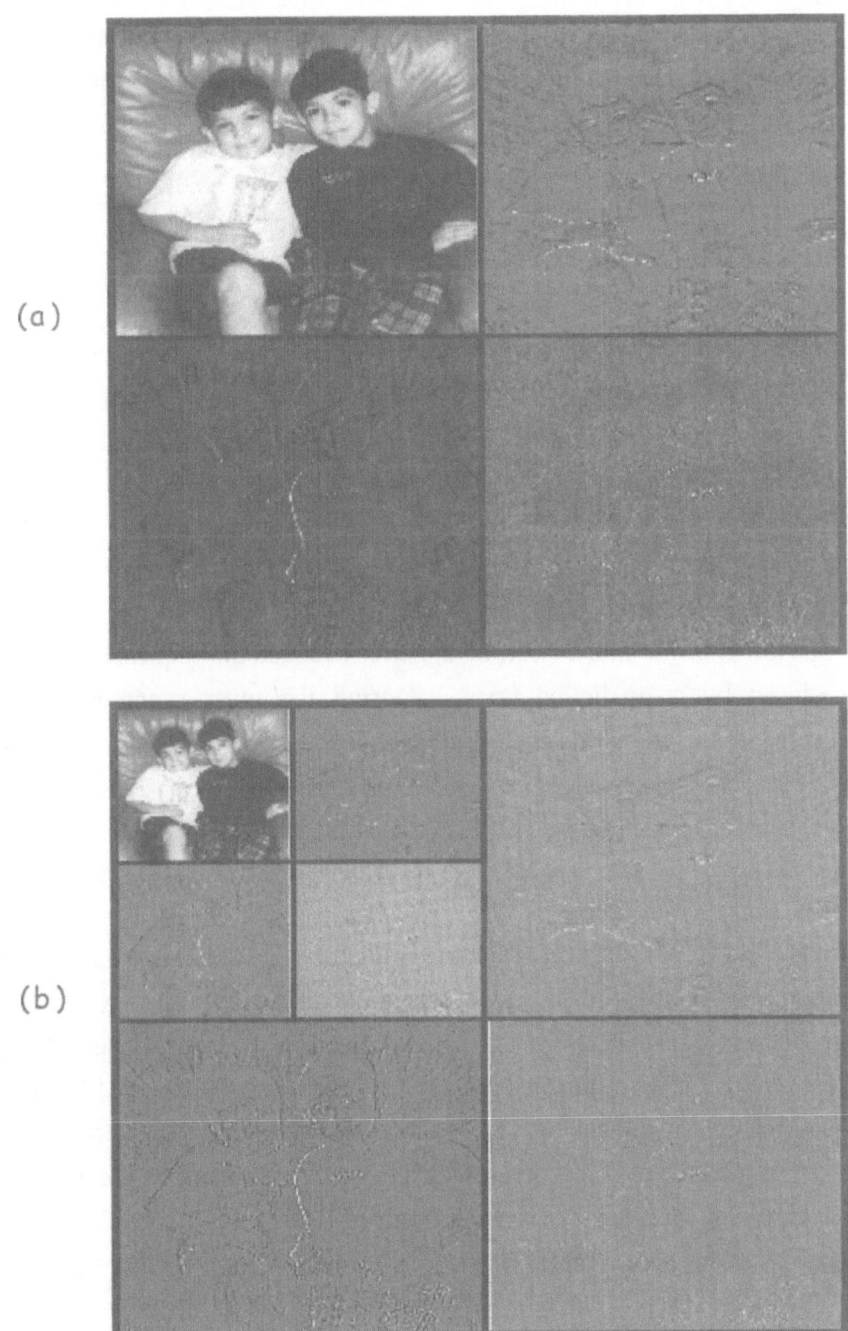

(a)

(b)

Figure 6.3.3 (a) First level decomposition of the image in Figure 6.3.2, (b) second level.

replace $h[n]$, forming a new transformed image with smaller first-order entropy. No estimation is subtracted from the sequence $l[n]$ because it forms the reduced-resolution image, which can be later transformed with the same method. If we define

$$\Delta l[n] = l[n-1] - l[n] \tag{6.3.6}$$

so that we will have zero mean estimation terms, and there is no need to subtract the mean from $e[n]$, the general form of the S+P image estimator is

$$h[n] = \sum_{i=-L_0}^{L_1} \alpha_i \Delta l[n+i] - \sum_{j=1}^{H} \beta_j h[n+j] \tag{6.3.7}$$

where the predictor coefficients are α_i, β_j.

Note that the index i can be negative because the sequence $l[n]$ is not replaced by a prediction error. During the inverse one-dimensional transform, the prediction can be added following a reverse order as follows:

$$h[n] = h_d[n] + (\bar{h}[n] + 1/2) \tag{6.3.8}$$

for $n = N/2-1, N/2-2, ...0$, so that the values of $h[n]$ required to calculate the prediction for the current value of n have already been recovered. The inverse one-dimensional S-transform is calculated after the sequence $h[n]$ is recovered. The two-dimensional S+P-transform is implemented by simply applying the one-dimensional S+P-transform sequentially to the columns and rows of the image. The order in which these transformations are performed is important, because the truncation operation makes the S+P-transform nonlinear. As an example, if the transformation was applied first to the columns and then to the rows, the inverse transformation must be applied first to the rows and then to the columns. In short, the inverse transformation algorithm is just like the transformation algorithm running backward.

We shall next discuss how to select the predictor coefficients. There are three basic schemes which have been proposed [Sai96]; namely minimum entropy, minimum variance, and frequency domain design. The coefficients that minimize the (first-order) entropy can be found with the Nelder-Mead simplex algorithm [Pre86], but their calculation requires a computational effort too large for practical applications. The coefficients that minimize the variance of $h_d[n]$ can be found by solving Yule-Walker equations [Kud92]. However, this common approach does not necessarily minimize the entropy of the S+P-transformed image, and numerical results have shown that even with high-order adaptive predictors, the minimum- variance schemes were inferior to fixed predictors designed in the frequency-domain. Thus, we will concentrate on frequency domain design in the remainder of this section.

If we disregard the truncations, it can be shown that the estimate $h_d[2n]$ can be regarded as the output of a noncausal FIR filter to the input sequence $e[n]$, which is undersampled by a factor of 2. The z-transform of the filter response is given by

$$F(z) = (z-1)\left[\frac{(z^{-1}+1)^2}{2}\sum_{i=-L_0}^{L_1}\alpha_i z^{2i} - \sum_{j=1}^{H}\beta_j z^{2j} - 1\right] \quad (6.3.9)$$

It is possible to have a noncausal response with predictive coding because the values of $l[n]$ are not replaced by the prediction.

Since most of the image energy is normally concentrated in the low frequencies, to reduce the variance we should select a filter with a strong attenuation in the low frequencies. However, due to the structure of the filter's z-transform response and the requirements for a reversible transformation, stronger attenuation in the low frequencies leads to a larger gain in the high frequencies. In theory, the choice for the best predictor depends on the image's characteristics: Smooth and noiseless images are better compressed using a filter with the largest attenuation in the low frequencies, whereas noisy and very detailed images require a filter with small gain at the high frequencies. It has been observed [Sai96] that the entropy has a low sensitivity to the predictor parameters. Thus, the predictor parameter choice is not as critical and it is possible to find predictors that are effective for a broad class of images (satellite images, medical, etc.).

Examples of these are available in the literature, which show that the difference between the S-transforms and S+P-transforms can be significant and that the S+P consistently yields better results. It can be argued that the entropy of the S+P-transform in some cases is not much smaller than obtained by JPEG. However, the capability of progressive transmission is quite important and should be taken into account. Compared to the HINT method, the S+P-transform is more advantageous because its reduced-resolution images have better quality and because it can be used for progressive fidelity transmission, as we will now discuss.

The progressive-resolution transmission schemes are easily implemented from the multiresolution transform because, in this case, the encoder just has to code the pixels beginning from the highest level of the pyramid. For entropy coding the S+P-transform, we can use the fact that there is a statistical dependence between pixels of the transformed image that cannot be further reduced by linear predictive methods, but that can be exploited during coding. In practice, we should also pay attention to the complexity of the coding methods; there are some components of the transformed image that cannot be efficiently compressed and may be transmitted uncoded. This fact was used to define one entropy-coding method in the JPEG still-picture compression standard. In JPEG's method, an integer value is decomposed in three parts: The length in bits, the sign, and a magnitude difference. The magnitude difference is the difference between the actual magnitude and the lowest magnitude in a particular predefined set of transform pixel magnitudes. The length, which is the number of bits needed to express the sign and this magnitude difference, is entropy coded. forming the variable-length code (VLC), and then the sign and the magnitude difference are transmitted uncoded in the variable-length integer (VLI) format (refer to the JPEG standard for more details). With this representation, there is a small loss due to the fact that the VLIs are not entropy coded, but with the advantage that the number of VLC symbols is small, which simplifies the entropy-coding process. In other words, with this representation, we can get bit rates near those that would be obtained if the complete integers were entropy coded, but with a smaller complexity. The same approach is used to entropy code the S+P-transform. However, to reduce the loss that must occur with the uncoded transmission of part of the numbers, slightly more complex integer coding methods have been proposed in which the number of a magnitude set is entropy coded first, then depending on its value, it is followed by the sign bit and the magnitude difference bits. In recent experiments, several different methods were employed to evaluate the true compression rates that can be achieved by the S+P-transform. One approach uses entropy coding similar to JPEG's standard for lossless compression, except that the S+P-transform replaces differential pulse code modulation and JPEG's fixed one-pass Huffman code is replaced by a Huffman code optimized for each image (two-pass encoding). Another method employs *set partitioning in hierarchical trees* (SPIHT), introduced in [Sai96a] for progressive fidelity transmission, combined with the S+P-transform and adapted for lossless compression.

Although good results have been obtained using all of these methods, it is difficult to present a comparative evaluation of different coding methods. For progressive fidelity transmission, it is possible to achieve good results using the method of set partitioning in hierarchical trees (SPIHT), which is in principle, similar to the *embedded zerotree wavelet* (EZW) algorithm [Sha93]. SPIHT is presently one of the most efficient methods known for lossy compression, both in terms of speed and compression. In addition, it has several other advantages, including being completely adaptive, simple to implement, and having the ability to produce a *fully embedded message* (a message corresponding to a rate Ro bits always forms the first Ro bits of any message with rate R1 > Ro). With embedded coding, at any point in the decoding process it is possible to recover the lossy version with distortion corresponding to the rate of the received message, which allows coding/decoding to exactly the desired rate or distortion. One important property of the SPIHT algorithm is that it codes information corresponding to individual bits of the transformed image, following a bit-plane ordering. Thus, it shares some characteristics of the well-known bit-plane binary coding methods and can be used for lossless compression. It also has some unique characteristics; the bits are not transmitted in the usual line-by-line order and tree structures are used in such a way that, with a single coded symbol, the decoder can infer that the all bits in a large region of a give n bit plane are zero.

The coding efficiency of the SPIHT algorithm comes from exploiting the self-similarity present in the wavelet multiresolution representations property also present in the S+P-transformed images. The only reason the S+P-transform cannot be used directly with SPIHT is that, for embedded lossy compression, the transmission priority given by the bit planes will minimize the mean squared error (MSE) distortion only when the transformation is unitary. The S+P-transform is not unitary, although we can achieve a good approximation.

For instance, if we use the relationships

$$l[n] = (e[2n] + e[2n+1])/\sqrt{2} \qquad\qquad (6.3.10)$$

$$h[n] = (e[2n] - e[2n+1])/\sqrt{2} \qquad\qquad (6.3.11)$$

then we obtain a unitary transformation. It is thus possibly to multiply the S+P-transform coefficients by integer values during the encoding process and obtain a unitary transformation (the same approach may be applied to the S-transform).

In the progressive fidelity transmission scheme, the decoder initially sets the transformed image to zero and updates its pixel values using the coded message. The decoder can decide at which rate to stop, and then it calculates the inverse S+P-transform to obtain a lossy version of the image. If it continues decoding to the end of the file, the image is recovered exactly. The SPIHT algorithm can also be used to code all bit planes to recover the image exactly. However, when it codes the least significant bits, its efficiency decreases, mostly in terms of speed and memory usage. This usually happens for a bit rate when the lossy version of the image is visually indistinguishable from the original and may be overcome using a hybrid approach in which SPIHT is used to code up to the third least significant bit and then a simplified version of the magnitude set encoding described earlier is used to code the remaining bit planes (for more details, see [Sai96]).   This change in the SPIHT method ensures that the coding process will disregard the parts of the bit planes that, due to scaling, are identically zero. Results obtained with this method have been shown to be slightly inferior (usually less than I dB) to those obtained with the SPIHT algorithm on a wavelet transform and are practically equal or superior to the EZW method and other much more complex coding methods like subband coding with adaptive vector quantization [Kim92].   The codec programs with the methods discussed above, including the S+P-transform. can be obtained via anonymous ftp from the host ipl.rpi.edu, directory pub/EW-Code, with instructions in the file FLEADME or via the Internet site with URL http://ipl.rpi.edu/SPIHT.

6.4   Embedded Zerotree Wavelets

As we have seen, the problem of obtaining the best image quality for a given compression rate can be quite complex.   While discussing the S+P-transform, reference was made to the embedded zerotree wavelet algorithm (EZW) as a point of comparison [Sha93]. The EZW algorithm merits additional attention, as it is a simple, yet remarkably effective, image compression algorithm, having the property that the bits in the image bit stream are generated in order of importance, yielding a fully *embedded code* (i.e., all encodings of the same image at lower bit rates are embedded in the beginning of the bit stream for the target bit rate).

An embedded code represents a sequence of binary decisions that distinguish a target image from the "null," or all gray, image. Since the embedded code contains all lower-rate codes "embedded" at the beginning of the bit stream, effectively the bits are "ordered in importance." Using an embedded code, an encoder can terminate the encoding at any point, thereby allowing a target rate or distortion metric to be met exactly. Typically, some target parameter, such as bit count, is monitored in the encoding process. When the target is met, the encoding simply stops. Similarly, given a bit stream, the decoder can cease decoding at any point and can produce reconstructions corresponding to all lower-rate encodings.

Embedded coding is similar in spirit to binary finite precision representations of real numbers. All real numbers can be represented by a string of binary digits. For each digit added to the right, more precision is added. Yet, the "encoding" can cease at any time and provide the "best" representation of the real number achievable within the framework of the binary digit representation. Similarly, the embedded coder can cease at any time and provide the "best" representation of an image achievable within its framework. Intuitively, for a given rate or distortion, a nonembedded code should be more efficient than an embedded code since it is free from the constraints imposed by embedding. However, work on embedded coding schemes have been motivated in part by other *universal coding* schemes that have been used for lossless data compression in which the coder attempts to optimally encode a source using no prior knowledge of the source. An excellent review of universal coding can be found in [Bel90].

In universal coders, the encoder must learn the source statistics as it progresses. In other words, the source model is incorporated into the actual bit stream. For lossy compression, there has been little work in universal coding. Typical image coders require extensive training for both scalar and vector quantization and generation of nonadaptive entropy codes, such as Huffman codes. However, it is also possible to develop a universal encoder by incorporating all learning into the bit stream itself; this approach is utilized in the EZW algorithm; our description of this algorithm follows the development of [Sha93].

The EZW algorithm is based on four key concepts, namely:

1. A discrete wavelet transform or hierarchical subband decomposition
2. Prediction of the absence of significant information across scales. by exploiting the self-similarity inherent in images
3. Entropy-coded successive-approximation quantization
4. Uiversal lossless data compression which is achieved via adaptive arithmetic coding.

The EZW algorithm contains the following features:

- A discrete wavelet transform which provides a compact multiresolution representation of the image.
- Zerotree coding which provides a compact multiresolution representation

of *significance maps* (binary maps indicating the positions of the significant coefficients). Zerotrees allow the successful prediction of insignificant coefficients across scales to be efficiently represented as part of exponentially growing trees.

- Successive aproximation which provides a compact multiprecision representation of the significant coefficients and facilitates the embedding algorithm.

- A prioritization protocol whereby the ordering of importance is determined, in order, by the precision, magnitude, scale, and spatial location of the wavelet coefficients. In particular, larger coefficients are deemed more important than smaller coefficients regardless of their scale.

- Adaptive multilevel arithmetic coding which provides a fast and efficient method for entropy coding strings of symbols and requires no training or prestored tables [Bel90].

- The EZW algorithm runs sequentially and stops whenever a target bit rate or a target distortion is met. A target bit rate can be met exactly, and an operational *rate-versus-distortion function* (RDF) can be computed point-by-point.

In order to better appreciate the EZW transform, let us first provide an overview of the ways in which wavelet theory and multiresolution analysis allow us to represent trends and anomalies in statistical data. This is important in image processing because edges (which represent anomalies in the spatial domain) represent important information, although they are only contained in a small number of the image samples. One of the oldest problems in statistics and signal processing is how to choose the size of an analysis window, block size, or record length of data so that statistics computed within that window provide good models of the signal behavior within that window. The choice of an analysis window involves trading the ability to analyze *anomalies* (signal behavior that is more localized in the time or space domain and tends to be wide band in the frequency domain) versus *trends* (signal behavior that is more localized in frequency but persists over a large number of lags in the time domain). To model data as being generated by random processes so that computed statistics become meaningful, stationary and ergodic assumptions are usually required which tend to obscure the contribution of anomalies.

One of the main contributions of wavelet theory and multiresolution analysis [Dau88, Mal89, Mal90, Rio91] is that it provides an elegant framework in which both anomalies and trends can be analyzed on an equal footing. Wavelets provide a signal representation in which some of the coefficients represent long data lags corresponding to a narrow-band, low-frequency range, and some of the coefficients represent short data lags corresponding to a wide-band, high-frequency range. Using the concept of scale, data representing a continuous trade-off between time (or space in the case of images) and frequency are available.

In image processing, most of the image area typically represents spatial *trends*, or areas of high statistical spatial correlation. However, *anomalies*, such as edges or object boundaries, take on a perceptual significance that is far greater than their numerical energy contribution to an image. Traditional transform coders, such as those using the DCT, decompose images into a representation in which each coefficient corresponds to a fixed-size spatial area and a fixed-frequency bandwidth, where the bandwidth and spatial area are effectively the same for all coefficients in the representation. Edge information tends to disperse so that many non zero coefficients are required to represent edges with good fidelity. However, since the edges represent relatively insignificant energy with respect to the entire image, traditional transform coders have been fairly successful at medium and high bit rates. At extremely low bit rates, however, traditional transform coding techniques, such as JPEG, tend to allocate too many bits to the "trends," and have few bits left over to represent "anomalies." As a result, blocking artifacts often result.

Wavelet techniques show promise at extremely low bit rates because trends, anomalies, and information at all scales in between are available. A major difficulty is that fine-detail coefficients representing possible anomalies constitute the largest number of coefficients, and, therefore, to make effective use of the multiresolution representation, much of the information is contained in representing the position of those few coefficients corresponding to significant anomalies.

The discrete wavelet transform forms the basis of the EZW algorithm; as discussed previously, it is identical to a hierarchical subband system, where the subbands are logarithmically spaced in frequency and represent an octave-band decomposition. To begin the decomposition, the image is divided into four subbands and critically subsampled, as previously shown in Figures 6.3.3a and 6.3.3b. Each coefficient represents a spatial area corresponding to approximately a $N/2 \times N/2$ area of the original image. The low frequencies represent a bandwidth approximately corresponding to frequencies between 0 and $\pi/2$, whereas the high frequencies represent the band between $\pi/2$ and π. The four subbands arise from separable application of vertical and horizontal filters. The subbands labeled LH, HL, and HH represent the finest-scale wavelet coefficients. To obtain the next coarser scale of wavelet coefficients, the subband LL is further decomposed and critically sampled in an iterative manner as shown in Figure 6.3.3b. The process continues until some final scale is reached.

Note that for each coarser scale, the coefficients represent a larger spatial area of the image but a narrower band of frequencies. At each scale, there are three subbands; the remaining lowest frequency subband is a representation of the information at all coarser scales. The issues involved in the design of the filters for the type of subband decomposition described above have been discussed in previous chapters and in many references [Dau88, Woo91, Zet90].

It is a matter of terminology to distinguish between a transform and a subband system, as they are two ways of describing the same set of numerical operations from different points of view. For example, let x be a column vector whose elements represent a scanning of the image pixels, and let X be a column vector whose elements are the array of coefficients resulting from the wavelet transform or subband decomposition applied to x. From the transform point of view, X represents a linear transformation of x which can be described by a transformation matrix, W, as

$$X = Wx$$

$$(6.4.1)$$

Although not actually computed this way, the effective filters that generate the subband signals from the original signal form basis functions for the transformation (the rows of W). Different coefficients in the same subband represent the projection of the entire image onto translates of a prototype subband filter, since from the subband point of view, they are simply regularly spaced different outputs of a convolution between the image and a subband filter. Thus, the basis functions for each coefficient in a given subband are simply translates of one another.

In subband coding systems, the coefficients from a given subband are usually grouped together for the purposes of designing quantizers and coders. Such a grouping suggests that statistics computed from a subband are in some sense representative of the samples in that subband. However, this statistical grouping once again implicitly deemphasizes the most significant anomalies or edges. We can use the terminology "wavelet transform" in this context to refer to the fact that each wavelet coefficient is individually and deterministically compared to the same set of thresholds for the purpose of measuring significance. Thus, each coefficient is treated as a distinct, potentially important piece of data regardless of its scale, and no statistics for a whole subband are used in any form. The result is that the small number of deterministically significant fine-scale coefficients are not obscured because of their statistical insignificance.

As a further note on terminology, the filters used to compute the discrete wavelet transform in this type of image coding may be based on the multi-tap symmetric quadrature mirror filters (QMF) discussed previously; the resulting transformation has also been called a QMF pyramid. As mentioned earlier, these filters offer good localization properties, and their symmetry allows for simple edge treatments. Additionally, using properly scaled coefficients, the transformation matrix for a discrete wavelet transform obtained using these filters is so close to unitary that it can be treated as unitary for the purpose of lossy compression. Since unitary transforms preserve some desirable properties of the signal, it makes sense from a numerical standpoint to compare all of the resulting transform coefficients to the same thresholds to assess significance.

An important aspect of low-bit-rate image coding is the coding of the

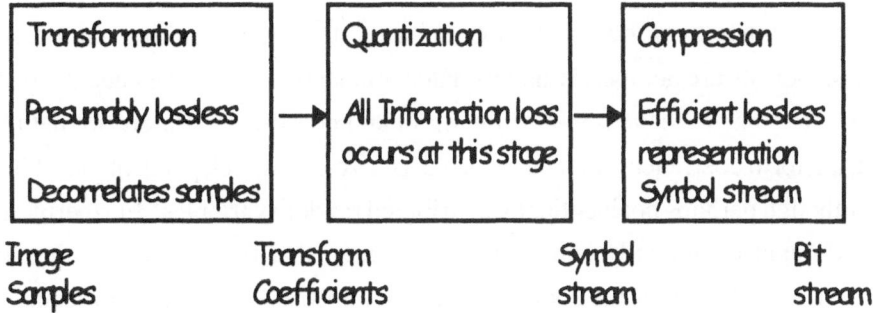

Figure 6.4.1 Block diagram of a generic transform encoder.

positions of those coefficients that will be transmitted as nonzero values. Using scalar quantization followed by entropy coding, in order to achieve very low bit rates, the probability of the most likely symbol after quantization (the zero symbol) must be extremely high. Typically, a large fraction of the bit budget must be spent on encoding the significance map, or the binary decision as to whether a sample, in this case a coefficient of a two-dimensional discrete wavelet transform, has a zero or nonzero quantized value. It follows that a significant improvement in encoding the significance map translates into a corresponding gain in compression efficiency. The position information contained in the significance map is important relative to the amplitude and sign information.

A typical low-bit rate image coder has three basic components: a transformation, a quantizer and compression, as shown in Figure 6.4.1. The original image is passed through some transformation to produce transform coefficients. This transformation is considered to be lossless, although, in practice, this may not be the case exactly. The transform coefficients are then quantized to produce a stream of symbols, each of which corresponds to an index of a particular quantization bin. Note that virtually all of the information loss occurs in the quantization stage. The data compression stage takes the stream of symbols and attempts to losslessly represent the data stream as efficiently as possible.

The goal of the transformation is to produce coefficients that are decorrelated. Ideally, we would like a transformation to remove all dependencies between samples. Assume for the moment that the transformation is doing its job so well that the resulting transform coefficients are not merely uncorrelated, but statistically independent. Also, assume that we have removed the mean and coded it separately so that the transform coefficients can be modeled as zero-mean, independent variables, although perhaps not identically distributed random variables. Furthermore, we might additionally constrain the model so that the probability density functions (PDF) for the coefficients are symmetric. The goal is to quantize the transform coefficients so that the entropy of the resulting distribution of bin indexes is small enough so that the symbols can be entropy coded at some target low bit rate, say, for example, 0.5 bits per pixel (bpp). Assume that the quantizers will be symmetric midtread, perhaps nonuniform, quantizers, although different symmetric midtread quantizers may be used for different groups of transform coefficients. Letting the central bin be index 0, note that because of the symmetry, for a bin with a nonzero index magnitude, a positive or negative index is equally likely. In other words, for each nonzero index encoded, the entropy code is going to require at least one bit for the sign. An entropy code can be desynched based on modeling probabilities of bin indices as the fraction of coefficients in which the absolute value of a particular bin index occurs. Using this simple model and assuming that the resulting symbols are independent, the entropy of the symbols H can be expressed as

$$H = -p \log_2 p - (1-p) \log_2(1-p) + (1-p)(1-H_N) \qquad (6.4.2)$$

where p is the probability that a transform coefficient is quantized to zero and H_N represents the conditional entropy of the absolute values of the quantized coefficients conditioned on them being nonzero. The first two terms in the sum represent the first-order binary entropy of the significance map, whereas the third term represents the conditional entropy of the distribution of nonzero values multiplied by the probability of them being nonzero. Thus, we can express the true cost of encoding the actual symbols as follows:

Total cost = Cost of significance map + Cost of nonzero values

$$(6.4.3)$$

It can be shown that no matter how optimal the transform, quantizer, or entropy coder, under very typical conditions, the cost of determining the positions of the few significant coefficients represents a significant portion of the bit budget at low rates and is likely to become an increasing fraction of the total cost as the rate decreases. By employing an image model based on an extremely simple and easy to satisfy hypothesis, it is possible to efficiently encode significance maps of wavelet coefficients.

To improve the compression of significance maps of wavelet coefficients, a new data structure called a *zerotree* is defined. A wavelet coefficient x is said to be insignificant with respect to a given threshold T if $|x| < T$. The zerotree is based on the hypothesis that if a wavelet coefficient at a coarse scale is insignificant with respect to a given threshold T, then all wavelet coefficients of the same orientation in the same spatial location at finer scales are likely to be insignificant with respect to T. More specifically, in a hierarchical subband system, with the exception of the highest-frequency subbands, every coefficient at a given scale can be related to a set of coefficients at the next finer scale of similar orientation. The coefficient at the coarse scale is called the *parent*, and all coefficients corresponding to the same spatial location at the next finer scale of similar orientation are called *children*. For a given parent, the set of all coefficients at all finer scales of similar orientation corresponding to the same location are called *descendants*. Similarly, for a given child, the set of coefficients at all coarser scales of similar orientation corresponding to the same location are called *ancestors*. For a QMF-pyramid subband decomposition, the parent-child dependencies are shown in Figure 6.4.2. With the exception of the lowest-frequency subband, all parents have four children. For the lowest-frequency subband, the parent-child relationship is defined such that each parent node has three children.

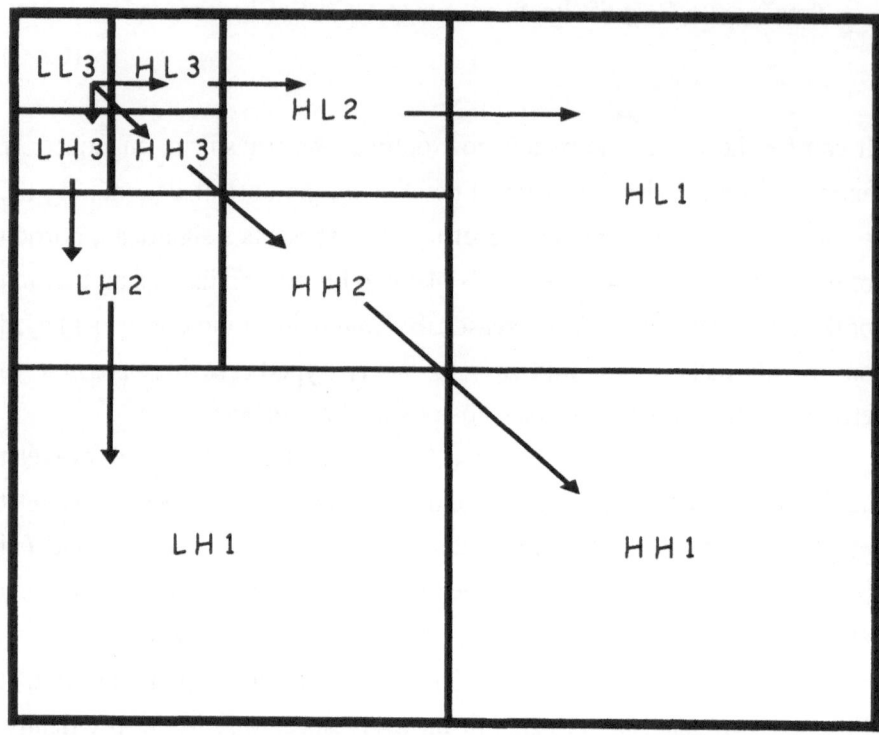

Figure 6.4.2 Parent-child dependencies of subbands; the arrows point from the subband of the parents to the subband of the children, with the lowest-frequency subband at the top left and the highest-frequency subband at the bottom right.

A scanning of the coefficients is performed in such a way that no child node is scanned before its parent. For an N-scale transform, the scan begins at the lowest-frequency subband, denoted as LLN, and scans subbands HLN, LHN, and HHN, at which point it moves on to scale $N-1$, etc. The scanning pattern for a three-scale QMF pyramid can be seen in Figure 6.4.3. Note that each coefficient within a given subband is scanned before any coefficient in the next subband.

Given a threshold level T to determine whether or not a coefficient is significant, a coefficient x is said to be an *element* of a zerotree for threshold T if itself and all of its descendants are insignificant with respect to T. An element of a zerotree for threshold T is a *zerotree root* if it is not the descendant of a previously found zerotree root for threshold T (it is not predictably insignificant from the discovery of a zerotree root at a coarser

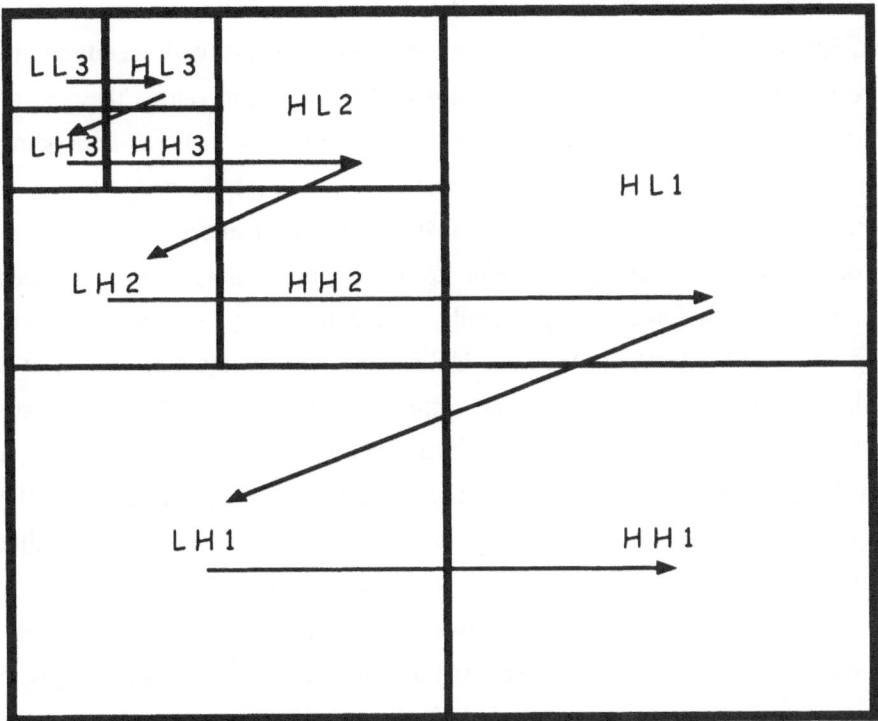

Figure 6.4.3 Scanning order of the subbands for encoding a significance map; parents must be scanned before children, and all positions in a given subband are scanned before the scan moves to the next subband.

scale at the same threshold). A zerotree root is encoded with a special symbol indicating that the insignificance of the coefficients at finer scales is completely predictable. The significance map can be efficiently represented as a string of symbols from a three-symbol alphabet which is then entropy coded. The three symbols used are: (1) zerotree root, (2) isolated zero, which means that the coefficient is insignificant but has some significant descendant, and (3) significant. When encoding the finest-scale coefficients, since coefficients have no children, the symbols in the string come from a two-symbol alphabet, whereby the zerotree symbol is not used.

In addition to encoding the significance map, it is useful to encode the sign of significant coefficients along with the significance map. Thus, in practice, four symbols are used: (1) *zerotree root*, (2) *isolated zero*, (3) *positive significant*, and (4) *negative significant*. This minor addition will be useful for embedding. Note that it is also possible to include two additional symbols such as "positive/negative significant but descendants are zerotrees," etc. At low bit rates, this addition often increases the cost of coding the significance map. Zerotree coding reduces the cost of encoding the significance map using self-similarity. Even though the image has been transformed using a decorrelating transform, the occurrences of insignificant coefficients are not independent events. More traditional techniques employing transform coding typically encode the binary map via some form of run-length encoding [Wal91]. Unlike the zerotree symbol, which is a single "termination" symbol and applies to all tree depths, run-length encoding requires a symbol for each run-length which much be encoded. A technique that is closer in spirit to the zerotrees is the *end-of-block* (EOB) symbol used in JPEG, which is also a "terminating" symbol indicating that all remaining coefficients in the block are quantized to zero.

To see why zerotrees may provide an advantage over EOB symbols, consider that a zerotree represents the insignificance information in a given orientation over an approximately square spatial area at all finer scales up to and including the scale of the zerotree root. Because the wavelet transform is a hierarchical representation, varying the scale in which a zerotree root occurs automatically adapts the spatial area over which insignificance is represented. The EOB symbol, however, always represents insignificance over the same spatial area, although the number of frequency bands within that spatial area varies. Given a fixed block size, there is exactly one scale in the wavelet transform in which if a zerotree root is found at that scale, it corresponds to the same spatial area as a block of the DCT. If a zerotree root can be identified at a coarser scale, then the insignificance pertaining to that orientation can be predicted over a larger area. Similarly, if the zerotree root does not occur at this scale, then looking for zerotrees at finer scales represents a hierarchical divide and conquer approach to searching for one or more smaller areas of insignificance over the same spatial regions as the block size. Thus, many more coefficients can be predicted in smooth areas where a root typically occurs at a coarse scale. Furthermore, the zerotree approach can isolate interesting nonzero details by immediately eliminating large insignificant regions from consideration. Note that this technique is quite different from previous attempts to exploit self-similarity in image coding [Pen91] in that it is far easier to predict insignificance than to predict significant detail across scales. The zerotree approach was developed in recognition of the difficulty in achieving meaningful bit-rate reductions for significant coefficients via additional prediction. Instead, the focus here is on reducing the cost of encoding the significance map so that, for a given bit budget, more bits are available to encode expensive significant coefficients.

A similar technique has been proposed known as the *LK coder* after the initials of its developers, Lewis and Knowles (LK) [Lew91, Lew92]. In the LK coder, a "tree" is said to be zero if its energy is less than a perceptually based threshold. Also, the "zero flag" used to encode the tree is not entropy coded. A promising technique representing a compromise between the EZW algorithm and the LK coder has also been described recently [Xio93].

The basic principle used thus far is that if a coefficient at a coarse scale is insignificant with respect to a threshold, then all of its descendants, as defined above, are also insignificant. This can be interpreted as an extremely general image model. One of the aspects that seems to be common to most models used to describe images is that of a *decaying spectrum*. For example, this property exists for both stationary autoregressive models and *non stationary fractal*, or *nearly 1/f models*, as implied by the name which refers to a generalized spectrum [Wom90]. The model for the zerotree hypothesis is even more general than "decaying spectrum," in that it allows for some deviations because it is linked to a specific threshold. Consider an example from [Sha93], where the threshold is 50, and we are considering a coefficient of magnitude 30, and whose largest descendant has a magnitude of 40. Although a higher-frequency descendant has a larger magnitude (of 40) than the coefficient under consideration (which is 30) (i.e., the "decaying spectrum" model is violated), the coefficient under consideration can still be represented using a zerotree root since the whole tree is still insignificant (magnitude less than 50). This holds true even though the "decaying spectrum model is not valid" (a higher-frequency descendant in this example has a larger magnitude than the coefficient under consideration). Thus, assuming that the more common image models have some validity, the zerotree hypothesis should be satisfied easily and extremely often. For those instances where the hypothesis is violated, it is safe to say that an informative, unexpected event has occurred, and we should expect the cost of representing this event to be commensurate with its self-information.

It should also be pointed out that the improvement in encoding significance maps provided by zerotrees is specifically not the result of exploiting any linear dependencies between coefficients of different scales that were not removed in the transform stage. In practice, the linear correlation between the values of parent and child wavelet coefficients has been found to be extremely small, implying that the wavelet transform is doing an excellent job of producing nearly uncorrelated coefficients. However, there is likely additional dependency between the squares (or magnitudes) of parents and children. Although this dependency is difficult to characterize in general for most images, even without access to specific statistics, it is reasonable to expect the magnitude of a child to be smaller than the magnitude of its parent. In other words, it can be reasonably conjectured (based on experience with real-world images) that had we known the details of the statistical dependencies and computed an "optimal" estimate, such as the conditional expectation of the child's magnitude given the parent's magnitude, that the "optimal" estimator would, with very high probability, predict that the child's magnitude would be the smaller of the two. Using only this mild assumption, based on an inexact statistical characterization, given a fixed threshold, and conditioned on the knowledge that a parent is insignificant with respect to the threshold, the "optimal" estimate of the significance of the rest of the descending wavelet tree is that it is entirely insignificant with respect to the same threshold (a zerotree). On the other hand, if the parent is significant, the "optimal" estimate of the significance of descendants is highly dependent on the details of the estimator whose knowledge would require more detailed information about the statistical nature of the image. Thus, under this mild assumption, using zerotrees to predict the insignificance of wavelet coefficients at fine scales given the insignificance of a root at a coarse scale is more likely to be successful in the absence of additional information than attempting to predict significant detail across scales.

The concept of predicting the insignificance of coefficients from low-frequency to high-frequency information corresponding to the same spatial localization is a fairly general concept and not specific to the wavelet transform configuration shown in Figure 6.4.2. Zerotrees are equally applicable to Quincunx wavelets [Sai93, Kov92, Sha92], in which case, each parent would have two children instead of four, except for the lowest frequency, where parents have a single child.

Also, a similar approach can be applied to linearly spaced subband decompositions, such as the DCT, and to other more general subband decompositions, such as wavelet packets [Coi92] and Laplacian pyramids [Bur83]. Of course, with the use of linearly spaced subbands, zerotree-like coding loses its ability to adapt the spatial extent of the insignificance prediction. Nevertheless, it is possible for zerotree-like coding to outperform EOB coding since more coefficients can be predicted from the subbands along the diagonal. For the case of wavelet packets, the situation is a bit more complicated, because a wider range of tilings in the space-frequency domain are possible; it may not always be possible to define similar parent-child relationships. Zerotree extensions such as these are interesting areas for further research.

So far, we have described a method of encoding significance maps of wavelet coefficients that seems to consistently produce a code with a lower bit rate than either the empirical first-order entropy or a run-length code of the significance map. The original motivation for employing successive approximation in conjunction with zerotree coding was that since zerotree coding was performing so well in encoding the significance map of the wavelet coefficients, it was hoped that more efficient coding could be achieved by zerotree coding more significance maps. Another motivation for successive approximation derives directly from the goal of developing an embedded code analogous to the binary representation of an approximation to a real number. Consider the wavelet transform of an image as a mapping whereby an amplitude exists for each coordinate in scale space. The scale-space coordinate system represents a coarse-to-fine "logarithmic" representation of the domain of the function; furthermore, successive approximation provides a coarse-to-fine, multiprecision "logarithmic" representation of amplitude information, which can be thought of as the range of the image function when viewed in the scale-space coordinate system defined by the wavelet transform. Thus, in a very real sense, the EZW coder generates a representation of the image that is coarse-to-fine in both the domain and range simultaneously.

To perform the embedded coding, *successive-approximation quantization* (SAQ) is applied. As will be seen, SAQ is related to bit-plane encoding of the magnitudes. The SAQ is related to bit-plane encoding of the magnitudes; it sequentially applies a sequence of thresholds to determine significance, where the thresholds are chosen so that

$$T_i = T_{i-1}/2 \tag{6.4.4}$$

The initial threshold T_0 is chosen so that $|X_j| < 2T_0$ for all transform coefficients. During the encoding (and decoding), two separate lists of wavelet coefficients are maintained. At any point in the process, the *dominant list* contains the coordinates of those coefficients that have not yet been found to be significant in the same relative order as the initial scan. This scan is such that the subbands are ordered, and within each subband, the set of coefficients are ordered. Thus, all coefficients in a given subband appear on the initial dominant list prior to coefficients in the next subband. The *subordinate list* contains the magnitudes of those coefficients that have been found to be significant. For each threshold, each list is scanned once.

During a dominant pass, coefficients with coordinates on the dominant list (those that have not yet been found to be significant) are compared to the threshold T_i to determine their significance and, if significant, their sign. This significance map is then zerotree coded. Each time a coefficient is encoded as significant (positive or negative), its magnitude is appended to the subordinate list, and the coefficient in the wavelet transform array is set to zero so that the significant coefficient does not prevent the occurrence of a zerotree on future dominant passes at smaller thresholds.

A dominant pass is followed by a subordinate pass in which all coefficients on the subordinate list are scanned and the specifications of the magnitudes available to the decoder are refined to an additional bit of precision. More specifically, during a subordinate pass, the width of the effective quantizer step size, which defines an uncertainty interval for the true magnitude of the coefficient, is cut in half. For each magnitude on the subordinate list, this refinement can be encoded using a binary alphabet with a "I" symbol indicating that the true value falls in the upper half of the old uncertainty interval and a "O" symbol indicating the lower half. The string of symbols from this binary alphabet that is generated during a subordinate pass is then entropy coded. Note that prior to this refinement, the width of the uncertainty region is exactly equal to the current threshold. After the completion of a subordinate pass, the magnitudes on the subordinate list are sorted in decreasing magnitude, to the extent that the decoder has the information to perform the same sort.

The process continues to alternate between dominant passes and subordinate passes where the threshold is halved before each dominant pass. (In principle, one could divide by other factors than 2, although this factor has an intuitive interpretation in terms of bit-plane encoding.)

In the decoding operation, each decoded symbol, both during a dominant and a subordinate pass, refines and reduces the width of the uncertainty interval in which the true value of the coefficient (or coefficients, in the case of a zerotree root) may occur. The reconstruction value used can be anywhere in that uncertainty interval. For minimum mean square error distortion, one could use the centroid of the uncertainty region using some model for the probability density function of the coefficients; another approach would be to simply use the center of the uncertainty interval as the reconstruction value. The encoding stops when some target-stopping condition is met, such as when the bit budget is exhausted. The encoding can cease at any time and the resulting bit stream contains all lower-rate encodings. Note, that if the bit stream is truncated at an arbitrary point, there may be bits at the end of the code that do not decode to a valid symbol since a codeword has been truncated. In that case, these bits do not reduce the width of an uncertainty interval or any distortion function. In fact, it is very likely that the first L bits of the bit stream will produce exactly the same image as the first $L+1$ bits which occurs if the additional bit is insufficient to complete the decoding of another symbol. Nevertheless, terminating the decoding of an embedded bit stream at a specific point in the bit stream produces exactly the same image that would have resulted had that point been the initial target rate. This ability to cease encoding or decoding anywhere is extremely useful in systems that are either rate constrained or distortion constrained. A side benefit of the technique is that an operational rate versus distortion plot for the algorithm can be computed.

Although the embedded coding system is considerably more general and more sophisticated than simple bit-plane encoding, consideration of the relationship with bit-plane encoding provides insight into the success of embedded coding. Consider the successive-approximation quantizer for the case when all thresholds are powers of 2 and all wavelet coefficients are integers. In this case, for each coefficient that eventually gets coded as significant, the sign and bit position of the *most significant binary digit* (MSBD) are measured and encoded during a dominant pass. By using successive approximation beginning with the largest possible threshold, where the probability of 0 is extremely close to 1, and by using, zerotree coding, whose efficiency increases as the probability of 0 increases, we should be able to code dominant bits with very few bits, since they are most often part of a zerotree. By factoring out a constant mantissa, M, the starting threshold T_0 can be expressed in terms of a threshold that is a power of 2

$$T_0 = M \cdot 2^E \tag{6.4.5}$$

where the exponent E is an integer, in which case the dominant and subordinate bits of appropriately scaled wavelet coefficients are coded during dominant and subordinate passes, respectively.

Note that the particular encoder alphabet used by the arithmetic coder at any given time contains either two, three, or four symbols depending whether the encoding is for a subordinate pass, a dominant pass with no zerotree root symbol, or a dominant pass with the zerotree root symbol. This is a real advantage for adapting the arithmetic codec. Since there are never more than four symbols, all of the possibilities typically occur with a reasonably measurable frequency. This allows an adaptation algorithm with a short memory to team quickly and constantly track changing symbol probabilities. This adaptivity accounts for some of the effectiveness of the overall algorithm. Contrast this with the case of a large alphabet, as is the case in algorithms that do not use successive approximation. In that case, it takes many events before an adaptive entropy coder can reliably estimate the probabilities of unlikely symbols. Furthermore, these estimates are fairly unreliable because images are typically statistically nonstationary and local symbol probabilities change from region to region.

The order of processing used in EZW implicitly defines a precise ordering of importance that is tied to several properties, namely (in order of importance) precision, magnitude, scale, and spatial location, as determined by the initial dominant list. The primary determination of ordering importance is the numerical precision of the coefficients. This can be seen in the fact that the uncertainty intervals for the magnitude of all coefficients are refined to the same precision before the uncertainty interval for any coefficient is refined further. The second factor in the determination of importance is magnitude. Importance by magnitude manifests itself during a dominant pass because prior to the pass, all coefficients are insignificant and presumed to be zero. When they are found to be significant, they are all assumed to have the same magnitude, which is greater than the magnitudes of those coefficients that remain insignificant. Importance by magnitude manifests itself during a subordinate pass by the fact that magnitudes are refined in descending order of the center of the uncertainty intervals (the decoder's interpretation of the magnitude). The third factor, scale, manifests itself in the a priori ordering of the subbands on the initial dominant list. Until the significance of the magnitude of a coefficient is discovered during a dominant pass, coefficients in coarse scales are tested for significance before coefficients in fine scales. This is consistent with prioritization by the decoder's version of magnitude since, for all coefficients not yet found to be significant, the magnitude is presumed to be zero. The final factor, spatial location, merely implies that two coefficients that cannot yet be distinguished by the decoder in terms of either precision, magnitude, or scale have their relative importance determined arbitrarily by the initial scanning order of the subband containing the two coefficients.

In one sense, this embedding strategy has a strictly non increasing operational distortion-rate function (for the distortion metric defined to be the sum of the widths of the uncertainty intervals of all of the wavelet coefficients). Since a discrete wavelet transform is an invertible representation of an image, a distortion function defined in the wavelet transform domain is also a distortion function defined on the image. This distortion function is also not without a rational foundation for low-bit-rate coding where noticeable artifacts must be tolerated, and perceptual metrics based on *just-noticeable differences* (JNDs) do not always predict which artifacts human viewers will prefer. Since minimizing the widths of uncertainty intervals minimizes the largest possible errors, artifacts, which result from numerical errors large enough to exceed perceptible thresholds, are minimized. Actually, as it has been described thus far, EZW is unlikely to be optimal for any distortion function. Despite the fact that without trial-and-error optimization, EZW is probably suboptimal; it is nevertheless quite effective in practice. Note also that using the width of the uncertainty interval as a distance metric is exactly the same metric used in finite-precision fixed-point approximations of real numbers. Thus, the embedded code can be seen as an "image" generalization of finite-precision fixed-point approximations of real numbers.

In a very similar approach, others references [Hua92] discuss a related approach to embedding, or ordering the information in importance, called *priority-position coding* (PPC). It has been shown that the entropy of a source is equal to the average entropy of a particular ordering of that source plus the average entropy of the position information necessary to reconstruct the source. Like the EZW algorithm, PPC implicitly defines importance with respect to the magnitudes of the transform coefficients. In one sense, PPC is a generalization of the successive-approximation method because PPC allows more general partitions of the amplitude range of the transform coefficients. On the other hand, since PPC sends the value of a significant coefficient to full precision, its protocol assigns a greater importance to the least significant bit of a significant coefficient than to the identification of new significant coefficients on the next PPC pass. In contrast, as a top priority, EZW tries to reduce the width of the largest uncertainty interval in all coefficients before increasing the precision further. Additionally, PPC makes no attempt to predict insignificance from low frequency to high frequency, relying solely on the arithmetic coding to encode the significance map. Also unlike EZW, the probability estimates needed for the arithmetic coder are derived via training on an image database.

Within the scope of this chapter, we have only been able to touch upon the large body of work that has been done in applying wavelet and subband methods to image processing and compression. For additional information, the reader is referred to the many excellent references in the literature.

6.5   References

[Abb97a]   A. Abbate, J. Frankel, and P. Das, Application of wavelet image processing for ultrasonic gaging, *Proc. 1997 SPIE Conference on Wavelets*, 1997.

[Ade87]   E. H. Adelson, E. Simoncelli, and R. Hingomni, Orthogonal pyramid transforms for image coding, *Proc. SPIE*, vol. 845, pp. 50-58, 1987.

[Ans88]   R. Ansari, H. Gaggioni, and D. J. LeGall, HDTV coding using a nonrectangular subband decomposition, *Proc. SPIE Conf. Visual Commun. Image Processing*, pp. 821-824, 1988.

[Bel90]   T. C. Bell, J. G. Cleary, and 1. H. Witten, *Text Compression*, Prentice-Hall, Englewood Cliffs, NJ, 1990.

[Bur83]   P.J. Burt and E.H. Andelson, The Laplacian pyramid as a compact image code, *IEEE Trans. Commun.*, vol. 31, pp. 532-540, 1983.

[Che84]   W. Chen, and W. Pratt, Scene adaptive coder, *IEEE Trans. Comm.*, vol. COM-32, pp. 225-232, 1984.

[Coi92]   R.R. Coifman and M. V. Wickerhauser, Entropy-based algorithms for best basis selection, *IEEE Trans. Inform. Theory*, vol. 38, pp. 713-718, 1992.

[Dau85]   J.G. Daugman, Uncertainty relation for resolution in space, spatial frequency, and orientation optimized by two-dimensional visual cortical filters, *J. Opt. Soc. Am.*, vol. A2, 1160-1169, 1985.

[Dau88]   I. Daubechies, Orthonormal bases of compactly supported wavelets, *Commun. Pure Appl. Math*, vol. XLI, pp. 909-996, 1988.

[DeV92]   R. A. DeVore, B. Lawerth, and B. J. Lucier, Image compression through wavelet transform coding, *IEEE Trans. Inform. Theory*, vol. 38, pp. 719-746, 1992.

[Eff92]   M. Effros. P. A. Chou. E. A. Riskin. and R. M. Gray, A progressive universal noiseless coder, *IEEE Trans. Inform. Theory*. vol, 40, pp. 108-117, 1994.

[Equ91]   W. Equitz and T. Cover, Successive refinement of information, *IEEE Trans. Inform. Theory*, vol. 37, pp. 269-275, 1991.

[Est77]   D. Estaban and C. Galand, Application of quadrature mirror filters to split band voice coding schemes, *Proc. ICASSP*, pp. 191-195, 1977.

[Gab46]   *D. Gabor, Theory of communication*, J. Inst. Elec. Eng., *vol. 93, pp.* 429-457, 1946

[Gha87]   H. Gharavi and A. Tabatabai, Application of quadrature mirror filtering to the coding of monochrome and color images, *Proc. ICASSP*, pp. 2384-2387, 1987.

[Hee90]   V. K. Heer and H.-E. Reinfelder, A comparison of reversible methods for data compression, *Proc. SPIE*, vol. 1233, pp. 354-365, 1990.

[Hua92]   Y. Huang, H. M. Driezen, and N. P. Galatsanos, Prioritized DCT for Compression and Progressive Transmission of Images, *IEEE Trans. Image Process.*, vol. 1, pp. 477-487, 1992.

[Jay84]   N. S. Jayant and P. Noll, *Digital Coding of Waveforms*, Prentice-Hall, Englewood Cliffs, N], 1984

[Kim92]   Y. H. Kim and L. W. Modestino, Adaptive entropy coded subband coding of images, *IEEE Trans. Image Process.*, vol. 1. pp. 31-48. Jan. 1992.

[Kud92]   G. R. Kuduvalli and R. M. Rangayyan, Performance analysis of reversible image compression techniques for high-resolution digital teleradiology, *IEEE Trans. Med Imaging*, vol.MI-1, pp. 430-445, 1992.

[Lew91]   A. S. Lewis and O. Knowles, A 64 kB/s video Codec using the 2-D wavelet transform, in *Proc. Data Compression Conf.*, Snowbird, Utah, IEEE Computer Society Press, 1991

[Lew92]   A.S. Lewis and O. Knowles, Image compression using the 2-D wavelet transform, *IEEE Trans. Image Process.*, vol. 1, pp. 244-250, 1992.

[Mal89]   S. G. Mallat, A theory for multiresolution signal decomposition: the wavelet representation, *IEEE Trans. on Pattern Analys. and Machine Intell.*, vol. 11m n. 7, pp. 674-693, 1989.

[Mal90] S. Mallat, Multifrequency channel decompositions of images and wavelet models, *IEEE Trans. Acoust. Speech Signal Process.*, vol. 37, pp. 2091-2110, 1990.

[Pen91] A Pentland and B. Horowitz, A practical approach to fractal-based image compression, *Proc. Data Compression Conf.*, Snowbird, Utah, 1991.

[Pre86] W. H. Press. B. P Flannery, S. A. Teukolsky. and W. T. Vetterling. *Numerical Recipes: The Art of Scientific Programming*, University Press, Cambridge, 1986.

[Rab91] M. Rabbani and P. W. Jones, *Digital Image Compression Techniques*, SPIE, Bellingham, WA, 1991.

[Rab92] M. Rabbani and P. W. Melnychuck., Conditioning contexts for the arithmetic coding of bit planes, *IEEE Trans. Signal Process.*, vol. 40, pp. 232-236. Jan. 1992.

[Ran88] S. Ranganath and H. Blume, Hierarchical image decomposition and filtering using the S-transform,. in *Proc. SPIE* , vol. 914. pp. 799-814, 1988.

[Rio91] O. Rioul and M. Vetterli, Wavelets and signal processing, *IEEE Signal Proc. Mag.*, pp. 14-38, 1991.

[Roo88] P. Roos, A. Viergever, and M. C. A. van Dijke, Reversible intraframe compression of medical images, *IEEE Trans. Med. Imaging*, vol. 7. pp. 328-336, Sept. 1988.

[Sai93] A. Said and W. A. Pearlman, Reversible image compression via multiresolution representation and predictive coding, in *Proc. SPIE*, vol. 2094. pp. 664-674, 1993.

[Sai96] A. Said and W. A. Pearlman, A new fast and efficient image codec based on set partitioning in hierarchical trees, *IEEE Trans. Circuits Syst. Video Tech.*, vol. 6. pp. 243-250, June 1996.

[Sai96a] A. Said and W. Pearlman, An image multiresolution representation for lossless and lossy compression, *IEEE Trans. Image Process.*, vol. 5, pp. 1303-1309, 1996.

[Sai96b] A. Said and W. Pearlman, A new, fast, and efficient image codec based on set partitioning in hierarchical trees, *IEEE Trans. Circuits and Syst. Video Tech.*, Vol. 6, p. 243-249, 1996.

[Say92] K. Sayood and K. Anderson, A differential lossless image compression scheme, *IEEE Trans. Signal Process.*, vol. 40. pp. 236-

241, 1992.

[Sha92]   J. M. Shapiro, An embedded wavelet hierarchical image coder, *Proc. IEEE Int. Conf. Acoust., Speech, Signal Processing*, 1992.

[Sha93]   J. M. Shapiro, Embedded image coding using zerotrees of wavelets coefficients, *IEEE Trans. Signal Processing*. vol. 41. pp. 3445-3462, Dec. 1993.

[Tak94]   S. Takamura and M. Takagi, Lossless image compression with lossy image using adaptive prediction and arithmetic coding, *Proc.of the Conference on Data Compression*, pp. 166-174, 1994.

[Tod85]   S. Todd, G. G. Langdon Jr., and J. Rissanen, Parameter reduction and context selection for compression of grey-scale images, *IBM J. Res. Dev.*, vol. 29, pp. 188-193, 1985.

[Tra87]   A.K. Tran, Liu, K. Tzou and E. Vogel, An efficient pyramid image coding scheme, *Proc. ICASSP*, pp. 744-747, 1987.

[Vai92]   J. Vaisey and A. Gersho, Image compression with variable block size segmentation, *IEEE Trans. Signal Process.*, vol. 2040, Aug. 1992.

[Vet90]   M. Vetterli, J. Kovačević, and D. J. LeGall, Perfect reconstruction filter banks for HDTV representation and coding, *Image Commun.*, vol. 2, pp. 349-364, Oct. 1990.

[Wal91]   G. K. Wallace, The JPEG still picture compression standard, *Comm. ACM.* vol. 34, pp. 30-44, Apr. 1991.

[Wal92]   G.G. Walter, A sampling theorem for wavelet subspaces, *IEEE Trans. Inform. Theory*, vol. 38, pp. 881-884, 1992.

[Wit87]   H. Witten, R. Neal, and J. G. Cleary, Arithmetic coding for data compression, *Comm. ACM*, vol. 30, pp. 520-540, June 1987.

[Wom90]   G. W. Womell, A Karhunen-Louve expansion for 1/f processes via wavelets, *IEEE Trans. Inform. Theory*, vol. 36, pp. 859-86 1, 1990.

[Woo91]   J. W. Woods ed., *Subband Image Coding*, Kluwer, Boston, 1991.

[Xio93]   Z. Xiong, N. Galatsanos, and M. Orchard, Marginal analysis prioritization for image compression based on a hierarchical wavelet de- composition, in *Proc. IEEE Int. Conf. Acoust., Speech, Signal Processing*, 1993.

[Zet90]   W. Zettler, I. Huffman, and D. C. P. Linden, Applications of compactly supported wavelets to image compression, *SPIE Image Processing Algorithms*, 1990.

Part III

Applications

Chapter 7

Applications of Wavelets to the Analysis of Transient Signals

7.1 Introduction

In the first two parts of this book, we have described some of the mathematical tools that can be used to analyze a transient signal. In particular, we have focused to two aspects of signal analysis:

- Continuous wavelet transform and wavelet frames. In this case, the emphasis was on describing a signal in terms of time and frequency evolution, at a cost of creating a redundancy in the representation.

- Fast wavelet transform and subband transform. In this case, a minimal representation of the signal is obtained to fulfill requirements such as compression, multirate processing, and noise robustness.

Wavelets and subband techniques are utilized in almost every field of science. Just by looking at any special issue on the subject or in the proceedings of a technical conference, we can easily realize that the wavelets are now becoming a standard analysis tools similar to the Fourier transform. Among the applications most reported in literature, we can find the following:

- Time-frequency analysis of acoustic, radar, music, ultrasound, and ECG and EEG biomedical signals [Abb95a, Fou94, Kad92, Uns96]

- Signal coding in digital communication [Ben94]

- Signal, image,and video compression algorithms [Abb96]

- Computer graphics and vision tools [Mal96]

- Analysis of fractals and turbulence [New93]
- Image processing for mapping the distant universe, etc. [Bij96]

Nevertheless, we can divide the applications into two areas: The first focused at understanding time-variant systems such as speech, ultrasound and biomedical signal analysis. The second subject area is focused more on the utilization of wavelets as a robust basis for increased compression ratios or multirate signal and image processing. The applications presented here are divided into these two subject areas. In the remainder of this chapter, we will focus on the utilization of the wavelet transform to characterize transient signals in ultrasonic inspection and in medical diagnostics. The purpose in this case is to understand and model the physical processes that create or alter a nonstationary signal. In the case of the ultrasonic testing, the resulting propagating models are used to extract the physical properties of the materials under inspection. We also discuss the application of wavelets for the analysis of EEG signals to understand the depth of anesthesia in a subject.

In Chapter 8, applications of wavelet and subband transform to the fields of digital communication and image analysis are presented. In the example, we discuss how multirate techniques are currently used in spread spectrum broad-band systems such as wireless and Internet. A progressive pattern recognition algorithm is also presented to discuss applications in mutiresolution search and multiscale edge detection.

A brief review of real-time hardware implementations of the wavelet transform with emphasis on digital VLSI and optical implementation can be found in Chapter 9.

It is not possible to cover every possible application in this book, and the ones given here are based on the authors' experience and research interests. Nevertheless, the concepts and results can be easily applied to many other problems in the fields of science and engineering. It is our aim to give the reader a flavor of the breath and wealth of information available, as well as covey some of the fundamental concepts of wavelet and subband applications.

7.2   Introduction to Time-Frequency Analysis of Transient Signals

Wavelet analysis can be used to study transient phenomena and create quantitative models of the physical systems involved in creating and altering such signals [Teo98]. In the following, the application of such techniques to ultrasound and biomedical EEG signals is discussed; however, many of the models and solutions described here are applicable to other areas involving wave propagation such as acoustic, seismology, or radar [Gui91, Gui96].

Let us begin by describing a dispersive channel propagation model and then its correspondent model in ultrasonic measurement system and components. In Figure 7.2.1, a block diagram of a dispersive channel system is shown. An input signal, which usually resembles an impulse function, is applied as an input to the system. From Fourier analysis, we know that the input signal is composed by many components of different frequency, all in phase, occurring at the same instant [Net95]. The analytic wavelet transform, AWT, of the input signal in Figure 7.2.1 is given in Figure 7.2.2a.

In a dispersive channel, the signals at different frequencies travel with different velocities [Ros99]. To characterize the channel, the plot of the group velocity as a function of the frequency is used, as plotted in Figure 7.2.1. This curve is commonly referred to as the dispersion curve [Net95].

The AWT of the output signal is given in Figure 7.2.2b. In this case, the energy of the signal is not all contained in a small window in time, but it has a different location on the time axis as a function of the instantaneous

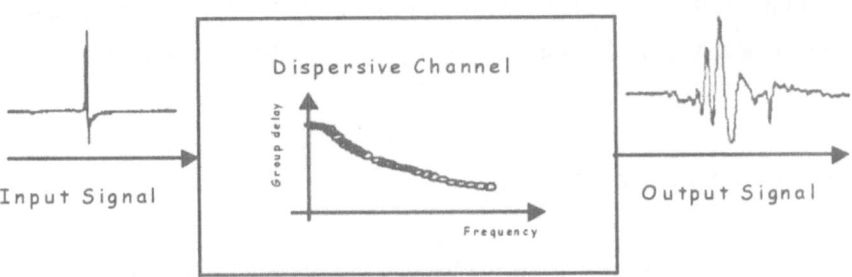

Figure 7.2.1 Block diagram of a dispersive channel system.

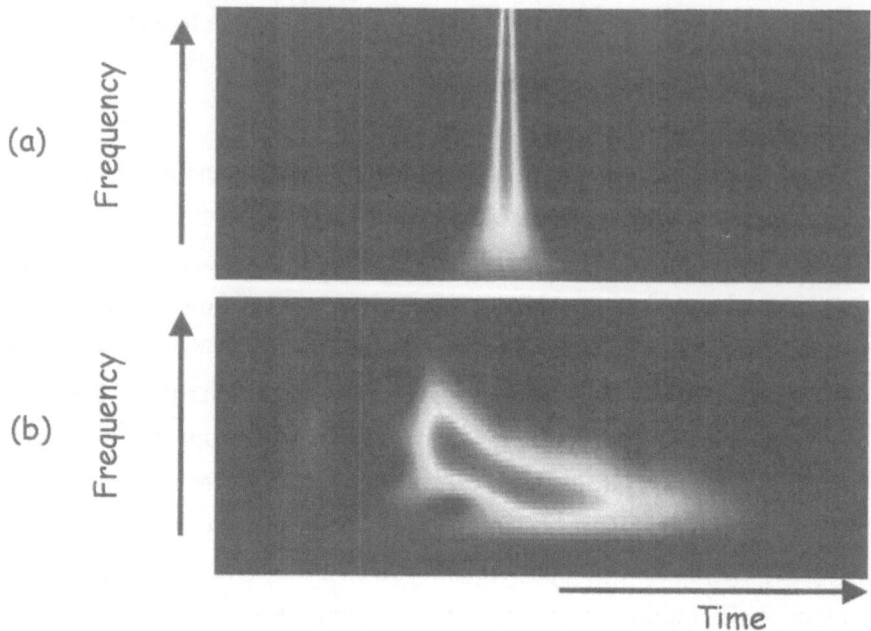

Figure 7.2.2 Magnitude of the AWT of the (a) input signal and (b) output signal in Figure 7.2.1. The time axes between the two AWT images are displaced to show both signals in the middle of the figure.

frequency. The concepts of analytic wavelet transform (AWT), group delay, and instantaneous frequency have been introduced in Chapter 3 [Szu92].

From the analysis of the wavelet transform of both signals, it is possible to extract the dispersion curve. In many applications, such as in ultrasonic, the dispersion curve is directly related to the mechanical characteristics of the material under analysis, and thus it is used to characterize the material. Many other examples of dispersive channels include air in radar, communication channels with multipath and interference, mechanical resonance systems in vibration analysis, etc. [Abb97, Vik67, Low95].

7.2.1 Ultrasonic Systems

A block diagram of a typical ultrasonic system is given in Figure 7.2.3. Briefly, a pulser is used to excite an ultrasonic stress wave in the material

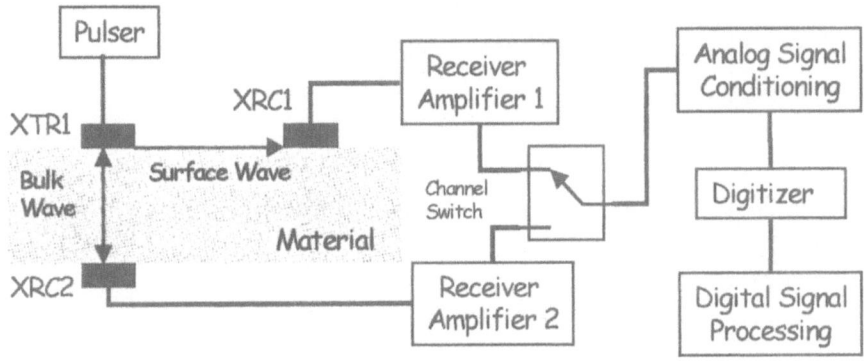

Figure 7.2.3 Block diagram of typical ultrasonic testing system.

using a piezoelectric transducer [Bra89]. The pulser generates repetitive short pulses, of the order of 10 to 100 ns in duration* and with a repetition period of 1 to 10 ms apart.

The electrical pulses drive a piezoelectric transducer that converts the electrical signal into mechanical energy that propagates in the material as ultrasound. The two receiving transducers perform the inverse operation of transforming mechanical energy into an electrical signal that is then amplified, filtered, and converted into a digital signal. The material is tested by looking at the propagation characteristics of ultrasonic waves propagating in the material (bulk waves) or along its surface (surface waves). The transducer (XTR1) is used to generate a sharp broad-band pulse in the frequency range between 250 kHz and 50 MHz, and different transducers, XRC1 or XRC2, are used to detect the surface or bulk wave, respectively. (See Figure 7.2.4) Analog signal conditioning is commonly an analog bandpass filter to eliminate aliasing and increase the dynamic range of the digitizer. Signals are sampled with resolutions of 8, 10, and 12 bits and sampling rates varying from 100 MHz to 1 GHz, depending on the application.

An example of a commercially available ultrasonic instrument is shown in Figure 7.2.5. The instrument can generate and detect the ultrasonic

---

* For reference 1 s = 1 second, 1 ns = 10^{-9} s, 1 μs = 10^{-6} s, 1 ms = 10^{-3} s.

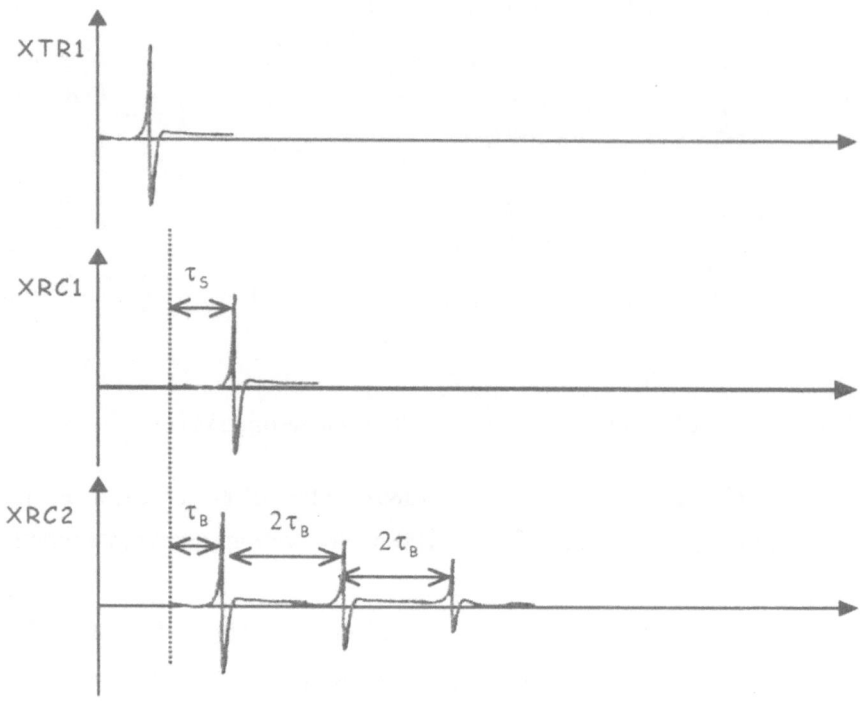

Figure 7.2.4 Typical ultrasonic signals associated with the generation (XTR1) and detection (XRC1 & XRC2).

signals and perform the digital signal processing in realtime, by means of a flash A/D converter and dedicated signal processing hardware and software. An example of a transducer used for ultrasonic testing is also shown in figure. The transducer and instrument shown in Figure 7.2.5 is commonly utilized to detect and characterize flaws in parts such as panels and other structures.

Typical signals obtained in ultrasonic systems are plotted in Figure 7.2.4. The delay between the XTR1 pulse and the XRC1 signal is associated with the propagation delay in the electrical channels and the travel time necessary for the ultrasonic wave to travel the distance between the two transducers on the surface of the material. For practical purposes, the electronic delay in the system is here not considered, as a lot of work has been done to minimize its effects by design or calibration.

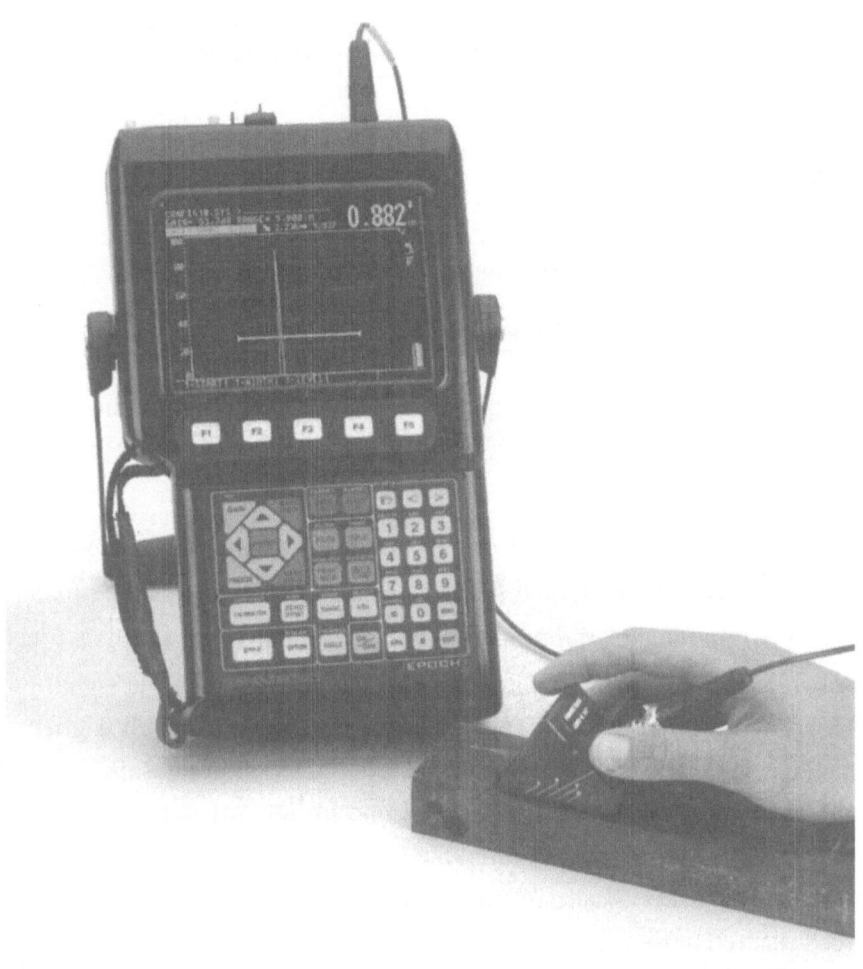

Figure 7.2.5 Example of a commercially available ultrasonic instrument with a typical transducer. Courtesy of Panametric Inc.

If x_S is the distance between the two transducer, then the delay τ_S is given by

$$\tau_S = \frac{x_S}{v_S} \qquad\qquad (7.2.1)$$

(a)

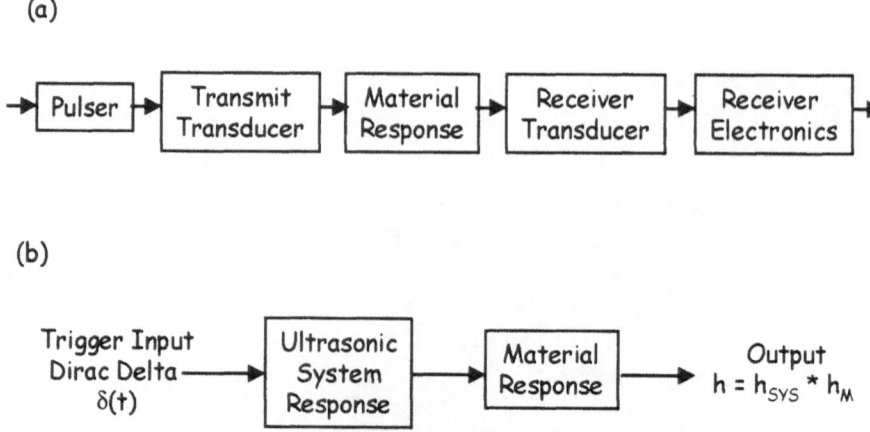

(b)

Figure 7.2.6 (a) Linear system equivalent of a typical ultrasonic testing system; (b) simplified block diagram with separation between the ultrasonic system response and the material response.

where v_S is the sound velocity of the surface ultrasonic wave. This equation is extremely important, as it links the signal characteristics (i.e., the group delay τ_S), to the material properties (i.e. the velocity v_S). If v_S is a function of frequency, as we will show in more detail later, then the signal group delay will be dispersive. The same can be said of the bulk signal from XRC2, with the added effect that multiple repetition of the same signal can be detected, as the signal travels back and forth between the two opposing surfaces. Of course, the amplitude of the signal is reduced as it travels through the material, as a result of attenuation. For this reason, the multiple echoes seen in Figure 7.2.4 show diminishing amplitude.

The ultrasonic system can be represented as a linear time-shift-invariant system, completely characterized by its time domain impulse response function or its frequency domain equivalent, the transfer function. As we are interested at measuring the effects of the propagation of the ultrasonic wave in the material, we can utilize a linear system block diagram, as in Figure 7.2.6a. For each block in the figure, we can define an impulse response, and the total response is thus given by [Kin87]

$$h(t) = h_E(t) * h_{RT}(t) * h_M(t) * h_{TT}(t) * h_P(t) \qquad (7.2.2)$$

where * represents the convolution operator and

- $h_E(t)$ is the impulse response of the receiving electronics;
- $h_{RT}(t)$ is the impulse response of the receiving transducer;
- $h_M(t)$ is the material's impulse response;
- $h_{TT}(t)$ is the impulse response of the transmitting transducer;
- $h_P(t)$ is the impulse response of the pulser.

As the electronics response can be calibrated out, the system in Figure 7.2.6a can be simplified in the block diagram in Figure 7.2.6b

$$h(t) = h_{SYS}(t) * h_M(t)$$

$$(7.2.3)$$

$$H(\omega) = H_{SYS}(\omega) H_M(\omega)$$

where $h_{SYS}(t)$ represents the ultrasonic system response.

Equation 7.2.3 shows us a good way to calibrate for the system response. If we apply the testing to a material of know characteristics, such as a steel standard of known velocity v_1 and attenuation A_1, then $h_M(t)$ is well defined and we have

$$H_M(\omega) = A_1 e^{-j\omega\tau_s}$$

$$H_{SYS}(\omega) = \frac{1}{A_1} H(\omega) e^{j\omega\tau_s}$$

$$(7.2.4a)$$

This is the case of the signal XRC1 in Figure 7.2.4. As the velocity v_1 of the material is know, the distance between XTR1 and XRC1 can now be determined as $x_0 = v_1 \tau_s$.

In the case of the signal XRC2, the function $H_M(\omega)$ is a little more complicated, as it represents multiple reflections of the ultrasonic wave within the material. In this case, we can write

$$H_M(\omega) = A_1 e^{-j\omega\tau_B} + \sum_{k=2}^{N} A_k e^{-j\omega(k-1)3\tau_B}$$

$$H_{SYS}(\omega) = H(\omega) H_M^{-1}(\omega)$$

$$(7.2.4b)$$

where τ_B is the time necessary to travel from one side to the other of the material.

For the sake of completeness, we should mention that the term $H_M(\omega)$ can be further decomposed into a series of response depending on the physical processes involved in the ultrasonic propagation. For the purpose of this book, we will only describe some of them as we deem necessary to better understand the application of wavelets.

To summarize this brief overview of ultrasonic measuring systems, we have defined a linear system equivalent model and separated the contributions to the overall system response in two groups, the first of which is only a function of the measuring system, and the second of which is only a function of the measured material system. Let us now see how we can apply wavelets to better characterize the second group of responses.

7.2.2   Ultrasonic Characterization of Coatings by the Ridges of the Analytic Wavelet Transform

Acoustic waves can be represented by a sum of harmonic signals, each having its own center frequency and amplitude modulation:

$$s(t) = \sum_{k=1}^{N} A_k(t) \cos(\omega_k t + \phi_k) \qquad (7.2.5)$$

where $\omega_k = 2\pi f_k$ is the angular frequency of the harmonic component, $A_k(t)$ is the low-frequency amplitude modulation, and ϕ_k is the phase shift between each component. It is to be noted that this modeling is extensively used in many other areas of real-time signal processing such as musical sound and voiced speech modeling. The same model applies for both the surface and the bulk ultrasonic wave configurations.

In the case of surface wave measurements on a uniform material, the detected ultrasonic signal resembles the initial ultrasonic pulse generated by XTR1, delayed by the time necessary for the wave to travel from the generation point to the detection point. For a composite structure, such as a coating deposited on a metallic part, the detected signals are more complex

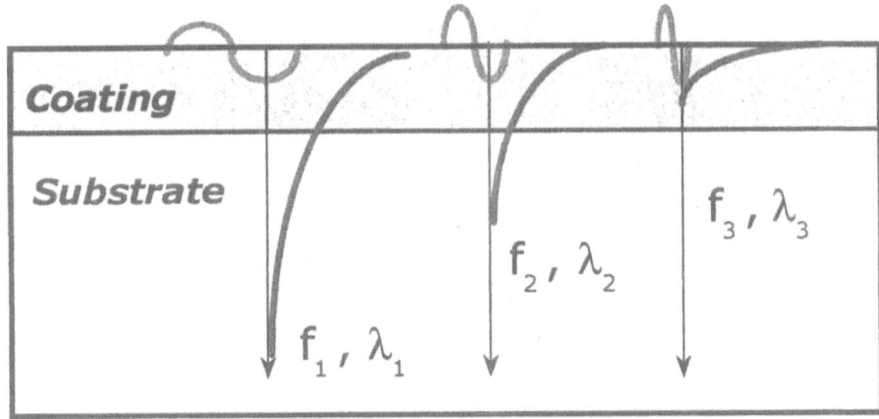

Figure 7.2.7   Model of a coated sample showing the relationship between wavelength and frequency.

in nature.   This complexity arises from the fact that the mechanical disturbance associated with the ultrasonic wave decays in amplitude away from the surface.   The rate of decay is proportional to the wavelength $2\pi\lambda = v/\omega$, where v is the wave velocity and ω is the angular frequency.   As the energy of the wave is mostly contained within a depth λ from the surface, the velocity v at which the wave propagates is a function of the properties of the material within a depth λ from the surface. Hence, the different harmonic components of the ultrasonic signal are affected by the characteristics of the material at different depths.   From the relationship between the wavelength and frequency, it is clear that higher-frequency components have a lower λ; hence, they are more sensitive to the properties of the surface layers [Lak95].

Assuming a model as the one in Figure 7.2.7, harmonic components with the higher-frequency content have smaller wavelengths and thus their propagation (i.e., their velocity), is mostly a function of the coating's properties.   As lower-frequency components penetrate more in depth, their velocity is affected by the properties of the substrate.   To summarize, different paths of travel of varying acoustic frequencies are present, and the velocity of the ultrasonic waves is now a function of the frequency.   This phenomenon is called dispersion, in which the original narrow pulse "spreads out" in time during propagation, and signals such as the ones shown in Figure 7.2.7 are obtained [Scr90, Abb95c].

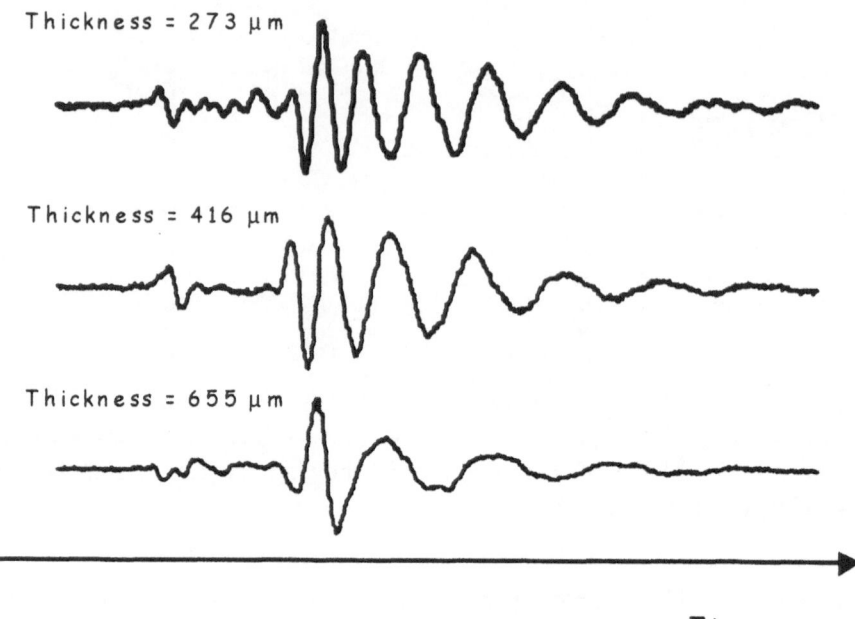

Figure 7.2.8 Ultrasonic surface wave signals traveling on coated steel samples. Notice the difference in dispersion for samples with different coating thicknesses.

The magnitude of signal dispersion depends on the thickness, elastic properties, and density of the coating as well as of the substrate, and thus can be used to obtain a signature of the coating layer and its underlying substrate. For example, in Figure 7.2.8, signals from samples with coatings of different thicknesses are plotted. As the coating thickness increases from 273 μm to 655 μm,** the signal shape is different. The effect of coating thickness on the dispersion of a surface wave is well defined by theoretical models, as shown in Figure 7.2.9, where the velocity is plotted as a function of frequency for coatings of different thicknesses [Bir95]. The figure can be explained as follows. At higher frequencies, the propagation velocity of the harmonic component of the surface wave is mostly a function of the coating velocity, which, in this case, is roughly 1000 m/s = 1.0 mm/μs. This is the value at which all the curves converge for high enough values of frequency.

** For reference, 1 m = 1 meter, 1 mm = 10^{-3} m, 1 μm = 10^{-6} m.

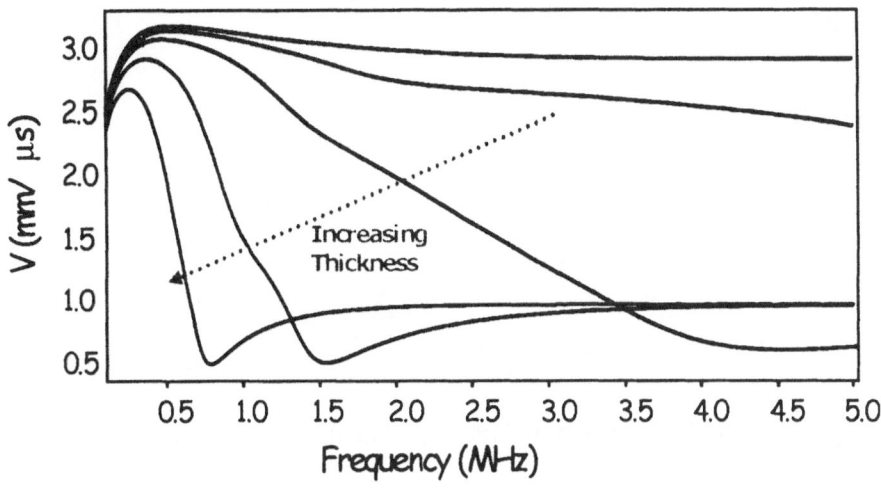

Figure 7.2.9 Theoretical calculation for the dispersion curve (velocity versus frequency) for steel samples with different coating thicknesses.

At the lower-frequency range, the effect of the substrate properties dominate and the curve tends to reach the value of the velocity in the steel substrate, which is 3.0 mm/μs. The transition between these two limits is defined by the thickness of the coating, as shown in Figure 7.2.9. As the thickness increases, the transition occurs at lower frequencies, consistent with the definition of the wavelength [Aus94].

To summarize, we can define a measurement algorithm to measure the properties of the coatings by comparing the experimental signals with theoretical models. A block diagram of the algorithm is shown in Figure 7.2.10. The algorithm can be divided into the following steps:

A. Acquire experimental signal

B. Perform time-frequency analysis of the signal to obtain the experimental group delay versus frequency dispersion curve

C. Compare the experimental dispersion curve with one generated using a theoretical model; perform a computer fit to determine the material's parameters for which the theoretical model best matches the experimental data.

Using this technique, it is possible to determine in real time the thickness and the elastic properties of the coating [Abb98, Kot94, Kot97].

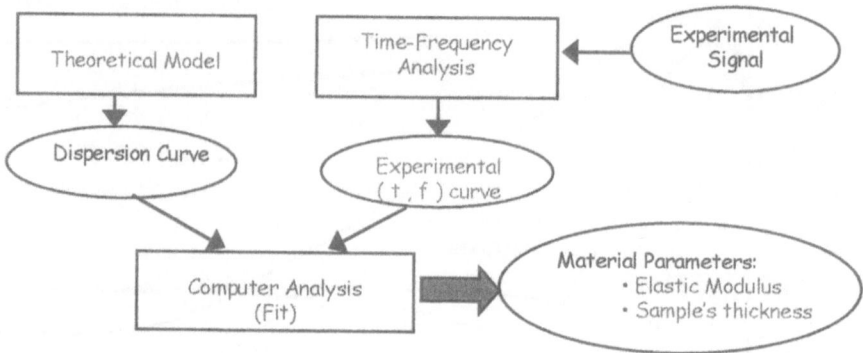

Figure 7.2.10   Block diagram of the algorithm discussed in the text.

Critical to this approach is obtaining a procedure to extract rapidly and accurately the dispersion curve from the experimental signals. This has been accomplished using the analytic wavelet transform and the concept of wavelet ridge [Gam81].

Many procedures have been developed to characterize a signal by some of the salient features of a specific transform, whether it is the short time Fourier or the wavelet transform. Ridges in the modulus of the transform determine regions in the transform domain with a high concentration of energy. For this reason, they are regarded as natural candidates for the characterization and reconstruction of the original signal. In order to measure the instantaneous frequencies of musical sounds from the local maxima of $P_s(\tau, \omega)$, Delprat, Escudié, Guillemain, Kronland-Martinet, Tchamitchian, and Torresani developed the concept of ridges [Kro91]. The mathematical details of these results are quite complex and beyond the scope of our present discussion; the reader is referred to the specific references. Instead, we will illustrate with a few examples [Abb98a].

In Chapter 3, we have introduced the concept of analytic wavelet transform (AWT) and have developed a closed-form approximation of the AWT for the special case in which the signal and the wavelet can be expressed using the product of two functions: (1) an harmonic modulation and (2) a lower-frequency amplitude modulation.

If we express the signal and the wavelet function as

$$s(t) = A(t)e^{j\varphi(t)}$$

$$\psi(t) = h(t)e^{j\omega_0 t}$$

(7.2.6)

then the wavelet transform can be expressed as:

$$W_S(a,b) \approx \frac{\sqrt{a}}{2} \cdot A(b)e^{-j\varphi(b)} H\left(a\left[\omega - \varphi'(b)\right]\right)$$

(7.2.7)

Please note that the quantities $A(b)$ and $\varphi'(b)$ represent the analytic amplitude and the instantaneous frequency, respectively, of the signal $s(t)$ at time $\tau = b$, where b represents the group delay [Kro87].

We make the following assumption regarding the window function $h(t)$:

- $h(t) = h(-t)$, i.e., the window is symmetrical around $t=0$
- $H(\omega)$ is a real symmetrical function
- $|H(\omega)| \leq H(0)$, for all values of $\omega \in \mathbb{R}$
- $H(0) = 1$. This condition ensures that the window function does not affect the amplitude of the wavelet transform.

The wavelet transform is thus given by the integral product between the signal $s(t)$ and the daughter wavelet functions. We define as T the time duration of the wavelet function, and if $A(b)$ does not change rapidly during the time interval aT, where a is the scaling coefficient, then we can derive an approximate expression for the scalogram, defined in Equation 3.4.42

$$P_W(\tau, \omega) = \left. \frac{|W_S(a,b)|^2}{a} \right|_{\left[\omega = \frac{\omega_0}{a}, \tau = b\right]} = \frac{1}{4} \cdot |A(\tau)|^2 \cdot \left| H\left(\frac{\omega_0}{\omega}[\omega - \varphi'(\tau)]\right)\right|^2$$

(7.2.8)

Given the conditions on the window function $h(t)$, $H(\omega)$ is maximum at $\omega = 0$; hence, the maximum of the scalogram will be at $\omega - \varphi'(\tau) = 0$ [i.e., at $\omega = \varphi'(\tau)$] [Kro88].

The above equation is based on the condition that the quantities $A(\tau)$ and $\varphi'(\tau)$ can be considered approximately constant in the time-frequency tile defined by the wavelet $\psi_{a,b}(t)$. If the wavelet function $\psi(t)$ has a bandwidth B_0, then, using Equation 7.2.6 and the definition of the scale $a = \omega_0/\omega$, we can calculate the bandwidth of the daughter wavelets at each frequency ω as

$$B(\omega) \;=\; \frac{B_0}{a} \;=\; \left(\frac{B_0}{\omega_0}\right)\omega \;=\; B_R\,\omega \qquad (7.2.9)$$

where $B_R = B_0 / \omega_0$ represents the relative bandwidth of the wavelet basis. Please note that given the scaling properties of the wavelet basis, the relative bandwidth B_R is constant and is thus a very useful quantity when dealing with wavelet analyses. If the wavelet function is constructed using Equation 7.2.6, then the bandwidth B_0 is the bandwidth of the window function $h(t)$. Assuming that $A(t)$ represents the low-frequency modulation of the signal and that the higher-frequency components are included in the phase term, then

$$\varphi'(t) \geq B(\omega) = B_R\,\omega \qquad (7.2.10)$$

Equation 7.2.10 imposes the condition that the instantaneous frequency of the signal must be larger than the bandwidth of the wavelet at the same frequency, thus ensuring that the signal amplitude in the tile can be considered constant. Equation 7.2.10 seems to suggest that we use a wavelet with the smallest possible bandwidth. Please remember that the time duration is inversely proportional; hence, we would lose the resolution in time [Kro91].

The scalogram at the position $[\tau, \varphi'(\tau)]$ in the time-frequency plane can be used to estimate the analytic amplitude of the signal $A(\tau)$

$$P_W\left(\tau,\varphi'(\tau)\right)\Big|_{Ridge\ Max} \approx \frac{1}{4}\,A^2(\tau) \qquad (7.2.11)$$

The time-frequency positions $[\tau, \omega_i(\tau)] = [\tau, \varphi'(\tau)]$ define the wavelet ridge points.

Next, let us consider the case of multispectral components in a signal. When the signal contains several spectral lines whose frequencies are sufficiently far apart, we want to utilize the wavelet ridges to determine the evolution in time of each spectral component. Let us consider a signal $s(t)$

$$s(t) = A_1(t)e^{j\varphi_1(t)} + A_2(t)e^{j\varphi_2(t)} \tag{7.2.12}$$

The wavelet transform is a linear operation; hence, we can write

$$W_s(a,b) = \frac{\sqrt{a}}{2} \left\{ A_1(t)\, e^{-j\varphi_1(b)} \cdot H(a[\omega - \varphi_1'(b)]) \right.$$

$$\left. + A_2(t)\, e^{-j\varphi_2(b)}\, H(a[\omega - \varphi_2'(b)]) \right\} \tag{7.2.13}$$

The two components can be resolved if the difference in instantaneous frequency between them is always larger than the bandwidth $B(\omega)$; that is,

$$\left| \varphi_1'(\tau) - \varphi_2'(\tau) \right| \geq B(\omega) = B_R\, \omega \tag{7.2.14}$$

The question now arises, How do we choose window functions for the determination of ridges? The measurement of instantaneous frequencies at ridge points is valid only if the size ΔT of the window is sufficiently small so that

$$| A'(\tau) | < \frac{1}{\Delta T} \sim B(\omega) \tag{7.2.15}$$

The first derivative of the amplitude must be sufficiently small, or the window width must be small. On the other hand, the frequency bandwidth $B(\omega)$ must also be sufficiently small to discriminate between different spectral components, resulting in the need for a trade-off between both constraints.

By using the ridges $[\tau, \varphi'(\tau)]$ of the wavelet transform, we have established a procedure to determine the relationship between the group delay

Figure 7.2.11 Ultrasonic surface wave signal acquired after propagation along the surface of a coated steel sample.

$\tau_g = \tau$ and the instantaneous frequency $\omega_i(\tau_g)$. As the distance x_0 between the XTR1 and XRC1 is calibrated, the group velocity v_g is given by $v_g = x_0 / \tau_g$. A experimental ultrasonic signal is plotted in Figure 7.2.11 and its correspondent AWT is given in Figure 7.2.12. The continuous line that tracts the maximum of the scalogram is also plotted. The line represents the wavelet ridge and it is used to calculate the dispersion curve plotted in Figure 7.2.13.

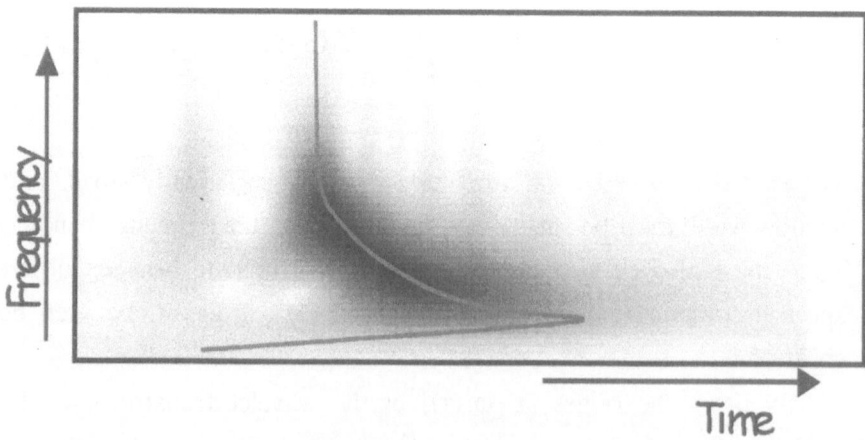

Figure 7.2.12 Analytic wavelet transform of the signal in Figure 7.2.11. The line over the image represents the calculated wavelet ridge.

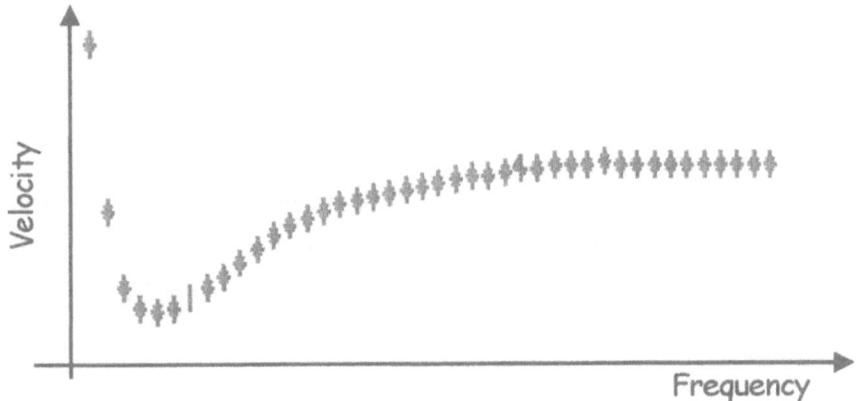

Figure 7.2.13   Surface acoustic wave velocity as a function of frequency, calculated using the points on the wavelet ridge.

Assuming for the signal a representation of the type of Equation 7.2.5, by definition the signal generated by the transmitting transducer (XTR1) has all the phase components $\phi_k = \phi_0$, $k = 1,..., N$. Equation 7.2.5 can thus be written as

$$H_0 = \sum_{k=1}^{N} A_k \, e^{j\omega_k \tau_k} \qquad (7.2.17)$$

where τ_k are the group delay and the $\omega_k = \omega_i(\tau_k)$ are the correspondent instantaneous frequencies.

If we define as $[\tau_k^0, \ \omega_k^0]$ the wavelet ridge of a signal acquired on a standard steel sample of known velocity $v_S = v_0$, and $[\tau_k^M, \ \omega_k^M]$ the wavelet ridge calculated for the material under test, we can calculate the material's response as

$$H_M(\omega) = \sum_{k=1}^{N} \frac{A_k^M}{A_k^0} e^{j\omega_k(\tau_k^M - \tau_k^0)} \qquad (7.2.18)$$

where we have inverted the function $\omega_k = f(\tau_k)$ into $\tau_k = g(\omega_k)$.

Comparing Equation 7.2.18 with the classical equation of wave propagation, we obtain

$$v(\omega_k) = \frac{x_0}{\tau_k^M - \tau_k^0}$$ (7.2.19)

The curve in Figure 7.2.12 is thus compared with theoretical models and the thickness and elastic properties of the coating are determined. In the next section, we will briefly describe how the algorithm depicted in Figure 7.2.10 was used to characterize the deposition of plasma spray coatings in real time.

7.2.3   Characterization of Coatings

Coatings are widely used to protect parts from aggressive environments. In applications such as land-based gas turbines, thermal barrier coatings (TBCs) are utilized to protect the turbine components from very high operating or firing temperatures. The TBCs are commonly applied by standard air plasma spray process, which is an open-loop operation with no feedback about the coating conditions during deposition. Unfortunately, on-line variations of the spray conditions, such as the continuous wearing of the torch hardware, can adversely affect the coating quality and create significant part-to-part variations. The standard method of evaluating coatings is destructive in nature; hence, these tests cannot be performed on each produced part. As a result, coated parts may not have the consistent quality and durability needed for many applications.

A new approach for the characterization of plasma sprayed coatings as deposited was developed utilizing surface ultrasonic waves and real-time digital signal processing to rapidly and nondestructively measure the thickness and the elastic modulus of TBC samples with thicknesses ranging from 100 to 700 μm and produced with different powder properties. The advantage of the proposed technique lies in its applicability to on-line

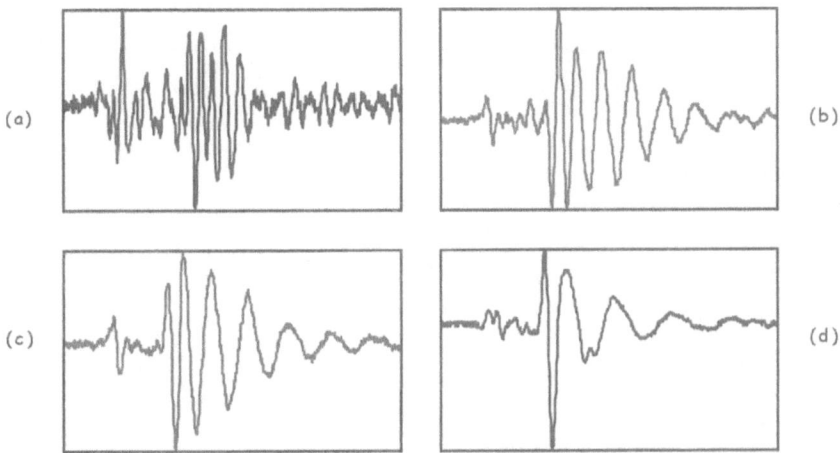

Figure 7.2.14 Ultrasonic signals acquired at different stages of the coating process showing the increase in thickness from (a) to (d).

measurements of thickness and elastic properties of coatings.   Currently, a measurement of the thickness and elastic modulus is obtained every 5 to 10 s using a PC computer without any dedicated digital signal processing (DSP) hardware.

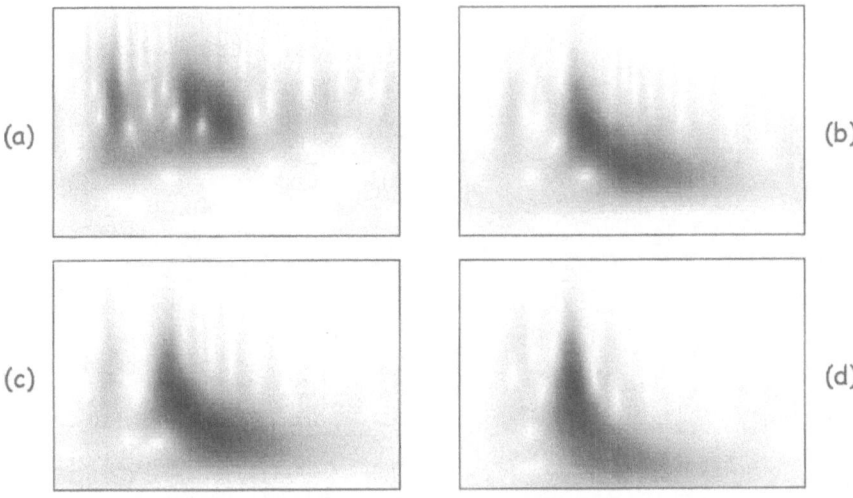

Figure 7.2.15 Analytic wavelet transforms of the ultrasonic signals in Figure 7.2.14.

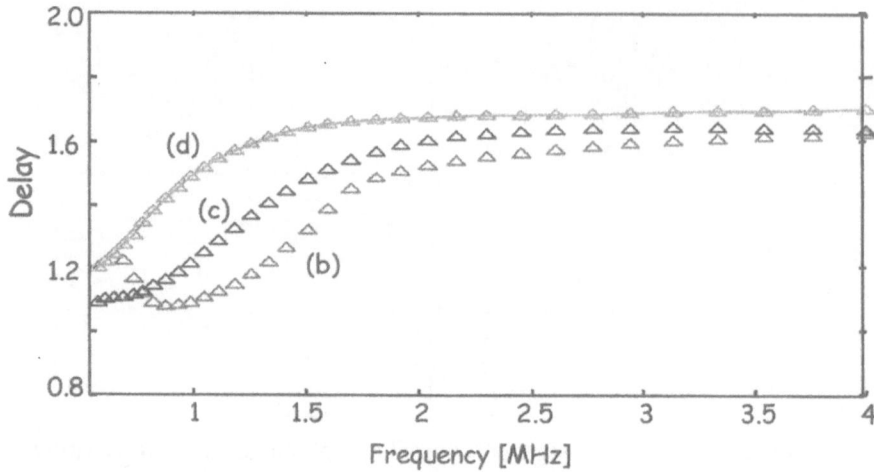

Figure 7.2.16 Dispersion curves for the ultrasonic signals (b), (c), and (d) in Figure 7.2.14.

The ultrasonic signals acquired at different stages of a coating process are plotted in Figures 7.2.14a to 7.2.14d. The signal in Figure 7.2.14a represents a sample with only a small bond coat deposited on it, and it is thus very different from the others. As the thickness is increased, the number of

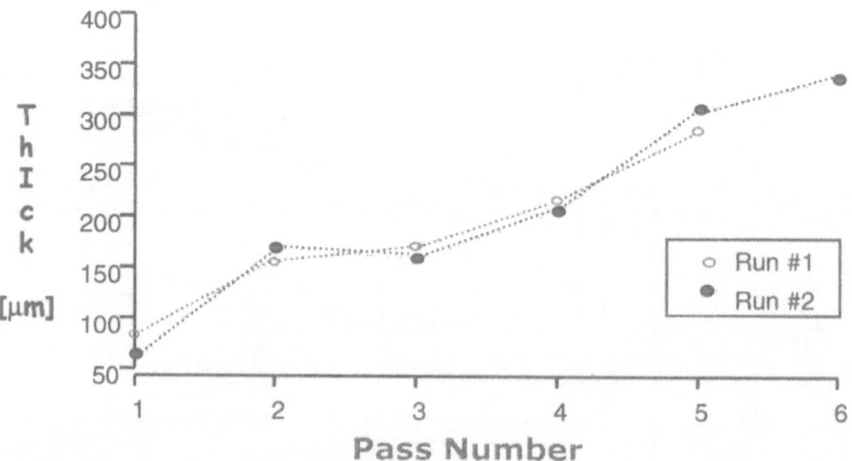

Figure 7.2.17 Real time determination of the coating thickness as a function of the pass number and for two different process run.

oscillations decreases, as more and more the harmonic components at higher frequencies travel with the same group velocity as the coating's. The correspondent spectrograms are plotted in Figure 7.2.15. Again, the effect of increasing coating thickness can be seen. The group delays as a function of the dispersion curves are plotted in Figure 7.2.16 for the ultrasonic signals (b), (c), and (d) in Figure 7.2.14. As the thickness of the coating increases, the frequency transition moves toward lower frequencies, consistent with the theoretical curves plotted in Figure 7.2.9. The comparison between the theory and experiment yields the coatings parameters [Abb98].

The output of the complete system is given in Figure 7.2.17, where the real-time measurement of the coating thickness is measured after each spray pass for two successive coating operations. The difference between the two curves is due both to the repeatability of the process from one spray operation to the next and the absolute accuracy of the measuring process, which is estimated to be of the order of 10 to 20 μm.

7.3. Biomedical Application of Wavelets: Analysis of EEG Signals for Monitoring Depth of Anesthesia

In a different application, the wavelet transform has been used to study the nonstationary information in the electroencephalograph (EEG) as an aid in determining the anesthetic depth [Abb96a]. The wavelet technique was utilized for the detection and spectral analysis of transient and background processes in the awake and asleep states. It can be observed that the response of both states before the application of the stimulus is similar in amplitude but not in spectral contents, which suggests a background activity of the brain. The brain reacts to the external stimulus in two different modes, depending on the state of consciousness of the subject. In the case of awake state, there is an evident increase in response, whereas for the sleep state, a reduction in this activity is observed. This analysis seems to suggest that the brain has an ongoing background process that monitors external stimulus in both the sleep and awake states [Abb96c].

The electroencephalograph (EEG) consists of an ensemble of electric potentials recorded continuously from a standard montage of scalp electrodes and is a direct monitor of the central nervous system (CNS). During surgery, the EEG is used to correlate the activity of the CNS to depth of anesthesia. Depth of anesthesia is related to the patient's ability to perceive and respond to noxious stimuli. Historically, anesthesiologists have used a variety of examination procedures to estimate awareness of a subject [Gho92]. The use of electroencephalography (EEG) to monitor depth of anesthesia was first suggested by Gibbs et al. [Gib37], who observed the correlation between the EEG patterns and levels of anesthesia induced by diethyl ether. With the development of fast and inexpensive computers, various signal processing techniques have been utilized to analyze and quantiify the signals both in the time or in the frequency domain. Klein et al. [Kle81] used the frequency and amplitude of the zeroth, first, and second derivative of the signal to monitor patients subjected to general anesthesia. A hierarchical clustering identification system based on autoregressive (AR) estimation of the EEG spectra was designed by Thomsen et al., Autoregressive (AR) estimation was used by Cerutti et al. in monitoring the influence of hypotension on the central nervous system during risky neurosurgery [Cer85]. A tenth-order AR model was also utilized by Sharma and Roy to train a neural network to predict the anesthetic depth [Sha92a].

In the frequency domain, Fourier analysis has been used to measure the spectral properties of the EEG. The spectral information is usually obtained in terms of the median frequency, spectral edge frequency (SEF), total power, and relative power in the individual frequency bands. These bands have been defined as Delta (1.0-3.5 Hz), Theta (3.5-7.5 Hz), Alpha (7.5-12.5 Hz), Beta1 (12.5-17.5 Hz), and Beta2 (17.5-25 Hz). In the Fourier transform, the signal is assumed to be stationary in the processing period; it is assumed that its spectrum does not change over time. It was shown by Koch et al. that the EEG spectral parameter SEF can lead to contradictory results since it is based on the assumption of stationarity [Koc94]. The authors observed that the SEF not only decreased when the patient was deeply anesthetized but also when surgically stimulated while under light anesthesia. Such occurrences have been labeled "paradoxical" arousal. For this reason, time-frequency techniques have been developed to study the spectral changes in the EEG. The most common technique is the short-time Fourier transform (STFT). It is obtained by sliding a window across the signal and computing the magnitude of the Fourier transform of the windowed segment with the assumption that the signal is stationary in the time interval defined by the window [Koe46]. The spectral resolution of the STFT can be improved by using a longer window. A longer window, however, leads to smearing of nonstationary data, whereas a short window leads to poor frequency resolution. Therefore, Nayak et al. studied the effect of noxious stimulus on the EEG by using a time-frequency spectral representation (TFSR) known as the Choi-Williams exponential distribution (ED). The TFSRs during stimuli induced wakefulness were localized in time and frequency with increased delta activity. In deeply anesthetized states, although the frequencies were in the delta band, the spectra were not localized. Although the results gave a broad picture of changes in the EEG, finer details could not be obtained due to the presence of cross terms. These cross terms carry redundant information and obscure important features of the signal [Nay94].

The ability of the wavelet transform (WT) to use different windows at different frequencies would tend to model the EEG in an efficient way. The windows at different frequencies have a constant Q factor $(Q^{-1} = \Delta f / f)$ where f is the center frequency and Δf is the spread of the window spectrum. Therefore, the stimuli-induced transients in the EEG can be studied more effectively by using the WT and lead to possible quantification of the EEG spectral information in relation to anesthetic depth [Vis91].

In many practical applications, as the one discussed in this chapter, the signal is digitized in discrete-time intervals. Furthermore, as for the STFT, the WT is higly redundant when the parameters a and b are continuous in value. Therefore, the WT is evaluated for a discrete number of points in the time-scale plane, corresponding to a discrete set of wavelets. Discretization of the time-scale parameters also leads to setting $a = 2^m$, and $b = n \cdot 2^m$, where it is assumed unity for the discrete sampling interval.

The EEG signals were obtained from mongrel conditioned dogs weighing approximately 20-25 kg. The dogs were anesthetized by orally intubating 2 mg/kg doses of methohexital (Brevital). Electrocardiograph leads were placed using needle electrodes, and four channels of EEG needle electrodes were applied. Arterial pressure was measured using a percutaneously placed femoral arterial line. Arterial pressure and ECG signals were analyzed using a Mennen Medical Horizon monitor, and the EEG signal was monitored by an Axon System Sentinel-4 EEG/EP monitor. Respiratory gases were monitored by a PPG Medspect mass spectrometer and a Criticare Systems POET capnograph. The capnograph provided a continuous visualization of the subject's breathing pattern. All signals were processed by a 486PC computer equipped with a Data Translation DT2801 analog-to-digital acquisition board.

The anesthetic agent (halothane) was increased from 0.2% to 1.2% in increments of 0.2% at time intervals of approximately 30 min. At the end of each interval, when the anesthetic level had stabilized, as determined by the mass spectrometer, the subjects were tested for depth of anesthesia. This test was accomplished by tail clamping for 30 s. This results in a supramaximal stimulus. At the time of tail clamping, the depth of anesthesia was estimated by observation of animal movement and changes in the respiration, heart rate, and mean arterial pressure. The depth was graded as either awake or asleep. At the conclusion of the experiment, the animal was awaken, extubated, and returned to the animal facility. This protocol was approved by the Institute Animal Care and Use Committee.

Typical signals acquired during this test are shown in Figure 7.3.1 for the case of awake (a) and asleep (b). Also, in the time window, the stimulus was applied as shown in the figure. The aim of this work was to obtain, by comparison between the signals prior and during the stimulus for awake and asleep, a quick EEG signal analysis procedure to determine the depth of anesthesia without having to apply the stimulus.

A Wavelet Transform Signal Processor was designed for investigation of the EEG signals. The system receives the EEG signal as input, $h(t)$ is the mother wavelet, and $W_x(a,b)$ is the WT of EEG. The WT technique was applied to the EEG signals. Two different mother wavelets were used for analysis of the EEG. In one case, the Morlet function was used as the mother wavelet to optimize the resolution of time-scale analysis. In the other

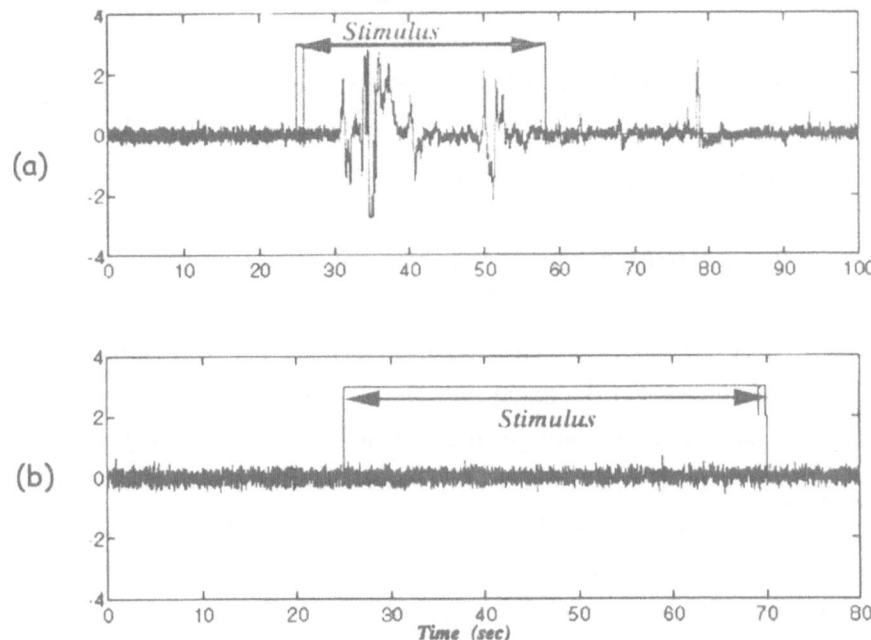

Figure 7.3.1 Typical EEG signals acquired before, during, and after the simulus for (a) awake and (b) asleep subjects.

case, the Mexican hat wavelet was used. This wavelet is commonly used in analyzing pulses in signals. Pulse detection consists of determining the presence or absence of a pulse and estimating its amplitude and arrival time even when the pulse is buried in time. Because the Q factor of the Mexican hat wavelet is equal to 1, the processes are depicted in different dilation ranges as long and short responses. Therefore, the Mexican hat approach with Q equal to 1 can be used to analyze the signal structure in the time domain.

The Mexican hat function was obtained by fitting the EEG response for the awake subject during stimulus and was used as the mother wavelet to detect the presence of spikes corresponding to the EEG obtained during light anesthesia. Thus, the WT transform can be interpreted as the correlation or affinity between the EEG signal and the brain response to stimulus, modeled by the Mexican hat. The EEG data before and during the period of tail clamp were analyzed.

7.3.1 Wavelet Spectral Analysis of EEG Signals

The spectral properties of the EEG signals were anayzed in the frequency range from dc to 25 Hz. The center frequency f_0 of $h_1(t)$ is 0.5 Hz therefore, for a generic dilation m, the center frequency of the associated filter is $0.5/2^m$. In order to cover the required range of frequencies, the coefficient m has to be as low as -6 ($f_m = 0.5/2^{-6} = 32$ Hz), and as high as $+1$ ($f_m = 0.5/2^1 = 0.25$ Hz). Therefore, the absolute bandwidth will vary from 6 to 0.05 Hz, for $m = -6$ and $m = 1$, respectively. The time-scale representations of the WT for the EEG obtained during the state of light (depth zero) and deep anesthesia (depth one) are given as density plots. The x and y-axes in the plot represent the dilation coefficient m and the time coefficient n, respectively. A negative dilation coefficient m represents compression of the mother wavelet and, thus, higher frequencies. The correspondence between dilation coefficient m and the frequency bands of interest is given in Table 1. The magnitude of the WT is plotted using a logarithmic gray scale in order to show contributions to the EEG signals from the various frequency bands. The white areas corresponds to higher values of the WT coefficients and the dark corresponds to the lower. Also in each case, depth zero and one, the magnitude of the WT coefficients were time-averaged over the prestimulus period and the stimulus period. These are plotted as a function of the center frequency of the filter associated to each daughter wavelet.

Table 1. Correspondence between the frequency bands and the dilation coefficients m.

Band	Frequency [Hz]	Dilation m
Delta (Δ, δ)	1.0 - 3.5	1.0 - 2.8
Theta (θ, Θ)	3.5 - 7.5	2.8 - 3.9
Alpha (A, α)	7.5 - 12.5	3.9 - 4.64
Beta1 (β1)	12.5 - 17.5	4.64 - 5.13
Beta2 (β2)	17.5 - 25	5.13 - 5.64

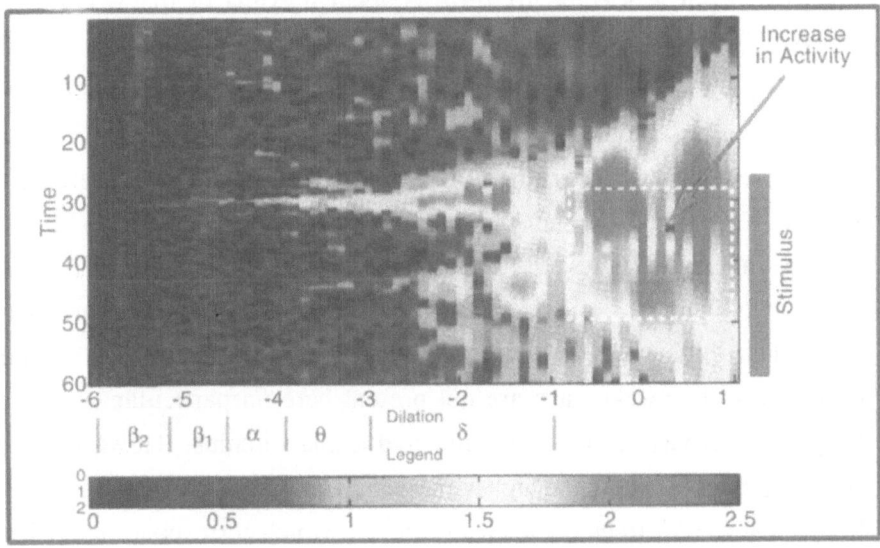

Figure 7.3.2 Wavelet transform of the EEG awake signal in Figure 7.3.1a.

Thirteen signals were processed using the WT, of which seven were classified as asleep and five as awake. A typical awake and sleep signal will be discussed in the following.

Figure 7.3.2 shows the density of plot of the time-scale representation

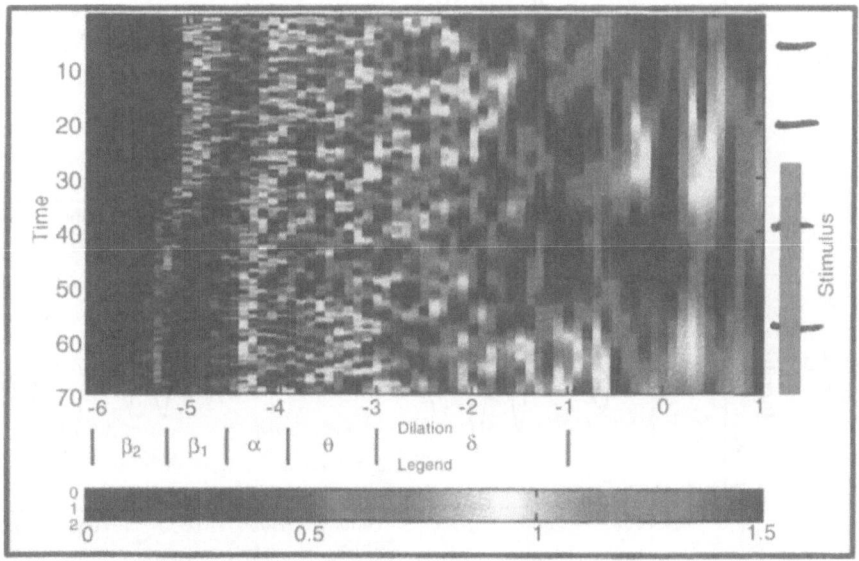

Figure 7.3.3 Wavelet transform ofthe EEG asleep signal in Figure 7.3.1b.

when the depth was equal to zero. A local maxima of the WT can be observed before the stimulus for $m = -5.5$ (22.6 Hz), which seems to shift to $m = -5.75$ (26.9 Hz) once the stimulus is applied. A somewhat lower density of values is observed during the stimulus in the range $m = -5$ to $m = -3$ (4-16 Hz), but it is difficult to quantify the variations. A decrease in amplitude is observed in the Alpha frequency range.

The time-scale representation of the WT for the sleep signal, depth one, is shown in Figure 7.3.3 as a density plot. The wavelet magnitudes are much lower than those given in Figure 7.3.3. The large changes in amplitude observed in the awake case are not present here, in particular the lower frequency activity seems to be unrelated to the stimulus. However some shifts can be observed in the higher-frequency ranges. In particular, the shift between $m = -3.7$ to $m = -4.1$ (6.5-8.6 Hz), which corresponds to a Theta to Alpha region shift, and $m = -4.7$ to $m = -5.1$ (13 to 17 Hz), in the Beta1 range. These observations are confirmed by the one-dimensional plots. A more subtle shift of values can be observed in the $m = -2$ to $m = -3$ range (2-4 Hz), which is the Delta region, but it is not possible to quantify the variations.

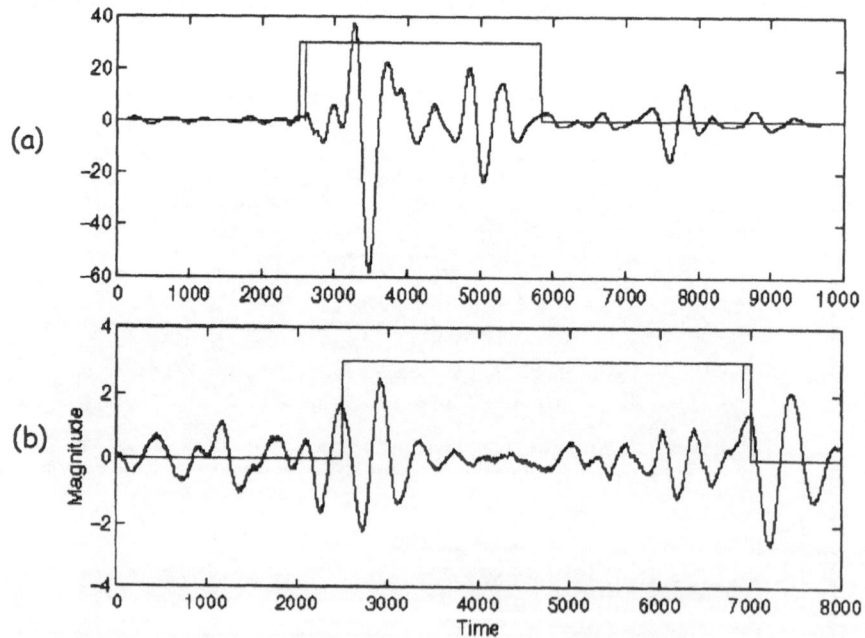

Figure 7.3.4 Wavelet processed EEG (a) awake and (b) asleep signals.

7.3.2   System Response Wavelet Analysis of EEG Signals

Figures 7.3.4a and 7.3.4b show the corresponding WT processed awake and sleep signals using the Mexican hat wavelet. It can be observed that the responses of both states before the application of the stimulus is similar in magnitude. However, the magnitudes are higher for depth-zero EEG when compared to those at depth one.

7.3.3   Discussion of Results

From the time-scale analysis of the EEG using the Morlet mother wavelet, it can be concluded that for the sleep case, a shift toward higher frequencies can be observed in the Beta1 region, which is analogous to the shift observed in the Beta2 region for the awake case. The average magnitudes of the pre stimulus WT coefficients corresponding to different frequencies for the

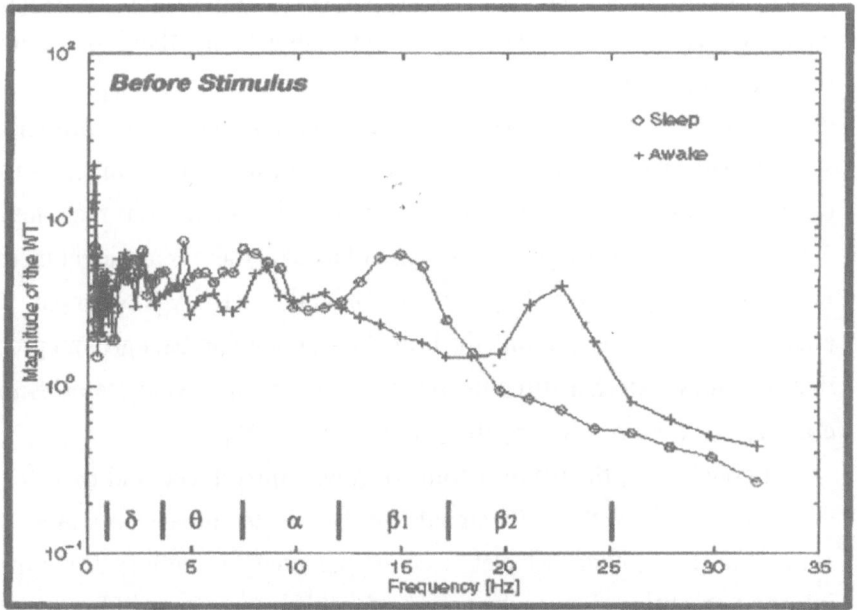

Figure 7.3.5 Average magnitude of the pre-stimulus wavelet coefficients as a function of frequency for asleep and awake EEG signals.

unstimulated sleep and awake states are plotted in Figure 7.3.5 [Abb96b, Abb96c]. The WT coefficients have roughly the same order of magnitude in both cases, but they show different spectra, especially in the higher-frequency range. The awake state is characterized by Beta2 (22.5 Hz) waves and lower-amplitude Alpha waves (7.5 Hz), whereas the sleep state is characterized by Beta1 (15 Hz) waves and slightly higher-amplitude Alpha-theta (5-8 Hz) waves. It has been suggested that alpha blocking can be related to the state of the patients, and thus can be utilized to evaluate the depth of anesthesia. Alpha blocking is a response to noxious stimulus under light-anesthesia conditions. Our results show the presence of Beta2 shift in addition to alpha blocking. During deep anesthesia there seems to exist a Beta1 shift. There also seems to be three distinct conditions in which the subject can be found: (1) sleep, with higher amplitude Alpha-Theta and Beta1 waves; (2) awake, with lower amplitude Alpha-Theta waves and significant Beta2 activity; and (3) stimulated awake state, with a large increase in amplitude in the low frequency range, and a shift in the main frequency in the Beta2 region.

The Beta2 shift in the case of depth-zero EEG and Beta1 shift associated with depth-one EEG could not be detected by the system analysis technique using the Mexican hat wavelet. This could suggest that the Beta frequency shifts are not due to external stimulus but due to some internal process. The difference in response at depth zero and the one due to the presence of stimulus suggests a background activity of the brain. The brain reacts to the external stimulus in two different modes depending on the state of consciousness of the subject. In the case of awake, there is a evident increase in response, whereas for sleep, a reduction in this activity is observed. This analysis seems to suggest that the brain has an ongoing background process that monitors external stimulus in both the sleep and awake states. Similar conclusions were reached by the authors in [Nay94].

In conclusion, the wavelet transform was introduced and used for the spectral analysis of the EEG signals in the awake and asleep states. The choice of the mother wavelet used in the analysis defines the type of information obtained by such analysis; for this reason, two different mother wavelets were utilized in this work. Using the WT with the Morlet wavelet, the awake and asleep states were characterized by the frequency contents. This approach is similar to the short-time Fourier transform. Differences in the frequency spectra between the two states were observed and are summarized as follows: (1) For the awake state, the stimulus causes an increase in the low-frequency energy; (2) for the asleep state, the stimulus causes the dog's EEG waves to shift in the awake direction, resulting in higher-frequency Alpha waves and a shift from Beta1 to Beta2 waves; (3) there exists a possibility of predicting the state of a subject by recording the changes in Alpha and Beta waves.

From these observations, there seems to be three distinct conditions with probably some transition states in between. These three states are as follows: (1) sleep, with the spectrum mainly localized in the Alpha-Theta and Beta1 regions; (2) awake, with the spectrum localized in the lower-frequency Alpha region and in the Beta2 range; and (3) stimulated awake state, with a dramatic increase in low-frequency energy on top of the regular awake frequency spectrum.

The second choice of the mother wavelet was to model the brain response to the external stimulus. Using the EEG signal in the awake state, a Mexican hat mother wavelet was constructed and used to study the EEG signals. Results show that there seems to be a constant background process that monitors internal and external stimuli. This background process is the same whether in the awake or sleep states if there is no supramaximal external stimulus. In the awake state, this external stimulus increases the system response; whereas in the asleep state the opposite seems to happen, resulting in a disruption of the background process. This approach could be used to model the brain functions using EEG data.

7.4   References

[Abb95b]   A. Abbate, J. Frankel, and P. Das, Wavelet transform signal processing applied to ultrasonics, *Proc. of the 1995 QNDE*, pp. 741-748, 1995.

[Abb95c]   A. Abbate, J. Frankel, and P. Das, Wavelet transform signal processing for dispersion analysis of ultrasonic signals, *Proc. of the 1995 IEEE International Ultrasonic Symposium*, vol. 1, pp. 751-755, 1995.

[Abb96]   A. Abbate, J. Frankel, R.W. Reed, and P. Das, Ultrasonic gauging and wavelet image processing for wear and erosion mapping, *Proc. of the 1996 QNDE*, vol. 18, 1996.

[Abb96a]   A. Abbate, J. Frankel, R.W. Reed, and P. Das, Wavelet image processing for wear and erosion mapping using an ultrasonic gauging technique, *Proc. 1996 Ultrasonic Symposium*, 1996.

[Abb96b]   A. Abbate and P. Das, Wavelet transform signal processing of electroencephalograph signals, US Army Tech. Rep. ARCCB-TR-96007, 1996.

[Abb96c]   A. Abbate, A. Nayak, J. Koay, R.J. Roy, and P. Das, Biomedical application of wavelets: analysis of electroencephalograph signals for monitoring depth of anesthesia, *Proc. 1996 SPIE Conf. on Wavelet Applications*, vol. 2762, pp. 412-423, 1996.

[Abb97]   A. Abbate, J. Koay, J. Frankel, S.C. Schroeder, and P. Das, Signal detection and noise suppression using a wavelet transform signal processor: Application to ultrasonic flaw detection, *IEEE Trans. Ulrason. Ferroelectrics Frequency Control*, vol. 44, n0. 1, pp. 14-26, 1997.

[Abb98]   A. Abbate, W. Russell, J. Goldmann, P. Kotidids, and C.C. Berndt, Nondestructive determination of thickness and elastic modulus of plasma spray coatings using laser ultrasonics, *Rev. Prog. QNDE*, vol. 18, pp. 373-380, 1998.

[Abb98a]   A. Abbate, D. Klimek, P. Kotidis, and B. Anthony, Analysis of dispersive ultrasonic signals by the ridges of the analytic wavelet transform, *Rev. Prog. QNDE*, vol. 18, pp. 703-710, 1998.

[Aus94]   J.D. Aussel and J.P. Monchalin, *Rev. Prog. QNDE*, Plenum Press, New York, pp. 535-542, 1994.

[Ben94]   J.J. Benedetto and D.M. Frasier eds., *Wavelets, Mathematics and Applications*, CRC Press, Boca Raton, FL, 1994.

[Bij96]   A. Bijaoui, E. Slezak, F. Rué, and E. Lega, Wavelets and the study of the distant universe, *Proc. IEEE*, vol. 84, n. 4, pp. 670-679, 1996.

[Bir95]   V. Biryukov, Y.V. Gulayev, V.V. Krylov, and V.P. Plessky, *Surface Acoustic Waves in Inhomogeneous Media*, Springer-Verlag, New York, 1995.

[Bra89]   D.E. Bray and R.K. Stanley, *Nondestructive Evaluation: A Tool in Design, Manufacturing and Service*, McGraw-Hill Book Company, New York, 1989.

[Cer85]   S Cerruti, D. Liberati, and P. Mascellani, Parameter extraction in EEG processing during riskful neurosurgical operations, *Signal Processing*, Vol. 9, pp. 25-35, 1985

[Fou94]   E. Foufoula-Georgiou and P. Kumar eds., *Wavelets in Geophysics*, Academic Press, New York, 1994.

[Gam81]   P.M. Gammel, Improved ultrasonic detection using analytic signal magnitude, *Ultrasonics*, pp. 73-76, 1981.

[Gam81a]   P.M. Gammel, Analogue implementation of analytic signal processing for pulse-echo systems, *Ultrasonics*, pp. 279-283, 1981.

[Gho92]   M.M. Ghoneim and R.I. Block, Learning and consciousness during general anesthesia, *Anesthesiology*, vol. 76, pp. 279-305, 1992.

[Gib37]   F.A. Gibbs, E.L. Gibbs, and W.G. Lenox, Effect of the electroencephalogram of certain drugs which influence nervous activity, *Arch. Intern. Med.*, vol. 60, pp. 154-166, 1937.

[Gui91]   P. Guillemain, R. Kronland-Martinet, and B. Martens, Estimation of spectral lines with the help of the wavelet transform. Applications in NMR spectroscopy, in *Wavelets and Applications*, edited by Meyer, Springer-Verlag, New York, 1991, pp. 38-60.

[Gui96]   P. Guillemain and R. Kronland-Martinet, Characterization of acoustic signals through continuous linear time-frequency representations, *Proc. IEEE*, vol. 84, n. 4, pp. 561-585, 1996.

[Kad92]   S. Kadambe and G.F. Boundreaux-Bartels, Applications of the wavelet transform for pitch detection of speech signals, *IEEE Trans. Inform. Theory*, vol. 38, pp. 917-924, 1992.

[Kin87]   G.S. Kino, *Acoustic Waves. Design, Imaging and Analog Signal Processing.* Prentice-Hall, Englewood Cliffs, NJ, 1987.

[Kle81]   F.F. Klein and D.A. Davis, The use of time domain analyzed EEg in conjnction with cardiovascular parameters for monitoring anesthetic levels, *IEEE Trans. Biomed. Eng.*, vol. 38, pp. 36-40, 1981.

[Koc94]   E. Koch, P. Bischoff, U. Pilchmeier, and J. S. Esch, Surgical stimulation induces changes in brain electrical activity during isoflurane/nitrous oxide anesthesia, *Anesthesiology*, vol. 80, pp.1026-1034, 1994.

[Koe46]   R. Koenig, H. Dunn, and L.Y. Lacy, The sound spectrograph, *J. Acoust. Soc. Amer.* vol. 18, pp. 19-49, 1946.

[Kot94]   P. Kotidis, J. Woodroffe, J. Shah, and T. Schultz, *Nondestructive Characterization of Materials*, edited by R.E. Green, Plenum Press, New York, 1994, vol. IV, pp. 21-28.

[Kot97]   P. Kotidis and D. Klimek, *Proceedings of the Eighth International Symposium on Nondestructive Characterization of Materials*, Boulder, CO, June 1997.

[Kro87]   R. Kronland-Martinet, J. Morlet, and A. Grossmann, Analysis of sound patterns through wavelet transforms, *Int. J. Pattern Recogn. Art.*, vol. 1, no. 2, pp. 273-302, 1987.

[Kro88]   R. Kronland-Martinet, The wavelet transform for analysis, synthesis and processing of speech and music sounds, *Computer Music J.*, vol. 12, pp. 11-20, 1988.

[Kro91]   R. Kronland-Martinet and A. Grossmans, Application of time-frequency and time-scale methods (wavelet transform) to the analysis, synthesis and transformations of natural sounds, *Representation of Musical Signals*, MIT Press, Cambridge MA, 1991, pp. 45-85.

[Lak95]   F. Lakestani, J.F. Coste, and R. Denis, *NDT&E Int.*, vol. 8, no. 5, pp. 171-178, 1995.

[Low95]   M. Lowe, *IEEE Trans. Ultrason. Ferroelectrics Frequency Control*, vol. 42, pp. 525-542, 1995.

[Mal96]   S. Mallat, Wavelets for a Vision, *Proc. IEEE*, vol. 84, no. 4, pp. 604-614, 1996.

[McE75]   J. McEwen, G.B. Ardenson, M. Low, and L. Jenkins, Monitoring the level of anesthesia by automatic analysis of spontaneous EEG activity, *IEEE Trans. Biomed. Eng.*, vol. 22, pp. 299-305, 1975.

[Nay94]   A. Nayak, R. J. Roy, and A. Sharma, Time-frequency spectral representation of the EEG as an aid in the detection of depth of anesthesia, *J. Acoust. Soc. Am.* vol. 22, pp. 501-513, 1994.

[Net95]   S. Nettel, *Wave Physics. Oscillations-Solitons-Chaos*, Springer-Verlag, New York, 1995.

[New93]   D.E. Newland, *An Introduction to random Vibrations, Spectral and Wavelet Analysis*, Wiley, New York, 1993.

[Ros99]   J.L. Rose, *Ultrasonic Waves in Solid Media*, Cambridge University Press, Cambridge, 1999.

[Scr90]   C.B. Scruby and L.E. Drain, *Laser Ultrasonics. Techniques and Applications*, Adam Hingler, IOP Publishing Ltd., Bristol, UK, 1990.

[Sha93a]   A. Sharma and R.J. Roy, Analysis of hidden nodes in a neural network trained for pattern classification of EEG data, *Proc. of the 15th Annual International Conf. of the IEEE EMBS*, 1993, vol. 1, pp. 252-253.

[Szu92]   R. Szu, B. Teller, and A. Lohynan, Causal analytical wavelet transform, *Opt. Eng.* vol. 31, pp. 1825-1829, 1992.

[Teo98]   A. Teolis, *Computational Signal Processing with Wavelets*, Birkhäuser, Boston, MA, 1998.

[Tho89]   C.E. Thomsen, K.N. Christensen, and A. Rosenfalck, Computerized monitoring of depth of anesthesia with isoflurane, *Br. J. of Anesthesia*, vol. 63, p. 36-43, 1989.

[Uns96]   M. Unser and A. Aldroubi, A review of wavelets in biomedical applications, *Proc. IEEE*, vol. 84, n. 4, pp. 626-638, 1996.

[Vik67]   I.A. Viktorov, *Rayleigh and Lamb Waves. Physical Theory and Applications.* Plenum Press, New York, 1967.

[Vis91]   R. Vishnoi and R.J. Roy, Adaptive control of closed circuit anesthesia, *IEEE Trans. Biomed. Eng.*, vol. 39, pp. 39-47, 1991.

Chapter 8

Applications of Subband and Wavelet Transform in Digital Communications

8.1 Introduction

Present-day communications systems are quite complex. In the early days of communication, one just needed a modulator for signal to be transmitted, a radio-frequency (r.f.) amplifier, and the antenna. On the receiving side, one needed a receiving antenna, a preamplifier, and a demodulator. Because of the wireless communication which involves networking, the complexity is steadily growing and it will continue to do so to accommodate wireless access to the Internet and multimedia. Figure 8.1.1 shows the block diagram of a digital communication system. Source encoding, also called data compression, removes the redundancy of the source by representing of the data using the smallest number of binary digits [Pro95]. In many cases, specially for secure Internet traffic, for an example, one uses encryption so that unauthorized detection of signal can be minimized or avoided. In contrast to the source coding which removes redundancy, channel coding introduces some redundancy in a predetermined fashion to improve the fidelity of the signal output. The channel has noise and interference, which causes error in the output. By introducing an n-bit codeword for a k-bit of information when $n>k$, the bit error rate (BER) can be drastically reduced. The digital modulator maps the binary output of the channel encoder to electrical signal waveforms. The modulator could be binary or M-ary, as also amplitude, phase, or frequency modulated or a combination of them.

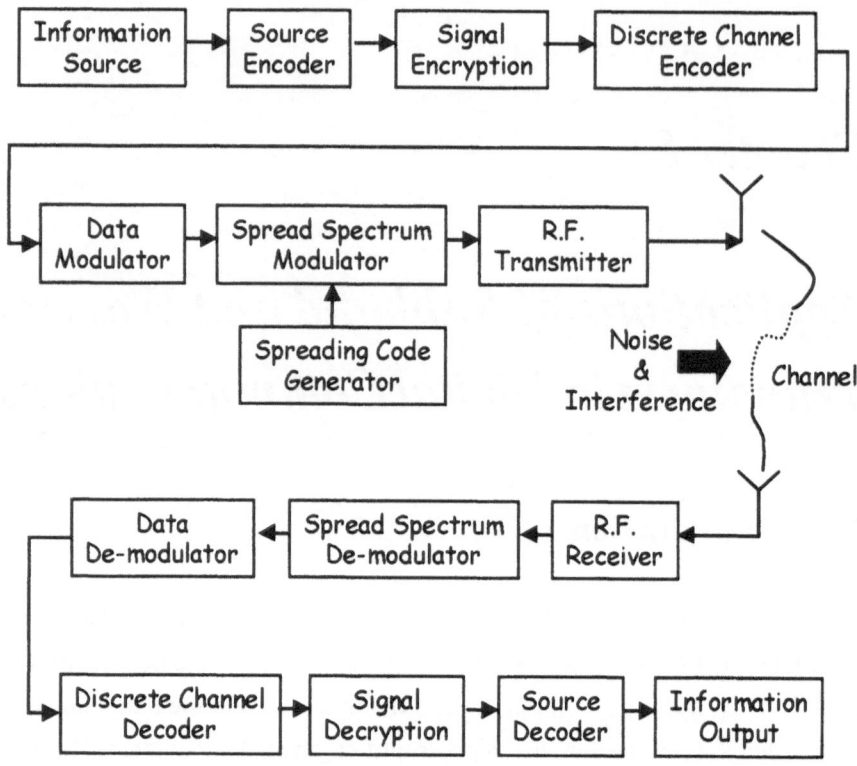

Figure 8.1.1 Block diagram of a digital communication system.

If the communication system is spread spectrum, then one uses a spread code generator to modulate the output of digital modulator to a higher-bandwidth signal [Het97, Med95a]. This is generally amplified and transmitted through the antenna. The receiver antenna receives corrupted and attenuated signal due to noise, multipath, fading due to interference, distance, etc. [Mil80]. The signal from the antenna is despread using spread spectrum code, data demodulated, channel decoded, decrypted, and source decoded to obtain the final output. As mentioned earlier, all blocks are not used all the time [Mil83]. The wavelet and subband filterbank structures have been used for modulation, source encoding, and spread spectrum code generation. These will be discussed in the next few sections [Aka99].

8.2   Applications in Spread Spectrum Communication Systems

In spread spectrum communication systems, digital modulation techniques are used which produce, at the output of the digital modulator, a signal whose bandwidth is much larger than the bandwidth of the input information. Typically, the ratio of the bandwidths can be 100 or 1000 and is generally referred to as the processing gain; the system is more resistant to interference.   In addition to the interference rejection property, spread spectrum systems have many other desirable features.   These are code-division multiple access, low-power spectral density, high-resolution ranging capability, and resistance to multipath fading.   The inherent interference rejection capability of a spread spectrum can be appreciated by considering Figure 8.2.1, used to find out the radius of effective communication in the presence of a jammer. If a non spread spectrum conventional system is used which requires at least 20 (signal to jammer ratio), then this radius is 18 km. However, if the S/J ratio of 10 is acceptable, then this radius is 48 km. Now, if one uses spread spectrum system with a processing gain of 30, then this radius increases from 48 km to 182 km, as shown in the Figure 8.2.1 [Mil83].

Although the spread spectrum communication system has inherent interference rejection capability due to processing gain, in many circumstances it is not enough specially in the presence of a high-power jammer. For this situation, one can augment the receiver performance by using transform domain processing, as mentioned in Section 8.1. Naturally, in place of Fourier transform, one can use a wavelet transform or subbands with better performance.   This particular application of interference suppression or excision has been the subject of many articles and we will review this particular application.

The price one pays for a spread spectrum receiver is its complexity. There are different kinds of spread spectrum system. The two most popular ones are direct sequence (DS) and frequency hopping. In the DS spread spectrum system, one modulates the data with a PN sequence, which has

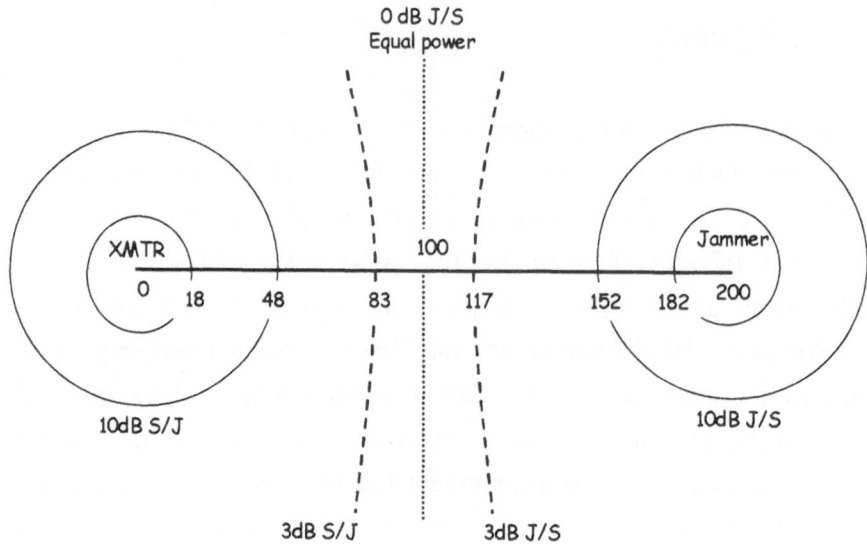

Figure 8.2.1   Inherent interference rejection capability of spread spectrum communication system leading to vast increase in the effective area of communication in the presence of jammer.

special properties. In place of using PN sequences, one can use the subband transform basis functions. Because DS receivers can be used for CDMA applications, one can use the subband filter generated codes for this multiple access purpose also. The spread spectrum signal has low-power spectral density. Thus, it can also be used when the communication system wants to stay undetected by the "enemy" intercept receiver. This low probability of intercept (LPI) property and the opposite one, the so-called intercept receiver when you want to detect a LPI receiver, can also be implemented using wavelet transforms [Res92].

8.2.1   Excision

Spread spectrum receivers have inherent protection against interference due to the processing gain. Actually, any level of interference protection can be provided by sufficiently spreading the signal to the required processing gain. The price for greater protection, however, is an increase in the bandwidth of the transmitted signal for a given data bandwidth. Practical considerations such as transmitter/receiver complexity and available frequency spectrum can serve to limit the reasonably attainable processing gain. As a result, it is beneficial to apply signal processing techniques to augment the processing gain of the spread spectrum signal itself, allowing greater interference protection without an increase in bandwidth. In general, these interference suppression techniques discriminate between the desired spread spectrum signal and the interference and work to suppress the interference. The processing can be performed in the time domain (e.g., adaptive transversal filtering), the spatial domain (e.g., adaptive array antennas), or the transform domain [Med95a, Het96].

The transform domain processing using wavelet transform and filter banks is the subject of interest in this section. When a signal is transformed or mapped to a different "space" and processed, the signal processing is said to have been done in the transform domain, or, in other words, that one is using transform domain processing. Note that this mapping should be unique and unambiguous and that an inverse mapping or transformation, which can return the signal to the time domain, should exist. The most widely used continuous-time transform is the Fourier transform, but there are many others of importance, such as the Fresnel, Hartley, Mellin, and Hilbert transforms, to name but a few. In communications and radar applications, particularly ones using spread spectrum techniques, transform domain processing can be utilized to suppress undesired interference and, consequently, improve performance. Here, the basic idea is to choose a transform such that the jammer or the undesired signal is nearly an impulse function in the transform domain, whereas the desired signal is transformed to a waveform that is very "flat" or "orthogonal" with respect to the transformed interference. A simple exciser, which sets the portions of the transform which are jammed to zero, can then remove the interferer without removing a significant amount of desired signal energy. An inverse transform then produces the nearly interference-free desired signal. Figure 8.2.2 is a block diagram of a transform domain excision system. The excision waveform usually takes only the values of zero and unity, resulting in the complete removal of portions of the transform that are determined to

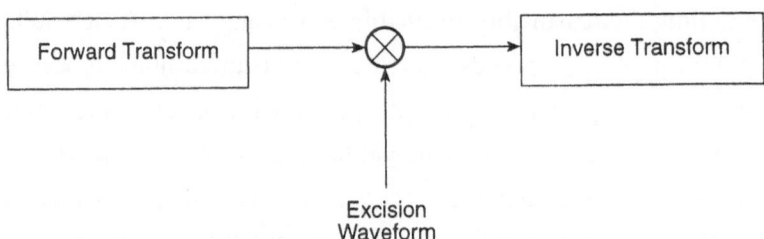

Figure 8.2.2 Block diagram of a transform domain excision system.

be "jammed." Figure 8.2.3 illustrates the excision process.

Most of the research and development related to the application of transform domain signal processing has been restricted to Fourier-transform-based applications. In fact, many practical implementations utilize short-time Fourier transforms, which observe only a segment of the input signal. In particular, discrete-time implementations typically use the fast Fourier transform (FFT), which is a computationally efficient implementation of its discrete counterpart, the discrete Forurier transform (DFT).

Under ideal circumstances, the transform domain representation of the

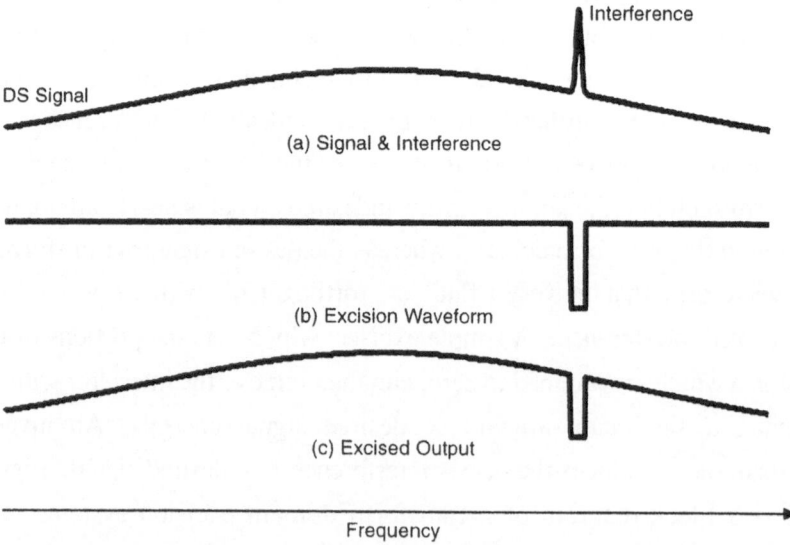

Figure 8.2.3 An illustration of the excision process.

interfering signal should appear as an impulse function. Implementing the short-time Fourier transform, however, requires the use of windowing functions to localize the input data in time. These windowing operations yield frequency domain representations that are characterized by undesired side lobes. As a result, when time domain windows are used, an excessive amount of interference energy may be detected and represented in neighboring spectral bins. The amount of energy contained in these side lobes is a direct function of the windowing function used; for example, a rectangular window yields very large side lobes due to its $\sin(x)/x$ frequency response, whereas Hamming, Bartlett, and other windows weight the data samples in such a manner that the size of the transform domain side lobes is significantly reduced at the cost of widening the main lobe. In an excision-based system, it is desirable to represent the interfering signal in as few transform domain bins as possible, in order to preserve the maximum amount of interference energy. Hence, it is advantageous to localize the interference energy to the main lobe of the frequency response and to keep the interference energy contained in the transform domain side lobes as small as possible. Although using non rectangular windows reduces the sides of the lobes, it requires the processing of overlapping segments of the input signal in order to get accurate reconstruction of the time signal, thus greatly increasing the computational requirements.

The idea of using the discrete wavelets transform (DWT) and related subband filter banks to suppress undesired interference in spread spectrum communication systems has been introduced in [Aka94, Her91, Het96a]. In this section, we will review the material. For details, the readers are referred to the original articles.

To evaluate the performance of the DS communications receivers of Figures 8.2.4 and 8.2.5, simulations were performed in which finite inpulse response (FIR) filters with coefficients derived by Daubechies were used in the para-unitary quadrature mirror filter (QMF) bank substructures. Note that this property of the receiver will be dependent on the type of filter chosen and its length. Actually, one can choose the optimum transform for particular situations. In those simulation studies, a 63-chip m-sequence was used to modulate a binary input data stream, thus producing a DS signal that

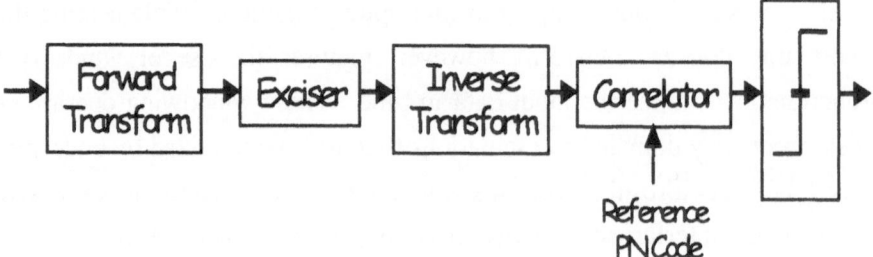

Figure 8.2.4 Block diagram of a DS communication receiver with subband domain excision.

was sampled 63 times per data bit. The resulting spread data signal was then transmitted over an Additive White Gaussian Noise (AWGN) channel with either additive narrow-band or pulse interference. At the receiver, five hierarchical levels of the two-channel perfect reconstruction (PR)-QMF bank substructures were used to generate a 32-band filter bank which mapped the time domain input into the subband transform domain. Full and dyadic subband tree structures using Daubechies' four-tap FIR filters were used to analyze the received data signal in the presence of continuous and transient interference, respectively.

The resulting bit error rate (BER) curves show the performance of the system as a function of the energy per data bit with respect to one-sided noise power spectral density ratio, E/N. Since interference energy is not considered in the calculation of E/N, the theoretical BER is equivalent to the performance of binary phase-shift keying in AWGN and serves as a benchmark of DS receiver performance to which subsequent results will be compared. When interference is present, the interference or jammer power

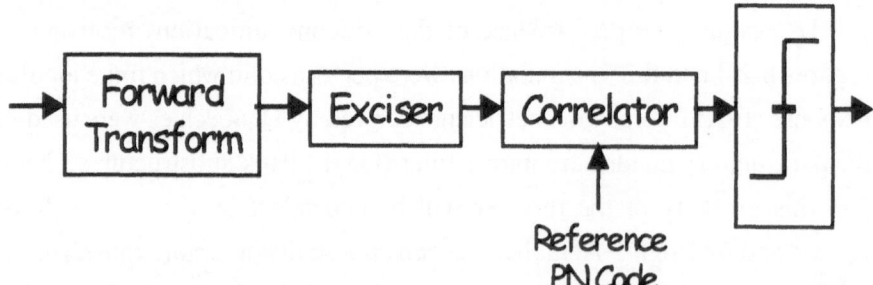

Figure 8.2.5 Block diagram of a DS communication receiver with subband transform domain excision and transform domain correlation.

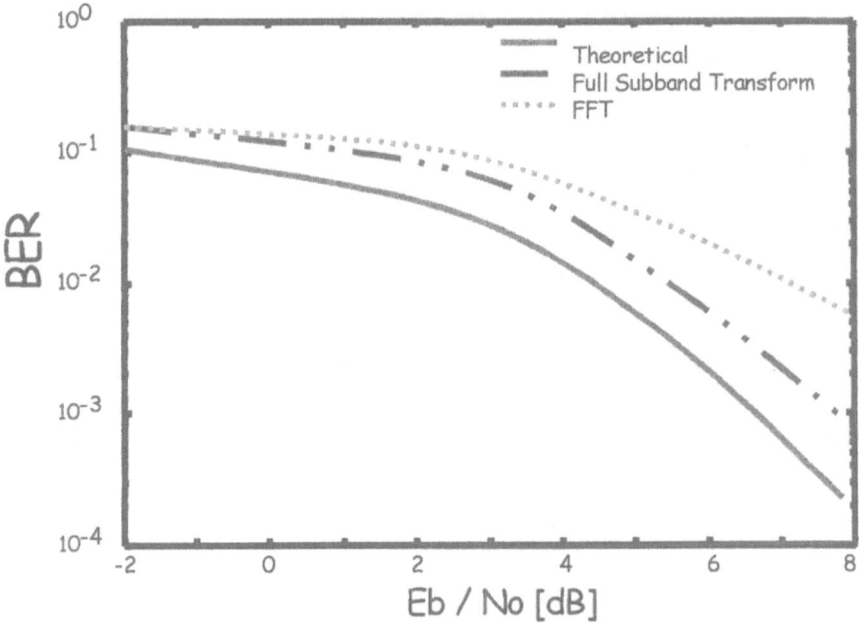

Figure 8.2.6 BER of full subband decomposition and FFT-based DS receiver compared with the theoretical curve. The performance was calculated in the presence of a continuous sinusoidal interference with eight bins excised.

is defined by the ratio of jammer power to DS signal power and is denoted as JSR.

The performance of DS transform domain receivers using the FFT and full subband decomposition in the presence of a continuous sinusoidal interferer having a JSR of 30 dB and a normalized frequency of 0.127 Hz (i.e., frequency normalized to a chip rate of unity) is shown in Figure 8.2.6. When no interference suppression technique is employed, the resulting BER is approximately 0.5, regardless of the E/N. For the results shown here, however, the interference is mitigated by excising the eight transform values with zero. As is evident from Figure 8.2.6 when these bins are removed the subband transform-based receiver yields a lower BER than the comparable FFT-based implementation. Since four-tap FIR filters derived by Daubechies were not specifically designed for interference suppression, receiver performance may be further improved by utilizing QMF banks with filters that are optimized for such applications.

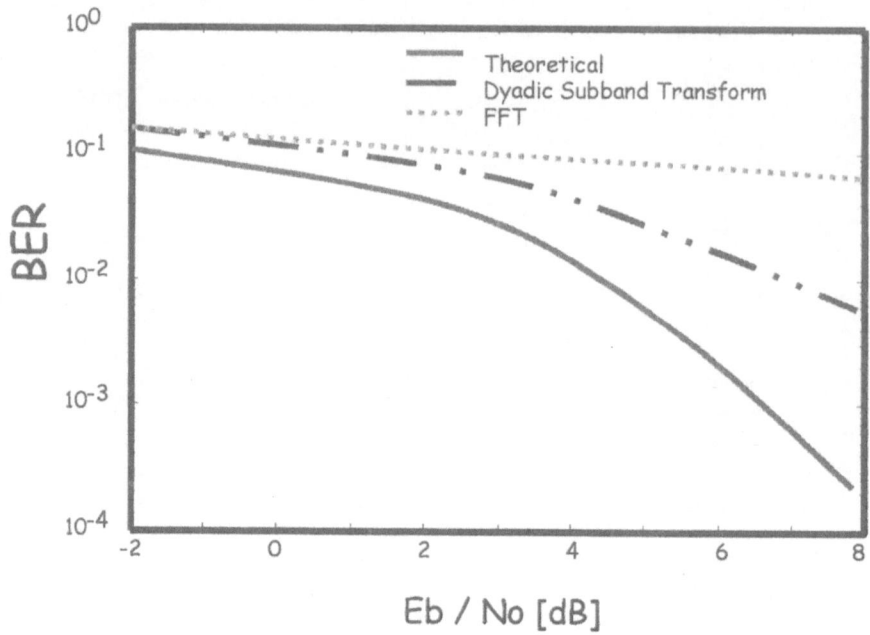

Figure 8.2.7 BER of dyadic subband decomposition and FFT-based DS receiver compared to the theoretical curve. The performance was calculated in the presence of a transient interference with eight bins excised.

Although transform domain excision using either full subband decomposition or FFT-based receivers may significantly improve receiver performance in the presence of continuous sinusoidal interference, their excision capability in the presence of a transient, or pulsed, interference source is far less effective. Some interference sources are typically characterized as relatively wide-band interferers which make the uniform spectral bands associated with the full binary subband tree and the FFT impractical in a transform domain excision system. As an alternative, transformation techniques that yield joint time-frequency resolution, such as the dyadic subband decomposition, may be more appropriate. For comparison, Figure 8.2.7 illustrates the relative performance of the DS receiver using the FFT and dyadic subband transform methods in the presence of a pulsed white noise jammer when 8 of the 64 transform domain bins are removed. In this case, the transient interference is represented by a white noise source with a duty cycle of 5% relative to a period of 70 samples, thus ensuring that the pulse duration is much less than that of the data bit, whereas the pulse period is slightly larger. In addition, the noise variance has been specified to yield an average jammer-to-signal ratio of 23.3 dB [Met94].

Despite the apparent improvement in receiver performance achieved using the dyadic subband transform, neither of these transformations techniques has shown the ability to significantly mitigate transient interference in a practical setting. In fact, since the number of subbands affected by the pulsed interferer increases directly with the pulse duty cycle, several subbands may be affected by the jammer and, hence, excision in the transform domain may be completely inappropriate. As a result, it may be necessary to utilize a transform domain technique related to the dyadic filter bank and DWT implementations which provides an optimal joint time-frequency representation of the received signal. In fact, it may even be necessary to suppress the transient interference directly in the time domain if the pulse duration is sufficiently short. Clearly, one must first determine the domain in which excision is most appropriate, if at all [Med97].

8.2.2 Adaptive Filter Bank Exciser

The approaches to interference suppression discussed so far utilize fixed filter bank structures and an exciser which simply replaces any transform bins which are considered to be primarily interference with zero. Although these systems are clearly effective at improving BER performance when interference is present, it seems likely that greater gains can be obtained by using processing systems that adaptively respond to the type of interference present at work to optimize some aspect of system performance. This section will discuss some of the approaches to making such an adaptive system and present some preliminary results obtained for one particular implementation [Med95].

There are a number of ways in which the filter bank structure can adapt to interference. One approach is to vary the tap weights of the filters in the system, with the goal of making the interference approximate an impulse function in the transform domain. This process is common in filter bank design and is formulated as the maximization of energy compaction. If taken from a transform point of view, the process is one of adapting the basis functions of the transform such that one basis function is essentially matched to the interference while other properties, such as orthonormality and perfect reconstruction, are maintained [Het95].

Another approach to adaptivity is to start following the full subband tree composed of two-channel para-unitary QMF filter banks, but instead of implementing the whole tree, make jammed/unjammed decisions at each level of the tree. When a signal at a particular level is found to be unjammed, that signal is not decomposed further but, rather, is passed directly to the synthesis section of the processor.

Conversely, any signal that is found to be jammed is decomposed further with the goal of isolating the interference to a smaller band. With this technique, the tree structure is determined by the interference and, by eliminating unnecessary decompositions, the computational load is reduced. Of course, the reduction in computational complexity is diminished somewhat by the requirement that jammed/ unjammed decisions must be made at each level of the tree rather than once at the point of excision. If the interference were a single sinusoid at zero frequency, the resulting tree would be the dyadic subband. A single sinusoid at a different frequency would produce a tree structure that is equivalent to the dyadic structure in complexity but which would provide the greatest resolution in the region of the interference rather that at low frequencies. Only in the case of strong interference spaced throughout the frequency band (e.g., multiple tone jammers) would the resulting tree structure approach the complexity of the full subband tree [Het95a].

A third approach to an adaptive system is to replace the simple two-level exciser with adaptable weights on each transform bin. Although the exciser itself is adaptive, it is constrained to either keeping or completely removing a transform bin (weighting by zero or one) and the excision decision is generally not based on a true optimization of a performance parameter such as output signal-to-interference ratio or BER. A better system could be formed by adaptively weighting each bin of the transform using continuously variable tap weights that are determined based on maximizing performance.

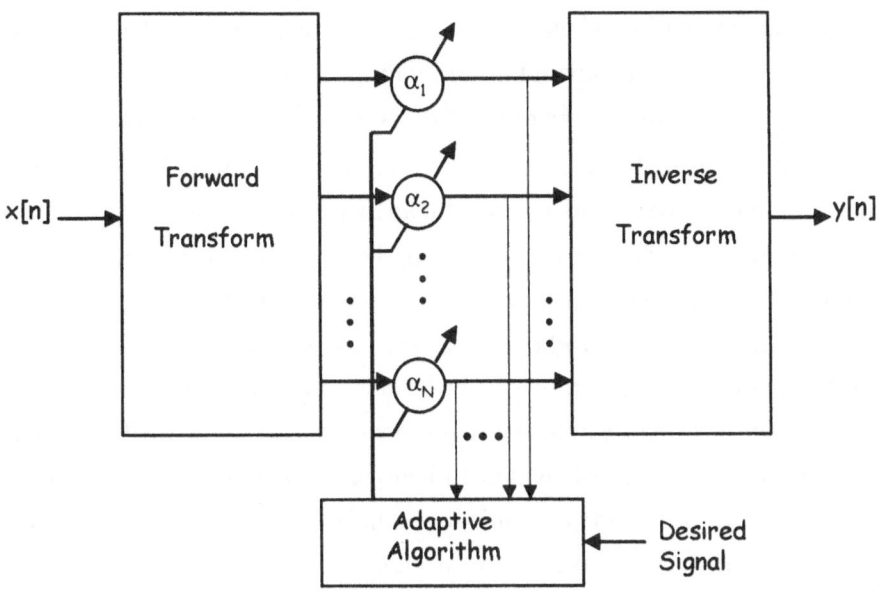

Figure 8.2.8 Transform domain adaptive filter.

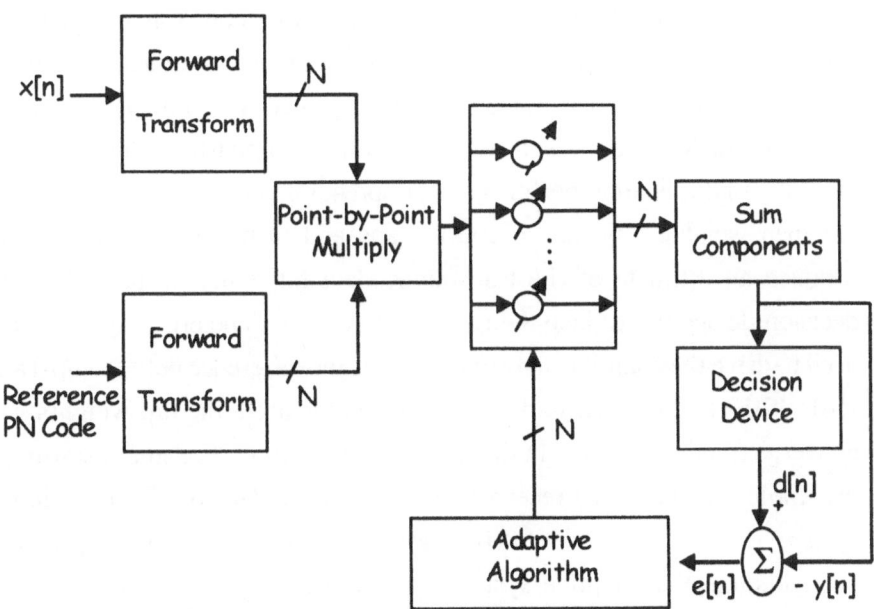

Figure 8.2.9 Adaptive transform domain DS receiver.

One possible structure used to perform this optimization is a transform domain adaptive filter, such as the one used shown in Figure 8.2.8. In this figure, the transform domain bins are multiplied by the adaptable weights, α_1, α_2,..., α_N, where N is the order of the transform. An adaptive algorithm works to minimize the difference between the weighted transform and some desired signal, generally using one of many well-known approaches such as the least-mean-square (LMS) algorithm, the recursive least-squares (RLS) algorithm, and their variations. Frequency domain adaptive filters incorporating these algorithms have been known for some time, and the extension of the wavelet transform domain is straightforward. Clearly, two keys to the performance of the system shown in Figure 8.2.8 are the adaptive algorithm and the desired signal. In many applications, the desired signal and the object of the adaptation is to minimize the difference between the weighted transform of the received signal $x[n]$ and this desired signal [Aka98].

For a DS spread spectrum receiver, it may be desirable to minimize the difference between the detected symbols and the output of the correlator. A system that does this and makes use of transform domain correlation is shown in Figure 8.2.9. In this system, as in the excision system depicted in Figure 8.2.5, no synthesis or reconstruction filter bank is required. The input signal and the reference PN spreading sequence are both processed by forward transforms and the results are multiplied point-by-point.

This multiplication performs the despreading operation. The products are then weighted by the adjustable taps and then summed in order to produce an estimate of the transmitted data bit which is passed to the decision device to produce the recovered symbol. The error signal that is used to drive the adaptive algorithm, $e[n]$, is the difference between $d[n]$ and $y[n]$. If the LMS or RLS adaptive algorithm is used, the tap weights will converge to minimize $e[n]$ in the mean squared sense. For initial startup, a preamble, known to the receiver, can be used in place of $y[n]$ to train the system, with a switch back to $y[n]$ taking place after convergence [Sau95].

Although the adaptive system of Figure 8.2.9 can produce a set of optimal weights, it has the drawback that it may not be able to react as quickly to changes in the input as the simple exciser. Most adaptive algorithms are either iterative, in which case they require time to converge to the optimal solution, or require information about the statistics of the input signal, which the receiver must estimate when the channel has unknown (and changing) properties. Therefore, if the interference is changing rapidly, the algorithm may not be able to track these changes and will be ineffective at suppressing the interference. Although some algorithms are better at tracking changing conditions than others, there is generally a trade-off between the tracking capability and the misadjustment noise, which is an indication of the variation of the weight values around the optimal values after convergence.

8.2.3   Transform-Based Low Probability of Intercept Receiver

The communications receivers discussed in the last two sections are receivers where intended parties communicate in the presence of intentional or unintentional interference or jamming. All the parties have access to the spreading or hopping sequence. In contrast, the intercept receiver is a "snooping" receiver that monitors the channel with the goal of detecting the presence of a signal. Once a signal is detected, it may direct a jammer to try to disrupt the communications channel or may work to extract information from the signal. Unlike the communications receiver, the intercept receiver does not have any a priori knowledge about the parameters of the transmitted signal (e.g., spreading sequence, chip rate, and data rate). Thus, intercept receivers simply try to detect the presence of a signal in the channel. The classical technique for detecting the presence of a DS signal is to use a total-power radiometer, which consists of a bandpass filter, a squaring device, and an integrator. The detected power of the received signals is compared to a threshold which is set using information about the receiver and channel noise. A decision on the presence of a DS signal is made when the measured detected power exceeds this threshold. The probability that the threshold is exceeded, given that a DS signal is present, is the probability of detection, P_d, whereas the probability that the threshold is exceeded in the absence of the signal is the probability of false alarm, P_{fa}. The performance of an intercept receiver can be characterized by plotting P_d versus P_{fa}, producing a receiver operating characteristic (ROC) for the detector [Het96a].

An intercept receiver can also be implemented in the transform domain. In this case, the total power in the transform domain is determined by squaring and summing the transform components. As in the time domain case, the result of the squaring and summing operations is compared to a threshold in order to make a decision on whether a signal is present or not. As for the case of the communications receiver, the performance of the intercept receiver can be improved by using excision. This has been studied in [Med95], where filter banks using Daubechies' coefficients have been used. The spread spectrum signal to be detected was a binary data bit. Each data bit is modulated using a 63-chip PN sequence and the resulting signal, sampled once per chip, is summed with AWGN and interference. In this case, performance results are presented as ROCs which provide P_d and P_{fa} for a given signal-to-noise ratio (SNR) and JSR, where SNR is defined as the ratio of signal power to AWGN. The threshold level decreases moving from left to right in the ROC, resulting in an increase in P_d and a corresponding decrease in P_{fa}.

The performance of an intercept receiver with excision using the full binary tree filter bank with 32 subbands is shown in Figure 8.2.10. For this figure, the ratio of the signal power to AWGN power is -7.0 dB. The narrow-band interference is a sinusoidal jammer with a normalized frequency of 0.0635 Hz and a JSR of 20 dB; excision is performed by removing the four or eight transform domain bins that have the largest magnitude. It is clear from Figure 8.2.10 that when no excision is performed, the probabilities of false alarm and detection are approximately equal, thus producing an unreliable and, in fact, "worst-case" determination of signal presence. Essentially, the probability that the threshold is passed is the same whether the signal is present or not because the jammer power is dominating the received signal. As frequency bins are excised, a more reliable indication of the presence of the DS signal is provided. The outermost curve represents the system's performance when no interference is present and serves as a benchmark for performance comparison.

It is important to keep in mind that in most transform domain excision applications, including this one, the performance benefits accompanying

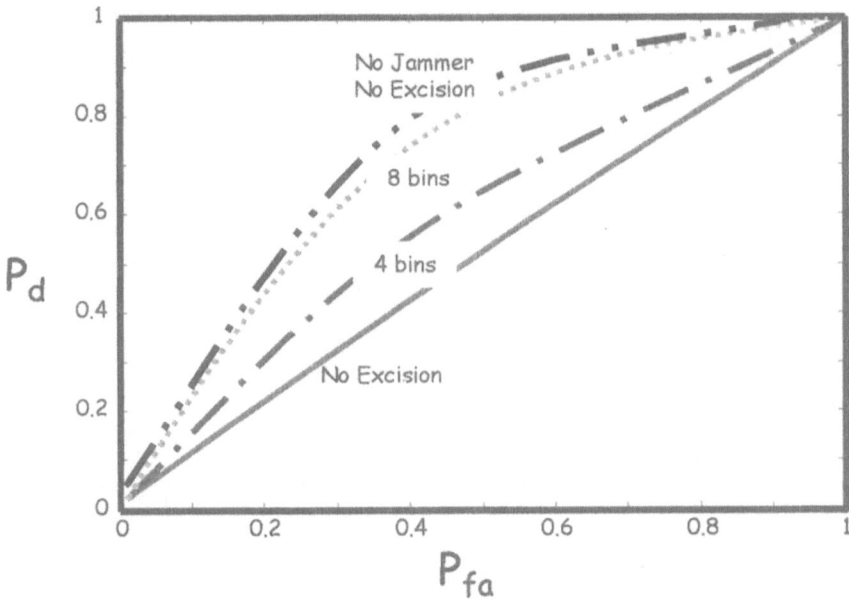

Figure 8.2.10 ROC using full subband decomposition and a varying number of bins excised in the presence of continuous sinusoidal interference.

excision are not without limit. In particular, if an excessive number of transform domain bins are removed, receiver performance will be significantly degraded since the excision of any bin removes both signal and interference energies. In fact, after a certain point, excising additional bins will remove more signal energy than interference energy, thereby worsening the overall system performance [Med95].

Figure 8.2.11 shows the receiver performance using a 32-subband transform with 8 subbands removed in the presence of a 20 dB sinusoidal jammer. For this system, normalized jammer frequencies of 0.0635, 0.127, and 0.254 Hz were tested. As expected, performance varied with jammer frequency due to aliasing effects within the filter bank and the characteristics of the spreading sequence. As with the communications receiver, aliasing and interference energy compaction within the subbands directly affect the subband transform's ability to isolate the undesired signal energy.

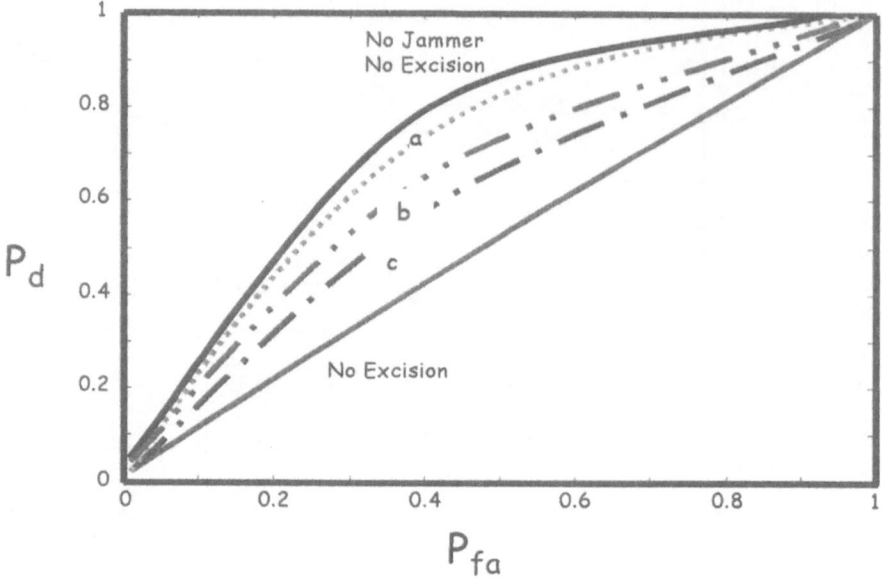

Figure 8.2.11 ROC of full subband decomposition and varying sinusoidal jammer frequency: $a = 0.0635$, $b = 0.1270$, $c = 0.2540$.

8.2.4 . Application of Multirate Filter Bank in Spreading Code Generation and Multiple Access

The most common spread spectrum block diagram in the simplified form is shown in Figure 8.2.12, where the spreading code is in general a *pn*-sequence, also called an *m*-sequence, where the *m* denotes the length of the code or the processing gain. The popularity of *m*-sequences is due to its ease of implementation using a shift register and very attractive noise like or randomness properties. For CDMA, one uses gold sequences, which has binary values and very low cross-correlation between codes. Most spreading sequences such as augmented *pn* codes and kasami sequences have also been considered.

The perfect reconstruction multirate filter bank discussed in Chapter 5 and shown in Figure 8.2.13 can also be used for spreading code generation. For the generation of the code, one uses the synthesis bank as shown in Figure 8.2.14. The unit vector contains only one nonzero element, and when applied to a particular subband, produces a unique waveform or spreading

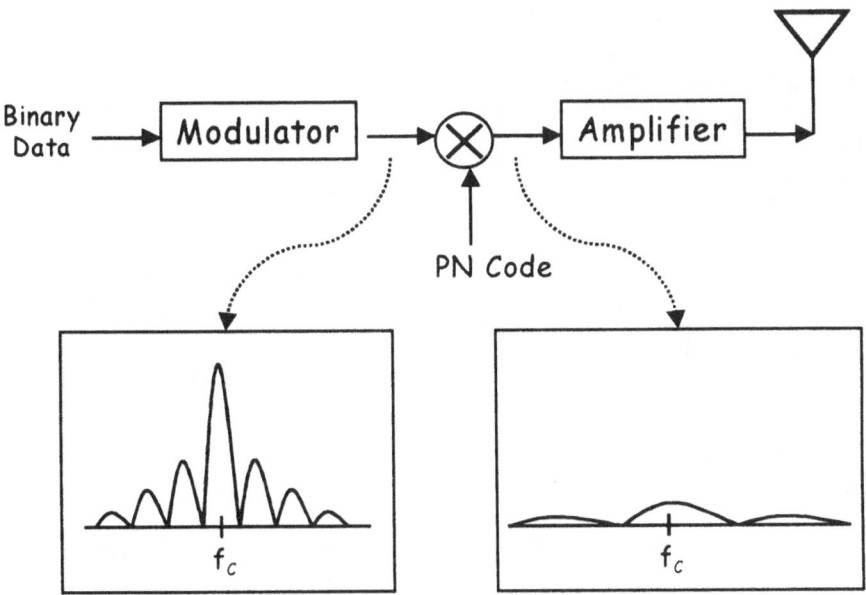

Figure 8.2.12 Block diagram of a simple DSS system.

sequence.   Note that the filterbank shown in Figure 8.2.13 can be implemented using a binary tree structure of M stages to generate 2^M spreading codes. The receiver structure is shown in Figure 8.2.15. The signal energy is isolated to a single band and the subband output is sent to a threshold device. Because the data bit polarity determines the polarity of spreading sequence, the demodulation can be easily performed. One

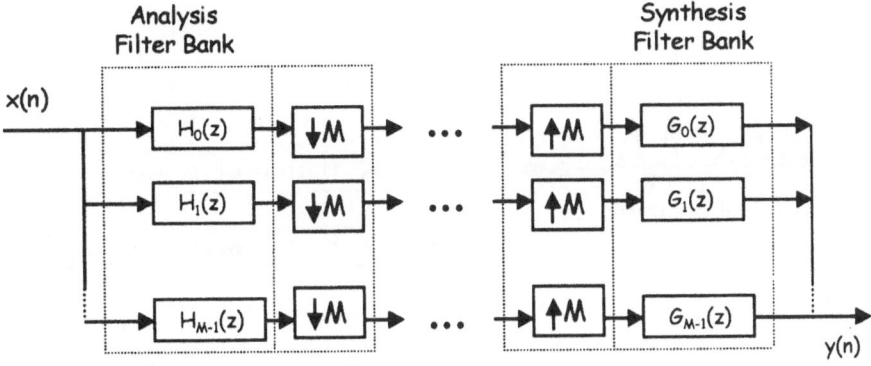

Figure 8.2.13 Block diagram of the perfect reconstruction multirate filter bank.

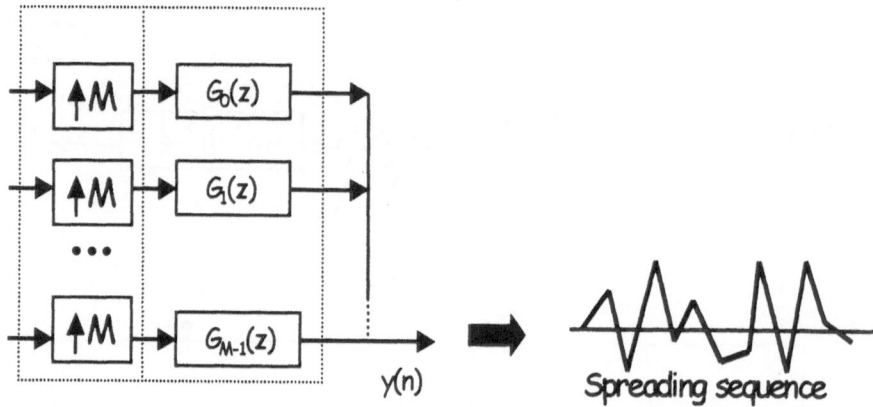

Figure 8.2.14 Block diagram of the filter bank structure used for the generation of the spreading code.

disadvantage of these spreading codes is that they are not binary valued. However, one has the freedom to choose the PR filter bank filter coefficients to optimize other properties such as improvement of transmission, reliability, and converters. For an example, consider the case for multipath interference mitigation. One can develop an objective function for this case to optimize the filter coefficient [Het95b].

Multirate filter bank-generated spreading codes can also be used for multiple access, as shown in Figure 8.2.16. Each user is assigned a particular subband or more subbands if robustness is necessary. The receiver uses the analysis filter banks, and if the synchronous system is used (i.e., no delay between codes), then the output signal appears only in the

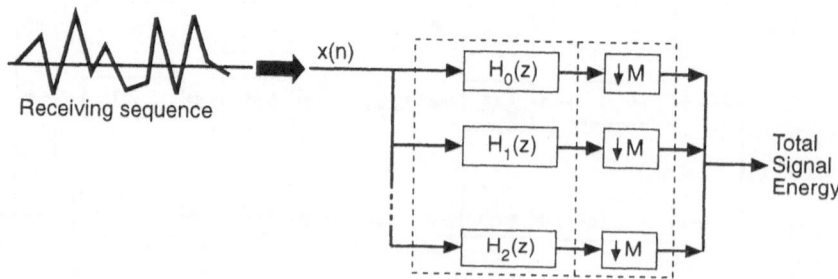

Figure 8.2.15 Block diagram of the receiving structure.

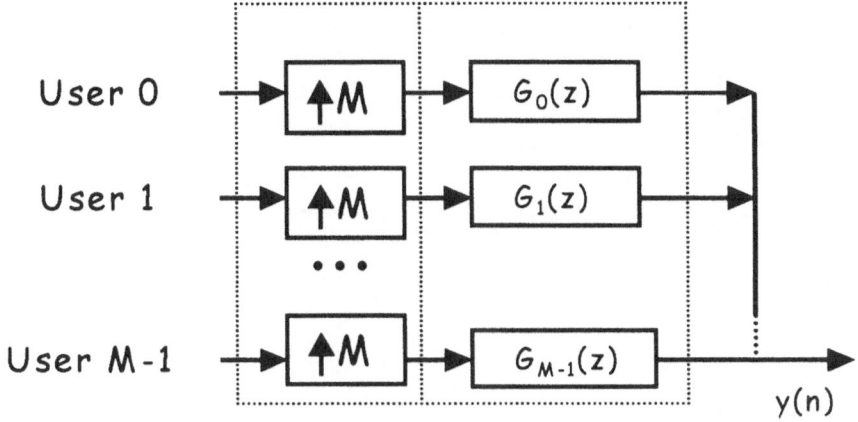

Figure 8.2.16 CDMA channel assignament using multirate filter banks.

particular subbands assigned. This, unfortunately, is not true for the asynchronous system, as it requires orthogonality for all possible delays, which is very difficult to achieve. However, if the delay spread is small, one can use the concept of constrained multivariable optimization to perform the filter design.

It has been shown that system BER performance in the presence of AWGN and no interference is identical to that achieved using binary phase-shift keying. Hence, in the absence of jammers and additional users, system performance using the wavelet-based spreading sequences is equivalent to that of traditional spread spectrum systems using PN and gold codes. Through simulation, additional BER results demonstrating the potential feasibility of applying the wavelet-based coding concept to asynchronous multiuser applications have also been generated. In these simulations, the AWGN channel and signal jamming were implemented at the base band and the spreading waveform energy was normalized such that it nearly matches the energy of a 63-chip DS system. The orthogonal wavelet basis was generated using four-tap FIR filters with wavelet coefficients derived by Daubechies. The full binary tree implementation of the filter bank was used with 5 stages of filters, resulting in 64 chips per message bit.

CDMA simulation results are shown in Figure 8.2.17. To generate these results, no restriction on the transmit start time was imposed on any of the additional users. In fact, each user on the channel was given a random start

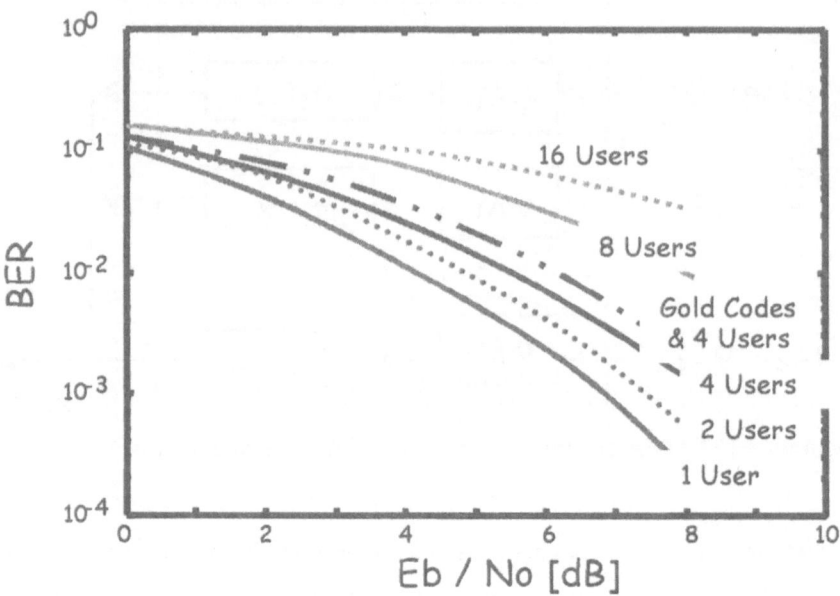

Figure 8.2.17 BER results for a CDMA system using subband spreading waveforms.

time which was uniformly distributed over the bit interval. Each data point in the figure is the average of the results from multiple simulations, each with different random start times. Obviously, if all of the users are synchronized, the orthogonality of the spreading codes ensures that the resulting BER will be equivalent to the theoretical rate achieved using binary phase-shift keying, BPSK, in an AWGN environment. For reference, the system performance when using 63-chip gold codes in a channel with 4 users is also shown in Figure 8.2.17. In such a case, the BER obtained using spreading codes based on Daubechies' four-tap FIR filters is slightly better than that achieved using gold codes.

Using anti jam waveform constraints, a new waveform design methodology that produces spreading codes which are particularly well suited to communications in the presence of various interferers has also been introduced. In fact, the BER simulation results shown in Figure 8.2.18 illustrate the potential use of such spreading sequences in a channel corrupted by narrow-band interference. To obtain these results, the subband filter bank basis functions were optimized with respect to an anti-jamming

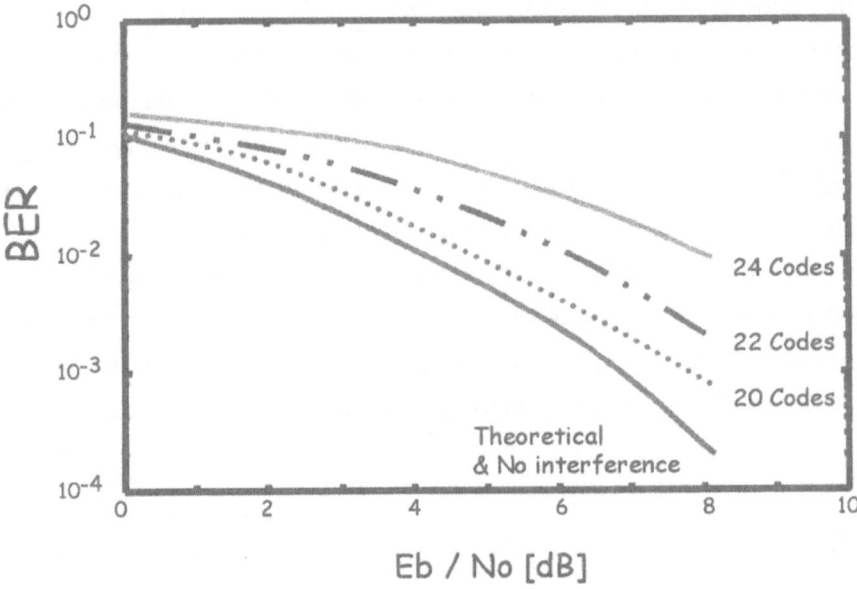

Figure 8.2.18 BER results obtained using a subband spreading function optimized for anti jam performance.

constraint to produce the maximum number of orthogonal spreading waveforms with minimal spectral content in the frequency range of interest, which, in this case, is 0.125 to 0.25 Hz, relative to a sampling frequency of unity.

More specifically, Figure 8.2.18 represents the system performance obtained in the presence of a 20-dB single-tone jammer at a normalized frequency of 0.21875 Hz. In accordance with the anti jam design constraint, four-tap FIR filters were designed and implemented in the hierarchical analysis and synthesis filter banks. Despite the existence of 32 possible spreading waveforms, which were obtained by extending the full binary subband tree structure to 5 levels, only 24 of the transform basis functions were sufficiently uncorrelated with the anticipated jammer; the remaining 8 waveforms are highly correlated with the interfering signal and, thus, not utilized. Figure 8.2.18 shows BER results obtained when the 20, 22, and 24 basis functions with the least correlation with the interference are used as the spreading sequence.

The anti jam constraint was used to produce filter bank spreading codes that allow multiple users to communicate in the presence of narrow-band interference.   In more general applications, similar constrains may be developed and used to derived an appropriate set of spreading functions that are well suited to a given application.   In fact, such constraints may be readily designed to yield covert LPI/D waveforms with spectral characteristics similar to those of the ambient noise.   Clearly, such a set of basis functions could easily generate a composite signal that would be very difficult to detect by conventional means.   In fact, using the wavelet transform, it has been shown that a variety of specialized waveforms may be produced that significantly improve communications covertness.

8.3   Modulation Using Filter Banks and Wavelets

Wavelet and subband transform yield orthogonal basis functions which can be used to modulate or spread data signals. Modulation can also performed including optimized filter banks, using an approach similar to the spreading code generation discussed in the last section. The optimization can be done using objective functions and constraints that improve overall performance.

One proposed modulation scheme optimized for additive Gaussian fractal noise channel has been discussed in [Wor92 and Wor96]. Fractal noise power spectra is proportional to $1/|f|^r$, where r can be between 0 and 2. Similar to the case of spreading code, the synthesis filter bank is used to modulate a block of pulse-amplitude-modulated data values $d_m(n)$. The composite transmitted waveform is given by

$$c(t) = \sum d_m(n) \, \psi_{m,n}(t)$$

$$(8.3.1)$$

where $\{\psi_{m,n}(t)\}$ is the set of the complete othonormal basis.

The original data are demodulated at the receiver using the corresponding analysis filter bank. Note that in the absence of any noise, perfect reconstruction of the transmitted is realized.

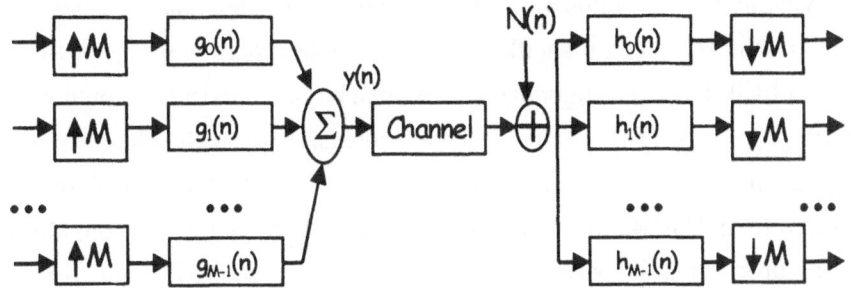

Figure 8.4.1 Basic block diagram of a multicarrier modulation-based transceiver.

8.4   Multitone Modulation

In multitone modulation, also referred to as multicarrier, the transmission channel is divided into a number of independent subchannels. This is done by utilizing a set of modulation line codes. The most common line codes are based on DFT. For this case, one can think of the channel as a combination of parallel brick wall subchannels for large number of tones. For this reason, this is also referred to as orthogonal frequency division multiplexing (OFDM) or Discrete Multitone (DMT) modulation. This modulation scheme is particularly useful when the channel power spectrum is distributed unevenly. The subchannels with better power levels are treated more favorably. Each subcarrier carries a different number of bits per symbol, commensurate with the subchannel attenuation. This modulation scheme is the backbone for the European T-DAB (Digital Audio Broadcasting) standard, and the ADSL (Asymmetric Digital Subscriber Line) communications standard which provides an efficient solution to the last mile problem (i.e., the connection to the home for high speed internet through the unshielded twisted pair copper cables) [Aka98].

The transmultiplexer previously discussed in Chapter 5 and shown in Figure 8.4.1 is the generic DMT communication system. The input discrete data stream is subdivided in M substreams and transmitted over its own orthogonal $g_k(t)$ subcarrier. Note that $g_k(t)$, with $k = 0$ to $M-1$ forms an orthogonal set. The data streams are modulated (QPSK for DAB) and the subsymbols $x_k(a)$ are obtained. The most common orthogonal basis sets are DFT based. However, it should be obvious to the reader that other wavelet or filter bank-based orthogonal basis sets can be generated which might have useful properties. This is a topic of very intensive research. Here, we only discuss one case which uses cosine-modulated filter banks and is generally referred to as DWMT modulation. In this multicarrier modulation, all the signal processing is performed on real signals. DWMT can also be considered as a combination of many carrierless single side-band modulation. The transmitted signals for DWMT are generated using a cosine-modulated synthesis filter bank. The kth subchannel response is given by

$$p_i(k) = 2v_i \cos[\pi(k+0.5)i/N + \phi_k]$$

$$\phi_k = (-1)^k \frac{\pi}{4} \tag{8.4.1}$$

where v_i is the prototype filter response with a baseband spectrum that satisfies the Nyquist criterion for transmission without any ISI (Inter Symbol Interference).

8.5   Noise Reduction in Audio and Images Using Wavelets

Detection and classification of signals in the presence of noise and interference is a critical issue in many areas of signal and image processing and analysis. A wide variety of applications involve signals with nonstationary or time-varying characteristics; examples include radar, sonar, communications, machine fault diagnostics, and biomedical and geophysical signal processing [Odg97]. The need for detection in such wide areas of applications has spurred a great deal of interest in time-frequency-based detection schemes [Abb97a, Str94].

If we define $y(t)$ the signal component we want to detect (information) and $n(t)$ the noise component, then the acquired signal $s(t)$ is given by

$$s(t) = y(t) + n(t) \tag{8.5.1}$$

Figure 8.5.1 Schematic representation of a noise reduction system using transform domain filtering.

In general, noise suppression methods in signal processing are based on representing the signal $s(t)$ in a manner that it is possible to separate the noise component from the information. A typical representation of the process of noise reduction is given in Figure 8.5.1. The filtering is performed in the transform domain, with the goal of the process to have $x(t)$ ~ $y(t)$, as defined in Equation 8.5.1.

Wiener filtering and, in general, Fourier domain smoothing algorithms have been used extensively as a tool to suppress noise, but recently methods based on the wavelet transform have become increasingly popular. The main advantage of wavelets lies in the additional "temporal" resolution of the transformed signal. In contrast to the Fourier transform, in the case of the DWT, the signal is decomposed into waves of finite length (i.e., into spatially localized waves). For this reason, whereas Fourier filtering affects all data points in the same manner, wavelets allow different parts of the signal frequency spectrum to be filtered individually. This, in principle, promises a considerably more refined and improved treatment of the problem. Seminal work in this subject has been developed by Donoho and Johnstone at Stanford University [Don92, Don93].

The Donoho-Johnstone approach begins with computing the discrete wavelet transform of the signal $s[n]$. The basic idea behind selective wavelet reconstruction is to choose a relatively small number of wavelet coefficients that contain most of the information. This is the fundamental assumption of the Donoho-Johnstone algorithm; virtually any signal s can be represented

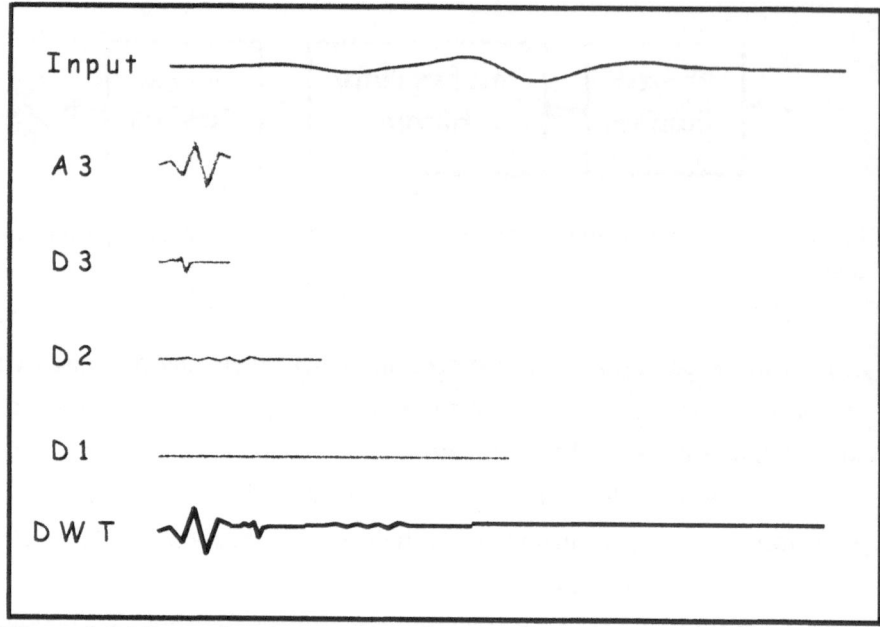

Figure 8.5.2 Input signal and wavelet coefficients A3, D3, D2, and D1 calculated using the Daubechies four-tap wavelet. The last signal on the bottom labeled DWT, represents a vector of size equal to the input signal and constructed concatenating sequentially [A3 D3 D2 D1].

by a small number of wavelet components at the various resolution levels of the wavelet decomposition, whereas the noise has components that cover the complete range.

For example, if we have a signal constituted of 64 discrete samples, a three-level wavelet decomposition will result in the decomposition of the signal into 4 components, D1, D2, D3, and A3 of length 32, 16, 8, and 8, respectively. An example is given in Figure 8.5.2. As it is clearly visible, higher decomposition numbers represent the output of successive stages of filtering and downsampling of the input signal. The last line in the figure is labeled DWT and represents the sequential concatenation of the previous subsampled signals. It is a representation similar to what is commonly adopted for images, and as in the two-dimensional case, the x axis no longer linearly corresponds to the original time axis.

For the case of a smooth signal, as the one in Figure 8.5.3, the higher-frequency details are very small. Hence, only a small number of wavelet

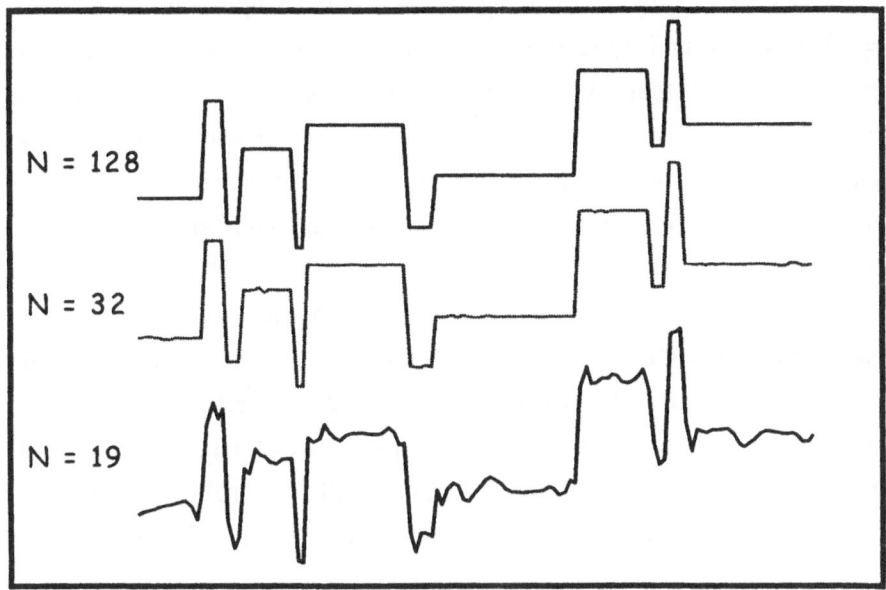

Figure 8.5.3 Reconstruction of blocks function [Don95] using different numbers of wavelet coefficients.

coefficients will be nonzero, as the wavelet basis are very well spatially localized.

Donoho and Johnstone [Don95] have pointed out that the wavelet coefficient W_{nm} is a measurement of the change in value of the signal in the time-frequency tile defined by the wavelet function $\Psi_{n,m}$, as extensively discussed in Chapters 3 and 4. If there is significant change, the correspondent wavelet coefficient W_{nm} will be large in magnitude, with a positive or negative sign. If not, W_{nm} will be small. As the dilation coefficient m increases, the time duration 2^{-m} decreases, resulting in a smaller number of coefficients affected by any transition. This is the reason why the wavelet transform is better suited to detect sharp transitions.

This concept of sparse wavelet representation of a smooth signal is extensively exploited both in noise reduction and signal compression algorithms based on wavelet transform domain processing [Don95]. For example, let us look at the signals in Figure 8.5.3, reconstructed using a different number of wavelet coefficients. The original signal was based on a signal described in [Don95] and available in the Matlab® Wavelet Toolbox [Mat96]. The reconstruction was performed using all the wavelet coefficients ($N = 128$), or by using the ones with the highest magnitude, ($N = 32$, $N = 19$), and by setting all the others to zero. What is clear from Figure 8.5.3 is that by maintaining about one-fourth of the wavelet coefficients, it is possible to reconstruct the signal with minimal error. Hence, the information is mostly contained in few coefficients, about 32 in our example, and the rest tend to be very small and thus more susceptible to noise.

By choosing a proper wavelet, for example, one that well correlates with the signal we want to detect, we can obtain large wavelet coefficient values when there is signal information, and much smaller coefficients when there is mostly noise. This is the fundamental principle behind the Donoho and Johnstone idea of wavelet thresholding, in which the signal reconstruction is achieved using only wavelet coefficients whose magnitudes are above a specified threshold value λ.

Using the threshold function $\delta_\lambda(x)$, we can thus write the reconstruction formula Equation 4.8.1:

$$x[k] = \sum_{m=0}^{M} \sum_{n=0}^{2^m-1} \delta_\lambda(W_{nm}) \cdot \psi_{n,m}[k] \quad , k = 1..K \qquad (8.5.2)$$

where K is the length of the original signal.

There are two possible schemes of thresholding:

- *Hard thresholding*

$$\delta_\lambda^H(x) = \begin{cases} x, & |x| > \lambda \\ 0, & \text{otherwise} \end{cases} \qquad (8.5.3)$$

- *Soft thresholding*

$$\delta_\lambda^s(x) = \begin{cases} x - \lambda, & x > \lambda \\ 0, & |x| < \lambda \\ x + \lambda, & x < -\lambda \end{cases} \qquad (8.5.4)$$

The soft threshold results in a reduction of the wavelet coefficient of the value λ, and for this reason, it is also known as wavelet shrinkage. Both operations are plotted in Figure 8.5.4.

The noise reduction algorithm proposed by Donoho and Johnstone is represented in Figure 8.5.5. After the wavelet transformation, the noise contribution is estimated and a proper threshold is chosen. The remaining wavelet coefficients are then used to reconstruct the original information signal $y(t)$ in Equation 8.5.1.

Regardless of the thresholding scheme chosen, the value λ used is critical. This can be verified in Figure 8.5.6, where white noise was added to the original signal. The noisy signal is also equivalent to the output of the system in Figure 8.5.5 when the threshold λ is set to zero. The magnitude of the wavelet coefficients of the noisy signal is plotted in Figure 8.5.7. The different thresholds are also plotted for comparison.

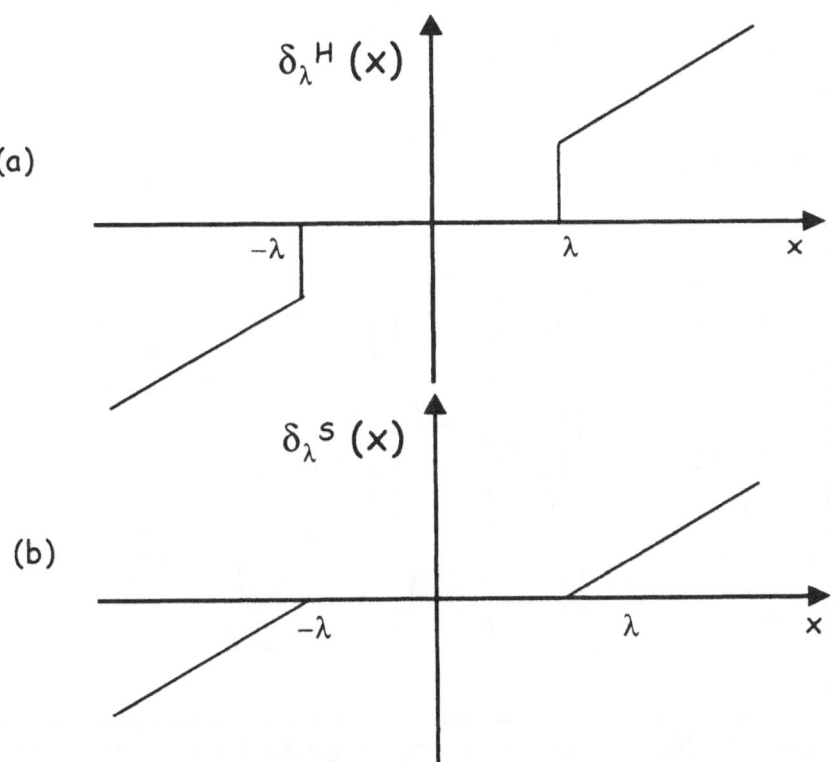

Figure 8.5.4 (a) Hard and (b) Soft threshold functions.

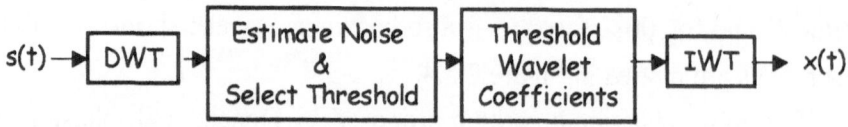

Figure 8.5.5 Block diagram of the wavelet noise reduction algorithm.

By choosing a very large value of λ, the number of coefficients that are utilized in the reconstruction is very small, $N = 13$ in the example. Since most of the higher-frequency terms are smaller in magnitude, this will result in a over smoothing of the signal, as seen in Figure 8.5.6 where one of the transitions is not properly reconstructed. On the other hand, if the threshold

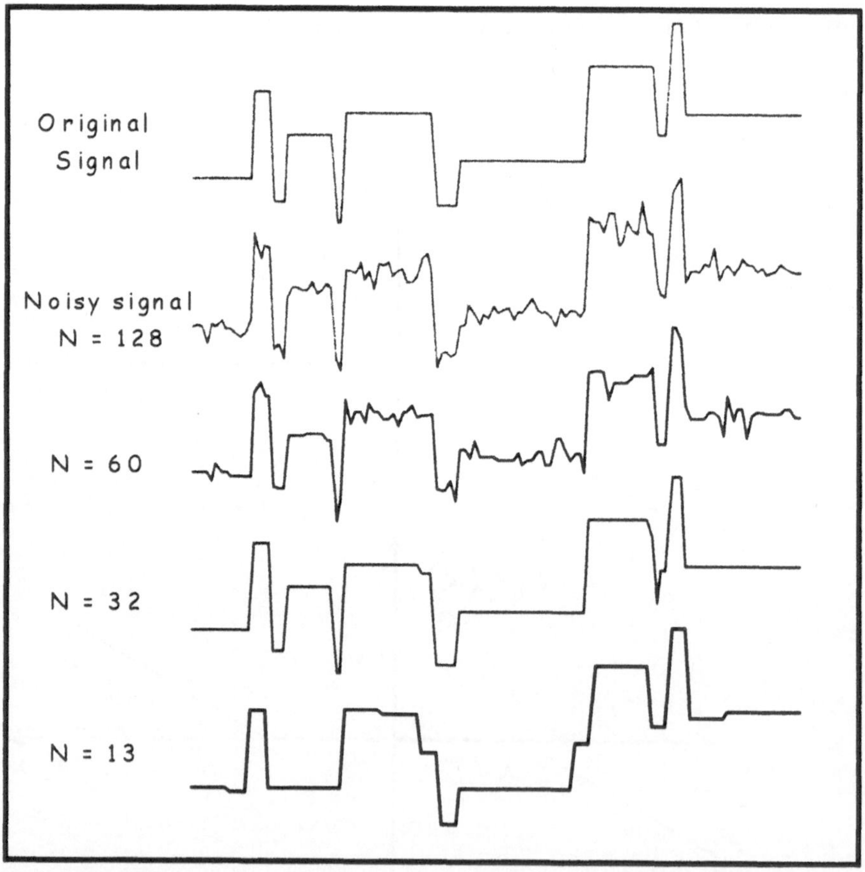

Figure 8.5.6 Noise was added to the original signal and different threshold levels were used to obtain the signals in the figure. N represents the number of coefficients not equal to zero after the thresholding operation.

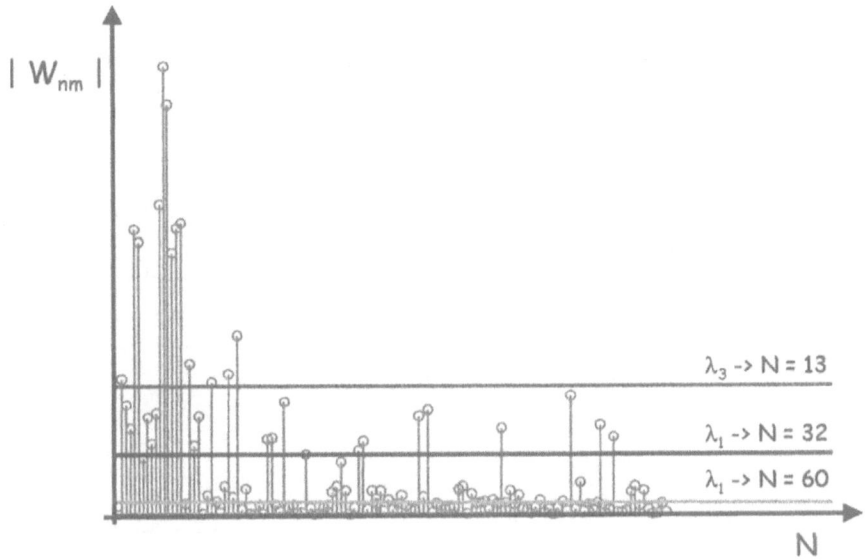

Figure 8.5.7 Magnitude of the wavelet coefficients of the noisy signal in Figure 8.5.6. The threshold values used to calculate the signals in Figure 8.5.6 are also plotted as straight lines, and labeled accordingly.

is set too low, as in the case of λ_1, then part of the noise is maintained and thus reconstructed at the output. This is the case of λ_1 with $N = 60$. For the example in Figure 8.5.6, the threshold λ_2 seems to be the optimal solution, as it achieves a good noise rejection performance without any major loss in spatial resolution.

There are various methods to choose the threshold λ, but they are moslty grouped as follows:

- Global threshold, in which the same value of λ is used across all decomposition levels m of the DWT. This is the case of the results shown in Figure 8.5.6.
- Level-dependent threshold, in which λ is a function of the decomposition level m.

Furthermore, in the case of global thresholding, it is typical to threshold only the coefficients at the higher resolution, as it expected that the noise coefficients have a higher contribution at the higher-resolution levels. This assumption is verified in Figure 8.5.8, where the magnitude of the wavelet

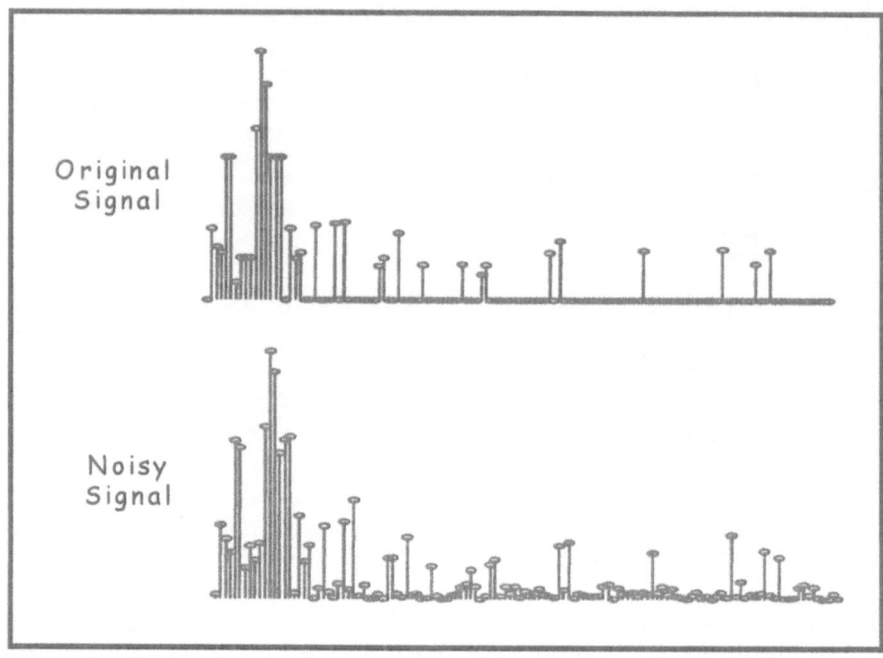

Figure 8.5.8 Comparison of the magnitudes of the wavelet coefficients of the original signal prior and after adding the noise.

coefficients of the signal with and without noise are plotted. This case clearly satisfies the fundamental assumptions of Donoho and Johnstone's approach: Most of the wavelet coefficients containing the signal information are in the lower resolution range, whereas the noise contribution to the wavelet coefficients is more spread out throughout the axis. Furthermore, by choosing a proper wavelet function, the wavelet coefficients due to the signal have a larger magnitude than the correspondent noise coefficients. This is critical, as it justifies the utilization of the nonlinear process of thresholding.

Hence, if we define J as the value of the scaling coefficient m below which most of the wavelet coefficients are strongly correlated to the signal, we can rewrite Equation 8.5.2 as follows:

$$x[k] = \sum_{m=0}^{J} \sum_{n=0}^{2^m-1} W_{nm} \cdot \psi_{nm}[k] + \sum_{m=J+1}^{M} \sum_{n=0}^{2^m-1} \delta_\lambda(W_{nm}) \cdot \psi_{nm}[k]$$

(8.5.5)

$$x[k] = a_J[k] + \sum_{m=J+1}^{M} \sum_{n=0}^{2^m-1} \delta_\lambda(W_{nm}) \cdot \psi_{nm}[k]$$

where $a_J[k]$ represents the approximation of the signal $s[k]$ obtained only by using the lower-resolution wavelet coefficients W_{nm}, with $m < J$. More details can be found in Section 4.13 of Chapter 4.

Using Equation 8.5.5, we can easily see the major difference between the discussed wavelet algorithm and more conventional low-pass smoothing algorithms. In the wavelet method, the high-frequency contribution due to the signal transients is retained by using the wavelet coefficients at higher resolution with the largest magnitude. In the low-pass filtering, all of the high-frequency information is eliminated indiscriminately.

The question is thus, How is the value of the threshold λ to be determined? The proper choice of the threshold must take into account at least two factors: The sample size K and the noise level σ^2.

Donoho and Johnstone have proposed a method to determine λ using an L^2 risk function. The resultant method is called *minimax thresholding* [Don94].

An empirical method to calculate the threshold was also presented by the authors based on the estimation of the standard deviation of the noise σ. This is the so-called *Visu-Shrink* method, in which

$$\lambda = \sqrt{\frac{2\log(K)}{K}} \sigma$$

(8.5.6)

where K is the length of the signal. As the visu-shrink rule results in a substantially larger value of λ, compared with the *minimax* calculation, it commonly results in a smoother estimate of the signal, whereas the minimax method does a better job at detecting sharp transitions.

We now need to develop a method to estimate the noise level σ. Assuming that the highest level of the wavelet coefficients consists mostly of the noise contribution, they can be used to estimate the noise level. In the absence of noise, these coefficients will be mostly zero; hence, we can utilize the median function to estimate σ:

$$\sigma = \frac{\text{median}(|W_{nm}|)}{0.6745} \tag{8.5.7}$$

In level-dependent thresholding, the wavelet coefficients at each level are utilized to estimate a threshold λ_m used to shrink the wavelet coefficients at that level. The estimation of λ_m is based on an unbiased estimate of risk called Stein's Unbiased Risk Estimate (SURE), which is currently the most popular data-dependent threshold selection procedure. The details of this method can be found in [Don95].

8.6   Audio/Video/Image Compression

One of the more widely publicized technical debates involves the decisions currently being reached about emerging standards for video compression. The achievement of high compression ratios with competitive technology is one of the key factors in the implementation of high-definition television (HDTV) systems; the practical and financial consequences are very great. The European telecommunications industry invested nearly a billion dollars in HDTV technology before deciding to follow the emerging North American standard. A number of possible video compression schemes have been considered; the Federal Communications Commission (FCC) has organized competition among several possible HDTV compression standards, and has urged cooperation among the leading contenders. In 1991, the Motion Pictures Experts Group (MPEG) held a competition in Japan among 30 companies with potential compression schemes; the best were judged to be windowed Fourier transforms, although several wavelet-based entries were high on the list as well. Although the current MPEG standard involves

Fourier-based analysis, the final decision has not yet been made; recent hardware developments, including those described in this chapter, and newer wavelet bases have not yet been evaluated, and the debate over the best compression standard continues [Woo91].

The main technical problem in video compression is motion estimation, or keeping the image updated in real time without retransmission of the entire image. Consider a video image to consist of a function $f(x,y,t)$ with x,y axes on the screen and the image changing with time. The function probably changes gradually; if we treat $f(x,y,t)$ as a sequence of still images to be compressed independently, the result is not very efficient. However, the direction of movement or change with t is unpredictable, and it is also inefficient to spend too much effort on extrapolation from a known image to the next likely position. A possible compromise is to encode every fifth or tenth image, and between them to work with the time differences between images - which contain less information and can therefore be compressed further. Starting from an image in which each pixel color is assigned a numerical shading between 0 and 255 ($256 = 2e8$), we have 8 bits for each color of red-green-blue. The goal is to compress these data and reduce transmission cost while increasing picture quality; the transmission bit rate is fixed by the channel capacity. Higher bit rates are required for improved resolution; typical compression ratios are about 100:1, with HDTV transmitted nominally at 24 Mbit/s and an extra 0.5 Mbit/s for each of four stereo audio channels. For the critical motion estimation step, a motion vector is computed for each region of the image, and the system transmits only the difference between this predicted image and the actual image (the motion-compensated residual). When the residual has too much energy (such as when there is a sudden scene change), the motion estimator is disabled and the most recent image is transmitted. Coding decisions are therefore based on the energy in different bands or on the size of the Fourier coefficients. Image quality, however, is still subjective; experts viewing standard images can judge acceptable differences that may not be apparent to the casual viewer. Thus, the final choice for HDTV may still differ from the MPEG standard and send a rougher picture at a lower bit rate. It has been shown, for example, that images reconstructed from Fourier and wavelet transforms after discarding 95 % of the coefficients in each case yield recognizable images, but the wavelet image is typically preferred [Str94].

Another potential application is compression of audio signals. An audio signal has fewer dimensions and as many as 512 frequency bands, divided into octaves of roughly equal energy. An active area of audio research involves taking advantage of psychoacoustic information about the human ear; the cochlea appears to have several critical bands per octave, which are key to a good reproduction. One goal of compression is the creation of a smaller CD disk; subband coding seems well suited to this application. Since music is basically sinusoidal and images have sharp edges, it would appear that wavelets are better for video compression and Fourier for audio; however, once again, a final decision has not yet been reached.

The most successful commercial application of wavelets to date has been digitizing of fingerprint information for the FBI. Over 30 million fingerprints are currently on file, with more being added daily. Comparing one to thousands of others is a daunting task; every improvement in matching technology leads to new matches and the solution of old crimes. Clearly, it is more efficient to digitize this database; new prints could then be electronically compared with those already on file. The key feature for matching fingerprints is the minute details of ridge endings and bifurcations. At 500 pixels per inch with 256 gray levels, each fingerprint card has about 10e7 bits of data; some form of compression is required [Str89].

The standard from the Joint Photographic Experts Group (JPEG) is Fourier based and unable to provide the 20:1 compression ratios required for this application; lines were broken within the image and impossible to match. The successful design uses wavelet compression techniques developed in association with Los Alamos National Laboratory and Yale University [Wic94]. This is probably only the first application where the use of wavelets provides performance not otherwise obtainable with conventional Fourier techniques.

As an example, we will now discuss a wavelet transform image decomposition technique utilized for compression and analysis of the ultrasonic images obtained by a Computerized Ultrasonic Gauging System (CUGS) [Abb96]. CUGS can generate very precise topographical maps of the outer and inner surfaces of tubes during various stages of manufacture and lifecycle of the parts.   Measurements of the tube dimensions are obtained with a resolution of 2.5 µm (0.0001 in.) and accuracies of the order of 10 µm (0.0004 in.) or better. A typical output of CUGS is an ultrasonic image, in which the horizontal and vertical axes represent the axial and angular positions of the part, respectively. Wavelet-based image analysis has been utilized to obtain representations of the same image with different resolutions and to enhance image features such as erosion pattern without the loss of localization [Abb96a].

A common problem of imaging systems is the storage and handling of a large quantity of data acquired. CUGS is currently utilized to gauge tubes as long as 7.5 m (25 ft) resulting in the acquisition of more than 10,000,000 data points, which require up to 40 Mbytes of disk memory on a computer. A wavelet transform image decomposition technique was thus developed for compression and processing of the ultrasonic images, resulting in a great decrease in time and resources needed to perform such operations [Abb96]. Wavelet-based image analysis has two distinct characteristics: multiresolution and high spatial localization [Jia96]. Multiresolution refers to the possibility of obtaining representations of the same image with different resolutions [DeC97b].

Multiresolution techniques have been used in computer vision for tasks such as object recognition and motion estimation as well as in image compression, with pyramid and subband coding. An important feature of such image compression techniques is their successive-approximation property: As more details (i.e., higher spatial frequency components), are added, higher resolution images are obtained. Furthermore, the high spatial localization properties of the filters used for the wavelet decomposition can also be utilized for the enhancement of features such as erosion pattern without the loss of localization, a problem commonly encountered in Fourier analysis. In short, besides good compression capabilities, the wavelet decomposition allows partial decoding of the images, which leads to subresolution approximations of the original image, which can be used to separate and enhance different features of the part. As the image is displayed at various resolutions, it is possible to enhance or diminish contributions

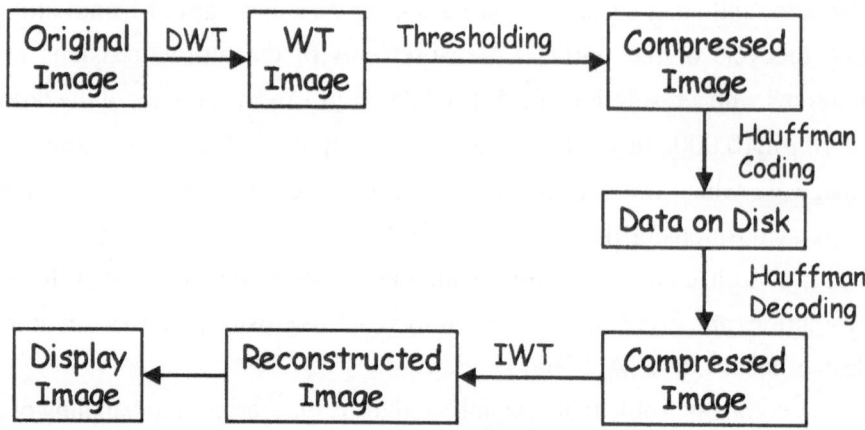

Figure 8.6.1 Block diagram of the wavelet-based compression scheme utilized for the ultrasonic imaging application described in the text.

from different physical atoms within the image. At a lower resolution, it is possible to determine properties of larger structures such as erosion patterns or gouges, whereas at finer resolutions, it is possible to enhance details such as pits or ridges.

The approach discussed here is utilized in many applications, including document and medical imaging, remote sensing, teleconferencing, and facsimile and data transmission [DeC97a].

The principle of the so-called *multiresolution visualization* is to utilize the least amount of data necessary to generate an usable approximation of the original data. The data are analyzed at different scales of resolution in which the higher scales present more details than the lower scales. This representation is thus used to reduce the size of the data to be visualized and thus the associated computational load. For example, if it is necessary to look at the tube in all its length, which is usually of the order of 5-10 m, it will not be useful to display features of the order of 1 mm. As the user requires further refinements (i.e., zooms into a specified area of the tube), more data are loaded into memory and a higher-resolution image is reconstructed and displayed. This is an equivalent process to the one performed by a microscope, in which the resolution is changed by looking

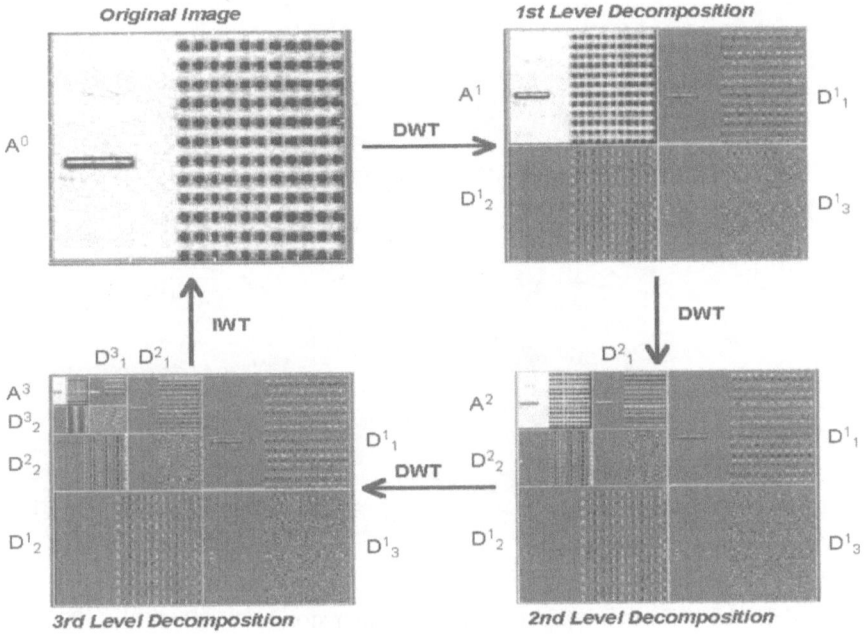

Figure 8.6.2 DWT of an ultrasonic image obtained using CUGS.

at different scales of the image. This concept is very well understood and utilized by cartographers when designing and reproducing maps.

The block diagram of this decomposition scheme and the subsequent reconstruction is shown in Figure 8.6.1, where the arrow labeled DWT represents the wavelet transform filter bank and the arrow labeled IWT represents its inverse operation.

The original ultrasonic image is processed using a discrete wavelet transform composed of a sequence of three filter banks. The output is an image of size $N{\times}N$ which has identical information content of the original image, nothing is lost, and perfect reconstruction is achieved. The reconstruction error is only due to the numerical precision of the filters utilized: in our implementation using MATLAB on a Silicon Graphics Indigo[2] Workstation, the maximum error of reconstruction is of $8{\times}10^{-11}$ in. An example of such decomposition, with its intermediate results, is given in Figure 8.6.2.

From wavelet theory, it can be shown that it is not necessary to retain

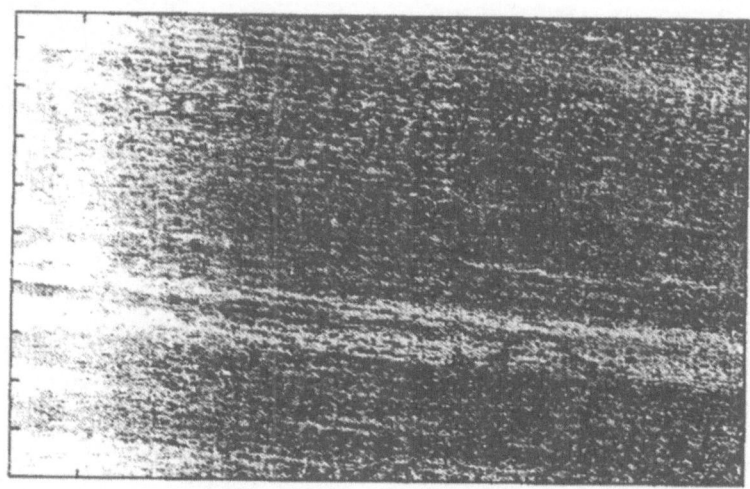

Figure 8.6.3 Ultrasonic image of the inside dimensions of a rifled tube. The white streaks show areas of reduced thickness due to erosion.

all of the $N{\times}N$ coefficients of the wavelet-transformed images. By thresholding the wavelet image, the number of coefficients needed to visualize the data is reduced, resulting in a reduction of the memory space necessary to store the ultrasonic images. The output image can be thresholded to reduce the amount of data necessary to represent the part. To retrieve the image, the inverse wavelet transform is performed to the resolution needed.

Other benefits of thresholding are the reduction of noise in the image as we have discussed in Section 8.5, and the enhancement of particular features. Wavelet transform has been utilized for feature enhancement and extraction, with promising results in many applications [Mal96, Sch96a].

The basic approach of image feature enhancement is given by these simple steps:

- Image decomposition with the forward wavelet transform
- Linear or non-linear processing of the wavelet coefficients
- Image reconstruction with the inverse wavelet transform.

Many tubes mapped using CUGS are internally rifled; that is they exhibit a periodic variation in the internal radius image. An image of one

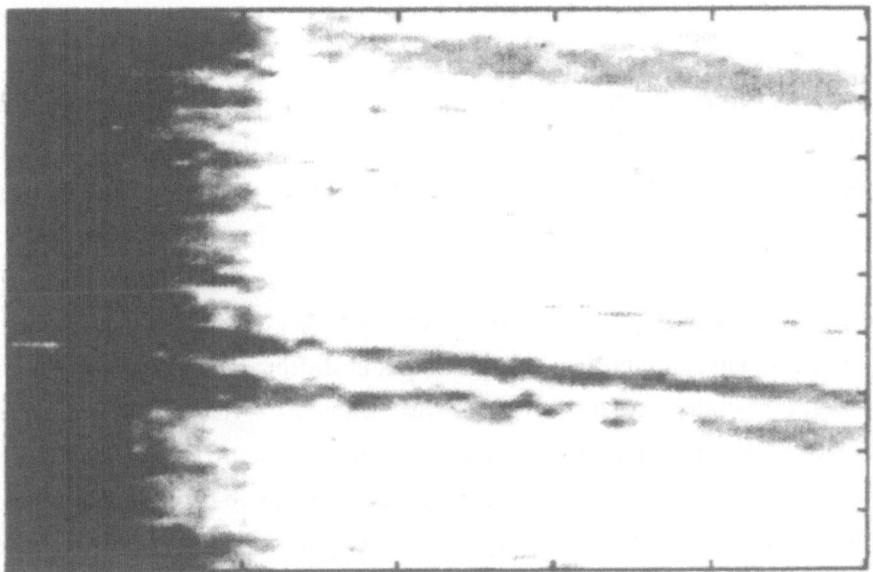

Figure 8.6.4 Ultrasonic image after wavelet processing, with the enhancement of the erosion pattern after removal of the higher-frequency components due to internal rifling.

of these tubes is given in Figure 8.6.3. In this case, an erosion pattern is visible by following the two white lines which transverse obliquely across the image. Wavelet analysis can be used to separate the different contributions of rifling and erosion from the image. By using the wavelet transform, the information regarding the rifling is mainly contained in the higher-resolution detail images D_2^1, D_2^2 and D_2^3, where the information of the erosion is mainly contained in A^3 and D_3^1, D_3^2, and D_3^3. By selectively setting to zero all the coefficients in the sub-images which mainly contain information about the rifling and by inverse transforming, we will obtain an image which contains only the erosion pattern, as shown in Figure 8.6.4. In this image, the variation in dimensions due to rifling is completely lost; in fact, the difference between this image and the one in Figure 8.6.3 represents the original rifling dimensions before erosion occurred.

8.7   Progressive Pattern Recognition

Wavelet-based systems have been used for many applications; for example, in the field of image processing wavelets have been applied to edge enhancement, feature extraction, multiresolution analysis, and target identification in the presence of noise or interference [Bar92, DeC96, Vet92]. The advantages which a wavelet approach offers over conventional Fourier analysis are related to the time-space localization properties of the daughter wavelets, which are produced by scaling and shifting the mother wavelet in order to vary the resolution of the wavelet transform. Recently, a scale and rotation invariant pattern recognition scheme was proposed, based on wavelet features of an image [DeC96a]. Because the two-dimensional nature of this problem lends itself to optical implementation, an extension of the joint transform correlator to a wavelet image processor (WIP) was also proposed [DeC96a, DeC97].

This approach takes advantage of the relationship between classic Huygens-Fresnel diffraction and wavelet transforms, since the free-space impulse response of the diffraction integral can be viewed as a two-dimensional wavelet kernel (although it violates the admissability condition). The WIP has been applied to various practical problems, including nondestructive testing based on ultrasound images and biomedical applications [Abb96, Abb96a]. Although this technique has been demonstrated to be very powerful at recognizing images distorted by white noise, it remains a very computationally intensive approach to the pattern recognition problem because it involves computing the cross-correlation between a target image and its wavelet transform.

A new technique has been proposed recently for dramatically reducing the computational time and improving the performance of this type of system. This approach is known as *progressive pattern recognition*, since it is based on correlating the target image with lower-resolution subbands of the wavelet transform, rather than with the complete wavelet transform of the image. The correlation improves as the resolution of the subband transform is increased; in this way, it is possible to identify the target image with sufficient detail using much lower resolution than the complete wavelet

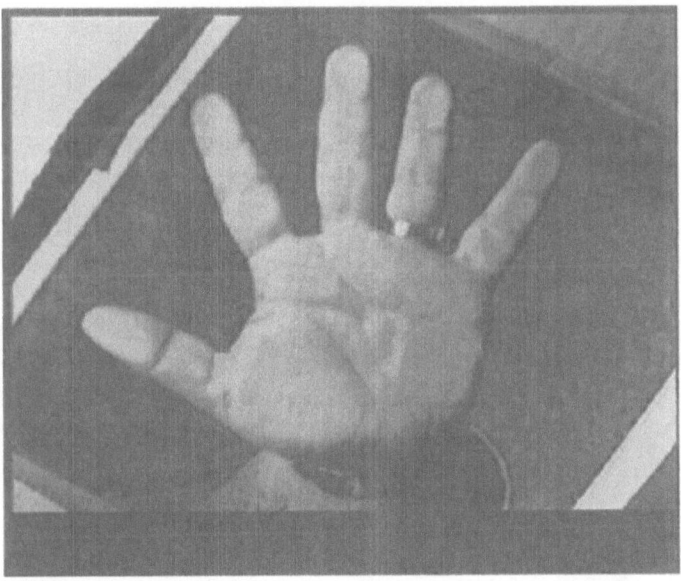

Figure 8.7.1 Handprint image used in our discussion.

transform. Conceptually, the approach is analogous to the way in which the human vision system perceives the world; for example, given a picture showing a bowl of fruit, it is very easy to tell the difference between an apple and an orange (or whether there is even an apple present in the scene), whereas further inspection might be required to tell what variety of apple is shown, Macintosh or Courtland. Similarly, the algorithm for progressive pattern recognition first computes the correlation between a desired image and a low-resolution subband of the wavelet transform; this result might be sufficient to determine whether the desired image is present, but not to provide any further details. Progressive correlation with the next-higher-resolution subbands will yield more detailed information about the image; it is often possible to obtain the desired result with significantly less computations than if the entire wavelet transform was used as a reference image.

The successive approximations of the hand image given in Figure 8.7.1 are plotted in Figure 8.7.2, where the quantity level refers to the size of the

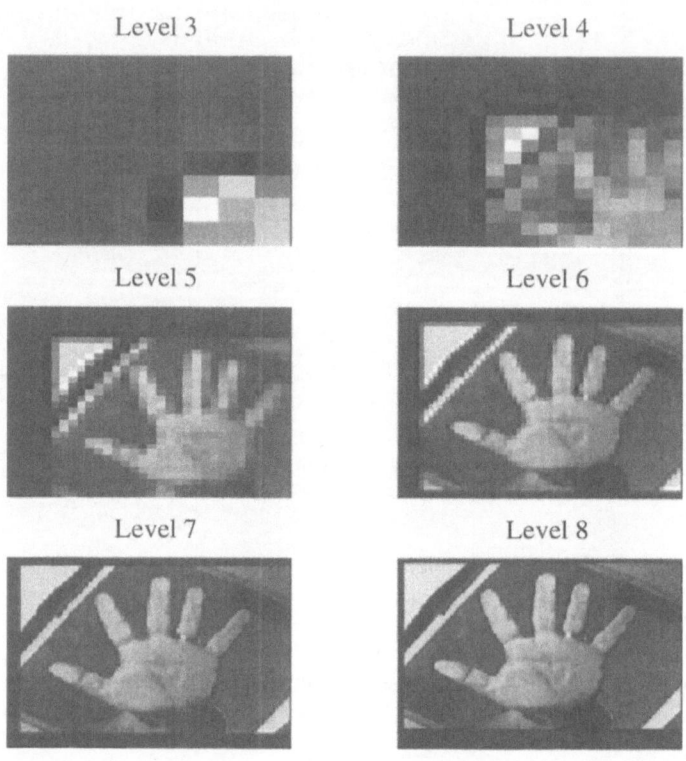

Figure 8.7.2 Successive levels of approximation of the hand image of Figure 8.7.1.

approximation image. For example, level 3 means a $2^3 \times 2^3 = 8 \times 8$ image. For reference, the original image is 512×512 pixels.

As it can be seen, the level-5 approximation already has enough detail to properly identify a left or right hand. Hence, we can utilize each approximation level to search through the library of possible patterns and eliminate possibilities, so that as we increase the resolution, we will have to search through a reduced library set. For this reason, this approach is referred to as the progressive pattern recognition scheme, as the library of possible patterns is progressively reduced at each level so that the highest-resolution comparison can be made to a much reduced set.

8.8   References

[Abb96]   A. Abbate, J. Frankel, R.W. Reed, and P. Das, Ultrasonic gauging and wavelet image processing for wear and erosion mapping, *Proc. 1996 QNDE*, vol. 18, 1996.

[Abb96a]   A. Abbate, J. Frankel, R.W. Reed, and P. Das, Wavelet image processing for wear and erosion mapping using an ultrasonic gauging technique, *Proc.1996 Ultrasonic Symposium*, pp. 705-707, 1996.

[Abb97a]   A. Abbate, J. Frankel, and P. Das, Application of wavelet image processing for ultrasonic gaging, *Proc. 1997 SPIE Conference on Wavelets*, 1997.

[Aka94]   A. N. Akansu and R.A. Haddad, *Multiresolution Signal Decomposition Transforms, Subbands, Wavelets*, Academic Press, New York, 1994.

[Aka98]   A.N. Akansu, P. Duhamel, X. Lin, and M. de Courville, Orthogonal transmultiplexers in communication: A review, *IEEE Trans. Signal Process.*, vol. 46, pp. 979-995, 1998.

[Aka99]   A.N. Akansu and M.J. Medley eds., *Wavelet, Subband and Block Transforms in Communications and Multimedia*, Kluwer, Boston, 1999.

[Bar92]   A. Baraniecki and S. Karim, Computational algorithms for discrete wavelet transforms, *Proc. SPIE*. vol. 1699, pp. 408-419, 1992.

[Chu92]   C.K. Chui, *An Introduction to Wavelets*, Academic Press, New York,1992.

[Das91]   P. Das and C. DeCusatis, *Acousto-optics: Fundamentals and Applications*, Artech House, Boston, 1991.

[DeC96]   C. DeCusatis, A. Abbate, and P. Das, Wavelet transform based image processing using acousto-optics correlators, *Proc. of 1996 SPIE Conf. on Wavelet Applications*, vol. 2762, pp. 302-313, 1996.

[DeC96a]   C. DeCusatis, A. Abbate, and P. Das, Scale and rotation invariant pattern recognition using a wavelet image processor, *Annual Meeting of the Optical Society of America*, 1996.

[DeC97]   C. DeCusatis, A. Abbate, D.M. Litynski, and P. Das, Wavelet image processing for optical pattern recognition and feature extraction, *Proc. SPIE*, vol. 3110, pp. 804-815, 1997.

[DeC97a]   C. DeCusatis, A. Abbate, and P. Das, Optical techniques for wavelet transform analysis, *Proc. Progress in Electromagnetics Research Symposium*, p. S19,1997.

[DeC97b]   C. DeCusatis, A. Abbate, and P. Das, Progressive pattern recognition using the wavelet transform, *Int. J. of Optoelectron.* vol. 11, p. 425-432, 1997.

[Don92]   D.L. Donoho and I.M. Johnstone, Nonlinear solution for linear-inverse problems by wavelet-vaguelet decomposition, Technical report 403, Stanford University, 1992.

[Don93]   D.L. Donoho, Nonlinear wavelet methods for recovery of signals, densities, and spectra from indirect and noisy data, *Proc. Appl. Math.*, vol. 47, pp. 173-205, 1993.

[Don95]   D.L. Donoho and I.M. Johnstone, Adapting to unknown smoothness via wavelet shrinkage, *J. Am. Statist. Assoc.*, vol. 90, pp. 1200-1224, 1995.

[Her91]   C. Herley and M. Vetterli, Linear phase wavelets: theory and design, *Proc. ICASSP*, vol. 3, pp. 2017-2020, 1991.

[Het95]   K. Hetling, G. Saulnier, and P. Das, Optimized filter design for PR-QMF based spread spectrum communications, *Proc. IEEE International Conference on Communications*, 1995, pp. 1350-1354.

[Het95a]   K. Hetling, G. Saulnier, and P. Das, Optimized PR-QMF based codes for multiuser communications, Proc. *SPIE*, vol. 2491, pp. 248-259, 1995.

[Het95b]   K. Hetling, G. Saulnier, and P. Das, PR-QMF based codes for multipath/multiuser communications, *IEEE Global Telecommunications Conference*, 1995.

[Het96]   K. Hetling, G. Saulnier, and P. Das, Performance of filter bank-based spreading codes for multipath/multiuser interference, in *SPIE Proceedings on Wavelet Applications*, 1996.

[Het96]   K. Hetling, *Multirate Filter Banks for Spread Spectrum Waveform Design*, PhD thesis, Rensselaer Polytechnic Institute, 1996.

[Het97]   K. Hetling, G. Saulnier, and P. Das, Performance of filter bank-based spreading codes for cellular and micro-cellular channels, in *SPIE Proceedings on Wavelet Applications*, 1997.

[Jia96]   J. Jian, C.Z. Chen, C.C. Chen, and G. Seethraraman, Image data compression methodologies using discrete wavelets, *Proc. SPIE*, vol. 2762, pp. 188-199, 1996.

[Joh95]   I.M. Johnstone and B.W. Silverman, Wavelet threshold estimators for data with correlated noise, Technical Report, Stanford University, 1995.

[Mal96]   S. Mallat, Wavelets for a vision, *Proc. IEEE*, vol. 84, n. 4, pp. 604-614, 1996.

[Mat96]   *Wavelet Toolbox User's Guide*, The Mathworks Inc., 1996.

[Med95]   M.J. Medley, *Adaptive Narrow-Band Interference Suppression Using Linear Transforms and Multirate Filter Banks*, PhD thesis, Rensselaer Polytechnic Institute, 1995.

[Med95a]   M.J. Medley, G.J. Saulnier, and P.K. Das, The application of wavelet-domain adaptive filtering to spread spectrum communications, *SPIE Proc. on Wavelet Applications for Dual-Use*, vol. 2491, pp. 233-247, 1995.

[Med97]   M.J. Medley, G.J. Saulnier, and P.K. Das, Narrow-band interference excision in spread spectrum systems using lapped transforms, *IEEE Trans. Commun.*, vol. 45, pp. 1444-1455, 1997.

[Met94]   M. Mettke, M.J. Medley, G.J. Saulnier, and P.K. Das, Wavelet transform excision using IIR filters in spread spectrum communication systems, *IEEE Global Communications Conference*, 1994, pp. 1627-1631.

[Mil80]   L.B. Milstein and P.K. Das, An analysis of a real-time transform domain filtering digital communication system, Part I: Narrowband interference rejection, *IEEE Trans. Commun.*, vol. 28, pp. 816-824, 1980.

[Mil83]   L.B. Milstein and P.K. Das, An analysis of a real-time transform domain filtering digital communication system, Part II: Wideband interference rejection, *IEEE Trans. on Commun.*, vol. 31, pp. 21-27, 1983.

[Odg97]   R.T. Odgen, *Essential wavelets for statistical applications and data analysis*, Birkhauser Press, Boston, 1997.

[Pro95]   J.G. Proakis, *Digital Communications*, 3rd ed., McGraw-Hill, New York,, 1995.

[Res92]   H. Resnikoff, Wavelets and adaptive signal processing, *Opt. Eng.* vol. 31, pp. 1229-1234, 1992.

[Sau95]   G. Saulnier, M.J. Medley, and P. Das, Wavelets and filter banks in spread spectrum communications, in *Subband and Wavelets Transforms*, edited by A.N. Akansu and N.J.T. Smith, Kluwer, Boston, 1995.

[Sch96a]   P. Schröder, Wavelets in computer graphics, *Proc. IEEE*, vol. 84, n. 4, pp. 615-625, 1996.

[Str89]   G. Strang, Wavelets and dilation equation: A brief introduction, *SIAM Rev.* vol. 31, pp. 614-627, 1989.

[Str94]   G. Strang, Wavelets, *American Scientist*, vol. 82, pp. 250-255, 1994.

[Vet92]   M. Vetterli and C. Herley, Wavelets and filter banks: theory and design, *IEEE Trans. Signal Proc.*, vol. 40, no. 9, 1992.

[Wic94]   M.V. Wickerhauser, *Adapted Wavelet Analysis from Theory to Software*, AK Peters, Wellesley, 1994.

[Woo91]   J. W. Woods. Ed., *Subband Image Coding*, Kluwer, Boston, 1991.

[Wor92]   G.W. Wornell and A.V. Oppenheim, Wavelet-based representations for a class of self-similar signals with applications to fractal modulation, *IEEE Trans. Inf.orm. Theory*, vol. 38, pp. 785-800, 1992.

[Wor96]   G.W. Wornell, Emerging applications of mutirate signal processing and wavelets in digital communications, *Proc. IEEE*, vol. 84, pp. 586-603, 1996.

Chapter 9

Real-Time Implementations of the Wavelet Transform

9.1   Digital VLSI Implementation

In this chapter, we will discuss various methods which can be used to realize wavelet and subband transforms for practical applications. It is impossible to give an all-inclusive description of every possible architecture for implementation of wavelet transforms; we will, instead, mention some significant results in digital electronic chip design, optical systems, and hybrid optical-electronic systems.

Since the wavelet transform is based on convolutions and can be formulated as a recursive operation, many existing software programs can be applied to simulate wavelet decomposition. The recent development of advanced computer-aided design (CAD) tools has made it practical to consider a large number of customized integrated circuit implementations of wavelet transforms. In designing VLSI hardware, the goal is to achieve complete circuit utilization and minimize the chip area required for fabrication. The chip area can be reduced by minimizing the number of processing elements, resistors, and interconnection wires. Certain types of processors, such as digital multipliers, consume more space and require more power than digital address, for example. Thus, we might also consider minimizing the number of multiplications required to implement the transform; this could also reduce the computation time in a programmable implementation. However, there are design trade-offs in a hardware implementation such as increased complexity, which may offset the area saved in this manner. Although there is currently no global optimization available for the digital design of wavelet processors, a number of architectures have been proposed which achieve near-theoretical performance using accepted benchmarking techniques. We will review several of these to illustrate the design principles involved; many variations are possible and may be found in the current literature.

Most digital implementations rely on the fast wavelet transform, first proposed by Mallat [Mal89]; this is the decomposition using quadrature mirror filters. Viewed in this way, the discrete wavelet transform is the multiresolution decomposition of an input sequence. The input signal $s(n)$ consists of N points, and the wavelet transform output is also a sequence of N points. The bandwidth of each channel is effectively reduced by a factor of 2 in this embodiment, and as discussed in prior chapters, the discrete wavelet transform is the information emerging from the high-pass filter channels. If N is the number of samples in the signal, a filter of length L requires LN multiply and add operations for the highest-frequency wavelet component. The next lower octave band requires $LN/2$, then $LN/4$ for the next band, and so on. Regardless of the depth of the binary tree structure, this geometric progression never exceeds $2LN$. Since most applications extract local information only, the filter is short and $L<N$; therefore, the wavelet transform is an $O(N)$ computation, ; that is, it requires on the order of N multiply/add operations. It is well known that the fast Fourier transform requires $O(N \log N)$ multiply/add operations; thus, the wavelet transform requires less computational time.

The wavelet transform exhibits frequency resolution which is lower at high frequencies and high at low frequencies; the time resolution is higher at high frequencies and lower at low frequencies. Consider decomposing the signal into J octave bands, so that $N=2^j$. The discrete wavelet transform (DWT) is then given by

$$W(n,j) = \sum_{m=0}^{2n} W(m,j-1)g(2n-m) \qquad (9.1.1)$$

$$W_H(n,j) = \sum_{m=0}^{2n} W(m,j-1) \, h(2n-m) \qquad (9.1.2)$$

where $W(n,0) = x(n)$ and the quadrature mirror filters are given by $h(n)$ and $g(n)$. Although this expression looks very different from the convolutions with a family of daughter wavelets given earlier, it is actually equivalent to these expressions. Implementation using QMF filter banks follows the pyramid algorithm proposed by Mallat for computing the DWT. This approach is to let $W(n,0)=x(n)$ for $i = 1$ to $\log N$, then compute the $(N/2)^j$ point convolution of $W(n, i\text{-}1)$ with $g(n)$; this will generate the output points.

Implementation of this algorithm is not as straightforward as it may seem. Since VLSI circuits are enormously complex, the asymptotic analysis of their performance has proven to be useful in studying design trade-offs. For example, it has been shown [Vis92] that the circuit area, A, required to implement the DWT and the period of circuit latency, T, are related and lower bounded by

$$AT^2 = O(N^2 \log^2 N) \qquad (9.1.3)$$

where N is the length of the input sequence. This is the same bound as the DFT implementation. In contrast, the discrete short-time Fourier transform is more difficult to compute; it is lower bounded by

$$AT^2 = O(N^2 M^2 \log^2(N_w + N)) \qquad (9.1.4)$$

where N_w is the size of the window, M is the number of time shifts, and N is now the number of frequencies we consider for the Fourier transform. It is generally accepted that a tight lower bound is a good metric against which to measure chip design and performance. These bounds provide a means to compare different digital designs. A theoretically optimal architecture for the DWT has been proposed, although it is not considered practical since it required complicated control; the data are tagged to indicate which octave band it belongs to, and the architecture performs on-the-fly data-dependent scheduling of the inputs during even and odd cycles of the clock phase. This controller has not yet been realized, and its design remains open to speculation [Vis92a].

It has been shown that a more practical implementation of the DWT involves a *systolic* architecture. The systolic processor concept was first suggested in [Kun81]; there are many processors, each of which receives input data, performs some short computations, and then passes the result along to the next processor. The system is clocked to maintain a regular floor of data (the term "systolic" is related to the physiological term *systole*, which refers to the rhythmic recurrent contractions of the heart and arteries). An advantage of this approach is that all processing blocks can be connected in parallel; thus, we need only $O(N)$ cycles to process N bits of data. Using the systolic model, a practical lower bound for the DWT is given by

$$AT^2 = N^2 N_w k^2 / r^3 = O(N^2 n_w k) \tag{9.1.5}$$

where N is the input sequence length, N_w is the filter length, k is the precision of the inputs and filter coefficients, and r is the input-output (I/0) rate. This assumes a practically achievable I/0 rate ($r > k$). We can now consider different DWT architectures and compare their performance against these benchmarks.

The recurrent nature of the DWT can be realized by cascading log N linear systolic arrays, as shown in Figure 9.1.1a [Szu92]. The operation of an individual cell is shown in Figure 9.1.1b; the *l*th row computes the *l*th octave of the output. Each row computes the convolution and the decimation by a factor of 2; in other words, each row implements the relationship

$$y(k) = \text{output} = \sum_{i=0}^{N_w - 1} x(i) w(2k - i) \tag{9.1.6}$$

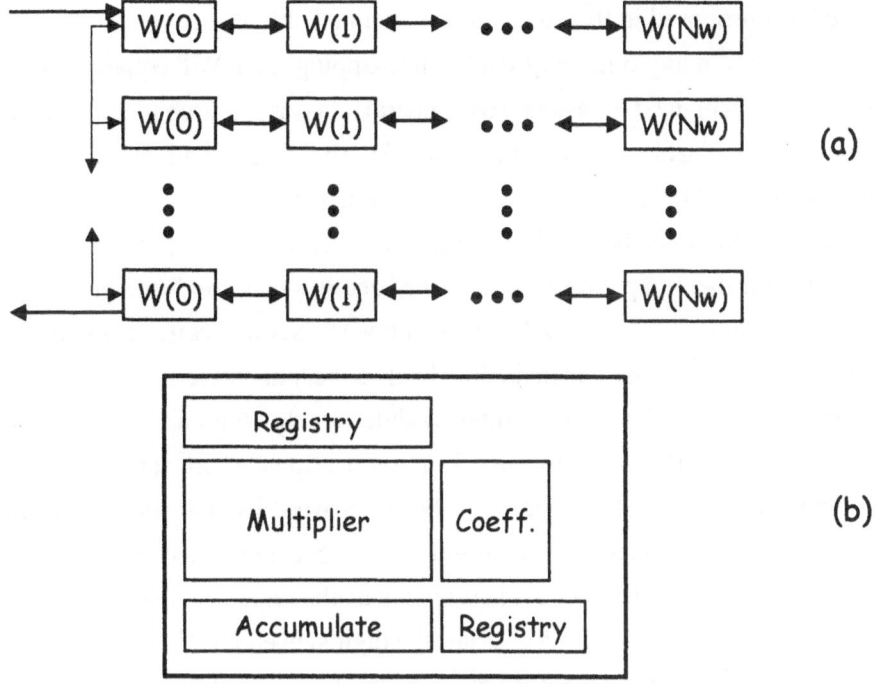

Figure 9.1.1 (a) Block diagram of a cascade of systolic arrays; (b) individual array cell.

where all inputs $x(i)$ and outputs $y(k)$ are provided through the leftmost cell of each row. Each stage gets inputs and produces outputs at half the rate of the previous stage; but since it also needs only half the number of inputs and outputs, the latency remains the same from stage to stage. Each building block contains an input register, multiplier and filter coefficients, summer, and output register. Either finite impulse response (FIR) or polyphase filters may be used, and polyphase filters will double the processor utilization. This implementation requires a relatively large area, which may not be justified by the gains in processor utilization. The architecture can be realized by using a routing network in which shift registers of J rows (J is the number of octaves) and N_w columns (the filter size) are used to route data among the processors. By routing data in this manner, the DWT can be computed in $2N$ clock cycles, and the total area is bounded by $O(N_w\,k\,j)$. A detailed design is given in [Vis92b]; performance is $\log N$ away from optimal, but this design is still practical to implement. A comparison of various architectures which have been proposed in the literature is given in Table 9.1.

There are many other approaches to mapping the DWT equations into VLSI. We will next consider two general classes of architectures first proposed by Parhi and Nishitani [Par93]: the folded and digital-serial architectures. The two approaches are independent and have different properties. First, consider the *folded architecture*, so called because it executes all the computations using only one low-pass and one high-pass filter. This approach has the advantage of low latency and can accommodate arbitrary word lengths. The main drawback is the long interconnection wires to wrap the output back to the filter modules (each interconnect is really a bus as wide as the word length). For applications where latency is not critical, hardware utilization can be improved by the *digital-serial architecture*. This processes a subset of the total number of bits in each cycle; the number of bits per cycle is known as the *digit size*. For a digit size of 1, and the architecture is bit serial; the digit size equals the word length, the architecture is bit parallel. The basic modules are digit-serial filtering blocks; the interconnections are local, which makes it easier to wire the design. A comparison of the two approaches is given in Table 9.2; note that the digital-serial also tends to consume less power for a given supply voltage and clock speed.

Table 9.1 Comparison of VLSI architectures for wavelet transforms

Architecture	Area on Chip	Latency	Period
Folded	$O(N_w\, k \log N)$	$2N+logN-4$	$2N$
Digit-serial	$O(N_w\, k \log N)$	$2N$	$2N$
Optimal	$O(N_w\, k)$	$O(N)$	$O(N)$
Hardware chip	$O(N_w\, k)$	$O(N \log N)$	$O(N \log N)$
Knowles architecture	$O(N_w\, logN(logN_w + loglogN))$	$O(N)$	$O(N)$

Table 9.2 Comparison of folded and digital-serial architectures for wavelet transforms

Property	Folded	Digital-Serial
Number of multipliers	16	14
Number of word-level registers	64	258
Latency in cycles	28	70
Power consumption	$16CV^2 f$	$9.76CV^2 f$
Interconnections	Complex	Simpler/local
Silicon area	Larger	Smaller
Constraints on word length	None	Multiples of 2^m
Constraint on speed	Less constrained	More constrained
Output pins or number of wires	Same	Same

An alternative proposal to optimizing VLSI circuits is to minimize the number of multipliers required, since these components are more complex, expensive, and power hungry than adders.

One proposal [Lew91] implements the Daubechies wavelet without using any multipliers for both forward and inverse transforms. This approach provides 8-bit precision and takes advantage of the fact that the Daubechies coefficients can be obtained by solving a series of algebraic expressions so that the multiplications can be calculated by adding the data to a shifted version of itself; once one of the coefficient products is obtained, the rest can be calculated using only additions and subtractions. Other approaches [Gan93, Sau92] take advantage of the matrix formulation, treating the DWT as a set of inner-product matrix operations. The matrix is often banded and has symmetry properties which simplify the computation. This approach is well suited to massively parallel architectures; by sharing memory storage among many processors the total number of processing elements is reduced

[Hoy92]. A fully parallel pipelined implementation would require $O(N)$ processing elements organized in $\log N$ stages, which are frequently underutilized. The design of application-specific integrated circuits (ASICs) using complementary metal oxide semiconductors (CMOS) has also been investigated. This approach offers the advantage of lower power consumption and higher speed approaching 50 million samples per second. The power consumption of a CMOS circuit is given by

$$P = \frac{1}{2}CV^2f \tag{9.1.7}$$

where C is the capacitance, V is the power supply voltage, and f is the clock frequency. We can see from this expression that if the DWT is implemented in bit serial fashion for an N-bit data word the clock frequency is N times higher than for a comparable data-path-intensive architecture. This affects both processing speed and power consumption. One example of a CMOS ASIC is the analog wavelet transform chip fabricated by Zilog Corp for applications in speech compression. Audio signals in the frequency range 180 Hz to 12 kHz may be compressed using this chip.

We have previously seen that the transfer function of a maximally decimated filter bank can be expressed in matrix notation. In order to realize these chips, the matrix size must be kept constant. This can be done with a number of algorithms; we will discuss one example based on the DWT.

An algorithm for construction of decimation and interpolation filter matrices was developed in [Vis92b]; the low- and high-pass filter matrices can be given by

$$L_{ij} = C_{2i-j+n-2}$$
$$H_{ij} = (-1)^{j+1}C_{j+1-2i} \tag{9.1.8}$$

where n is the number of filter coefficients, i is the row number, j is the column number, N is the input signal length. The number of rows is half the number of columns, and the number of columns equals N. These matrices have certain properties; the sum of coefficients in any row of L_{ij} should equal 2, and in any row of H_{ij}, it should equal 0.

The two matrices are orthonormal (their inner products are zero) and they must satisfy the identity conditions

$$LL^* = \mathbb{I}, \quad HH^* = \mathbb{I}, \quad LL^* + HH^* = \mathbb{I} \qquad (9.1.9)$$

where $\mathbb{I}$ is the identity matrix. L and H are sparsely populated and contain many zeros; they are thus often easier to compute than corresponding fast Fourier transform (FFT) matrices. However, the matrix size grows proportionally to the signal length. An alternate approach for generating fixed size matrices would be

$$L_{ij} = C_{2i-j}$$
$$H_{ij} = (-1)^{j+1} C_{j-2i+n-1} \qquad (9.1.10)$$

where the result is always a matrix of size $(n-i)$ by n. Orthogonality is not necessarily met by this solution and must be insured by manipulating the input data. This approach has been successfully applied to decomposition of a 128×128 pixel image; further work is necessary to devise an efficient hardware implementation.

9.2   Optical Implementation

In this section, we will discuss the implementation of wavelet transforms using optical signal processing techniques. There are many features of wavelet transforms which are well suited to an optical implementation; the transform is inherently two-dimensional (2-D) and parallel, it requires real-time implementation for many applications, and it is well suited for image processing. Since the basic building block of the wavelet transform is the

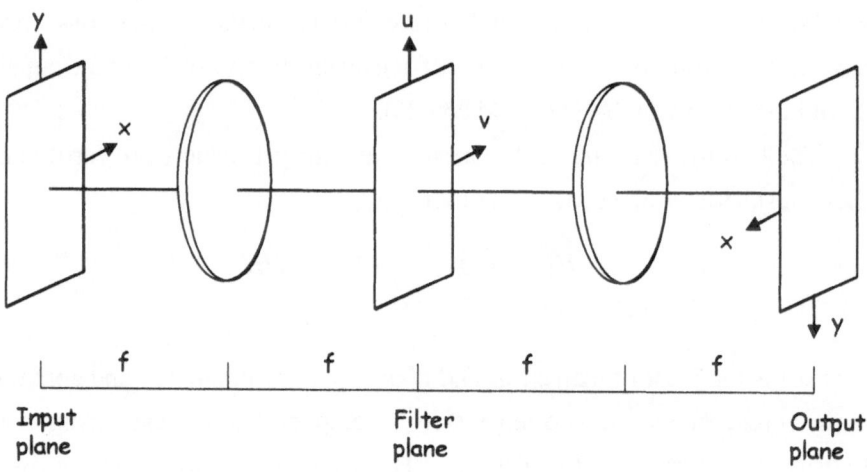

Figure 9.2.1 Block diagram of the Vander Lugt 4f optical correlator.

convolution, we will first present an overview of optical convolvers and correlators. We will then examine several proposed architectures for optical implementation of the wavelet transform.

Many fundamentals of optical data processing were developed in the early 1960s [Cut60]. The first optical correlator/convolver was the classic holographic matched filter proposed by Vander Lugt. The basic architecture is shown in Figure 9.2.1, a so-called *4f system* because its total length is four times the focal length of the lenses used. An object function in the form of a transparency is placed at the front focal plane of a lens; the lens performs a 2-D spatial Fourier transform of the input image at the back focal plane. A matched filter is placed in this plane, also as a transparency. The overlap of the matched filter function and the input function is equivalent to an optical multiplication, and multiplication in the Fourier domain is equivalent to convolution or correlation in the time domain. A second lens produces the inverse Fourier transform, with its coordinates reversed in space, at the output focal plane. This output represents the correlation between the input and matched filter. Since the matched filter is, in general, a complex function containing both amplitude and phase information, Vander Lugt proposed a holographic implementation for the filter. Early filters were recorded on thick emulsions of photographic film; the approach was limited, since the filter function could not be easily changed. In the presence of white noise, this architecture produces the optimal output signal-to-noise ratio. The success of this approach created a demand for more practical two-dimensional spatial light modulators, which are, in general, difficult and expensive to fabricate. Several alternatives exist today, including the liquid-crystal television (LCTV), smart pixels, and magneto-optic spatial light modulator (MOSLM). Acousto-optic devices may also be used as light modulators in an optical correlator.

Using a Vander Lugt correlator, a non-real-time 2-D optical wavelet transform has been implemented. The input image and the Fourier transform of a single daughter wavelet were stored on film. To correlate the input with multiple daughter wavelets, this approach required that daughter wavelets be loaded sequentially into the processor, with each daughter wavelet recorded on a separate film plate, This approach was probably the first successful optical implementation, but it is not practical for most applications. An improved approach replaces the film with a programmable spatial light modulator, such as a MOSLM. The operation of a magneto-optic device is illustrated in Figure 9.2.2; the device is based on the Faraday effect, which causes the plane of polarization of light to rotate when passing through a magnetic field [Pin91]. Incident light is polarized and then passes through a magneto-optic thin film; this film is divided into picture elements, or pixels, each of which can be addressed electronically under computer control. By addressing pixels in this manner, it is possible to change their magnetization state.

As shown in Figure 9.2.2, when the magnetic field is oriented in one direction, the linearly polarized light is rotated clockwise. When the field is reversed, the plane of polarization of the light is rotated counterclockwise. A second polarizer placed behind the device is oriented to allow only one linear polarization to pass through. Thus, pixels which have rotated the light to line up with the second polarizer appear light, whereas the other pixels rotate light so that it is blocked by the second polarizer and they remain dark. It is also possible to force the pixels into a third intermediate state, in which they are partially magnetized in both directions; this has the effect of allowing some fraction of the light to pass through, and the pixels appear

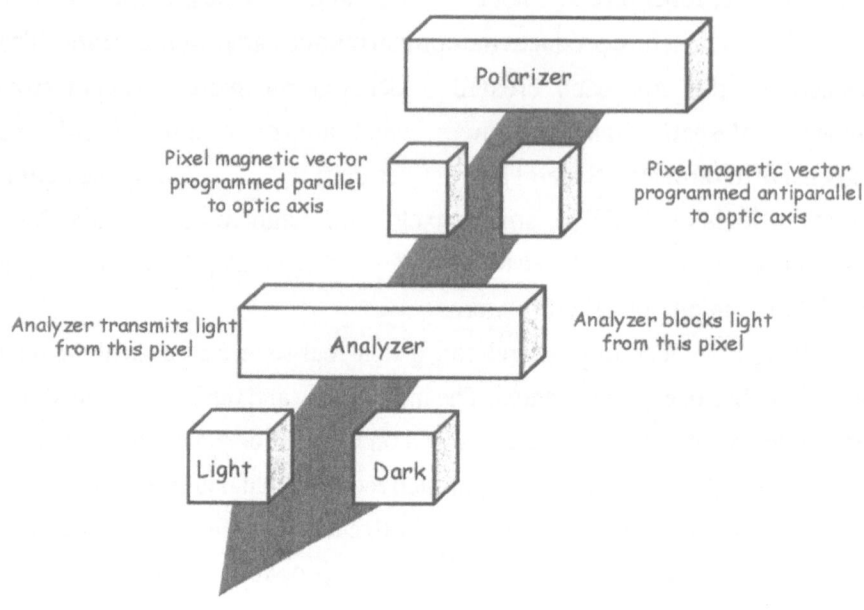

Figure 9.2.2 Operating principle of a magneto-optic spatial light modulator (MOSLM).

gray (in between the light and dark states). Arrays of 256×256 pixels are commercially available.

By inserting a MOSLM in the Fourier plane of a Vander Lugt correlator, the correlator becomes programmable. Daughter wavelets can be written to different rows of the MOSLM, and an input image can be correlated against all the daughters simultaneously. The output correlation peaks can be detected using a photodetector or charge-coupled device (CCD) camera. Alternately, the MOSLM can be used to provide the input images, and the wavelets can be recorded using modem holography techniques such as thermoplastic holograms, which are created electrically. The apparatus of Figure 9.2.3 was used to implement a wavelet transform correlator. Because the MOSLM has three stable states, it is well suited for implementation of the piecewise continuous Haar wavelet, which has only three possible values: 0 and ± 1.

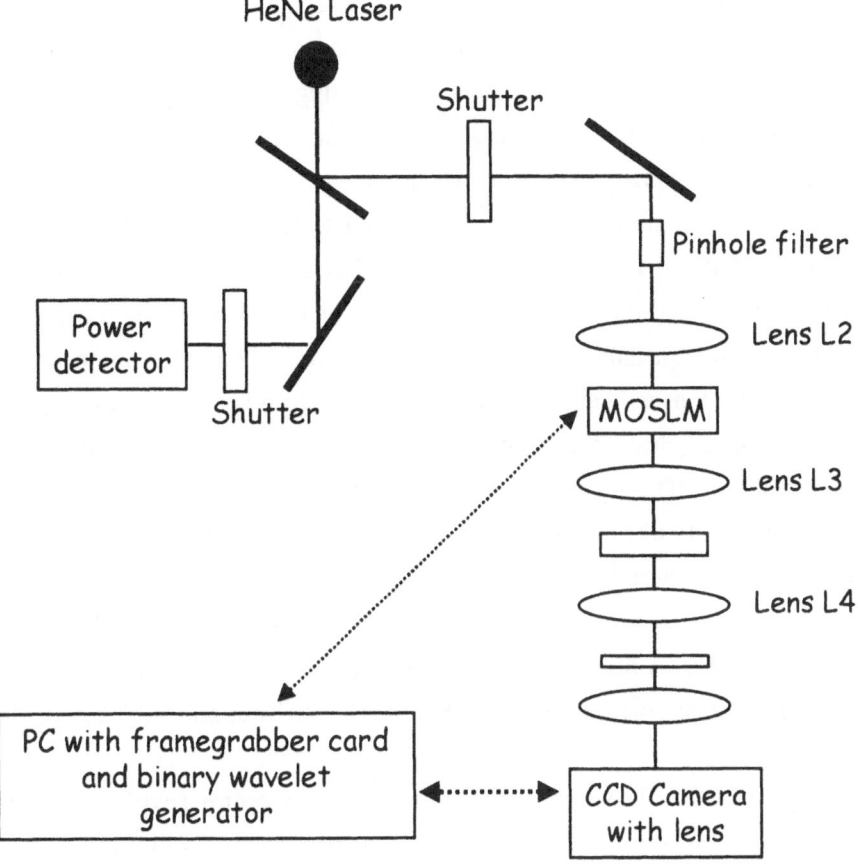

Figure 9.2.3 Design of a programmable optical correlator using MOSLM.

Although the Haar wavelet is not very useful for image compression as a basis, correlation of this wavelet with an image results in edge detection and enhancement. This is an important area for pattern recognition, image segmentation, and image classifiers. In the approach shown in Figure 9.2.3, the MOSLM under computer control generates a Haar wavelet image, which is illuminated by a helium neon laser beam (the laser is pinhole spatially filtered, and optical power is monitored by a beamsplitter, shutter, and power detector). The Fourier-transformed wavelet is imaged onto a thermoplastic holographic camera, which contains the image to be studied. Diffracted light output from the hologram represents the correlation between the Haar wavelet and the image; this is detected and displayed by a CCD camera and frame grabber card in a PC. Digital correlation with horizontal and vertical edges in the target image was successfully demonstrated.

A sinc approach has been demonstrated using a liquid-crystal light valve (LCLV) as the spatial light modulator for an optical correlator [She92]. Although this work treats more complex wavelets such as Morlet and Mexican hat, many of these popular wavelets are real-valued symmetric functions. Therefore, they can be represented as simple intensity transmission masks in a spatial light modulator. By using more complex coding techniques, it is possible to incorporate negative and complex values in the wavelet functions as well; we will not go into detail here. By recording the proper holographic filters, it is possible to implement multichannel correlators. This approach takes advantage of the free space interconnectivity of optical light beams to avoid the interconnect and wiring problems encountered in VLSI implementations. Since this architecture can present a joint space frequency representation of a signal, it is also known as an N^4 correlator. A number of wavelets have been implemented in real time using this processor; output is limited by the diffraction efficiency of the transparencies.

Whereas the classic optical matched filter is optimal with respect to output signal-to-noise ratio, at least in the presence of white noise, it is sensitive to rotation and scale changes of the input image. Many potential solutions to this problem have been suggested; among the most powerful is the joint transform correlator architecture [DeC97], which is a method for performing correlation without the use of a reference filter function. The basic architecture is shown in Figure 9.2.4; the input consists of two images offset from the optic axis by a distance x_0:

$$\text{input} = f(x - x_0, y) + g(x + x_0, y) . \tag{9.2.1}$$

When the system is illuminated, the electric field of the light in the focal plane of the first lens is simply the Fourier transform of the input; this operation contains the product of the Fourier transforms of the images. By using a second lens to invert the Fourier transform, the cross-correlation can be obtained:

$$I_{\text{output-plane}} = f(x,y) \otimes f(x,y) + g(x,y) \otimes g(x,y)$$
$$+ f \otimes g \, \delta(x - 2x_0) + f \otimes g \, \delta(x + 2x_0)$$

(9.2.2)

The desired cross-correlation is centered about the points $+2x_0$ and $-2x_0$ as indicated by the delta functions; either term may be used. This architecture has been applied to image correlation [Kir90].

It is also possible to implement the wavelet transform through recursive reuse of a 2-D wavelet correlator. Instead of dilating the wavelet, the signal is dilated. This takes advantage of the relationship between bandwidth and sampling. For example, shrinking the signal by a factor of k is equivalent to expanding the wavelet by a factor of k, which reduces the bandwidth by a half. Since the density of the points sampled by the intensity detector remains constant, the number of points sampled becomes directly proportional to the bandwidth. Figure 9.2.5 shows a Mach-Zender configuration. Instead of dilating the wavelet, the signal is dilated. In this setup, all of the lenses are used to perform Fourier transforms. The input passes through lens L1, causing the Fourier transform (FT) of the input to appear at plane R1. The Fourier transform then passes through lens L3, which has a focal length that is $1/a$ times as long as the focal length of lens L1. The inverted input image is reduced in scale by a factor of A and appears at plane P2. Lens L2 takes

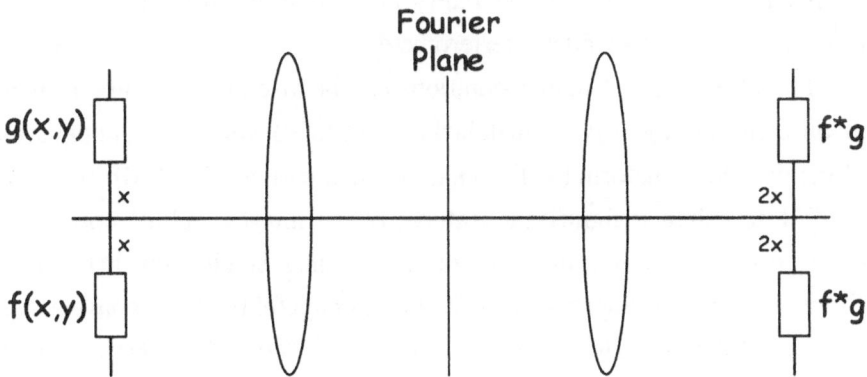

Figure 9.2.4 Optical joint transform correlator architecture.

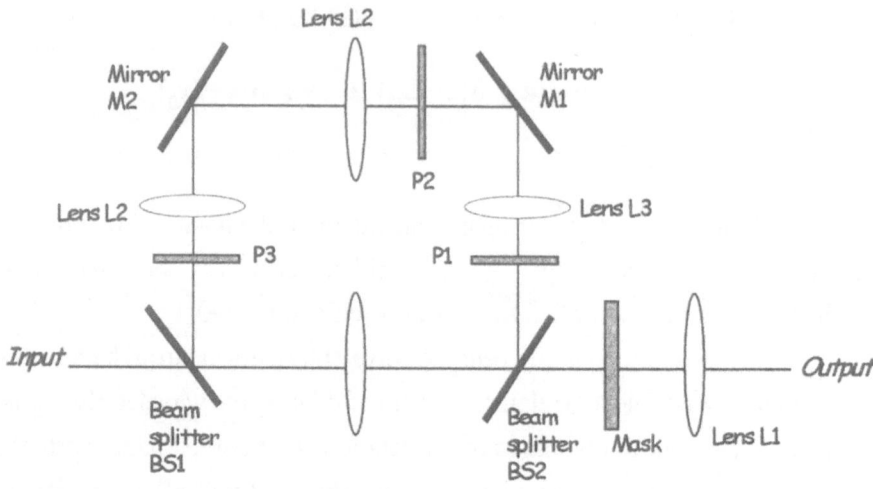

Figure 9.2.5 Mach-Zender configuration for a wavelet correlator.

the Fourier transform of the image at plane P2. The FT will appear at M2, which allows us to tilt mirror M2 to create a phase gradient to cause a noninverted input image that is shifted in position to appear at plane P3. (This prevents all of the dilated components from appearing on top of each other.) The cycle then repeats itself. However, each time, some of the light escapes at BS2 and passes through the mask and a second lens LI. The correlated input appears at the output. If the mask was a holographic optical element, different wavelets could be encoded onto the mask. An alternative approach might be to use the Fabry-Perot interferometer configuration instead of the Mach-Zender interferometer.

The above optical implementations can be also applied without many modifications to the wavelet matched filter (WMF), which is an application of the wavelet transform (WT) to pattern recognition. The WMF performs the WT to enhance significant features of the images and the correlation between two wavelet transform coefficients in a single step. It has been shown to have the improved discrimination capability of the conventional matched spatial filter (MSF) against unknown objects and is relatively robust against noise.

9.2.1   Matrix Processing and Neural Networks

We have previously mentioned the use of both adaptive filtering and matrix processing to implement wavelet transforms; it follows that both optical matrix processors and neural network architectures can be used for the same purpose. A method of adoptively generating wavelet templates for neural networks has been proposed in [Szu92]. A *neural network* is basically a network of processing nodes, each of which has a simple input/output rule (often a sigmoid function or similar nonlinear response) and is connected to all other nodes in the network. Many neural net architectures have been studied for different applications; they derive their processing power from the ability to update the weights of connections between nodes to strengthen or weaken certain output responses as the input changes. We have shown that a signal $s(t)$ can be approximated by a sum of daughter wavelets. This approximation can be mapped onto the neural net architecture, where the artificial neurons contain wavelet nonlinearities; the weights, shifts, and scaling parameters can be optimized by existing methods such as the *least mean squares* (LMS) algorithm. This approach has been modeled as a classifier for images or phonemes used in speech recognition.

The matrix formulation of wavelets can be applied to various optical matrix processor architectures, as well. The fundamental optical matrix processor was first proposed by Goodman et al. [Goo78] and is illustrated in Figure 9.2.6. A two-dimensional matrix is represented as a transmission mask, and an input vector is formed by modulating the light intensity of a vertical column of lasers or LEDs. The inputs are spread vertically across the transparency by a cylindrical lens; this realizes multiplication between the input vector components and the matrix elements. A second cylindrical lens focuses the light horizontally, performing an optical summation of the partial products in the matrix columns; in this way, the vector-matrix product is formed. Output values are taken from a horizontal linear detector array. This is a fully parallel vector-matrix multiplication; in some sense, it is an optimal design because processing occurs as fast as light can traverse the system. Processing speed is therefore limited by the input data rate and noise levels at the detector. This architecture for inner-product processing is very similar

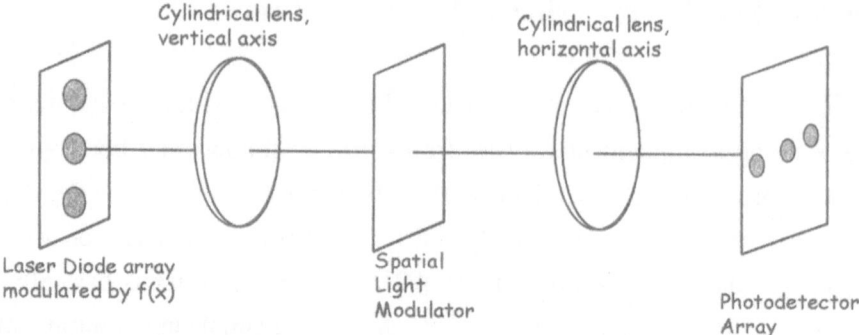

Figure 9.2.6   Block diagram of an optical matrix processor.

to the Vander Lugt optical correlator design; by using a programmable mask and extending the input to two dimensions, matrix-matrix multiplications can be achieved. Because all quantities are handled as discrete points in space, this design is called a *space multiplexed* matrix processor. An alternative is the *time-space multiplexed* processor, which is basically the same architecture except that a point source is used and modulated with successive vector components in time. A one-dimensional mask is used to perform successive vector multiplications. The output is obtained from a one-dimensional device which accumulated charge proportional to the input light intensity until all the vector products are formed; the result is then clocked out of the detector array. The advantage to this approach is the use of only one-dimensional devices; however, processing speed is slower because of the clocking required to process an $N{\times}N$ matrix in N clock cycles. A third approach is the *time-frequency* multiplexed processor, in which different vector components are frequency multiplexed onto the one-dimensional spatial light modulator to perform parallel processing of the vector products. Many variations of these designs have been reported, including efficient approaches using integrated optical devices which reduce the optical matrix processor to the dimensions of a VLSI chip.

9.2.2   Acousto-Optic Devices

Because the wavelet transform is essentially a convolution operation, it is well suited to implementation using *acousto-optic* (AO) devices, especially devices based on *surface acoustic waves* (SAW). In such devices, a radio frequency (r.f.) signal is applied to a piezoelectric transducer on a crystal such as quartz (if the material itself is piezoelectric, transducers can be fabricated directly on the crystal using standard photolithographic techniques). When the transducer is excited by an r.f. signal, it generates an acoustic wave or ultrasound signal which propagates through the crystal and causes periodic perturbations in the refractive index. This forms an optical phase grating in the material, which is capable of diffracting light just like a conventional diffraction grating. The diffracted light is modulated by the acoustic signal, and the modulated light passing through the material can be measured with a photodetector. In this way, the AO device acts as a one-dimensional spatial light modulator. If the acoustic signals are confined near the surface of the material, it is called a surface acoustic wave device; the advantage of a SAW device is that the interaction between light and ultrasound is stronger if the energy of the acoustic wave is concentrated near the crystal surface. The acoustic wave propagates through the crystal, passing through the incident optical beam. If we recall the graphical interpretation of a convolution in which one signal is inverted, moved across another, and the product signal integrated at each point, then we can see that the physical structure of an AO device lends itself to implementation of the convolution integral. There are a large number of devices and materials which can be employed for this purpose, and several different types of diffraction which can occur [Das93]; in this section, we will limit our attention to an overview of the basic convolver and correlator architectures, then discuss a few specific examples to illustrate the possible implementations.

Since the convolution integral can be performed over either space or time, there are two basic classes of optical convolvers; the *time integrating convolver* (TIC) and *space integrating convolver* (SIC). The difference is illustrated by the TIC and the SIC of Figure 9.2.7. In the TIC, one of the signals to be convolved, $f(t)$, is used to modulate an optical point source; the other, $g(t)$, excites the AO transducer in a Bragg diffraction cell. The acoustic signal propagates across the optical beam with some delay, t, that is characteristic of the AO material and the velocity of sound propagation in

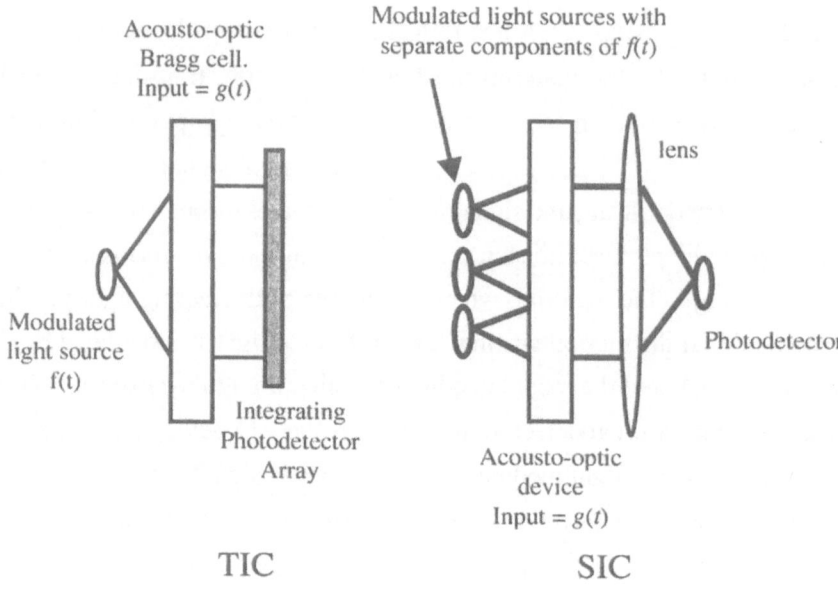

Figure 9.2.7 Block diagram of the TIC and SIC integrated convolvers.

that material. The product of the two signals is formed as the acoustic beam passes through the optical beam; this product falls on an integrating photodetector, which accumulates a signal from the partial products until the matrix multiplication is complete. The convolver output is given by

$$y(t) = \int_0^T f(\tau) \, g(t_0 - \tau) \, d\tau \qquad (9.2.3)$$

where the integration occurs over the time interval 0 to T. The output will be a single point of the total convolution function for each detector element in the TIC. In contrast, the SIC uses a broad optical beam to illuminate the entire aperture of the AO device; the optical beam may consist of multiple point sources, each modulated with a different component of the input signal $f(t)$. The entire convolution is calculated, not just a single point, and the output is integrated spatially by a lens placed after the AO device. The processing time is much shorter than a TIC and is given by the time required for the AO signal to propagate the length of the AO device. Also, the SIC performs the entire convolution integral, whereas the TIC produces samples of the convolution output for each detector element. However, the integration time is shorter for an SIC, being limited to about 10 μs. By simply allowing the detected light to accumulate, the TIC can achieve integration times of several seconds; thus the time-bandwidth product of a TIC can be much larger. Clearly, there are application-dependent trade-offs to consider when choosing an integration scheme.

There are many possible variations on these basic architectures; for example, a pair of counterpropagating acoustic waves could be used to provide the input signals, rather than using one acoustic wave and a modulated optical signal. Relative movement between the two signals is the only requirement for realizing a convolver architecture, but there are many permutations on this design. Actually, the relative displacement principle is also used by other types of devices including the *acoustic charge transport* (ACT) device and the *charge-coupled device* (CCD). lthough we will not discuss these devices in detail, they function on the same principle of displacing one input signal relative to another, and they can be employed in place of AO devices for many of the architectures we will discuss. We have already seen that the convolvers may be either space or time integrating, SAW or bulk ultrasound, and use either one or two acoustic delay lines. The light source may be coherent or incoherent (laser or LED), since it is used to probe the acoustic signal in the AO device. Depending on the choice of optical source, the detection scheme may be simple *intensity detection* or *heterodyning (interferometric detection)*, which increases the linear dynamic range of the processor but also places coherence length requirements on the optical source. The AO device is inherently a one-dimensional light modulator; however, several devices may be stacked to form an array of either horizontal or vertical components and build up a two-dimensional processor. Yet another option is the so-called *memory correlator*, in which one signal is stored in the AO cell using photorefractive memory techniques, so that multiple inputs can be correlated against a single reference signal.

The use of AO convolvers specifically to implement the wavelet transform was first demonstrated by Szu et al. [Szu92] using an architecture similar to the classic 4f correlator. This approach can potentially implement the wavelet transform in real time. Collimated laser light illuminates the AO cell, and a lens computes the Fourier transform of the input signal, $s(t)$, which is then multiplied by a fixed holographic mask. This mask stores the causal Fourier transform of a wavelet kernel along the x axis, with the scale factor a varying in the vertical direction. Another lens inverts the Fourier transform for each dilation on the y axis, and the light intensity of the wavelet transform of $s(t)$ is detected with a photodiode (square law detection) at the output plane. Several other proposals for the implementation of wavelet transforms using AO convolvers have been published [DeC95], including some recent architectures which implement perfect reconstructon filters using AO devices [DeC97, DeC97a]. In a setup similar to Szu's, the wavelet and inverse wavelet can also be combined into a single step. Filtering can be achieved by putting a mask over the filter bank to block out or cut down certain frequency bands. To simplify implementation, discrete wavelets that cover the bandwidth of the signal should be used. The filter can also be made variable in real time (as is the case with a computer-controlled light valve), which allows for the possibility of adaptive filtering. This would probably require some electronic connection between the detector and the mask. Similar to the 2-D case, it is also possible to recursively reuse the wavelet correlator rather than use a filter bank. Although it is not necessary, one might also dilate the signal rather than the wavelet.

Acousto-optic devices can also be used to implement both *finite impulse response* (FIR) and *infinite impulse response* (IIR) filters. When an acoustic signal propagates along the AO device, several optical beams can pass through the signal at different points. If the light intensity is modulated in proportion to a set of filter coefficients, then the diffracted light from each beam represents a delayed, weighted sample of the acoustic signal. The AO filter outputs may be detected using photodiode arrays or similar means. The AO device can also be used as a spatial light modulator in an optical matrix processor; the structure of an AO device lends itself particularly to systolic architectures. One of the earliest attempts to use AO devices for systolic

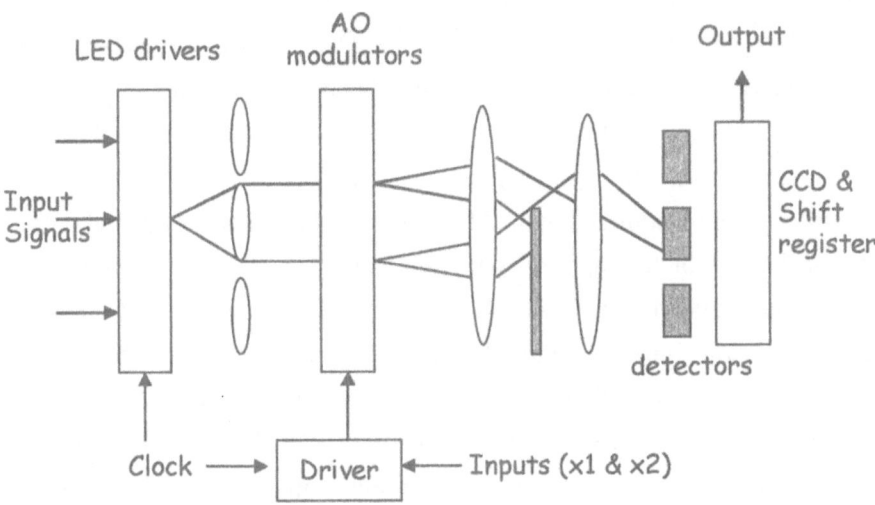

Figure 9.2.8 Systolic array processor design.

matrix processing was proposed by Caulfield et al. [Cau81], and is illustrated in Figure 9.2.8. Matrix elements are introduced as modulated LED intensities, and vector components are supplied to the AO device in the proper sequence. Note that both inputs are clocked, so the acoustic signal is synchronized with the modulated light input. In other words, the acoustic signal propagates along the device until it is aligned with one of the LEDs; this diode is then pulsed to supply the matrix element corresponding to the acoustic signal vector. The AO device acts as a spatial light modulator to perform analog multiplication, which generates the matrix partial products; these are summed by accumulation at a CCD which is clocked at the same rate as the rest of the system. Note that the capacity of the system can be increased by using multiple transducers in the AO device or by frequency multiplexing different matrix components on the same acoustic signal. Since the angle of diffracted light is proportional to the acoustic signal frequency, these components can be spatially separated in the detector plane. The diffracted light is also frequency modulated by the AO cell, so it is possible to filter out different frequency components at the detector array or use heterodyne detection schemes.

An architecture similar to the above one could be used as a two-channel filter bank, a recursive element in the binary subband tree implementation of the DWT. For example, four taps (like Daubechies' wavelet) can be implemented by an array of four laser diodes. (Although not shown in Figure 9.2.8, each diode has its own lenslet.) The AO cell modulates the diode light by $x(n)$, the input signal. Two cylindrical lenses would act as summers of the modulated light. The detector would just sample the output at half the rate of the input signal. A computer would store the signal and then recursively reuse the QMF bank. Of course, this implementation would not be real time, but an integrated optic version with computer chip control might be developed to allow us to cascade these banks together. It might also be possible to have rows of diodes and an array of detectors. Each row of diodes could be used to display a different signal. The AO cell would modulate the diode light by the tap coefficients. Since all the clock speeds are integer multiples of the clock speed of the input signal and the tap coefficients are the same at each stage, a common AO cell could be used. Alternately, one might stack up the AO cells. Then the diodes would display the tap coefficients. Each AO cell would modulate the diode light with a different signal.

The AO systems discussed so far are analog in nature and therefore less accurate than a digital system (although their processing speed is much higher). Data throughput can be greatly increased with these approaches because no analog-to-digital conversion is required and because of the real time optical processing elements. Highly accuracy versions of the AO matrix processor have been developed with the equivalent of 10-bit accuracy, as opposed to the typical 5- or 6-bit accuracy of analog matrix processors. Using commercially available hardware, a linear dynamic range of 60-70 dB can be achieved, making these types of processors suitable for realizing FIR filters. A fundamentally different approach involves the use of analog processors to perform calculations that can be represented in digital form. For matrix processing, we can take advantage of the fact that multiplying two numbers is equivalent to convolving their digital representations on a digit-by-digit basis. This is known as the *digital multiplication by analog convolution* (DMAC) algorithm; it is a powerful method for increasing the accuracy of matrix calculations without changing the hardware implementation. To perform a multiplication operation with this method, each value is first converted into twos complement form (binary form with an appropriate sign bit). This is the minimum number of bits required to represent the number without truncation or rounding. Performing a bit-by-bit convolution of the resulting values yields the result in mixed binary form; each digit is weighted by a power of 2, but the value of a digit may be greater than 2. This is necessary because the arithmetic carry operation cannot be performed optically. The results can then be converted back into the original number base. Since it is relatively easy to perform convolutions optically using AO devices, DMAC is a powerful algorithm; any AO convolver can also be used for matrix processing. This approach was first developed by Whitehouse and Spieser [Whi77] and first implemented by Psaltis [Psa80]. Many other authors have investigated this approach, including Guifoyle [Gui89], who combined the systolic processor of Caulfield with the high accuracy of the DMAC algorithm.

The resulting architecture is a digital version of the optical systolic array processor and is known as the *systolic acousto-optic binary convolver* (SAOBIC). This system uses parallel AO devices oriented horizontally and vertically to modulate an incident beam of light and to provide the spatial mask in the Fourier plane. To process an N bit signal requires N parallel AO devices and N photodetectors in the output plane. Using currently available technology, if we assume a 500 MHz bandwidth AO device with a 250 detector array, then the architecture can perform multiplication of a 250×250 matrix by a 250 element vector in about 40 μs, with 20-bit accuracy; this corresponds to a throughput rate of 1.5×10^9 operations per second. Following an extensive analysis of this architecture, it has been concluded that the design places significant restrictions on the complexity and accuracy of the electronic post processors. An alternative approach is a fully digital optical computer with programmable, arbitrary bit length. The architecture is very similar to the SAOBIC processor, except that instead of parallel AO devices, a single device with frequency multiplexed inputs is used for the input and Fourier planes. By using AO devices fabricated from tellurium dioxide, an anisotropic material, up to 64 parallel acoustic channels can be supported in a single device. The architecture performs an optical Boolian vector-matrix multiplication; specifications for the general purpose architecture are given in [Gui89].

Using a 10 ns clock, the device can perform 100 million 62-bit matrices per second. The theoretical limits of this system have not yet been reached; external electronics will probably determine the ultimate limits of this design. Various other fully digital optical systems have been proposed, many of which have yet to be fully explored as potential implementations of the wavelet transform.

9.2.3 Other Optical Implementations

As noted earlier, the wavelet transform may be implemented using any type of convolver or correlator architecture. Hybrid optoelectronic systems have been proposed which attempt to take advantage of the relative strengths of both optics and electronics, as well as all optical parallel processing schemes. Optical systems are particularly well suited for image processing because of their inherent two-dimensional nature and fast processing speed. Many possible architectures have been demonstrated, including interferometric systems such as the Smartt interferometer [Zha94] and the photorefractive interferometric correlator [DeC98]. However, most optical implementations are variants of the classic 4f optical correlator and matched filter first proposed by Vander Lugt, in which the Fourier transform of the desired wavelet filter is implemented at the transform plane of an optical correlator. This type of architecture requires holographic techniques to implement the filters, which are, in general, complex functions so that both amplitude and phase information must be preserved. This method has been used to realize a non-real-time wavelet transformer, with each daughter wavelet recorded separately on a holographic filter and loaded sequentially into the optical path. Computer-generated holographic filters have been used in a similar approach as part of an automated pattern recognition system or image segmenter. In order to realize a real-time wavelet transformer, many different types of spatial light modulators have been proposed for this application. There is a general lack of availability for high-speed, programmable, two-dimensional spatial light modulators; however, experimental demonstrations have successfully used magneto-optic devices [Bur92], liquid-crystal light valves [She92], phase conjugate mirrors [Yu94], and acousto-optic devices [DeC96].

In particular, acousto-optic (AO) devices are well suited to the optical implementation of wavelet and subband transforms. We have previously discussed a hybrid optical architecture for implementation of the wavelet transform; this same approach can be extended to progressive pattern recognition. The proposed architecture uses acousto-optic finite impulse response filters with electronic feedback to realize a QMF. The discrete input signal (rows or columns of an image) are applied to the parallel array of acousto-optic cells, whereas the filter coefficients are implemented as an array of intensity-modulated light-emitting diodes (LEDs). As the input signals propagate along the acousto-optic delay lines, they are sampled optically by pulsing the LED array; the outputs, which constitute a partial wavelet transform of the input signal, are summed by a lens onto a detector array. By sampling the output at half the clock rate of the date (to implement downsampling by a factor of 2), the desired wavelet output is obtained and stored in a digital computer. The computer can recursively process the signal, so that rows of the image are processed first followed by columns, for a direct optical implementation of the subband tree decomposition.

Another possible architecture eliminates the need for a two-dimensional spatial light modulator by incorporating multiplexing techniques. Previous publications [DeC96, DeC97] have described acousto-optic image correlators, which use only a single acousto-optic cell to realize the correlation between an $M \times M$ pixel reference image $g(x, y)$ and a larger $N \times N$ input image, $f(x, y)$, where $N > M$. In this design, both images are described as a sequence of sequentially scanned rows. Each row of $g(t, m)$ temporally modulates an element of an LED array, which contains one LED for each row of the reference image. The modulated light is focused onto the input aperture of an acousto-optic cell, which is modulated by the other image $f(t,n)$. The diffracted light is reimaged onto a CCD array detector such that each LED is focused onto a separate row of the detector. The CCD thus contains the correlation between one row of the input image and all rows of the reference image. To obtain the remaining partial correlations, the CCD is clocked down one row and the process is repeated; the CCD accumulates and sums the partial correlations between a row of the reference and input image and the correlation between the previous rows of the reference and input images. The sum of partial one-dimensional correlations is equivalent to a single two-dimensional correlation and is well suited to implementation of the progressive correlations on subband features of the image. In other words, the system acts like a series of parallel one-dimensional time-integrating correlators; the sum of partial correlations is equivalent to the two-dimensional correlation of the original images. The processing speed of this system is limited by the clock rate of the CCD and the input memory readouts. It can be shown that this architecture processes images at a frame rate given by

$$f_m \frac{N}{M+N} \tag{9.2.2}$$

where f_m is the frequency of the master clock driving the LED array, AO device input, and CCD. In order to increase system throughput without using a faster clock, it is possible to reuse the information in the AO device several times before a new line of the input image is read out. With this modification, each LED illuminates a separate row of the CCD; these rows are now separated by several blank rows, shielded from illumination by a mask. When an input image line enters the AO device, it is correlated against all the rows of the reference image and the result is stored on the exposed row of the CCD. The CCD can then be clocked down one row before the information in the AO cell has moved no more than a few pixels. A second reference image can then be read into the LEDs and correlated against the same information in the AO cell. This result is stored on the exposed row of the CCD, whereas the previous partial correlation is protected by the mask. In this way, several reference images can be correlated against the same input image; the number of blank rows between open areas of the CCD corresponds to the number of reference images used and the number of times the data in the AO cell are reused. After the first row of the input image has been correlated against all the reference images, the CCD is once again positioned to accumulate the partial correlation between the second line of the input image and the first reference image. This approach represents an improvement in processing speed of about 8.8 times over the previous system.

The advantage of image correlators is their ability to perform a large number of 1-D correlations in parallel using only a 1-D AO device with moderate bandwidth requirements. Using a 30-50 MHz clock, it is possible to achieve better than 1000 frames per second throughput. Performance is not otherwise possible without the use of very high frequency CCDs (in which case accuracy is limited by electrical cross-talk) or by using a 2-D spatial light modulator. The use of image correlators to implement real-time wavelet transforms has also been proposed. One-dimensional wavelet filter banks can be encoded as two-dimensional images: Instead of correlating two images, it is possible to implement the wavelet transform of multiple signals in parallel, greatly reducing the processing time.

Another option is the use of so-called "smart pixel" devices, a relatively new technology which closely integrates silicon electronic circuitry and optical devices on a common chip [Sho97]. A two-dimensional array of smart pixels, which potentially includes optical sources and modulators, detectors, and electronic gain or signal processing functions, could provide both electrical and optical inputs and outputs. There have been many proposed fabrication methods for smart pixels, including monolithic integration, direct epitaxy, epitaxial liftoff, and other hybrid techniques. In particular, liquid crystal on silicon (LCOS) is a hybrid approach with tremendous potential for smart pixel applications. Electronic circuitry is first fabricated on a conventional complementary metal oxide (CMOS) silicon substrate, including regularly spaced metal contact pads on the chip surface. A cover glass coated with a transparent conductive material such as indium tin oxide (ITO) is then placed on polysilicon spacers and mounted on top of the silicon circuitry (alternately, the cover glass may be mounted atop the natural contours of the processed circuitry). This creates a cavity between the cover glass and the silicon circuitry, which is filled with liquid-crystal material. A thin layer of obliquely evaporated silicon monoxide or rubbed polyvinyl alcohol is deposited on the ITO; this makes contact with the liquid-crystal material and induces spatial alignment of the liquid-crystal molecules. The liquid crystal is thus in contact with both the alignment layer beneath the conductive coating on the glass and the metal pads on the silicon circuitry. An electric field between these two defines individual pixels at the location of the metal pads. If the pixels are then illuminated, modulation of the electric field results in a corresponding modulation of the phase, intensity, or polarization of the optical signal. In practice, a polarized optical beam is imaged through the liquid crystal onto the metal pads, which also act as high-reflectivity mirrors for the modulated optical beam. In the case of photodetectors, the liquid crystal is not dynamically modulated and the incident optical beam is simply absorbed by the silicon substrate. Because the smart pixels operate as a spatial light modulator in a reflective mode, the 4f correlator is modified slightly, but the basic architecture remains the same. For an initial proof of concept demonstration, an 8×8 pixel array has been experimentally demonstrated using a 2.0 μm silicon CMOS process on a single chip 6900 μm × 6800 μm square. By using a combination of acousto-optic modulators, smart pixels, and diffractive optical elements, it is possible to compress the wavelet transform architecture into a compact device with no moving parts. This design holds the potential to construct a "black box" optical wavelet transform processor which does not require adjustment of the focal planes and could be integrated into larger signal processing systems for practical applications.

Recently, a rotation-invariant pattern recognition scheme based on wavelet features was proposed for use in a joint transform correlator [DeC96]. An extension of this technique can be used to provide both scale and rotation invariance in a wavelet-based image processor (WIP). The use of wavelets offers several advantages to such a processor, compared with conventional approaches. For example, it is possible to optimize the correlation performance over a wide range of image scales and rotation angles through proper choice of the mother wavelet. By using features of the image in the wavelet domain, we can easily form a composite reference image which allows for rotation-invariant pattern recognition. Using the probability density function (PDF), or histogram of the input image, we can optimize scale invariance and improve performance in the presence of noise;

this optimization also allows us to choose a single scaling parameter, reducing the complexity of the wavelet transform from four dimensions to only two dimensions. Finally, since this is an analog approach, a possible implementation involves using acousto-optic image correlators; this makes it possible to realize two-dimensional optical processing using only a single, one-dimensional input device with appropriate multiplexing schemes.

In this chapter, we have covered many of the fundamental approaches to implementation of wavelet transforms using both optics and electronics. As this is a rapidly developing field, the reader is encouraged to consult the wide array of technical literature available in this area for recent developments and a review of wavelet theory fundamentals.

9.3   References

[Bur92]   T.J. Burns, Optical Haar wavelet transform, *Optical Engineering* Vol. 31, pp. 1852-1858, 1992.

[Cau81]   B. Caufield, W.T. Rhodes, M. Foster, and S. Horwity, Optical implementation of systolic array processing, *Opt. Comm.,*. vol. 40, p. 86, 1981.

[Cut60]   L.J. Cutrona, E. Leith, N. Palerrno, and L.J. Poreello, Optical data processing and filtering systems, *IRE Trans. Inform. Theory*, vol. IT-6, p. 386, 1960.

[Das91]   P. Das and C. DeCusatis, *Acousto-optics: Fundamentals and Applications*, Artech House, Boston, 1991.

[Das93]   P. Das and C. DeCusatis, A review of acousto-optic image correlators, *Proc. SPIE 5th Annual School Seminar on Acousto-Optics and Applications*, vol. 1844, pp. 33-48, 1993.

[DeC93]   C. DeCusatis and P. Das, Acousto-optic implementation of wavelet transforms, *OSA Annual Meeting Tech. Digest*, vol. 16, p. 208, 1993.

[DeC94]   C. DeCusatis, P. Das, and J. Koay, Perfect reconstruction wavelets using acousto-optic correlators, *Proc. OSA Annual Meeting*, 1994.

[DeC95]   C. DeCusatis, J. Koay, D.M. Litynski, and P. Das, The wavelet transform: fundamentals, applications, & implementation using acousto-optic correlators, *Proc. SPIE*, vol. 2643, pp. 17-37,1995.

[DeC96]   C. DeCusatis, A. Abbate, and P. Das, Wavelet transform based image processing using acousto-optics correlators, *Proc. SPIE Conf. on Wavelet Applications*, vol. 2762, pp. 302-313, 1996.

[DeC96a]   C. DeCusatis, A. Abbate, and P. Das, Scale and rotation invariant pattern recognition using a wavelet image processor, *Annual Meeting of the Optical Society of America*, 1996.

[DeC96b]   C. DeCusatis, J. Koay, and P. Das, Hybrid optical implementation of discrete wavelet transforms: a tutorial, *Opt. Laser Tech,*. vol. 28, no. 2, pp. 51-58, 1996.

[DeC97]   C. DeCusatis, A. Abbate, D.M. Litynski, and P. Das, Wavelet image processing for optical pattern recognition and feature extraction, *Proc SPIE*, vol. 3110, pp. 804-815, 1997.

[DeC97a]   C. DeCusatis, A. Abbate, and P. Das, Optical techniques for wavelet transform analysis, *Proc. Progress in Electromagnetics Research Symposium*, p. S19, 1997.

[DeC97b]   C. DeCusatis, A. Abbate, and P. Das, Progressive pattern recognition using the wavelet transform, *Int. J. Optoelectron.*, vol. 11, p. 425-432, 1997.

[DeC98]   C. DeCusatis and P. Das, Acousto-photorefractive holographic interferometric correlator for progressive pattern recognition using wavelet transforms, *Proc. SPIE*, vol. 3470, pp. 214-225, 1998.

[Gan93]   S. Ganesan,S. Mahalingam, and S. Nagabhushana, VLSI synthesis of a programmable DWT chip for the optimal choice of a prototype wavelet, *IEEE Computer*, pp. 127-131,1993.

[Goo78]   J.W. Goodman, A.R. Dias, and L.M Woody, Fully parallel, high speed incoherent optical method for performing discrete Fourier transforms *Opt. Lett.*, vol. 2, pp.1-3, 1978.

[Gui89]   P.S. Guilfoyle, Acousto-optic digital computer fundamentals, *Proc. IEEE Ultrasonics Symp.* pp. 481-486, 1989 .

[Hoy92]   J. D. Hoyt and H. Wechsler, The wavelet transform - a CMOS VLSI ASIC implementation, *Proc. 11th IAPR Int. Conf. on pattern recognition, vol. IV conf. D: Architectures for Vision and Pattern Recognition*, pp. 19-22, 1992.

[Kir90]   J. Kirsch, Light efficient joint transform correlator, *Proc. SPIE Optical Information Processing Systems and Architectures*, vol. 1347, pp. 147- 154, 1990.

[Kun81]   E.T. Kung, Why systolic architectures?, *IEEE Computer*, vol. 15, p. 37, 1981.

[Lew91]   A.S. Lewis and G. Knowles, VLSI architecture for 2-D Daubeches wavelet transform without multipliers, *Elec. Lett.* vol. 27, pp. 171-173, 1991.

[Mal89]   S. G. Mallat, A theory for multiresolution signal decomposition: the wavelet representation, *IEEE Trans. Pattern Anal. Machine Intell.*, vol. 11, n. 7, pp. 674-693, 1989.

[Par93]   K. Parhi and T. Nishitani, VLSI architectures for discrete wavelet transforms, *IEEE Trans. VLSI Systems*, vol. 1, pp. 191- 202, 1993.

[Pin91]   S.D. Pinski, Optical image segmentation using wavelet correlation, *SPIE Proc. Intelligent Robots and Computer Vision X*, vol. 1608, pp. 414-423, 1991.

[Psa80]   D. Psaltis, Accurate numerical computation by optical convolution, *Proc. SPIE*, vol. 262, p. 151, 1980.

[Sau92]   M. Sauer and J. Gotze, A VLSI architecture for fast wavelet computations, *Proc. IEEE SP Intl. Symp. On Time and Frequency Scale Analysis*, pp. 407-410, 1992.

[She92]   Y. Sheng et.al., Optical N implementation of a two-dimensional wavelet transform, *Opt. Eng.*, vol. 31, pp. 1859-1864, 1992.

[Sho97]   B.L. Shoop, A.H. Sayles, G.P. Dudevoir, D.A. Hall, D.M. Litynski, and P. K. Das, Smart pixel based wavelet transformation for wideband radar and sonar signal processing, *Proc. SPIE*, vol. 3078, pp. 415-423, 1997.

[Szu92]   H. Szu, B. Telfer, and S. Kadambe, Neural network adaptive wavelets for signal representation and classification, *Opt. Eng.*, vol. 31, p. 1907-1916, 1992.

[Vis92]   M. Vishwanath and R. Ownes, Discrete wavelet transforms in VLSI, *IEEE Int. Conf on App. Specific Array Processors*, pp. 218-229, 1992.

[Vis92a]   M. Vishwanath and R. Owens, An AT 2 lower bound for wavelet transforms in VLSI, *IEEE Int. Conf on App. Specific Array Processors*, pp. 649-652,1992.

[Vis92b]   M. Vishwanath, R. Owens, and M. Irwin, An efficient systolic architecture for QMF filter bank trees, *IEEE Workshop on VLSI Signal Processing*, 1992.

[Whi77]   H.J. Whitehouse and D. Spiezer, Linear signal processing architectures, *Aspects of Signal Processing with Emphasis on Underwater Acoustics*, edited by G. Tacconi, vol. 2, Reidel, Hingham, MA, 1977.

[Yu94]   F.T.S. Yu and G. Lu, Short time Fourier transform and wavelet transform with Fourier domain processing, *Appl. Optics* vol. 33, pp. 5262-5270, 1994.

[Zha94]   Y. Zhang et al., Optoelectronic wavelet processors based on Smartt interferometry, *Appl. Optics* 33, pp. 5279-5266, 1994.

Appendix

A   Fourier Transform

In many applications the quantity of interest is a transient signal s(t0 that characterizes the behavior of a system in the time domain. The pulse is transformed in the frequency domain using the Fourier transform:

$$S(\omega) = \frac{1}{\sqrt{2\pi}} \int_{-\infty}^{+\infty} s(t) e^{-j\omega t} dt$$

where the angular frequency is $\omega = 2\pi f$, and f is the frequency. Physically, $S(\omega)$ represents the magnitude and phase of the harmonic component of $s(t)$ with frequency ω and it is thus a complex function:

$$S(\omega) = |S(\omega)| \, e^{j\Phi(\omega)} = S_R(\omega) + j \, S_I(\omega) \quad .$$

The inverse Fourier transform is defined as

$$s(t) = \frac{1}{\sqrt{2\pi}} \int_{-\infty}^{+\infty} S(\omega) e^{j\omega t} d\omega$$

For a real signal, we have S($-\omega$)=S*(ω). The above equation states that the negative frequency components are the complex conjugate of the positive frequency components. This means that:

$$|S(-\omega)| = |S(\omega)|$$

$$\Phi(-\omega) = \Phi(\omega)$$

$$S_R(-\omega) = S_R(\omega)$$

$$S_I(-\omega) = -S_I(\omega)$$

A1   Properties of the Fourier Transform

Convolution of real signals

Given two real signals $f(t)$ and $g(t)$ with:

$$f(t) \leftrightarrow F(\omega)$$

$$g(t) \leftrightarrow G(\omega)$$

then

$$y(t) = f(t) * g(t) = \int_{-\infty}^{+\infty} f(\tau)\, g(t-\tau)\, d\tau = \int_{-\infty}^{+\infty} g(\tau)\, f(t-\tau)\, d\tau$$

$$y(t) \leftrightarrow Y(\omega) = F(\omega)\, G(\omega)$$

• **Symmetry**

$$f(t) \leftrightarrow F(\omega)$$

$$F(t) \leftrightarrow f(-\omega)$$

• **Shifting**

$$f(t) \leftrightarrow F(\omega)$$

$$f(t-\tau_0) \leftrightarrow F(\omega)\, e^{-j\omega\tau_0}$$

- Scaling

$$f(t) \leftrightarrow F(\omega)$$

$$f(at) \leftrightarrow \frac{F(\omega/a)}{a}, \quad \text{for } a > 0$$

- Conjugation

$$f^*(t) \leftrightarrow F^*(-\omega)$$

$$f^*(-t) \leftrightarrow F^*(\omega)$$

- Differentiation

$$f^{(n)}(t) = \frac{d^n f(t)}{dt^n} \leftrightarrow (-j\omega)^n S(\omega)$$

$$\text{if } f^{(m)}(t) \rightarrow 0, \quad \text{as } |t| \rightarrow \infty, \quad \text{for } m = 0, 1, .., n-1$$

- Integration

$$\int_{-\infty}^{t} f(\tau)d\tau \leftrightarrow \pi F(0)\, \delta(\omega) - \frac{F(\omega)}{j\omega}$$

$$\pi f(0)\, \delta(t) + \frac{f(t)}{jt} \leftrightarrow \int_{-\infty}^{\omega} f(\xi)\, d\xi$$

- Convolution

$$\int_{-\infty}^{+\infty} s_1^*(\tau)\, s_2(t-\tau)d\tau \leftrightarrow S_1(\omega)\, S_2(\omega)$$

$$s_1(t)\, s_2(t) \leftrightarrow \frac{1}{2\pi} \int_{-\infty}^{+\infty} S_1(\xi)\, S_2(\omega - \xi)\, d\xi$$

- Correlation

$$\int_{-\infty}^{+\infty} s_1^*(\tau)\, s_2(t+\tau) d\tau \;\leftrightarrow\; S_1^*(\omega)\, S_2(\omega)$$

$$s_1^*(t)\, s_2(t) \;\leftrightarrow\; \frac{1}{2\pi} \int_{-\infty}^{+\infty} S_1^*(\xi)\, S_2(\omega+\xi) d\xi$$

- Parseval s Theorem

$$\int_{-\infty}^{+\infty} s_1^*(\tau)\, s_2(\tau) d\tau \;\leftrightarrow\; \frac{1}{2\pi} \int_{-\infty}^{+\infty} S_1^*(\omega)\, S_2(\omega) d\omega$$

If $s(t) = s_1(t) = s_2(t)$, then

$$\| s \|^2 = \int_{-\infty}^{+\infty} |s(\tau)|^2 d\tau \;\leftrightarrow\; \frac{1}{2\pi} \int_{-\infty}^{+\infty} |S(\omega)|^2 \, d\omega$$

B   Discrete Fourier Transform

In most modern systems, the signals are digitized and processed by computers. In this section, we examine the consequences of dealing with a sampled signal and consider how the Fourier transform can be efficiently computed using the discrete samples.

To begin, let us rewrite the Fourier transform using the frequency f instead of the angular frequency ω:

$$S(f) = \int_{-\infty}^{+\infty} s(t)\, e^{j2\pi ft}\, dt$$

$$s(t) = \int_{-\infty}^{+\infty} S(f)\, e^{-j2\pi ft}\, df \;.$$

Now, we consider the discrete (sampled) values of the signal $s(t)$ at times t_k = $k\,\Delta t$, where $k = 0, \pm 1, \pm 2, .., \infty$. We also define the sampling frequency $f_S = 1/\Delta t$. Hence, we can write:

$$s(t_k) = \int_{-\infty}^{+\infty} S(f) e^{-j2\pi f\, t_k}\, df$$

but for $t_k = k\,\Delta t = k\,/\,f_S$:

$$s(t_k) = \int_{-\infty}^{+\infty} S(f) e^{-j2\pi \frac{f}{f_S} k}\, df \ .$$

We can choose to separate the integral into a combination of smaller integrals of width f_S :

$$s(t_k) = \sum_{m=-\infty}^{+\infty} \int_{m f_S}^{(m+1) f_S} S(f) e^{-j2\pi \frac{f}{f_S} k}\, df$$

but since the complex exponential function is periodic of period 2π [i.e., for any integer n, $\exp(-j2\pi n) = 1$], we can rewrite the exponential function in the previous equation as

$$e^{-j2\pi \frac{f}{f_S} k} = e^{-j2\pi \frac{f}{f_S} k}\, e^{-j2\pi k m}$$

$$e^{-j2\pi \frac{f}{f_S} k} = e^{-j2\pi \frac{(f-m f_S)}{f_S} k} \ .$$

By using the change of variable $u = f - m f_S$, we thus have

$$\int_{m f_S}^{(m+1) f_S} S(f) e^{-j2\pi \frac{f}{f_S} k}\, df = \int_{0}^{f_S} S(u + m f_S) e^{-j2\pi \frac{u}{f_S} k}\, du \ .$$

It follows that

$$s(t_k) = \sum_{m=-\infty}^{+\infty} \int_0^{f_s} S(u+mf_s)e^{-j2\pi\frac{u}{f_s}k}\,du$$

If we again change the variable to $f = u$ and we define the periodic function

$$S_P(f) = \sum_{m=-\infty}^{+\infty} S(f+mf)\ ,$$

then we have

$$s(t_k) = \int_0^{f_s} S_P(f)\,e^{-j2\pi\frac{f}{f_s}k}\,df\ .$$

Since the function $S_P(f)$ is a periodic function in f with a period f_s, it can be expanded into a Fourier series as

$$S_P(f) = \sum_{l=-\infty}^{+\infty} C_l\,e^{j2\pi\frac{f}{f_s}l}$$

with

$$C_l = \frac{1}{f_s} \int_0^{f_s} S_P(f)e^{-j2\pi\frac{f}{f_s}l}\,df$$

but we can compare this equation with the expression of $s(t_k)/f_s$ and we finally have

$$S_P(f) = \frac{1}{f_s} \sum_{k=-\infty}^{+\infty} s(t_k)\,e^{j2\pi\frac{f}{f_s}k}\ .$$

This equation is extremely important as it shows that we can obtain the $S_P(f)$ using the discrete samples of $s(t)$. The function $S_P(f)$ differs from the original $S(f)$ by the fact that $S_P(f)$ is constructed by sum of multiple shifted versions of $S(f)$. The shift is given by the sampling frequency f_S.

The definition of $S_P(f)$ thus introduces the concept of *aliasing*. If $S(f)$ is band-limited [i.e., $S(f) = 0$ for $|f| > f_{\text{MAX}}$], then we must impose $f_S \geq f_{\text{MAX}}$. This is the well-known Nyquist sampling criterion, which ensures that a true replica of the original signal can be recovered by low-pass filtering $S_P(f)$ in the frequency domain and then inverting in the time domain.

Now, let us suppose that we also sample the signal in the frequency domain at $f_n = n\,\Delta f$, $n = 0, \pm 1, \pm 2, ..$, where $\Delta f = 1\,/\,T$ and T is the total time interval covered by the sampled signal. Please note that $T \gg \Delta t = 1\,/\,f_S$, as previously discussed.

Using the expression $f_n = n\,\Delta f$, we can write

$$S_P(f_n) = \frac{1}{f_S} \sum_{k=-\infty}^{+\infty} s(t_k)\, e^{j2\pi \frac{f_n}{f_S} k} = \frac{1}{f_S} \sum_{k=-\infty}^{+\infty} s(t_k)\, e^{j2\pi \frac{n}{T f_S} k} \quad .$$

However the product $T f_S$ is an integer N (i.e., $N = T f_S$), and since $\exp(j2\pi kn/N)$ is a periodic function with period N:

$$S_P(f_n) = \frac{T}{N} \sum_{k=0}^{N-1} s_P(t_k)\, e^{j2\pi \frac{n}{N} k}$$

where we have defined a periodic function in the time domain:

$$s_P(t_k) = \sum_{k=-\infty}^{+\infty} s(t_k + kT) \quad .$$

We can now introduce the concept of *time domain aliasing* by choosing a T large enough that the replicas of $s(t)$ for $|k| > 1$ do not overlap in time.

Given that aliasing is negligible, we now have an equation that links the sampled values of the signal $s(t)$ to the sampled values of its frequency components:

$$S_P(f_n) = \frac{T}{N} \sum_{k=0}^{N-1} s_P(t_k)\, e^{j2\pi \frac{nk}{N}}$$

$$s_P(t_k) = \frac{1}{T} \sum_{n=0}^{N-1} S_P(f_n)\, e^{-j2\pi \frac{nk}{N}}$$

with

$t_k = k/f_S$, $k = \{0,1,..,N-1\}$, f_S = sampling frequency
$f_n = n/T$, $n = \{0,1,..,N-1\}$, T = total signal duration

In the absence of aliasing, the discrete Fourier transform provides an explicit way of calculating sampled values of the Fourier transform $S(f)$ directly from sampled values of the signal $s(t)$, and vice versa.
In the following we will write:

$$s(k) = s(t_k) = s[k] \quad \text{and} \quad S(n) = S(f_n) = S[n]$$

B1.   Properties of the Discrete Fourier Transform

- <u>Symmetry of real discrete-time signals</u>

$$\text{Re}\{S(N-n)\} = \text{Re}\{S(n)\}$$
$$\text{Im}\{S(N-n)\} = -\text{Im}\{S(n)\}$$

where Re and Im represent the real and imaginary components of the DFT, respectively.

- **Even symmetry signals**

If $s(n)$ is an even signal around the zero [i.e., $s(n) = s(-n)$], then

$$S(n) = \sum_{k=0}^{N-1} s(k) \cos\left(\frac{kn}{N}\right)$$

- **Odd symmetry signals**

If $s(n)$ is an odd signal around the zero, [i.e., $s(n) = - s(-n)$], then

$$S(n) = -j \sum_{k=0}^{N-1} s(k) \sin\left(\frac{kn}{N}\right)$$

- **Parseval s theorem**

$$\sum_{k=0}^{N-1} s^2(k) = \frac{1}{N} \sum_{n=0}^{N-1} |S(n)|^2$$

- **DFT of delta function**

$$s(k) = \delta(k) \leftrightarrow S(n) = 1$$

- **Circular cross-correlation between two data series**

The circular cross-correlation of two finite length sequences $s_1(k)$ and $s_2(k)$, each of length N, is defined to be

$$r_{s_1 s_2}(l) = \frac{1}{N} \sum_{k=0}^{N-1} s_1(k)\, s_2(k+l)\,, \quad l = 0,\ldots,N-1$$

can be calculated using the DFTs of the two sequences $S_1(n)$ and $S_2(n)$:

$$r_{s_1 s_2}(l) = \frac{1}{T} \sum_{n=0}^{N-1} S_1^*(n)\, S_2(n) e^{-j2\pi \frac{n}{N} l}$$

C   z-Transform

A discrete-time signal is represented as a sequence of numbers:

$$s(n) = s(nT) = s[n] = s_n.$$

The z-transform of the signal is given by

$$S(z) = Z[s(n)] = \sum_{n=-\infty}^{+\infty} s(n)\, z^{-n}\,.$$

The inverse z-transform allows us to recover the discrete-time sequence. The z-transform is very useful in digital signal processing, as it is used to describe the impulse response of digital filters. In general, the z-transform is expressed as a power series:

$$S(z) = \sum_{n=0}^{+\infty} s(n)\, z^{-n}$$

$$= s(0) + s(1)z^{-1} + s(2)z^{-2} + s(3)z^{-3} + \ldots$$

where the coefficients of the series are exactly the sampled values of the series.

The above expression describes an infinite response [i.e., the impulse response of an IIR (infinite impulse response) filter]. If the signal has a finite length N, we can utilize the more explicit FIR (finite impulse response) expression:

$$S(z) = \frac{a_0 + a_1 z^{-1} + a_2 z^{-2} + .. + a_N z^{-N}}{b_0 + b_1 z^{-1} + b_2 z^{-2} + .. + b_M z^{-M}} .$$

D   Orthogonal Representation of Signals

We will now describe in more detail the concepts of orthogonal representation of signals and the orthonormal discrete wavelet transform. The concept of orthogonal signal representations should be familiar to those with a basic background in the fundamentals of linear algebra and matrix theory. The basic idea is to represent a signal $s(t)$ as a linear combination of other functions:

$$s(t) = \int F(a) u(a,t) \, da$$

where $\{u(a,t)\}$ represents the set of basis functions and $F(a)$ represent the "expansion coefficients, given by

$$F(a) = \int s(t) \, u^*(a,t) \, dt .$$

An example of orthogonal signal representation is the well-known Fourier transform:

$$u(a,t) = e^{j\omega t}\Big|_{\omega = a}$$

$$s(t) = \int F(\omega)e^{j\omega t}\, d\omega$$

$$F(\omega) = \int s(t)e^{-j\omega t}\, dt \quad .$$

The term $F(a_0)$ thus represents the weighting of the function $u(a_0,t)$ in the expansion of $s(t)$. While this example was given in terms of continuous time functions, the same principle applies to discrete signals. If we utilize a discrete number of coefficients a_n, the previous integral becomes a summation:

$$s(t) = \sum_{n=1}^{\infty} c_n\, u_n(t)$$

$$c_n = \int u_n^*(t)\, s(t)dt \quad .$$

In both cases, a reduced representation of the signal is obtained if the basis functions are *orthonormal*; that is, if they satisfy the following relationship:

$$\int u_m^*(t)\, u_n(t)\, dt = \delta_{nm}$$

$$\sum_{n=1}^{\infty} u_n^*(t)\, u_n(t_0) = \delta(t - t_0) \quad .$$

The classical approach of expanding signals in a series such as those shown above has its limitations; in particular, there are no local Fourier series expansions that have a high degree of localization in both time and frequency. The first alternative to this approach is known as the *Haar basis* where, in addition to time-shifting, one uses scaling instead of modulation in order to obtain an orthonormal basis for $L^2(\mathbb{R})$. We would like to find a set of orthonormal continuous-time functions $\{\varphi_k(t)\}$ such that a signal $s(t)$ can be expressed as a linear combination of them:

$$s(t) = \sum_{k=0}^{\infty} \left\langle \varphi_k(u), s(u) \right\rangle \varphi_k(t)$$

$$\left\langle \varphi_k(u), s(u) \right\rangle = \int_{-\infty}^{+\infty} \varphi_k^*(u) f(u) du \quad .$$

In other words, the signal $s(t)$ can be written as the sum of its orthogonal projections onto the basis vectors $\{\varphi_k(t)\}$. There are two conditions which the set of basis functions must satisfy. First, the set must satisfy the orthogonality condition

$$\left\langle \varphi_k(u), \varphi_l(u) \right\rangle = \delta_{kl} = \delta(k-l) \quad .$$

Second, the set of basis functions must be complete. We can illustrate both of these properties with the well-known Fourier series expansion. We begin with a periodic function $f(t+T) = f(t)$ which can be written as a linear combination of sine and cosine functions or complex exponentials

$$f(t) = \sum_{k=-\infty}^{+\infty} F(k) e^{+j2\pi kt/T}$$

where the $F(k)$ terms represent the Fourier coefficients

$$F(k) = \frac{1}{T} \int_{-T/2}^{T/2} e^{-j2\pi kt/T} f(t)\, dt \quad .$$

It is easy to verify that this set is orthogonal and complete, since it obeys the relationship

$$\left\langle e^{j2\pi kt/T}, e^{j2\pi lt/T} \right\rangle_{[-T/2, T/2]} = T\, \delta(k-l) \quad .$$

Unfortunately, this expansion has a very slow decay in time, which limits the temporal resolution and makes for an inaccurate representation of some types of signals. It could be more useful to construct a series expansion based on wavelets, which have a higher resolution in time.

If we are free to choose the mother wavelet which forms the basis of such an expansion, it is natural to ask what desirable features this basis function should have. First, it should allow a characterization of the functions of interest in as simple a manner as possible. Second, it should have good localization properties in both time and frequency (i.e., appropriately fast decay in both domains). Third, it should be invariant under certain elementary operations (e.g., time-shifting). Fourth, it should be continuous, differentiable and, in general, be a smooth well-behaved function. Finally, we can attempt to choose a function such that its zero and higher-order moments have desirable properties. Some of these requirements conflict with each other and, ultimately, the application at hand will greatly influence the choice of the basis.

Bibliography

Books

[Aka94]   A. N. Akansu and R.A. Haddad, *Multiresolution Signal Decomposition Transforms, Subbands, Wavelets*, Academic Press, NY, 1994.

[Aka99]   A.N. Akansu and M.J. Medley, eds., *Wavelet, Subband and Block Transforms in Communications and Multimedia*, Kluwer, Boston, MA, 1999.

[Bel90]   T. C. Bell, J. G. Cleary, and 1. H. Witten, *Text Compression*, Prentice-Hall, Englewood Cliffs, NJ, 1990.

[Ben94]   J.J. Benedetto and D.M. Frasier, eds., *Wavelets, Mathematics and Applications*, CRC Press, Boca Raton, FL, 1994.

[Bir95]   V. Biryukov, Y.V. Gulayev, V.V. Krylov, and V.P. Plessky, *Surface Acoustic Waves in Inhomogeneous Media*, Springer-Verlag, NY, 1995.

[Bra89]   D.E. Bray and R.K. Stanley, *Nondestructive Evaluation: A Tool in Design, Manufacturing and Service*, McGraw-Hill Book Company, NY, 1989.

[Chu92]   C.K. Chui, *An Introduction to Wavelets*, Academic Press, NY, 1992.

[Chu95]   C.K. Chui, *Wavelets: Theory, Algorithms, and Applications*, Academic Press, NY, 1995.

[Coh95]   L. Cohen, *Time-Frequency Analysis*, Prentice-Hall, Englewood Cliffs, NY, 1995.

[Cro83]   R.E. Crochiere and L.R. Rabiner, *Multirate Digital Signal Processing*, Prentice Hall, Englewood Cliffs, NJ, 1983.

522   Bibliography

[Dah97]   W. Dahmen, A. Kurdila, and P. Oswald, *Multiscale Wavelet Methods for Partial Differential Equations*, Academic Press, NY, 1997.

[Das91]   P. Das and C. DeCusatis, *Acousto-Optics: Fundamentals and Applications*, Artech House, Boston, MA, 1991.

[Dau92]   I. Daubechies, *Ten lectures on Wavelets*, SIAM, Philadelphia, 1992.

[Dud84]   D.E. Dudgeon and R.M. Mersereau, *Multidimensional Digital Signal Processing*, Prentice-Hall, Englewood Cliffs, NJ, 1984.

[Fou94]   E. Foufoula-Georgiou and P. Kumar, eds., *Wavelets in Geophysics*, Academic Press, NY, 1994.

[Hay88]   S. Haykin, *Digital Communications*, Wiley, NY, 1988.

[Ife93]   E.C. Ifeachor and B.W. Jervis, *Digital Signal Processing. A Practical Approach*, Addison-Wesley, Reading, MA, 1993.

[Jay84]   N. S. Jayant and P. Noll, *Digital Coding of Waveforms*, Prentice-Hall, Englewood Cliffs, NJ, 1984.

[Kai95]   G. Kaiser, *A Friendly Guide to Wavelets*, Birkhäuser, Boston, 1995.

[Kin87]   G.S. Kino, *Acoustic Waves. Design, Imaging and Analog Signal Processing*. Prentice-Hall, Englewood Cliffs, NJ, 1987.

[Kot94]   P. Kotidis, J. Woodroffe, J. Shah, and T. Schultz, Testing of materials using laser ultrasound, *Nondestructive Characterization of Materials*, Vol. IV, edited by R.E. Green, Plenum Press, NY, 1994, pp. 21-28.

[Kro91]   R. Kronland-Martinet and A. Grossmans, Application of time-frequency and time-scale methods (wavelet transform) to the analysis, synthesis and transformations of natural sounds, *Representation of Musical Signals*, MIT Press, Cambridge MA, 1991, pp. 45-85.

[Lak95]   F. Lakestani, J.F. Coste, and R. Denis, *NDT&E International*, Elsevier Science Ltd, Amsterdam, 1995, vol. 15, pp. 171-178.

[Mal98]   S. Mallat, *A Wavelet Tour of Signal Processing*, Academic Press, San Diego, 1998.

[Mar91]   R.J. Marks II, *Introduction to Shannon Sampling Theory*, Springer-Verlag, NY, 1991.

[Mat96]   The Mathworks Inc., *Wavelet Toolbox User's Guide*, Natick, MA, 1996.

[Mey89]   Y. Meyer, Orthonormal wavelets, in *Wavelets: Time-Frequency, Methods and Phase Space*, edited by J.M. Combes, A. Grossman, and Ph. Tchamitchain, Springer-Verlag, NY, 1989.

[Mey93]   Y. Meyer, *Wavelets. Algorithms and Applications*, translated by R.D. Ryan, SIAM, Philadelphia, 1993.

[Net95]   S. Nettel, *Wave Physics. Oscillations-Solitons-Chaos*, Springer-Verlag, NY, 1995.

[New93]   D.E. Newland, *An Introduction to Random Vibrations, Spectral and Wavelet Analysis*, Wiley, NY, 1993.

[Odg97]   R.T. Odgen, *Essential Wavelets for Statistical Applications and Data Analysis*, Birkhäuser Press, Boston, MA, 1997.

[Opp89]   A.V. Oppenheim and R.W. Schafer, *Digital Signal Processing*, Prentice-Hall, Englewood Cliffs, NJ, 1989.

[Pap77]   A. Papoulis, *Signal Analysis*, McGraw-Hill, NY, 1977.

[Pap84]   A. Papoulis, *Probability, Random Variables, and Stochastic Processes*, McGraw-Hill, NY 1984.

[Pra91]   W. K. Pratt, *Digital Image Processing*, Wiley, NY, 1991.

[Pra97]   L. Prasad and S.S. Iyengar, *Wavelet Analysis with Applications to Image Processing*, CRC Press, Boca Raton, FL, 1997.

[Pre86]   W. H. Press. B. P Flannery, S. A. Teukolsky, and W. T. Vetterling, *Numerical Recipes: The Art of Scientific Programming*, Cambridge University Press, Cambridge, UK, 1986.

[Pro88]   J.G. Proakis and D.G. Manolakis, *Introduction to Digital Signal Processing*, Macmillan Publ. Co., NY, 1988.

[Pro95]   J.G. Proakis, *Digital Communications*, 3rd ed., McGraw-Hill, NY, 1995.

[Rab78]   L.R. Rabiner and R.W. Schafer, *Digital Signal Processing of Speech Signals*, Prentice-Hall, Englewood Cliffs, NJ, 1978.

[Rab91]   M. Rabbani and P. W. Jones, *Digital Image Compression Techniques*, SPIE, Bellingham, WA, 1991.

[Ros99]   J.L. Rose, *Ultrasonic Waves in Solid Media*, Cambridge University Press, Cambridge, MA, 1999.

[Rus92]   M.B. Ruskai, G. Beylkin, R. Coifman, I. Daubechies, S. Mallat, Y. Meyer, and L. Raphael, eds., *Wavelets and their Applications*, Jones and Bartlett Publ., 1992.

[Sch96]   L.L. Schumaker and G. Webb, eds., *Recent Advances in Wavelet Analysis*, Academic Press, NY, 1994.

[Scr90]   C.B. Scruby and L.E. Drain, *Laser Ultrasonics. Techniques and Applications*, Adam Hingler, IOP Publishing Ltd., Bristol, UK, 1990.

[Str96]   G. Strang and T. Nguyen, *Wavelets and Filter Banks*, Wellesley-Cambridge Press, Cambridge, UK, 1996.

[Sto96]   E.J. Stollnitz, T.D. Derose, D.H. Salesin, *Wavelets for Computer Graphics, Theory and Applications*, Morgan Kaufmann Publishers, Inc., San Francisco, 1996.

[Sut97]   B.W. Suter, *Multirate and Wavelet Signal Processing*, Academic Press, NY, 1997.

[Teo98]   A. Teolis, *Computational Signal Processing with Wavelets*, Birkhäuser, Boston, MA, 1998.

[Vai93]   P.P. Vaidyanathan, *Multirate Systems and Filter Banks*, Prentice-Hall, Englewood Cliffs, NJ, 1993.

[Vet95]   M. Vetterli and J. Kovačević, *Wavelets and Subband Coding*, Prentice-Hall, Englewood Cliffs, NJ, 1995.

[Vik67]   I.A. Viktorov, *Rayleigh and Lamb Waves. Physical Theory and Applications*, Plenum Press, NY, 1967.

[Wic94]   M.V. Wickerhauser, *Adapted Wavelet Analysis from Theory to Software*, AK Peters, Wellesley, UK, 1994.

[Woo91]   J. W. Woods, ed., *Subband Image Coding,* Kluwer, Boston, 1991.

Journal Articles and Other Publcations

[Abb94]   A. Abbate, J. Koay, J. Frankel, S.C. Schroeder, and P. Das, Application of wavelet transform signal processor to ultrasound, *Proc. of the 1994 IEEE International Ultrasonic Symposium*, 1994, pp. 1147-1152.

[Abb95]   A. Abbate, M. Doxbeck, and P. Das, Applications of wavelet transform in signal processing, *Proc. of the International Conference on Signal Processing Applications and Technology*, 1995, pp. 652-655.

[Abb95a]   A. Abbate, Wavelet transform applied to ultrasonics, US Army Tech. Rep. ARCCB-TR-95013, 1995.

[Abb95b]   A. Abbate, J. Frankel, and P. Das, Wavelet transform signal processing applied to ultrasonics, *Proc. of the 1995 QNDE*, 1995, pp. 741-748.

[Abb95c]   A. Abbate, J. Frankel, and P. Das, Wavelet transform signal processing for dispersion analysis of ultrasonic signals, *Proc. of the 1995 IEEE International Ultrasonic Symposium*, 1995, vol. 1, pp. 751-755.

[Abb96]   A. Abbate, J. Frankel, R.W. Reed, and P. Das, Ultrasonic gauging and wavelet image processing for wear and erosion mapping, *Proc. of the 1996 QNDE*, 1996.

[Abb96a]   A. Abbate, J. Frankel, R.W. Reed, and P. Das, Wavelet image processing for wear and erosion mapping using an ultrasonic gauging technique, *Proc.1996 Ultrasonic Symposium*, 1996, pp. 705-707.

[Abb96b]   A. Abbate and P. Das, Wavelet transform signal processing of electroencephalograph signals, US Army Tech. Rep. ARCCB-TR-96007, 1996.

[Abb96c]   A. Abbate, A. Nayak, J. Koay, R.J. Roy, and P. Das, Biomedical application of wavelets: analysis of electroencephalograph signals for monitoring depth of anesthesia, *SPIE Proceedings*, vol. 2762, pp. 412-423, 1996.

[Abb97]   A. Abbate, J. Koay, J. Frankel, S.C. Schroeder, and P. Das, Signal detection and noise suppression using a wavelet transform signal processor: application to ultrasonic flaws detection, *IEEE Trans. on Ulrason. Ferroelectr. Frequency Control*, vol. 44, no. 1, pp. 14-26, 1997.

[Abb97a]   A. Abbate, J. Frankel, and P. Das, Application of wavelet image processing for ultrasonic gaging, *Proc. 1997 SPIE Conference on Wavelets*, 1997.

[Abb98]   A. Abbate, W. Russell, J. Goldmann, P. Kotidids, and C.C. Berndt, Nondestructive determination of thickness and elastic modulus of plasma spray coatings using laser ultrasonics, *Rev. Prog. QNDE*, vol. 18, pp. 373-380, 1998.

[Abb98a]   A. Abbate, D. Klimek, P. Kotidis, and B. Anthony, Analysis of dispersive ultrasonic signals by the ridges of the analytic wavelet transform, *Rev. Prog. QNDE*, vol. 18, pp. 703-710, 1998.

[Ade87]   E. H. Adelson, E. Simoncelli, and R. Hingomni, Orthogonal pyramid transforms for image coding, *Proc. SPIE*, vol. 845, pp. 50-58, 1987.

526   Bibliography

[Ahm95]   F. Ahmed, M.A. Karim, and M.S. Alam, Wavelet transform based correlator for the recognition of rotationally distorted images, *Opt. Eng.*, vol. 34, pp. 3187-3192, 1995.

[Aka98]   A.N. Akansu, P. Duhamel, X. Lin, and M. de Courville, Orthogonal transmultiplexers in communication: A review, *IEEE Trans. Signal Process.*, vol. 46, pp. 979-995, 1998.

[And93]   J.C. Anderson, A wavelet magnitude analysis theorem, *IEEE Trans. Signal Process.*, vol. 41, pp. 3541-3543, 1993.

[Ans88]   R. Ansari, H. Gaggioni, and D. J. LeGall, HDTV coding using a nonrectangular subband decomposition, *Proc. SPIE Conf. Visual Commun. Image Processing*, 1988, pp. 821-824.

[Aus94]   J.D. Aussel and J.P. Monchalin, *Rev. Progr. QNDE*, edited by D.O. Thompson and D.E. Chimenti, Plenum Press, NY, 1994, pp. 535-542.

[Bar92]   A. Baraniecki and S. Karim, Computational algorithms for discrete wavelet transforms, *Proc. SPIE,* vol. 1699, pp. 408-419, 1992.

[Ben90]   J.J. Benedetto and W. Heler, Irregular sampling and the theory of frames, *Math. Note*, vol. 10, pp. 103-125, 1990.

[Ben91]   A. Benveniste, Multiscale signal processing: from QMF to wavelets, *Int. Workshop on Algorithms and Parallel VLSI Architectures*, Elsevier, Amsterdam, 1991, pp.71-76.

[Bij96]   A. Bijaoui, E. Slezak, F. Rué, and E. Lega, Wavelets and the study of the distant universe, *Proc. IEEE*, vol. 84, n. 4, pp. 670-679, 1996.

[Blo94]   P. Block, S. Rogers, and D. Ruck, Optical wavelet transforms from computer generated holography, *Appl. Opt.*, vol. 33, pp. 5275-5278, 1994.

[Bur83]   P.J. Burt and E.H. Andelson, The Laplacian pyramid as a compact image code, *IEEE Trans. Commun.* vol. COM-31, pp. 532-540, 1983.

[Bur92]   T.J. Burns, Optical Haar wavelet transform, *Opt. Eng.*, vol. 31, pp. 1852-1858, 1992.

[Cas84]   D.P. Casasent, Unified synthetic discriminant function computational formulations, *Appl. Opt.*, vol. 23, pp. 1620-1627, 1984.

[Cau81]   B. Caufield, W.T. Rhodes, M. Foster, and S. Horwity, Optical implementation of systolic array processing, *Opt. Comm.*, vol. 40, p. 86, 1981.

[Cer85]   S Cerruti, D. Liberati, and P. Mascellani, Parameter extraction in EEG processing during riskful neurosurgical operations, *Signal Process.*, vol. 9, pp. 25-35, 1985.

[Che84]   W. Chen and W. Pratt, Scene adaptive coder, *IEEE Trans. Commun.*, vol. COM-32, pp. 225-232, 1984.

[Cho89]   H.I. Choi and W.J. Williams, Improved time-frequency representation of multicomponent signals using exponential kernels, *IEEE Trans. Acoust. Speech Signal Proc.*, vol. 37, pp. 862-871, 1989.

[Coh89]   L. Cohen, Time-frequency distributions - A review, *Proc. IEEE*, vol. 77 n. 7, pp. 941-981, 1989.

[Coh92]   A. Cohen, I. Daubechies, and J.C. Feauveau, Bi-orthogonal bases of compactly supported wavelets, *Commun. Pure Appl. Math.*, vol. 45, pp. 485-560, 1992.

[Coh93]   L. Cohen, The scale representation, *IEEE Trans. Signal Process.*, vol. 41, pp. 3275-3292, 1993.

[Coh96]   A. Cohen and J. Kovačević, Wavelets: the mathematical background, *Proc. IEEE*, vol. 84, n. 4, pp. 514-522, 1996.

[Coi92]   R.R. Coifman and M. V. Wickerhauser, Entropy-based algorithms for best basis selection, *IEEE Trans. Inform. Theory*, vol. 38, pp. 713-718, 1992.

[Cut60]   L.J. Cutrona, E. Leith, N. Palerrno, and L.J. Poreello, Optical data processing and filtering systems, *IRE Trans. Info. Theory,* vol. IT-6, p. 386, 1960.

[Das93]   P. Das and C. DeCusatis, A review of acousto-optic image correlators, *Proc. SPIE 5ᵗʰ Annual School Seminar on Acousto-optics and Applications*, vol. 1844, pp. 33-48, 1993.

[Dau88]   I. Daubechies, Orthonormal bases of compactly supported wavelets, *Commun. Pure Appl. Math*, vol. XLI, pp. 909-996, 1988.

[Dau89]   I. Daubeches and J. Lagarias, Two scale differential equations, 11 local regularity, infinite products of matricies, and fractals, AT&T Bell Labs Tech. Report, 1989.

[Dau90]   I. Daubechies, The wavelet transform, time-frequency localization and signal analysis, *IEEE Trans. Inform. Theory*, vol. 36, pp. 961-1005, 1990.

[Dau91]   I. Daubechies, The wavelet transform: a method for time-frequency localization, in *Advances in Spectrum Analysis and Array*

Processing, edited by S. Haykins, Prentice Hall, Englewood Cliffs, NJ, 1991, pp. 366-417.

[Dau85]   J.G. Daugman, Uncertainty relation for resolution in space, spatial frequency, and orientation optimized by two-dimensional visual cortical filters, *J. Opt. Soc. Am.*, vol. A2, 1160-1169, 1985.

[DeC93]   C. DeCusatis and P. Das, Acousto-optic implementation of wavelet transforms, *OSA Annual Meeting Tech. Digest* vol. 16, p. 208, 1993.

[DeC94]   C. DeCusatis, P. Das, and J. Koay, Perfect reconstruction wavelets using acousto-optic correlators, *Proc. OSA Annual Meeting*, 1994.

[DeC95]   C. DeCusatis, J. Koay, D.M. Litynski, and P. Das, The wavelet transform: fundamentals, applications, & implementation using acousto-optic correlators, *Proc. SPIE*, vol. 2643, pp. 17-37,1995.

[DeC96]   C. DeCusatis, A. Abbate, and P. Das, Wavelet transform based image processing using acousto-optics correlators, *Proc. of 1996 SPIE Conf. on Wavelet Applications*, vol. 2762, pp. 302-313, 1996.

[DeC96a] C. DeCusatis, A. Abbate, and P. Das, Scale and rotation invariant pattern recognition using a wavelet image processor, *Annual Meeting of the Optical Society of America*, 1996.

[DeC96b] C. DeCusatis, J. Koay, and P. Das, Hybrid optical implementation of discrete wavelet transforms: a tutorial, *Opt. Laser Tech.* vol. 28, no. 2, pp. 51-58, 1996.

[DeC97]   C. DeCusatis, A. Abbate, D.M. Litynski, and P. Das, Wavelet image processing for optical pattern recognition and feature extraction, *Proc. SPIE,* vol. 3110, pp. 804-815, 1997.

[DeC97a] C. DeCusatis, A. Abbate, and P. Das, Optical techniques for wavelet transform analysis, *Proc. Progress in Electromagnetics Research Symposium,*1997, p. S19.

[DeC97b] C. DeCusatis, A. Abbate, and P. Das, Progressive pattern recognition using the wavelet transform, *Int. J. Optoelectron.*, vol. 11, p. 425-432, 1997.

[DeC98]   C. DeCusatis and P. Das, Acousto-photorefractive holographic interferometric correlator for progressive pattern recognition using wavelet transforms, *Proc. SPIE,* vol. 3470, pp. 214-225, 1998.

[DeV92]   R. A. DeVore, B. Lawerth, and B. J. Lucier, Image compression through wavelet transform coding, *IEEE Trans. Inform. Theory*, vol. 38, pp. 719-746, 1992.

[Djo97]   I. Djokovic and P.P. Vaidyanathan, Generalized sampling theorems in mutiresolution subspaces, *IEEE Trans. Signal Process.,* vol. 45, pp. 583-599, 1997.

[Duf52]   R.J. Duffin and A.C. Schaeffer, A class of nonharmonic Fourier series, *Trans. Amer. Math. Soc.,* vol. 72, pp. 314-366, 1952.

[Eff92]   M. Effros. P. A. Chou, E. A. Riskin, and R. M. Gray, A progressive universal noiseless coder, *IEEE Trans. Inform. Theory,* vol, 40, pp. 108-117,1994.

[Equ91]   W. Equitz and T. Cover, Successive refinement of information, *IEEE Trans. Inform. Theory,* vol. 37, pp. 269-275, 1991.

[Est77]   D. Estaban and C. Galand, Application of quadrature mirror filters to split band voice coding schemes, *Proc. International Conference on Acoutsics, Speech and Signal Processing ICASSP,* 1977, pp. 191-195.

[Far98]   G.D. Farney and G. Ungerboeck, Modulation and coding for liner Gaussian channels, *IEEE Trans. Inform. Theory,* vol. 44, pp. 2384-2415, 1998.

[Fey90]   E. Feysz, Optical wavelet transform of fractal aggregates, *Phys. Rev. Lett.,* vol. 64, pp 7745-7748, 1990.

[Fla90]   P. Flandrin, F. Magand, and M. Zakharia, Generalized Target Description and Wavelet Decomposition, *IEEE Trans. Acoust. Speech Signal Process.,* vol. 38, pp. 350-362, 1990.

[Fra97]   J. Franca, A. Petraglia, and S.K. Mitra, Multirate analog digital systems for signal processing and conversion, *Proc. IEEE,* vol. 85, pp. 242-262, 1997.

[Fre93]   M. Freeman, Wavelets: signal representations with important advantages , *Opt. Photonics News,* vol. 4, pp. 8-14, 1993.

[Gab46]   D. Gabor, Theory of communication, *J. Inst. Electr. Eng.,* vol. 93, pp. 429-457, 1946.

[Gam81]   P.M. Gammel, Improved ultrasonic detection using analytic signal magnitude, *Ultrasonics,* pp. 73-76, 1981.

[Gam81a]   P.M. Gammel, Analogue implementation of analytic signal processing for pulse-echo systems, *Ultrasonics,* pp. 279-283, 1981.

[Gan93]   S. Ganesan,S. Mahalingam, and S. Nagabhushana, VLSI synthesis of a programmable DWT chip for the optimal choice of a prototype wavelet, *IEEE Computer,* pp. 127-131,1993.

[Gha87]   H. Gharavi and A. Tabatabai, Application of quadrature mirror filtering to the coding of monochrome and color images, *Proc. ICASSP*, pp. 2384-2387, 1987.

[Gho92]   M.M. Ghoneim and R.I. Block, Learning and consciousness during general anesthesia, *Anesthesiology*, vol. 76, pp. 279-305, 1992.

[Gib37]   F.A. Gibbs, E.L. Gibbs, and W.G. Lenox, Effect of the Electroencephalogram of certain drugs which influence nervous activity, *Arch. Intern. Med.*, vol. 60, pp. 154-166, 1937.

[Gou84]   P. Goupillaud, A. Grossmann, and J. Morlet, Cycle-octave and related transforms in seismic signal analysis, *Geoexploration* vol. 23, pp. 85-102, 1984.

[Goo78]   J.W. Goodman, A.R. Dias, and L.M Woody, Fully parallel, high speed incoherent optical method for performing discrete Fourier transforms *Opt. Lett.,* vol. 2, pp.1-3, 1978.

[Gro84]   A. Grossman and J. Morlet, Decomposition of Hardy functions into square integrable wavelets of constant shape, *SIAM J. Math. Anal.*, vol. 15, no. 4, pp. 723-736, 1984.

[Gro86]   A. Grossmann, J. Morlet, and T. Paul, Transforms Associated to square Integrable Group Presentations II. Examples, *Am. Inst. Henry Poincare*, vol. 45, no. 3, pp. 293-309, 1986.

[Gro87]   A. Grossmann, M. Holschneider, R. Kronland-Martinet and J. Morlet, Detection of abrupt changes in sound signals with the help of wavelet transforms, in *Inverse Problems*, Academic Press, NY, 1987, pp. 289-306.

[Gro89]   A. Grossmann, R. Kronland-Martinet, and J. Morlet, Reading and understanding continuous wavelet transforms, in *Wavelets*, edited by J.M. Combes, A. Grossmann and Ph. Tchamitchian, Springer-Verlag, Berlin, 1989, pp. 2-20.

[Gui84]   P.S. Guilfoyle, Systolic acousto-optic binary convolver, *Opt. Eng.* vol. 23, p. 20, 1984.

[Gui89]   P.S. Guilfoyle, Acousto-optic digital computer fundamentals, *Proc. IEEE Ultrasonics Symp.,* 1989, pp. 481-486.

[Gui91]   P. Guillemain, R. Kronland-Martinet, and B. Martens, Estimation of spectral lines with the help of the wavelet transform. Applications in NMR spectroscopy, in *Wavelets and Applications*, edited by Meyer, Springer-Verlag, NY, 1991, pp. 38-60.

[Gui96]   P. Guillemain and R. Kronland-Martinet, Characterization of acoustic signals through continuous linear time-frequency representations, *Proc. IEEE*, vol. 84, no. 4, pp. 561-585, 1996.

[Har41]   G.H. Hardy, Notes of special systems of orthogonal functions - IV: The orthogonal functions of Whittaker s series, *Proc. Camb. Phil. Soc.*, vol. 37, pp. 331-348, 1941.

[Har92]   J. Hartung, Architecture for real-time implementation of three dimensional subband video coding, *IEEE Inter. Conf. on Acoustics, Speech and Signal Processing*, 1992, vol. 3, pp. 225-229.

[Hee90]   V. K. Heer and H.-E. Reinfelder, A comparison of reversible methods for data compression, in *Proc. SPIE-Med. Imaging V*, vol. 1233, pp. 354-365, 1990.

[Her91]   C. Herley and M. Vetterli, Linear phase wavelets: theory and design, *Proc. ICASSP*, vol. 3, pp. 2017-2020, 1991.

[Her93]   C. Herley, J. Kovačević, K. Ramchandranard, and M. Vetterli, Tilings of the time-frequency plane: construction of arbitrary orthogonal bases and fast tiling algorithms, *IEEE Trans. Signal Process.*, vol. 41, pp. 3341-3359, 1993.

[Her93a]   C. Herley and M. Vetterli, Wavelets and recursive filter banks, *IEEE Trans. Signal Process.*, vol. 41, pp. 2536-2556, 1993.

[Hes96]   N. Hess-Nielsen and M.V. Wickerhauser, Wavelets and time-frequency analysis, *Proc. IEEE*, vol. 84, n. 4, pp. 523-540, 1996.

[Het95]   K. Hetling, G. Saulnier, and P. Das, Optimized filter design for PR-QMF based spread spectrum communications, *Proc. of the IEEE International Conference on Communications*, 1995, pp. 1350-1354.

[Het95a]   K. Hetling, G. Saulnier, and P. Das, Optimized PR-QMF based codes for multiuser communications, *SPIE Proceedings on Wavelet Applications for Dual Use*, volume 2491, pp. 248-259, 1995.

[Het95b]   K. Hetling, G. Saulnier, and P. Das, PR-QMF based codes for multipath/multiuser communications, *IEEE Global Telecommunications Conference*, 1995.

[Het96]   K. Hetling, G. Saulnier, and P. Das, Performance of filter bank-based spreading codes for multipath/multiuser interference, *SPIE Proceedings on Wavelet Applications*, 1996.

[Het96]   K. Hetling, *Multirate Filter Banks for Spread Spectrum Waveform Design*, PhD thesis, Rensselaer Polytechnic Institute, 1996.

[Het97]   K. Hetling, G. Saulnier, and P. Das, Performance of filter bank-based spreading codes for cellular and micro-cellular channels, in *SPIE Proceedings on Wavelet Applications*, 1997.

[Hoy92]   J. D. Hoyt and H. Wechsler, The wavelet transform - a CMOS VLSI ASIC implementation, *Proc. 11th IAPR Int. Conf. on Pattern Recognition, Vol. IV Conf. D: Architectures for Vision and Pattern Recognition*, 1992, pp. 19-22.

[Hsu82]   Y.N. Hsu and H.H. Arsenault, Optical pattern recognition using circular harmonic expansion, *Appl. Opt.*, vol. 21, pp. 4016-4019, 1982.

[Hua92]   Y. Huang, H. M. Driezen, and N. P. Galatsanos, Prioritized DCT for compression and progressive transmission of images, *IEEE Trans. Image Process.*, vol. 1, pp. 477-487, 1992.

[Jia96]   J. Jian, C.Z. Chen, C.C. Chen, and G. Seethraraman, Image data compression methodologies using discrete wavelets, *Proc. SPIE*, vol. 2762, pp. 188-199, 1996.

[Jer77]   A.J. Jerri, The Shannon sampling theorem - Its various extensions and applications: A tutorial review, *Proc. IEEE*, vol. 65, pp. 1565-1596, 1977.

[Kad92]   S. Kadambe and G.F. Boundreaux-Bartels, Applications of the wavelet transform for pitch detection of speech signals, *IEEE Trans. Inform. Theory*, vol. 38, pp. 917-924, 1992.

[Kha93]   M.R.K. Khansari and A. Leon-Garcia, Subband decomposition of signals with generalized sampling, *IEEE Trans. Signal Process.*, vol. 41, n. 12, pp. 3365-3376, 1993.

[Kho97]   R. Khoni-Poorfard, L.B. Lim, and D.A. Johus, Time-interleaved oversampling A/D converters: Theory and practice, *IEEE Trans. Circuits Syst. II: Analog Digital Signal Process.*, vol. 44, pp. 634-645, 1997.

[Kim92]   Y. H. Kim and L. W. Modestino. Adaptive entropy coded subband coding of images. *IEEE Trans. Image Process.*, vol. 1. pp. 31-48. Jan. 1992.

[Kir90]   J. Kirsch, Light efficient joint transform correlator, *Proc. SPIE Optical Information Processing Systems and Architectures*, vol. 1347, pp. 147- 154, 1990.

[Kle81]   F.F. Klein and D.A. Davis, The use of time domain analyzed EEG in conjnction with cardiovascular parameters for monitoring anesthetic levels, *IEEE Trans. Biomed. Eng.*, vol. 38, pp. 36-40, 1981.

[Koc94]   E. Koch, P. Bischoff, U. Pilchmeier, and J. S. Esch, Surgical stimulation induces changes in brain electrical activity during isoflurane/nitrous oxide anesthesia, *Anesthesiology*, vol. 80, pp.1026-1034, 1994.

[Koe46]   R. Koenig, H. Dunn, and L.Y. Lacy, The Sound spectrograph, *J. Acoust. Soc. Am.*, vol. 18, pp. 19-49, 1946.

[Kot97]   P. Kotidis and D. Klimek, *Proceedings of the Eighth International Symposium on Nondestructive Characterization of Materials*, Boulder, CO, June 1997.

[Kov92]   J. Kovačević and M. Vetterli, Nonseparable multidimensional perfect reconstruction filter banks and wavelet bases, *IEEE Trans. Inform. Theory*, vol. 38, pp. 533-555, 1992.

[Kro87]   R. Kronland-Martinet, J. Morlet, and A. Grossmann, Analysis of sound patterns through wavelet transforms, *Int. J. Pattern Recogn. Art.*, vol. 1, no. 2, pp. 273-302, 1987.

[Kro88]   R. Kronland-Martinet, The wavelet transform for analysis, synthesis and processing of speech and music sounds, *Computer Music J.*, vol. 12, pp. 11-20, 1988.

[Kud92]   G. R. Kuduvalli and R. M. Rangayyan, Performance analysis of reversible image compression techniques for high-resolution digital teleradiology, *IEEE Trans. Med Imaging*, vol. 1, pp. 430-445, 1992.

[Kun81]   E.T. Kung, Why systolic architectures?, *IEEE Computer*, vol. 15, p. 37, 1981.

[Le87]   P. Le, D.Y. Zang, G.D. Xu, and C.S. Tsai, An integrated optical digital correlator module using acousto-optic and electro-optic Bragg diffraction in LiNbO3, *Proc. IEEE Ultrasonics Symp.*, 1987, pp. 467-470.

[Lew91]   A. S. Lewis and O. Knowles, A 64 kB/s video Codec using the 2-D wavelet transform, *Proc. Data Compression Conf.*, Snowbird, Utah, IEEE Computer Society Press, Los Alamitos, CA, 1991.

[Lew91a]   A.S. Lewis and G. Knowles, VLSI architecture for 2-D Daubeches wavelet transform without multipliers, *Elec. Lett.* vol. 27, pp. 171-173, 1991.

[Lew92]   A.S. Lewis and O. Knowles, Image compression using the 2-D wavelet transform, *IEEE Trans. Image Process.*, vol. 1, pp. 244-250, 1992.

[Low95]   M. Lowe, *IEEE Trans. Ultrason. Ferroelect. Frequency Control*, vol. 42, pp. 525-542, 1995.

[Mac94]   R.P. MacDonald, Optical wavelet transform for fingerprint identification, *Proc. SPIE*, vol. 2237, pp. 302-313, 1994.

[Mal89]   S. G. Mallat, A theory for multiresolution signal decomposition: the wavelet representation, *IEEE Trans. Pattern Anal. Machine Intell.*, vol. 11, no. 7, pp. 674-693, 1989.

[Mal90]   S. Mallat, Multifrequency channel decompositions of images and wavelet models, *IEEE Trans. Acoust. Speech Signal Process.*, vol. 37, pp. 2091-2110, 1990 .

[Mal96]   S. Mallat, Wavelets for a Vision, *Proc. IEEE*, vol. 84, n. 4, pp. 604-614, 1996.

[McA93]   A. McAuley and J. Wang, Optical wavelet transform classifier with positive real Fourier transform wavelets, *Opt. Eng.*, vol. 32, pp. 1333-1339, 1993.

[McE75]   J. McEwen, G.B. Ardenson, M. Low, and L. Jenkins, Monitoring the level of anesthesia by automatic analysis of spontaneous EEG activity, *IEEE Trans. Biomed. Eng.*, vol. 22, pp. 299-305, 1975.

[Med95]   M.J. Medley, *Adaptive Narrow-Band Interference Suppression Using Linear Transforms and Multirate Filter Banks*, PhD thesis, Rensselaer Polytechnic Institute, 1995.

[Med95a]   M.J. Medley, G.J. Saulnier, and P.K. Das, The application of wavelet-domain adaptive filtering to spread spectrum communications, *SPIE Proceedings on Wavelet Applications for Dual-Use*, vol. 2491, pp. 233-247, 1995.

[Med97]   M.J. Medley, G.J. Saulnier, and P.K. Das, Narrow-band interference excision in spread spectrum systems using lapped transforms, *IEEE Trans. Commun.*, vol. 45, pp. 1444-1455, 1997.

[Met94]   M. Mettke, M.J. Medley, G.J. Saulnier, and P.K. Das, Wavelet transform excision using IIR filters in spread spectrum communication systems, *IEEE Global Communications Conference*, 1994, pp. 1627-1631.

[Mey89a]   Y. Meyer, Wavelets and operators, *Proc. Special Year in Modern Analysis, Urbana 1986-87*, Cambridge University Press, Cambridge, 1989.

[Mil80]   L.B. Milstein and P.K. Das, An analysis of a real-time transform domain filtering digital communication system - Part I: Narrowband interference rejection, *IEEE Trans. Commun.*, vol. 28, pp. 816-824, 1980.

[Mil83]   L.B. Milstein and P.K. Das, An analysis of a real-time transform domain filtering digital communication system - Part II: Wideband interference rejection, *IEEE Trans. Commun.*, vol. 31, pp. 21-27, 1983.

[Nay94]   A. Nayak, R. J. Roy, and A. Sharma, Time-frequency spectral representation of the EEG as an aid in the detection of depth of anesthesia, *J. Acoust. Soc. Am.,* vol. 22, pp. 501-513, 1994.

[Ngu89]   T.Q. Nguyen and P.P. Vaidyanathan, Two channel perfect reconstruction FIR QMF structures which yield linear-phase analysis and synthesis filters, *IEEE Trans. Acoust. Speech Signal Process.*, vol. 37, pp. 676-690, 1989.

[Nyq28]   H. Nyquist, Certain topics in telegraph transmission theory, *Trans. Am. Inst. Electr. Eng.*, vol. 47, pp. 617-644, 1928.

[Pap77a]   A. Papoulis, Generalized sampling expansion, *IEEE Trans. Circuit Syst.,* vol. 24, pp. 652-654, 1977.

[Par93]   K. Parhi and T. Nishitani, VLSI architectures for discrete wavelet transforms, *IEEE Trans. VLSI syst.*, vol. 1, pp. 191- 202, 1993.

[Pen91]   A Pentland and B. Horowitz, A practical approach to fractal-based image compression, *Proc. Data Compression Conf.*, Snowbird, Utah, IEEE Computer Society Press, Los Alamitos, CA, 1991.

[Pet92]   A Petraglia and S.K. Mitra, High speed A/D conversion incorporating a QMF bank, *IEEE Trans. Instrum. Measurement,* vol. 41, pp. 427-431, 1992.

[Pin91]   S.D. Pinski, Optical image segmentation using wavelet correlation, *SPIE Proc. Intelligent Robots and Computer Vision*, vol. 1608, pp. 414-423, 1991.

[Poc88]   E. Pochapsky and D. Casasent, Optical linear heterodyne matrix-vector processor, *Proc. SPIE,* vol. 886, pp. 158-170, 1988.

[Psa80]   D. Psaltis, Accurate numerical computation by optical convolution, *Proc. SPIE,* vol. 262, p. 151, 1980.

[Rab92]   M. Rabbani and P. W. Melnychuck, Conditioning contexts for the arithmetic coding of bit planes, *IEEE Trans. Signal Process.*, vol. 40, pp. 232-236, 1992.

[Ram96]   K. Ramchandran, M. Vetterli, and C. Herley, Wavelets, subband coding, and best bases, *Proc. IEEE*, vol. 84, n. 4, pp. 541-560, 1996.

[Ran88]   S. Ranganath and H. Blume. Hierarchical image decomposition and filtering using the S-transform. in *Proc. SPIE-Med. Imaging*, vol. 914. pp. 799-814, 1988.

[Res92]   H. Resnikoff, Wavelets and adaptive signal processing, *Opt. Eng.* vol. 31, pp. 1229-1234, 1992.

[Rio91]   O. Rioul and M. Vetterli, Wavelets and signal processing, *IEEE Signal Proc. Mag.*, pp. 14-38, 1991.

[Rio92]   O. Rioul and P. Flandrin, Time-scale energy distributions: A general class extending wavelet transforms, *IEEE Trans. Signal Process.*, vol. 40, pp. 1746-1757, 1992.

[Rio92a]   O. Rioul and P. Duhamel, Fast Algorithms for Discrete and Continuous wavelet Transforms, *IEEE Trans. Inform. Theory*, vol. 38, n. 2, pp. 569-586, 1992.

[Rio93]   O. Rioul, A Discrete-time Multiresolution theory, *IEEE Trans. Signal Process.*, vol. 41 n. 8, pp. 2591-1606, 1993.

[Rob95]   M.C. Robini et al., Application of the wavelet packet transform to flaw detection in ultrasound B-scans, *Proc. IEEE Ultrasonics Symp.*, 1995, pp.1-4.

[Roo88]   P. Roos, A. Viergever, and M. C. A. van Dijke, Reversible intraframe compression of medical images, *IEEE Trans. Med. Imaging*, vol. 7, pp. 328-336, 1988.

[Ros82]   W.E. Ross, Two-dimensional magneto-optic spatial light modulator for signal processing, *SPIE Proc. Real Time Signal Processing*, vol. 341, pp. 191-198, 1982.

[Sai93]   A. Said and W. A. Pearlman, Reversible image compression via multiresolution representation and predictive coding, *Proc. SPIE*, vol. 2094. pp. 664-674, 1993.

[Sai96]   A. Said and W. A. Pearlman, A new fast and efficient image codec based on set partitioning in hierarchical trees, *IEEE Trans. Circuits Sys. Video Tech.*, vol. 6, pp. 243-250, 1996.

[Sai96a]   A. Said and W. Pearlman, An image multiresolution representation for lossless and lossy compression, *IEEE Trans. Image Process.*, vol. 5, pp. 1303-1309, 1996.

[Sai96b]   A. Said and W. Pearlman, A new, fast, and efficient image codec based on set partitioning in hierarchical trees, *IEEE Trans. Circuits Syst. Video Tech.*, vol. 6, pp. 243-249, 1996.

[San95]   S.D. Sandberg and M.A. Tzannes, Overlapped discrete multitone modulation for high speed copper wire communications, *IEEE J. Select. Areas Comm.* vol. 13, pp. 1570-1585, 1995.

[Sau92]   M. Sauer and J. Gotze, A VLSI architecture for fast wavelet computations, *Proc. IEEE SP Intl. Symp. on Time and Frequency Scale Analysis*, 1992, pp. 407-410.

[Sau95]   G. Saulnier, M.J. Medley, and P. Das, Wavelets and Filter Banks in Spread Spectrum Communications, in *Subband and Wavelets Transforms*, edited by A.N. Akansu and N.J.T. Smith, Kluwer, Boston, 1995.

[Say92]   K. Sayood and K. Anderson, A differential lossless image compression scheme. *IEEE Trans. Signal Process.*, vol. 40. pp. 236-241, 1992.

[Sch96a]   P. Schröder, Wavelets in computer graphics, *Proc. IEEE*, vol. 84, n. 4, pp. 615-625, 1996.

[Sel99]   I.W. Selesnick, Interpolating multiwavelets bases and the sampling theorem, *IEEE Trans. Signal Process.*, vol. 47, pp. 1615-1621, 1999.

[Sha49]   C.E. Shannon, Communication in the presence of noise, *Proc. IRE* vol. 37, pp. 10-21, 1949.

[Sha92]   J. M. Shapiro, An embedded wavelet hierarchical image coder, *Proc. IEEE Int. Conf. Acoust., Speech, Signal Processing*, 1992.

[Sha93]   J. M. Shapiro, Embedded image coding using zerotrees of wavelets coefficients, *IEEE Trans. Signal Process.*, vol. 41. pp. 3445-3462, Dec. 1993.

[Sha93a]   A. Sharma and R.J. Roy, Analysis of hidden nodes in a neural network trained for pattern classification of EEG data, *Proc. of the 15th Annual International Conf. of the IEEE EMBS*, 1993, vol. 1, pp. 252-253.

[She92]   Y. Sheng, Optical N implementation of a two-dimensional wavelet transform, *Opt. Eng.*, vol. 31, pp. 1859-1864, 1992.

[Sho97]   B.L. Shoop, A.H. Sayles, G.P. Dudevoir, D.A. Hall, D.M. Litynski, and P. K. Das, Smart pixel based wavelet transformation for wideband radar and sonar signal processing, *Proc. SPIE,* vol. 3078, pp. 415-423, 1997.

[Sho98]   B.L. Shoop, P. Das, and M. Litynski, Improved resolution of photonic A/D conversion using oversampling techniques, *Proc. SPIE,* vol. 3463, pp. 192-199, 1998.

[Smi84]   M.J. Smith and T.P. Barnwell III, A procedure for designing exact reconstruction filter banks for tree structured subband coders, *Proc. IEEE Intl. Conf. ASSP,* 1984, pp. 1-4.

[Smi86]   M.J. Smith and T.P. Barnwell, III, Exact reconstruction for tree-structured subband coders, *IEEE Trans. Acoust. Speech Signal Process.,* vol. 34, pp. 434-441, 1986.

[Str89]   G. Strang, Wavelets and dilation equation: a brief introduction, *SIAM Rev.,*vol. 31, pp. 614-627, 1989.

[Str94]   G. Strang, Wavelets, *Am. Scientist,* vol. 82, pp. 250-255, 1994.

[Szu92]   R. Szu, B. Teller, and A. Lohynan, Causal analytical wavelet transform, *Opt. Eng.,*vol. 31, pp. 1825-1829, 1992.

[Szu92a]   H. Szu, B. Telfer, and S. Kadambe, Neural network adaptive wavelets for signal representation and classification, *Opt. Eng.,*vol. 31, p. 1907-1916, 1992.

[Tak94]   S. Takamura and M. Takagi, Lossless image compression with lossy image using adaptive prediction and arithmetic coding. in *Proc. Conf. on Data Compression,* 1994, pp. 166-174.

[Tho89]   C.E. Thomsen, K.N. Christensen, and A. Rosenfalck, computerized monitoring of depth of anesthesia with isoflurane, *Br. J. Anesthesia,* vol. 63, p. 36-43, 1989.

[Tod85]   S. Todd. G. G. Langdon. Jr., and J. Rissanen, Parameter reduction and context selection for compression of grey-scale images. *IBM J. Res. Dev.* vol. 29, pp. 188-193, 1985.

[Tra87]   A.K. Tran, Liu, K. Tzou and E. Vogel, An efficient pyramid image coding scheme, *Proc. ICASSP,* pp. 744-747, 1987.

[Tsa79]   C.S. Tsai, Guided wave acousto-optic Bragg modulators for wideband integrated optical communication and signal processing, *IEEE Trans. Circuits Syst.,* vol. CAS-26, p. 1072, 1979.

[Uns95]   M. Unser, A general Hilbert space framework for the discretization of continuous signal processing operators, *Proc. SPIE Conf. on*

Wavelet Applications in Signal and Image Processing, 1995, pp. 51-61.

[Uns96]   M. Unser and A. Aldroubi, A Review of wavelets in biomedical applications, *Proc. IEEE*, vol. 84, n. 4, pp. 626-638, 1996.

[Uns00]   M. Unser, Sampling - 50 years after Shannon, *Proc. IEEE*, vol. 88, pp. 569-587, 2000.

[Vai87]   P.P. Vaidyanathan, Quadrature mirror filters banks, M-band extensions and perfect-reconstruction technique, *IEEE ASSP Mag.*, vol. 4, pp. 4-20, 1987.

[Vai88]   P.P. Vaidyanathan and P.Q. Hoang, Lattice structures for optimal design and robust implementation of two-band perfect reconstruction QMF banks, *IEEE Trans. Acoust. Speech Signal Process.*, vol. 36, pp.81-94, 1988.

[Vai90]   P.P. Vaidyanathan, Multirate digital filters, filterbanks, polyphase networks and applications: A tutorial, *Proc. IEEE,* vol. 78, pp. 56-93, 1990.

[Vai92]   J. Vaisey and A. Gersho, Image compression with variable block size segmentation, *IEEE Trans. Signal Process.*, vol. 2040, 1992.

[Van64]   A. VanderLugt, Signal detection by complex spatial filtering, *IEEE Trans. Inform. Theory,* vol. IT-10, p. 2, 1964.

[Vel98]   S.R. Velzquez, T.Q. Nguyen, and S.R. Broadstone, Design of Hybrid filterbanks for analog/digital conversion, *IEEE Trans. Signal Process.*, vol. 46, pp. 956-967, 1998.

[Vet86]   M. Vetterli, Filter banks allowing perfect reconstruction, *Signal Process.*, vol. 10, no. 3, pp. 219-244, 1986.

[Vet87]   M. Vetterli, A Theory of Multirate Filter Banks, *IEEE Trans. Acoust. Speech Signal Process.*, vol. 35, pp. 356-372, 1987.

[Vet89]   M. Vetterli and D. LeGalli, Perfect reconstruction FIR filter banks: Some properties and factorizations, *IEEE Trans. Acoust. Speech Signal Process.*, vol. 37, pp. 1057-1071, 1989.

[Vet90]   M. Vetterli, J. Kovačević, and D. J. LeGall, Perfect reconstruction filter banks for HDTV representation and coding, *Image Commun.*, vol. 2, pp. 349-364, Oct. 1990.

[Vet90a]   M. Vetterli and C. Herley, Wavelets and filter banks: relationships and new results, *Proc. ICASSP,* vol. 3, pp. 1723-1726, 1990.

[Vet92]   M. Vetterli and C. Herley, Wavelets and Filter Banks: Theory and Design, *IEEE Trans. Signal Process.*, vol. 40 no. 9, 1992.

[Vis91]   R. Vishnoi and R.J. Roy, Adaptive control of closed circuit anesthesia, *IEEE Trans. Biomed. Eng.*, vol. 39, pp. 39-47, 1991.

[Vis92]   M. Vishwanath and R. Ownes, Discrete wavelet transforms in VLSI, *IEEE Int. Conf. App. Specific Array Processors*, 1992, pp. 218-229.

[Vis92a]   M. Vishwanath and R. Owens, An AT 2 lower bound for wavelet transforms in VLSI, *IEEE Int. Conf App. Specific Array Processors*, 1992, pp. 649-652.

[Vis92b]   M. Vishwanath, R. Owens, and M. Irwin, An efficient systolic architecture for QMF filter bank trees, *IEEE Workshop on VLSI Signal Processing*, 1992.

[Wal91]   G. K. Wallace, The JPEG still picture compression standard, *Comm. ACM.*, vol. 34, pp. 30-44, Apr. 1991.

[Wal92]   G.G. Walter, A sampling theorem for wavelet subspaces, *IEEE Trans. Inform. Theory*, vol. 38, pp. 881-884, 1992.

[Whi77]   H.J. Whitehouse and D. Spiezer, Linear signal processing architectures, *Aspects of Signal Processing with Emphasis on Underwater Acoustics*, edited by G. Tacconi, Reidel, Hingham, MA, 1977.

[Wit87]   H. Witten, R. Neal, and J. G. Cleary, Arithmetic coding for data compression, *Comm. ACM*, vol. 30, pp. 520-540, June 1987.

[Wom90]   G. W. Womell, A Karhunen-Louve expansion for 1/f processes via wavelets, *IEEE Trans. Inform. Theory*, vol. 36, pp. 859-861, 1990.

[Wor92]   G.W. Wornell and A.V. Oppenheim, Wavelet-based representations for a class of self-similar signals with applications to fractal modulation, *IEEE Trans. Inform. Theory*, vol. 38, pp. 785-800, 1992.

[Wor96]   G.W. Wornell, Emerging applications of mutirate signal processing and wavelets in digital communications, *Proc. IEEE*, vol. 84, pp. 586-603, 1996.

[Xia93]   X.G. Xia and Z. Zhang, On sampling theorem, wavelets and wavelet transforms, *IEEE Trans. Signal Proc.*, vol. 41, pp. 3524-3535, 1993.

[Xio93]   Z. Xiong, N. Galatsanos, and M. Orchard, Marginal analysis prioritization for image compression based on a hierarchical wavelet *de-composition*, Proc. IEEE Int. Conf. Acoust., Speech, Signal *Processing*, 1993.

[Yu94]   F.T.S. Yu and G. Lu, Short time Fourier transform and wavelet transform with Fourier domain processing, *Appl. Opt.,* vol. 33, pp. 5262-5270, 1994.

[Zet90]   W. Zettler, I. Huffman, and D. C. P. Linden, Applications of compactly supported wavelets to image compression, *SPIE Image Processing Algorithms,* 1990.

[Zha94]   Y. Zhang, Optoelectronic wavelet processors based on Smartt interferometry, *Appl. Opt.,* vol. 33, pp. 5279-5266, 1994.

[Zib93]   M. Zibulski and Y.Y. Zeevi, Oversampling in the Gabor Scheme, *IEEE Trans. Signal Process.,* vol. 41, pp. 2679-2687, 1993.

Index